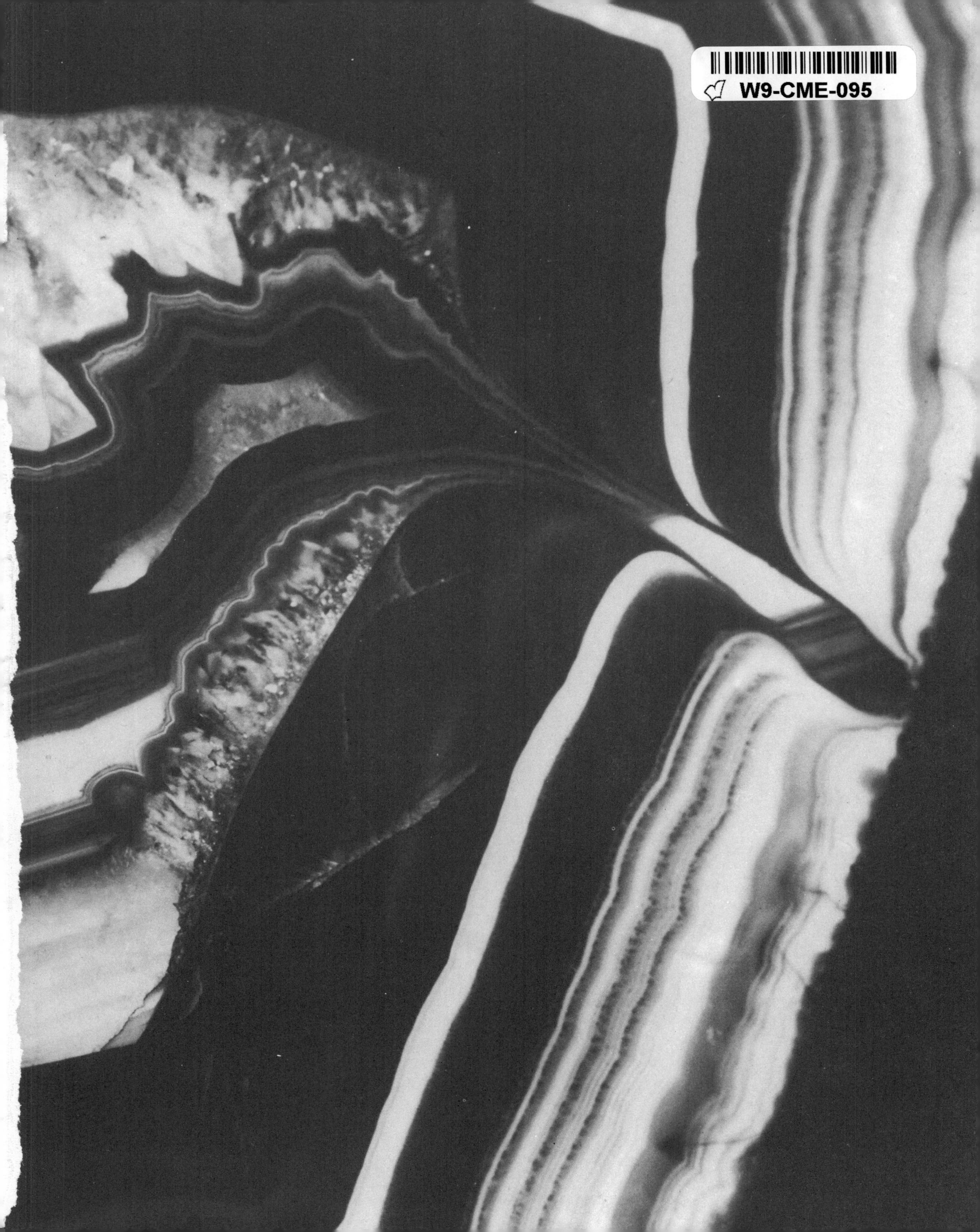

The ENCYCLOPEDIA of MINERALS and GEMSTONES

The ENCYCLOPEDIA of MINERALS and GEMSTONES

Edited by Michael O'Donoghue

G. P. PUTNAM'S SONS New York

Opposite: Polished variscite slice

Preceding pages: Crystals of amazonite

End papers: Section through a geode, showing various types of chalcedony

Associate Editor: Christopher Cooper

SBN: 399-11753-9
Library of Congress Catalog Card Number: 76-1419
Printed in Italy by IGDA, Novara

Contents

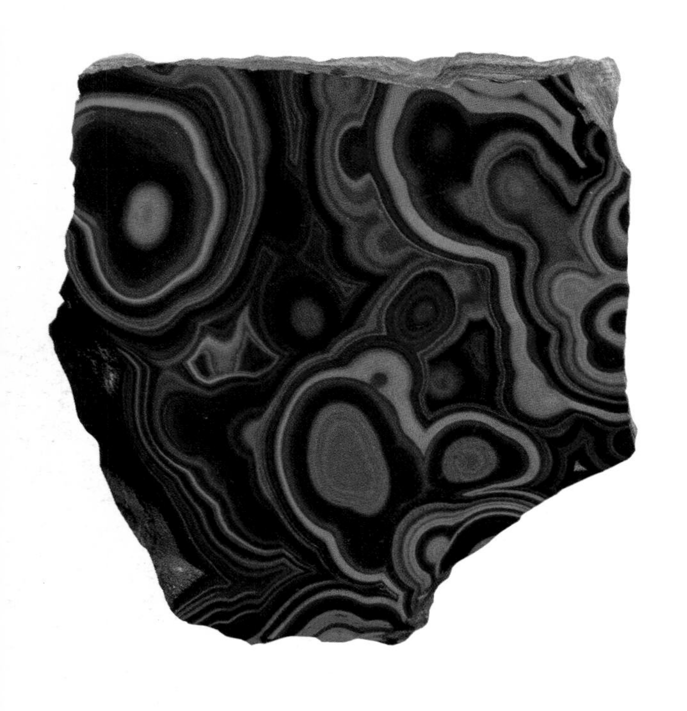

Introduction

Left: Magnificent hexagonal crystals of chalybite, an iron carbonate. The triangular 'growth shadows' that are visible represent the form of the crystals at an earlier stage in their development

This book has several purposes: to introduce readers to the beauty of the world of minerals and fashioned gems; to lead those who have already made a first acquaintance with mineralogy a little deeper into the subject; to provide information about minerals that are not commonly found; and to give guidance to those who wish to fashion stones suitable for jewellery from the specimens that they find. But if the book leads only one reader to fall in love with minerals it will have been well worth while.

What is the outstanding feature presented by a mineral or a gemstone to the casual observer? In many cases it is its colour. This, like some products of living things, such as the wings of butterflies, can show an incomparable purity and brightness. In this connection, one great advantage that readers of books today possess that their counterparts of fifty years ago lacked is the high quality of colour reproduction. However, the representations of minerals that appeared in older books could be quite sufficient to fire the imagination—and that was this author's experience.

Our present-day love of minerals was doubtless shared by early man, who must have noticed brightly coloured pebbles on the ground or on the beds of streams. These objects were put to use as simple ornaments as well as tools or weapons on account of their comparative hardness or toughness. It is possible that pearls were recovered from salt and fresh waters and similarly used. But centuries elapsed before a serious attempt was made to classify minerals or record their occurrence. Indeed, especially noteworthy minerals were likely to be regarded as having a divine, rather than a natural, origin.

Man learnt to work stone before he learnt to work metal, since most metals need the intermediate process of smelting, whereas stone can be dug and roughly shaped quite quickly. But serious attempts to classify minerals date only from the Greeks and the Roman Empire. Aristotle, living from 384 to 322 BC, and Theophrastus, living from about 370 to 287 BC, both mention researches made by earlier scholars, so that we can place the beginning of mineralogical study before 400 BC. Aristotle's work *Meteorologica* includes remarks on minerals and metals. *Concerning Minerals*, by Theophrastus, is the first treatise still surviving that deals with minerals alone. In about 44 BC the Greek historian Diodorus Siculus wrote *Bibliotheca Historica*, part of which tells of minerals known to his contemporaries from the areas of Greek civilization. Probably the best-known classical writer on stones is Pliny the Elder, living from 23 to 79 AD; his large *Natural History* contains 37 separate books, four of which deal with minerals of various kinds, including gems. Other writers of the time concentrated on the magical and medical properties of stones, a semi-mystic approach to gems that still survives today.

The post-Roman era was not productive of much scientific research, though some was undertaken by Arab scholars, among whom Avicenna is one of the most celebrated. But the revival of learning in the Middle Ages led to an increase in writings of all kinds. Some of the mineralogical writing is purely fanciful, especially from the early part of the period, when the books known as 'lapidaries' were fashionable. Many of these are speculative in the extreme and add little to knowledge. Works from this period include those by Marbodus, Albertus Magnus and Camillus Leonardus. These are charming if insubstantial to modern minds.

By the sixteenth century work of a higher value was taking place. Georg Bauer (Georgius Agricola, 1494–1555), produced several notable books, including *De Re Metallica* and *De Natura Fossilium*. The latter is sometimes hailed as an early textbook since in it minerals are arranged in an elementary classification based on colour and weight. As in the sister science of botany, with its elaborate taxonomy (classification), old systems have given way to new and minerals have been re-allocated to different places in them, but on the whole Agricola's work gives some idea of how things could be done. In the following century, de Boodt, who wrote *Gemmarum et Lapidum Historia*, dated 1609, is one of those who treat minerals seriously enough for their work to have some significance today.

One of the problems in any science is to so describe specimens that other workers who do not have similar specimens before them will be able to gain some idea of their appearance. Some of the best descriptions of minerals were given in the eighteenth century by the German geologist Werner (1750–1817) who was skilled in accurate description; his work *On the External Characteristics of Fossils* was published in 1774. Later writers worked in a similar way and there has been no change in the general basis behind classification systems since the development of chemical testing methods began.

Above: A woodcut from Agricola's great treatise on metals and their extraction. Miners are seen labouring in the shafts and tunnels of a sixteenth-century mine

When we visit a modern mineral collection (the correct term is 'cabinet'), the chances are that it will be arranged in the order devised by the great American mineralogist James Dwight Dana (1813–1895). This system, based on chemical classes, owes a great deal to the work of Berzelius (1779–1848), a Swede. Dana's *System of Mineralogy* appeared in 1837 and the present revision, still incomplete, began in 1944.

Work is going on perpetually in mineralogy as in all the other natural sciences, and this means that new discoveries are constantly being made. Some of these discredit existing minerals, identifying them with others instead of naming them as separate species. New minerals are

arabic⁹ grosse viriditatis. ʒ nõ trāslucid⁹ sicut smarald⁹. ꝛ ē mollis ¶ Fert ꝙ ꝟtutē hēat custo diēdi gestantē se a nocuis ca sibꝫ. Et sup cuna bula infantiū. Hoc aūt in lapidario suo enax

Capitulum. lxxxiij

Enphites. Alber. Mēphites ē lapis m a ciuitate egipti ꝙ mēphis vocat. ꝗ ꝺꝛ vt ignis calere. virtute quidē nō a ctu

Operationes

¶ Hic tritus et in aqua mixtus in potu detur vrendis vel secandis. quia inducit insensibili tatē ne sentiat cruciandus. Hoc idē ait Enax in lapidario suo.

Capitulum. lxxxiiij

m Edus. Dyas. Medus lapis ē niger ꝗ trit⁹ emittit croceū colorē. Alb. Lapis medus vel medo ꝺꝛ ꝙ a regiōe medo rū fert vbi ples iueniunt Sunt aūt hui⁹ spēs due vnus niger ꝛ alter viridis

Operationes

¶ Alb. Hui⁹ ꝟtutē ē dicūt ꝯ vetere podagrā. oculoꝝ cecitatē. ꝛ nefreticā passionē. etiā refoue re ꝺꝛ fessos ꝛ debiles ¶ Fert etiā ꝙ fragmenta nigri si in aq̄ ca. missa fuerit. ꝛ se ꝗs ī aq̄ illa la uerit incurrit mēbroꝝ excoriatōeꝫ. ꝛ si ex ea bi berit pibit vomen do ¶ Dyas. Est aūt duplicis potētie. s. bone ꝛ male. nā trit⁹ in cote medicia li cū lacte muliebri (ꝗ sel' creauit pmo pruꝑue rū) ꝛ iunct⁹ albugies curat oculoꝝ. ꝛ oēm illo rū doloꝛē etiā si despatus fuerit medꝫ podagri cos ¶ Eodem mō cū lacte ouis ꝗ sel' genuit ma sculū sanat. Renes ꝛ anhelit⁹ ꝛ iterioꝛes ptes corpis Hoc medicamē ppatū hēas in vase vi treo vel argenteo ꝛ pfice ieiun⁹. ad vindictā ꝗꝫ supra cotē ꝛ da cui vis. ꝛ laue faciem et exicca bit. ¶ Si quis autem ex eo biberit. pulmoneꝫ vomet ¶ O pꝫ aūt bibēti paululū dare vt non satiet. Inuent⁹ ē aūt iuxta fluuiū fasin.

Capitulum. lxxxv

m Ogra. Sera. Mogra ex ea est ques dam que defert ex terris lincos. ꝛ melioꝛ ex ea est grauis densa cuius coloꝛ nō est diuersus. ꝛ ē sicut coloꝛ cpatis aut sāguinis nō bnis lapidē. Et qñ ponit in aqua dissoluit facile et crescit. Illa ꝟo mogra ꝗ vtunt carpen

Above: A page from a late fifteenth century treatise on minerals by an anonymous writer

Above right: James Dwight Dana, the great American mineralogist

frequently discovered, however, and new methods of testing are devised. This work is reported in such journals as *Mineralogical Magazine* and *American Mineralogist* and from time to time a new textbook appears. Keeping up with the literature is very difficult and a number of small periodicals devoted to the literature of the subject have been published in recent years, with the purpose of improving communication over a very complex field. Most of the important current literature appears in the bibliography of this book.

The study of the gem materials, today universally known as gemmology, is more closely linked with the science of mineralogy than with any other, since nearly all gem materials are minerals. However, gemmologists borrow testing methods from chemistry and physics, too. In past years those concerned with the testing of gemstones could rely to a large extent on experience when attempting to identify a synthetic gem set in jewellery. But now that so many new materials are making their appearance, there is no substitute for scientific testing.

The gemmologist is called upon to make commercial assessments that are outside the field of the laboratory mineralogist. But the two professions are able to help one another a great deal; the gemmologist can make available stones that are of superlative quality compared with the usual mineralogical specimen; the mineralogist can identify new materials and those that defy the reasonably simple testing methods used by gem-testing laboratories.

Both mineralogist and gemmologist assist the lapidary in a variety of ways. Both can tell him which materials will make the best finished stones; they can warn him of such dangers as perfect cleavages, and advise him of attractive optical effects. From his side, the lapidary may notice features of gem material that only come to light when the stones are ground and polished.

All those who have been concerned with *The Encyclopedia of Minerals and Gemstones* hope that those who read it will share the enthusiasm possessed by its authors. The object of books is to give pleasure as well as to inform; we hope that we have achieved both ends.

The Chemistry of Minerals

Left: Prismatic azurite crystals on a matrix of highly discoloured limonite. Minerals are almost infinitely varied in their colours, textures and compositions. Yet they are built up from less than a hundred naturally occurring elements

You may collect, enjoy and in many cases identify minerals without any knowledge of their chemistry. However, to ignore modern theories of the nature of atoms and how they combine to form spectacular and complex crystals is to miss contact with a fascinating subject. The purpose of this chapter is to provide a simple account of some of the basic principles of chemistry, in intelligible language. If this objective is attained, you should be able to understand the formulae and some technical terms which abound in the Mineral Kingdom section of this book.

THE NATURE OF THE ATOM

What is an atom? It is the smallest particle of an *element* that retains the essential properties of the latter. Elements are the simplest substances that may be isolated by chemical means. There are a hundred or so known and each is composed of atoms that, chemically, are identical (there can be variations of weight among them, discussed later in the section on isotopes). Let us take an example. Gold is an element; it can combine with other elements, as in alloys, but cannot be refined further into other more fundamental metals. If you wished to hoard gold, you could start with a single atom. This would be a scrap with the tiny diameter of 1.5 angstroms. (The angstrom, symbol A, is a hundred-millionth of a centimetre, or about a two-hundred-millionth of an inch.) If you now gave this atom to a nuclear physicist, he could split it up in a particle accelerator or nuclear reactor, but the resulting fragments would no longer be gold.

There are three main types of sub-atomic particle within atoms. Two of these, *protons* and *neutrons*, have approximately similar masses and cluster together at the centre of the atom in the *nucleus*. Each proton has a small positive electrical charge, whereas neutrons have none. Particles of the third type, *electrons*, make up a swarm orbiting the nucleus. The mass of an electron is nearly two thousand times less than that of a proton but, nevertheless, it carries a negative electrical charge equal to the proton's positive charge. Atoms are not electrically charged overall, because they contain equal numbers of protons and electrons. Most of an atom's mass is in its nucleus; in contrast, most of its volume is occupied by the electron swarm.

The studies of modern physics have shown that, despite their smallness, electrons are rather complicated. For instance, some of their properties are more like those of waves than particles.

Fortunately, however, these additional factors are inessential to the explanation of the basic features of mineral chemistry. For our present purposes we can stick to what is known as the Rutherford–Bohr theory of atomic structure. This may be described as the 'solar system analogy', in which the atomic nucleus is the sun and the electrons are orbiting planets. Planets remain at approximately fixed distances from the sun and do not wander freely around the solar system. Similarly, the electrons in the cloud around an atomic nucleus keep to fixed orbits. Let us now consider how these sub-atomic particles are combined to form the atoms of each element.

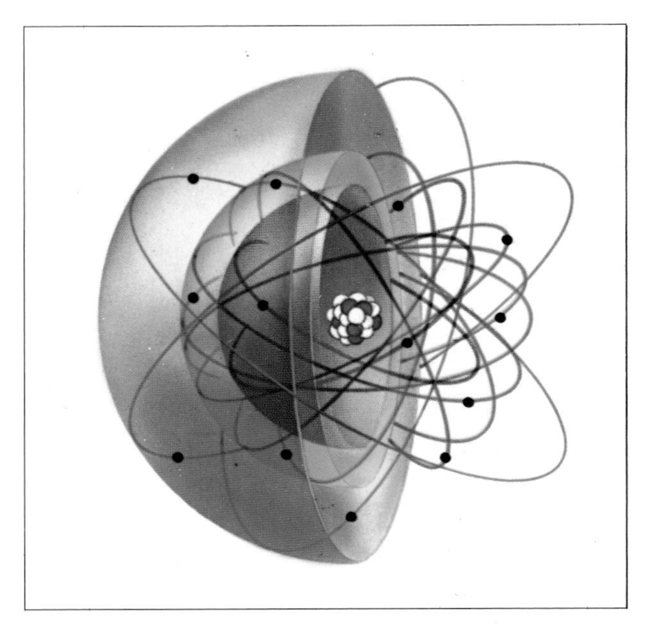

Left: The structure of the atom. The nucleus consists of positively charged protons (red) and uncharged neutrons (white). Round it circle the negatively charged electrons. Their orbits are grouped into 'shells' symbolized here by the coloured spheres

The nucleus

Each element has a characteristic number of protons in its nucleus, balanced electrically by an equal number of orbiting electrons. The number of protons (known as the *atomic number*) varies from 1 in the simplest atom, that of hydrogen, to slightly over 100 in the heaviest elements known. Neutrons accompany the protons in the nuclei of all atoms except hydrogen. The number of neutrons need not be the same in all atoms of an element. For instance, the atom of strontium always contains 38 protons, but has from 46 to 50 neutrons. Each of these sub-species, with its particular number of neutrons, is known as an *isotope* of strontium.

The *atomic weight* of an element was originally defined as the weight of one of its atoms compared with that of an atom of hydrogen. When the complications introduced by the occurrence of isotopes were more clearly understood, the standard unit of atomic weight was redefined in such a way that the common isotope of carbon, with 6 protons and 6 neutrons in its nucleus, has an atomic weight of exactly 12. Because most naturally occurring elements are composed of two or more isotopes, each with differing nuclear masses, their atomic weights (which, by convention, means their *average* weights) are not exact whole numbers. This may be seen in the accompanying table, which also lists the symbols used by chemists for each element, a kind of scientific shorthand.

The elements

element	symbol	atomic number	atomic weight	element	symbol	atomic number	atomic weight
Aluminium	Al	13	26.98	Neodymium	Nd	60	144.24
Antimony	Sb	51	121.75	Neon	Ne	10	20.18
Argon	Ar	18	39.95	Nickel	Ni	28	58.71
Arsenic	As	33	74.92	Niobium	Nb	41	92.91
Barium	Ba	56	137.34	Nitrogen	N	7	14.01
Beryllium	Be	4	9.01	Osmium	Os	76	190.20
Bismuth	Bi	83	208.98	Oxygen	O	8	16.00
Boron	B	5	10.81	Palladium	Pd	46	106.40
Bromine	Br	35	79.91	Phosphorus	P	15	30.97
Cadmium	Cd	48	112.40	Platinum	Pt	78	195.09
Caesium	Cs	55	132.91	Potassium	K	19	39.10
Calcium	Ca	20	40.08	Praseodymium	Pr	59	140.91
Carbon	C	6	12.01	Protactinium	Pa	91	231.04
Cerium	Ce	58	140.12	Radium	Ra	88	226.02
Chlorine	Cl	17	35.45	Radon	Rn	86	222
Chromium	Cr	24	52.00	Rhenium	Re	75	186.2
Cobalt	Co	27	58.93	Rhodium	Rh	45	102.90
Copper	Cu	29	63.54	Rubidium	Rb	37	85.47
Dysprosium	Dy	66	162.50	Ruthenium	Ru	44	101.07
Erbium	Er	68	167.26	Samarium	Sm	62	150.35
Europium	Eu	63	151.96	Scandium	Sc	21	44.96
Fluorine	F	9	19.00	Selenium	Se	34	78.96
Gadolinium	Gd	64	157.25	Silicon	Si	14	28.09
Gallium	Ga	31	69.72	Silver	Ag	47	107.87
Germanium	Ge	32	72.59	Sodium	Na	11	22.98
Gold	Au	79	196.97	Strontium	Sr	38	87.62
Hafnium	Hf	72	178.49	Sulphur	S	16	32.06
Helium	He	2	4.00	Tantalum	Ta	73	180.95
Holmium	Ho	67	164.93	Tellurium	Te	52	127.60
Hydrogen	H	1	1.00	Terbium	Tb	65	158.92
Indium	In	49	114.82	Thallium	Tl	81	204.37
Iodine	I	53	126.90	Thorium	Th	90	232.04
Iridium	Ir	77	192.20	Thulium	Tm	69	168.93
Iron	Fe	26	55.85	Tin	Sn	50	118.69
Krypton	Kr	36	83.80	Titanium	Ti	22	47.90
Lanthanum	La	57	138.91	Tungsten	W	74	183.85
Lead	Pb	82	207.19	Uranium	U	92	238.03
Lithium	Li	3	6.94	Vanadium	V	23	50.94
Lutetium	Lu	71	174.97	Xenon	Xe	54	131.30
Magnesium	Mg	12	24.31	Ytterbium	Yb	70	173.04
Manganese	Mn	25	54.94	Yttrium	Y	39	88.91
Mercury	Hg	80	200.59	Zinc	Zn	30	65.37
Molybdenum	Mo	42	95.94	Zirconium	Zr	40	91.22

The electrons

Electron orbits fall into groups called *shells.* There can be up to seven shells in an atom, depending on the number of electrons. They are labelled alphabetically from K to Q outwards. Each shell can contain any number of electrons

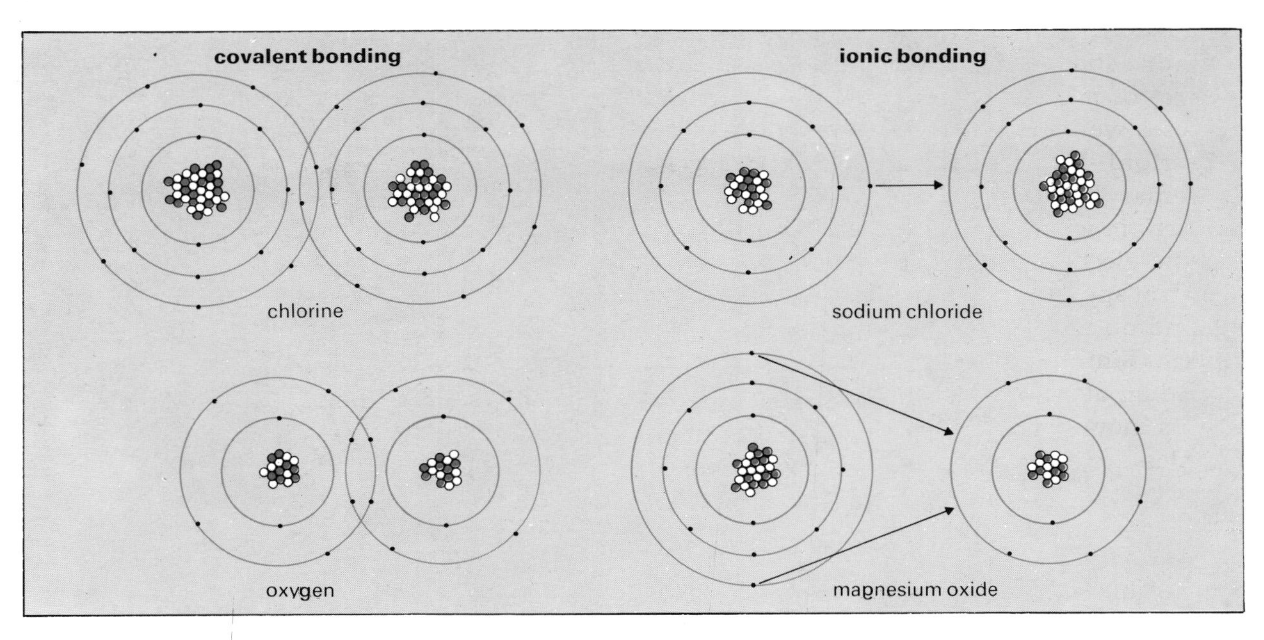

Right: In the covalent bonding of two atoms of chlorine one electron from each is shared. In oxygen, two from each atom are shared. In the ionic bonding of sodium chloride sodium gives its spare outermost electron to the chlorine, which then has a stable electron configuration. In magnesium oxide, magnesium gives two electrons

Valencies of some simple and complex ions

	cations					anions			
valency	5+	4+	3+	2+	1+	1−	2−	3−	4−
	P	Si	Al	Be	H	F	O	N	C
		Sn	Cr	Mg	Li	Cl	S	P	(SiO_4)
		Pb	Fe	Ca	Na	Br	Se	As	
		Ti	Mn	Sr	K	I	Te	Sb	
		Mn	Ti	Ba	Rb	(OH)	(CO_3)	(BO_3)	
		Zr	V	Fe	Cs	(NO_3)	(SO_3)	(PO_3)	
		W		Co	Cu		(SeO_3)	(AsO_3)	
		Pt		Ni	Ag		(Al_2O_3)	(SbO_3)	
				Mn	Au		(SO_4)	(PO_4)	
				Zn	(NH_4)		(CrO_4)	(AsO_4)	
				Cd			(MoO_4)	(VO_4)	
				Hg			(WO_4)		
				Pb			(SeO_4)		
				Cu					

up to a maximum. For instance, the limit is 2 for K, 8 for L, 18 for M and 32 for N. The lighter elements, those with up to 18 protons in their nuclei, fill their electron shells systematically outwards. Thus, hydrogen has 1 nuclear proton, attended by a single orbiting electron in its K shell. Helium, the element with 2 protons, has a complete K shell containing 2 electrons. Lithium, with 3 protons, has a full K shell plus a solitary electron in L. The simple shell-filling pattern continues up to argon, with 18 protons. Thereafter, elements of higher atomic number show a confusing tendency to allow electrons into an outermost shell before inner shells are filled. We need not worry about the details of this here.

Next we must consider a peculiar group of elements known as the *inert gases*. The first of these is helium, with 2 electrons that fill its K shell. The others all have 8 electrons in their outermost shell. These are the rare gases neon, argon, krypton, xenon and radon. They are called inert because it is practically impossible to combine them chemically with any other elements. This fact gave rise to the idea that chemical stability results from outermost electron shells having the same contents as those of the inert gases. From this has evolved the basic principle governing our understanding of how elements combine: that atoms link by swapping or sharing electrons so that their electron clouds can become identical to those of the inert gases.

Let us look at this process of chemical combination a little more closely. Where electron swapping is involved, it is easiest to think of it in two distinct stages. First, the atoms of an element gain or lose electrons until they reach the configuration of the nearest inert gas. In doing so the atoms become electrically charged. They are then known as *ions*. Ions with gained electrons are negatively charged and are known as *anions*, whilst ions with lost electrons are positively charged and are called *cations*.

The number of electrons which an atom must gain or lose in order to reach the electronic configuration of the most similar inert gas is known as its *valency*. Let us take some examples. The elements sodium (Na) and fluorine (F) have one more and one less electron, respectively, than the inert gas neon (Ne). Their ions are written thus: Na^+ and F^-. Both ions are *monovalent*; Na^+ is a cation and F^- is an anion. Magnesium (Mg) has two more electrons than neon. By losing both of these it becomes Mg^{2+}, a *divalent* cation.

Elements such as Na and Mg, with electronic configurations close to those of inert gases, can only form one sort of ion. Others, such as carbon (C) and silicon (Si), have atomic structures midway between two inert gases. Thus, carbon can either gain 4 electrons to become C^{4-} and resemble neon, or lose 4 electrons and become C^{4+}, resembling helium. Both situations occur in nature and carbon may be an anion or a cation.

A final complication affects elements whose atoms have incomplete inner electron shells. Many of these can form ions with several valencies. For example, iron (Fe) may occur as divalent Fe^{2+} or trivalent Fe^{3+}, depending on various circumstances. A list of the ions and valencies of elements forming common minerals is given in the table on the previous page.

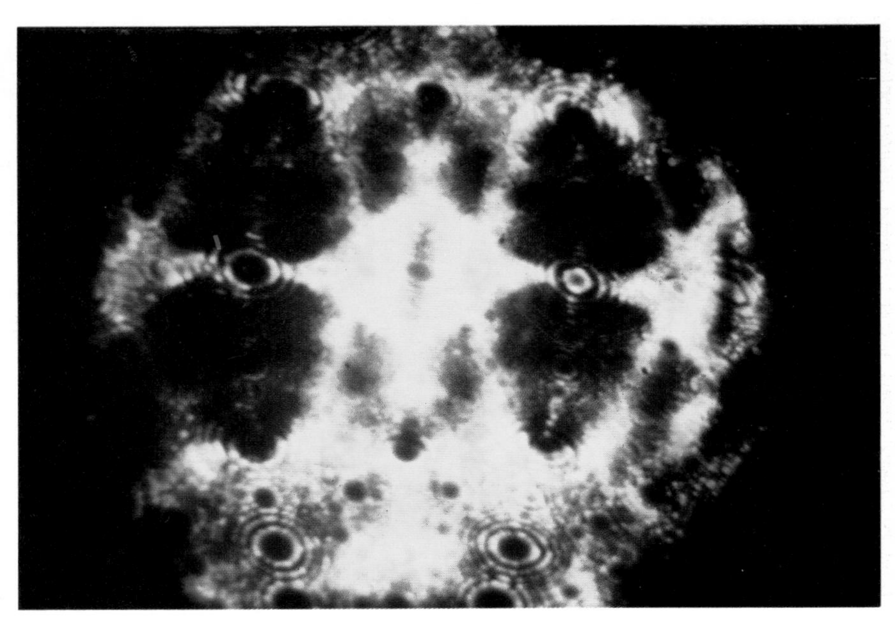

Above: The atomic pattern of a crystal of the metal ruthenium. This photograph, made with a field-ion microscope, shows a tiny area of the crystal's surface magnified five million times. Each bright spot represents a single ruthenium atom

The linking of atoms

Electrons are so tiny that the cloud they form around an atomic nucleus is largely empty space. In view of this it is reasonable to wonder why the nuclei of adjacent atoms do not collide and fuse together to form larger atoms. The answer lies in the electrical charges of their protons, which keep them comparatively far apart by mutual repulsion. As a result, when two atoms link together they do so by means of interactions of their outermost electrons.

There are three main ways in which atoms may be attached, or *bonded*. The *ionic bond* is the commonest type found in minerals. In its simplest form it arises between two elements that can form ions of equal and opposite valency. The 'surplus' electron, or electrons, of one atom become separated from it and fill the vacancies in the outermost shell of the other. The pair of ions thus formed are held together by electrical forces. The bond can also link groups of ions, such as a divalent cation with two univalent anions. Some examples are illustrated on the preceding page.

The *covalent bond* is an electron-sharing arrangement. Two or more atoms coming close together can join in this way by sharing electrons so that all atoms effectively have 8 in their outermost shells.

Finally, the metals, such as iron, copper and nickel, have a distinctive type of *metallic bond* in which the ions of the element pack closely together and leave their outermost loosely-bound electrons free to move around comparatively independently.

Substances possessing the different types of bonding acquire distinctive features from the differing ways in which their atoms are linked. For instance, metals may be beaten into thin sheets or drawn into wires because metallic bonding allows their constituent atoms to glide easily over each other. Similarly, the ability of metals to conduct electricity lies in their mobile clouds of loosely held electrons. Covalent and ionic bonds are both strong but the former are established in specific directions around an atom, whereas the latter are non-directional. Just to confuse the issue, it is common for bonds to be intermediate in character between these two extremes. Solids containing ionic or covalent bonding are generally relatively rigid, so that they fracture rather than deform when hit with a hammer.

We must next consider the products of atomic bonding. What are they called? If the linked atoms are of more than one element, a chemical *compound* is formed. This is distinguished easily from a mere mixture of the same elements, because the requirement that a bonded structure should be electrically neutral guarantees that its component atoms are combined in a fixed ratio.

The linkage of elements in a compound is not necessarily all of one type. For instance, it is common for sub-groups of atoms to be bonded strongly together in a predominantly covalent way, forming what are known as *complex ions* or *radicals*. An example of this is the tetrahedral arrangement of 4 oxygen atoms around 1 of silicon to give a complex anion, which may be written in chemists' shorthand as $(SiO_4)^{4-}$. These clusters of atoms bond together ionically with either single atoms or other groups in a very wide variety of mineral structures.

Let us use the $(SiO_4)^{4-}$ complex anion to

Above: Three ice crystals. The detailed shape of every ice crystal differs from that of all others, but all show a hexagonal (sixfold) symmetry

illustrate the next point. $(SiO_4)^{4-}$ must be bonded to a cation or cations with a total valency of 4, in order to produce an electrically balanced compound. Two atoms of the divalent metal magnesium (Mg) would be appropriate and, indeed, these are linked with an (SiO_4) group in the mineral forsterite. The chemical shorthand for forsterite is Mg_2SiO_4. This is known as its *formula* and you will find a similar one given for every mineral described in the Mineral Kingdom section of this book. When a formula is written in this ultra-abbreviated form the reader is expected to work out several things for himself. For instance, the addition of information about valencies and a bracket around the distinctive (SiO_4) group would result in forsterite symbolized thus: $(Mg_2)^{4+}(SiO_4)^{4-}$. It is an interesting exercise to re-write some of the formulae in the reference section of this book in the same way.

ATOMIC STRUCTURES

Having discussed the nature of atoms and how they link together, we are now in a position to consider the ways in which these fundamental scraps of matter are built into minerals. At this point is is necessary to restrict our scope a little. The term 'minerals' is often taken to include naturally occurring organic materials—that is, those formed from living or once living things. These include coal, oil and gas, as well as coral, pearl, jet and so on. Here and throughout this book the term is used in its stricter sense of 'inorganic natural materials' (though some organic materials of interest to the gem-fashioner are listed in an appendix to the Mineral Kingdom).

Solids, liquids and gases

You certainly do not have to be a paid-up member of the chemists' union to know that a compound such as water (H_2O) may occur in solid, liquid or gaseous forms, depending on its temperature. Heating a material causes its atoms to vibrate faster, so that the bonds between them become broken. Thus, in steam the independent H_2O groups, called *molecules*, fly around freely. In water they are loosely linked, whereas in ice the bonding becomes rigid, forming a crystalline solid. In the chapter of this book called The Crystalline State it is explained that crystals are distinguished from other types of solid by possessing symmetry on both the atomic and visible scale. This feature is not very apparent in ice on a pond or fresh from a refrigerator. The best way to see it is to catch a snowflake (also made of ice). You will find that it has nearly perfect hexagonal (sixfold) symmetry, though its detailed shape can be almost infinitely varied.

You can find specimens of non-crystalline materials without the trouble of going out of doors. Glass is an important example. This artificial material is produced by cooling or *quenching* a molten mixture of minerals so rapidly that the loose interatomic bonding of the liquid is 'frozen' before it has time to become organized into crystals. Glassy rocks (remember that a rock is a mixture of minerals) are not uncommon as rapidly quenched products of volcanoes. There are not many kinds of glassy mineral, however. One interesting example is maskelynite, which is formed when a meteorite strikes a crystal of the common mineral plagioclase, shaking its bonds until they are totally disorganized. This mineral is extremely rare on the earth, although comparatively common on the moon.

There is a range of minerals whose structures fall between the regimentation of the atoms in crystals and the disorder of those in glasses. These are known as *amorphous* compounds. One of the best examples of them is the gemstone opal, a form of the compound SiO_2 with some attached water. The distinction between such minerals and those composed of aggregates of extremely tiny crystals is of course ill-defined.

Coordination polyhedra

A crystal may be likened to a brick wall, in which a large number of simple units (the bricks) are bonded together to make a final product with a complicated-looking pattern. Strictly speaking, the fundamental particles within crystals are ions. However, if we equate them with bricks, we run into a snag which would tease the patience of a master brick-layer; ions are essentially spherical in shape and occur in a large range of sizes. Nature resolves this architectural problem by packing ions together in two stages. Firstly, ions of diverse sizes combine into small groups. These groups can be imagined to be bounded by simple solid shapes with flat faces, called *coordination polyhedra*. These imaginary polyhedra can be packed together, building up the crystal.

If you assemble a collection of rigid spherical objects, such as oranges, billiard balls, marbles and peas, you can work out the shapes of the polyhedra yourself. Let us start with a small sphere (a pea) surrounded by large ones (oranges). Remember that these are meant to be ions and must therefore all touch, in order to form bonds. The best fit of oranges around a pea is three around one. Draw a diagram showing this, join up the centre points of the oranges, and you have a *triangle*. This is the shape of the very common complex anion $(CO_3)^{2-}$, called the *carbonate* ion. If you now choose a slightly larger small sphere, you will find that the number of oranges that fit around it increases to four and that these are at the corners of a pyramid-like *tetrahedron*. This is the configuration of the principal building block in the earth's crust, the complex *silicate* anion, $(SiO_4)^{4-}$. As the ion within the polyhedron becomes still larger, the number of ions around it increases to six and these are arranged at the corners of an *octahedron*, resembling two pyramids joined base-to-base. Finally, as the 'small' ion in the centre becomes comparable in size with those surrounding it, the 'large' ions come to number eight and arrange themselves at the corners of a *cube*.

Ions of equal size can link together to form a crystal without first assembling into polyhedra. Like bricks in a wall, they can be bonded together in a number of slightly different patterns, all of which are known as varieties of *close packing*.

The concepts of coordination polyhedra and close packing are applicable principally to crystals with ionic or metallic bonding, because these are non-directional links between crystals. However, as the oriented bonds in covalent structures usually point in the same directions as the principal axes of the polyhedra, these fundamental atomic groupings can usually be recognized in covalent-bonded minerals. For instance, blende (ZnS) is covalent but may be thought of as consisting of interpenetrating close-packed arrays of zinc and sulphur atoms.

Structural variation: polymorphism

Heat and pressure are powerful opposing influences upon the way in which the atoms of an element or compound combine to form crystalline minerals. When a material gets hotter its atoms vibrate faster, so that they fit best into loosely-packed crystal structures. Pressure compresses the electron clouds of the atoms, encouraging them to form dense crystals. For example, the compound SiO_2 can crystallize in five different major structures known as *polymorphs*, depending on the pressure and temperature. These are cristobalite (found only in volcanic lavas), tridymite, quartz, coesite and stishovite (found only in the impact areas of meteorites, but predicted to occur at depths of more than 400 km in the earth). Stishovite is nearly twice as dense as cristobalite, despite their identical chemical compositions.

The minerals diamond and graphite are

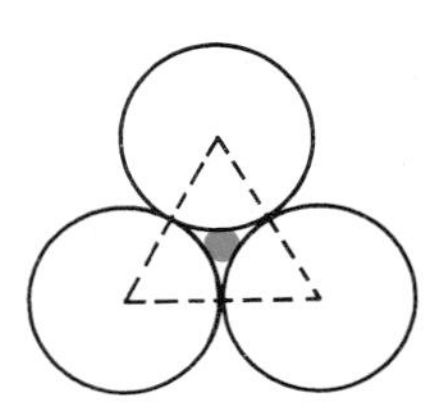

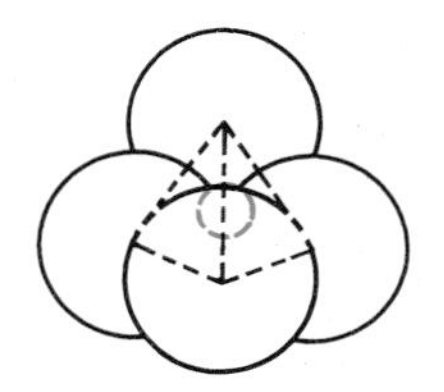

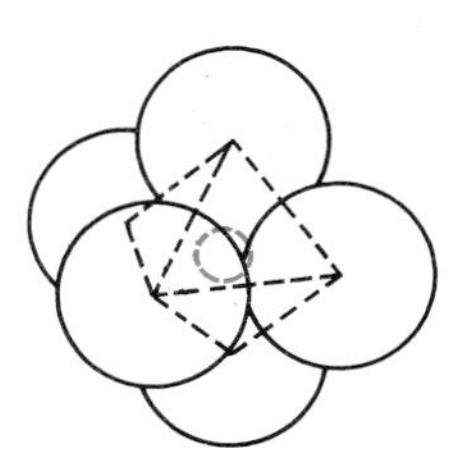

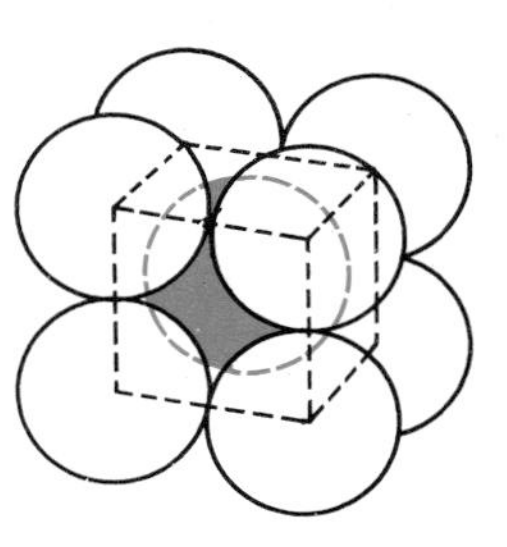

Above: The basic coordination polyhedra: triangle, tetrahedron, octahedron and cube

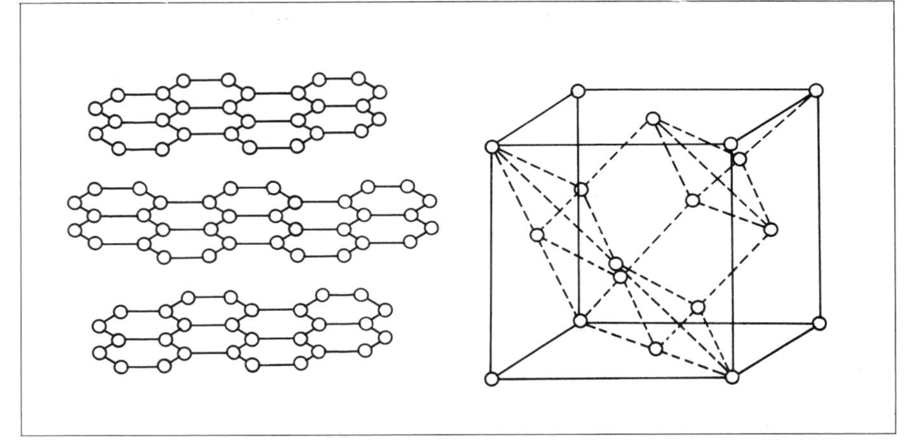

Below: The atomic structures of diamond (left) and graphite. Both materials are pure carbon, but diamond is hard while graphite is soft and easily cleaved

Above: Thin section of dunite seen through crossed polarizers. Individual grains are revealed in different colours. Dunite is a rock consisting entirely of members of the olivine series

another pair of strikingly different polymorphs. Both are composed entirely of the element carbon. In graphite the carbon atoms are linked into sheets which are almost completely free to slide over each other. The resulting soft, black, greasy-feeling mineral is used for lubrication and in pencil leads. In diamond, the high-pressure polymorph, the carbon atoms are held closely together by a three-dimensional network of rigid tetrahedrally oriented covalent bonds. The resulting mineral is our immensely hard, tough, dense, glittering 'best friend'. Diamonds form at great depths in the earth and only reach the surface because they are transported upwards very rapidly in volcanic pipes.

Identical structures: isomorphism

You will remember that the previous discussion of coordination polyhedra emphasized the importance of the size of ions in determining the crystal structure adopted by a mineral. Many pairs of ions are similar in size and can substitute for each other in a mineral without changing its structure. Let us take an example. The minerals merwinite, $Ca_2CaMg(SiO_4)_2$, and brianite, $Na_2CaMg(PO_4)_2$, have identical crystal structures: they are thus *isomorphous*. Brianite is a *phosphate* because it contains the tetrahedrally coordinated complex anion $(PO_4)^{3-}$. Notice that, because the phosphorus ion P^{5+} is quinquivalent and the oxygen ion O^{2-} is divalent, the $(PO_4)^{3-}$ group is trivalent; the $(SiO_4)^{4-}$ group in merwinite is tetravalent because the silicon ion, Si^{4+}, itself tetravalent, is combined with the divalent oxygens.

It follows that the cations in each merwinite complex have to possess 8 positive charges in all, and the brianite cations have to possess only 6, in order to achieve electrical balance. Accordingly, a monovalent Na^+ ion in brianite replaces two of the divalent Ca^{2+} ions in merwinite. By the way, if you can understand the last paragraph you have grasped most of the contents of this chapter!

Despite the identity of their crystal structures, there is a complete chemical gap between merwinite and brianite. The former is almost devoid of phosphorus and the latter of silicon. However, some isomorphous pairs of minerals show a different chemical relationship. For instance, let us take the pure magnesium and iron members of the group of minerals known as olivines. These are called forsterite, Mg_2SiO_4, and fayalite, Fe_2SiO_4. If we melt a mixture of these minerals and allow the melt to cool, the crystals that form are no longer of forsterite and fayalite but of a single olivine containing both Mg and Fe. You can think of this as the forsterite and fayalite dissolving in each other; hence its name, *solid solution*. A complete series of olivines intermediate between forsterite and fayalite exists in nature. In fact solid solution is extremely common in most mineral groups.

At this point we must return to the subject of chemical formulae to see how these cope with solid solutions. The method is simple; our magnesium-iron olivine is written $(Mg,Fe)_2SiO_4$. If chemical analysis shows that the olivine contains, say, 75 percent of the forsterite component, this may be written into the formula as $(Mg_{0.75}Fe_{0.25})_2SiO_4$.

The next aspect of solid solution we must think about is the effect of temperature upon it. This is quite dramatic, because the faster thermal vibration of atoms at high temperatures causes crystal structures to 'stretch' and become more loosely bonded than at low temperatures. In this expanded state it is possible for the structure to incorporate ions that would be rather too large for their sites if the temperature were lower. Sure enough, if a high-temperature solid solution of this type is now cooled, the oversize ions are squeezed out and the crystal unmixes into two separate mineral species. These usually remain closely intergrown in grains or layers which are so small that they can only be seen with the aid of a microscope. This unmixing process is known as *exsolution*.

The final variety of solid solution to note here is a rather peculiar one. Despite all that was discussed previously about the importance of electrical balance amongst the anions and cations in a crystal structure, some minerals defy this restriction to a limited extent. For instance, pyrrhotine has the formula $Fe_{1-x}S$, where x ranges from zero to 0.125 in various specimens. The missing Fe^{2+} cations are not replaced by other kinds of cation; they simply leave little holes, rather quaintly known as *vacancies*, in the structure. Pyrrhotine's natural magnetism is due to these vacancies.

CLASSIFICATION OF MINERALS

It is a common failing of the classifiers of naturally occurring materials that they attempt to force their schemes to be what they are not. The classificatory systems of such people, both ancient and modern, are all part of the deep-rooted instinct of mankind to seek an aesthetically satisfactory comprehensive order in nature. The classifier is on a course tangential to that of the day-to-day work of science. This requires no more of classifications than that they should place the material (minerals, rocks, plants or whatever) in a series of pigeon holes, convenient for use in museums or, for that matter, in the Mineral Kingdom section of this book!

Classification by chemical composition has now superseded earlier and simpler systems. The order of the Mineral Kingdom's entries is based on a scheme proposed by the eminent British mineralogist M H Hey. This scheme groups together minerals that have the same *anion* (simple or complex). The system works well unless you want to do something like comparing the various minerals of an element that always forms cations, such as sodium or copper. To find these minerals you will have to rummage through all the anionic groups.

The next few pages contain some general notes about the chemistry and crystal structures of the principal mineral groups in the Mineral Kingdom. Their order has been re-arranged here to correspond as closely as possible with their abundance in the earth's *crust*, the outermost 20 km or so of our planet. Oxygen is much the most common element by weight. But its anion is considerably larger than those of the other common elements, such as silicon and aluminium. As a result, the *volumetric* abundance of oxygen in the earth's crust is an amazing 94 percent. This means that, when you contemplate a mighty mountain range, you are essentially looking at a vast heap of oxygen! If you could disregard the laws of chemistry for a moment and remove all the atoms of elements other than oxygen from your view, the remaining 'oxygen mountains' would be almost imperceptibly smaller than their rocky predecessors. We now look at the commonest mineral groups.

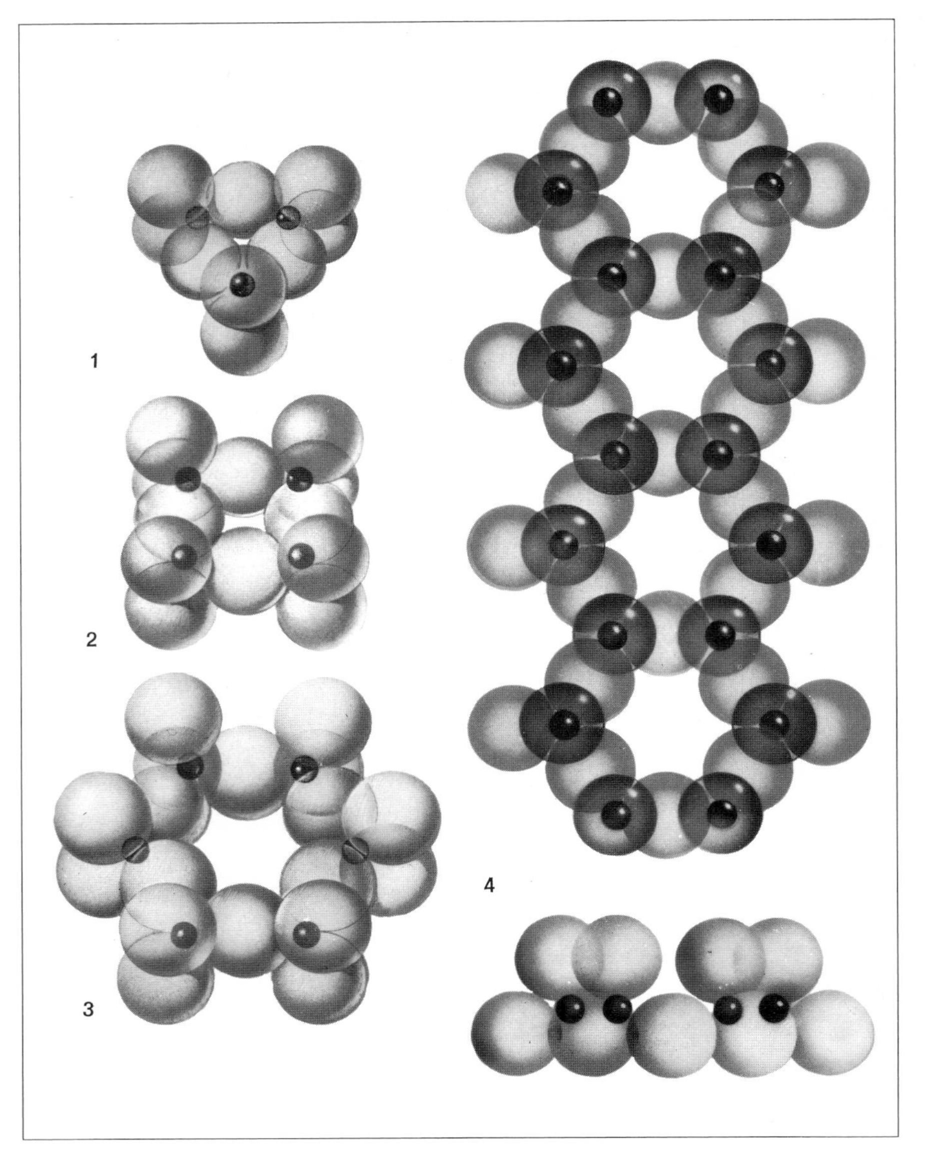

Above: Silicate structures. (1) Si_3O_9; (2) Si_4O_{12}; (3) Si_6O_{18}. These are all ring silicates, which tend to form very hard crystals. (4) A double chain silicate, Si_4O_{11}, which forms moderately hard crystals

Right: A SiO_4 tetrahedron (top) found in minerals such as olivine; and a double tetrahedron, Si_2O_7, found in, for example, melilite

Silicates. The element silicon has a crustal abundance next to that of oxygen, with which it readily combines to form the tetrahedrally coordinated complex silicate anion $(SiO_4)^{4-}$. Of the major groups of minerals the silicates are the most abundant in the earth's crust. They show

Right: Magnesite crystals growing on quartz, a framework silicate

a great variety of physical appearance and properties because the SiO_4 tetrahedra can join together in several distinctive ways. The tetrahedra can occur as separate units (unlinked), pairs, rings, chains, double chains, sheets or frameworks.

In silicates with *independent tetrahedra* the SiO_4 groups are held tightly together by interspersed cations. Two examples are olivine, $(Mg, Fe)_2 SiO_4$, and garnet, $X_3Y_2Si_3O_{12}$ (X is a divalent cation such as Ca^{2+}, Mg^{2+} and Fe^{2+}; Y is a trivalent ion such as Al^{3+}, Fe^{3+} or Ti^{3+}). Members of this group tend to form equidimensional (not elongated) crystals that are dense and hard.

There are very few silicates with *pairs* of SiO_4 tetrahedra but their close relatives, the *ring silicates*, are a little more common. The crystal symmetry of a ring silicate depends on the number of SiO_4 groups in the ring. Thus beryl, $Be_3Al_2Si_6O_{18}$, contains six SiO_4 tetrahedra in its rings and has hexagonal symmetry. As the diagram shows, the tetrahedra share oxygen anions at the corners that touch; this is why the ratio of oxygens to silicons is reduced from 4 to 1 for a single tetrahedron to 3 to 1 for the rings. The rings tend to stack on each other, producing elongate crystals that are moderately dense and very hard.

The *chain* and *double-chain silicates* are closely related, forming the prominent mineral groups known as the pyroxenes and amphiboles, respectively. A typical example is the pyroxene diopside, $CaMgSi_2O_6$, in which 8-coordinated Ca and 6-coordinated Mg cations bind together the SiO_4 chains. Chain and double-chain silicates have moderate hardness and variable density, the latter being low if the mineral is rich in Mg and high if it is rich in Fe or Mn (manganese). Two interesting chain silicates are jadeite, $NaAlSi_2O_6$, a pyroxene that forms the ornamental stone called jade, and tremolite, $Ca_2Mg_5Si_8O_{22}(OH)_2$, an amphibole sometimes occurring in a fibrous form, which is one of the varieties of commercial asbestos.

In *sheet silicates* the SiO_4 tetrahedra are linked by three of their corners into continuous layers which are bound together by cations. The principal minerals in this group are the micas, talc and the clays. As the bonding between sheets is much less strong than that within them, these minerals are all comparatively soft and split easily into very thin wafers. This is why talc (as a crystal or in talcum powder) has a characteristic 'greasy' feel, exactly like that of the sheet-structured mineral graphite, which we discussed in the section above on polymorphism. Another economically important example of this ease of splitting is provided by muscovite, $KAl_4Si_3O_{10}(OH)_2$, a mica which has long been used to provide efficient electrical and thermal insulation in the form of very thin sheets and remains valuable for this purpose.

The *framework silicates* have their SiO_4 tetrahedra linked together at all four corners to give a rigid three-dimensional network. Unless another element is substituted for some of the silicon, this network is electrically balanced without the addition of further cations and forms the common oxide mineral quartz, SiO_2, and its rarer polymorphs. In other framework silicates, such as the feldspars and zeolites, between a quarter and a half of the Si^{4+} ions in the framework are replaced by Al^{3+}. In these *aluminosilicates* other cations, such as Na^+, K^+ and Ca^{2+}, are incorporated into the framework to make up the required number of positive charges. A typical example is provided by the feldspar albite, $NaAlSi_3O_8$. The feldspars, which are the principal aluminosilicate group, are also by far the most common minerals in the earth's crust, forming slightly more than half of it (by weight or volume). The relatively rigid, partially covalent Si–O bonding in framework silicates causes them to have comparatively loosely packed structures. The resulting minerals combine great hardness with moderately low densities.

Oxides and hydroxides. A mineral is an oxide when its anion is oxygen alone. If its sole anion is the complex $(OH)^-$, it is a *hydroxide*. The oxides are quite a widespread mineral group, which occur in almost every type of crustal rock. In the many oxides that have cations considerably smaller than the O^{2-} anion the structure of the oxygens is essentially close-packed and gives rise to dense, compact minerals, which are in many cases extremely hard. Corundum, Al_2O_3, is a good example; two of its coloured varieties are the gemstones ruby and sapphire. The very important ore minerals hematite, Fe_2O_3, and cassiterite, SnO_2, a tin oxide, are representative of the many oxides of great economic value. A large subgroup of oxides, called the spinels, have the general formula XY_2O_4, where X is a divalent and Y a trivalent cation. An example is chromite, $FeCr_2O_4$, the principal ore of chromium. Hydroxide minerals are mostly formed by the reaction of oxides in rocks with water. They are characteristically soft and in the form of aggregates of very tiny crystals.

Phosphates. These minerals contain the complex anion $(PO_4)^{3-}$ which resembles the silicate radical closely except that, because phosphorus is quinquivalent, the PO_4 group is only trivalent. Like silicates, PO_4 tetrahedra can occur separately in crystal structures or link into pairs or rings, but not the more complex networks. Although phosphorus is the eleventh most abundant element in the earth's crust, its minerals are not very prominent. The commonest and most widespread is apatite, $Ca_5(PO_4)_3(OH, F, Cl)$, which is a minor constituent of many igneous rocks. Animal bone is composed of calcium phosphate, and sedimentary deposits of fossil bones, recrystallized to very fine-grained aggregates of apatite, are the world's main sources of phosphorus for use in fertilizers.

Carbonates. Carbon is only twelfth in abundance amongst crustal elements but the carbonates, which contain the complex anion $(CO_3)^{2-}$, are a prominent group of minerals. The reason for this is that many common forms of marine life, such as corals, extract dissolved carbonate ions from sea water and incorporate them into their skeletons or shells. Vast accumulations of the hard parts of such organisms become buried in the ever-thickening layers of sediment on the sea floor and become the rocks we know as limestones. The predominant carbonate mineral in limestone is calcite, $CaCO_3$, though dolomite, $CaMg(CO_3)_2$, is equally important locally. As you may remember, the CO_3 group is triangular. This imparts threefold symmetry to the crystals

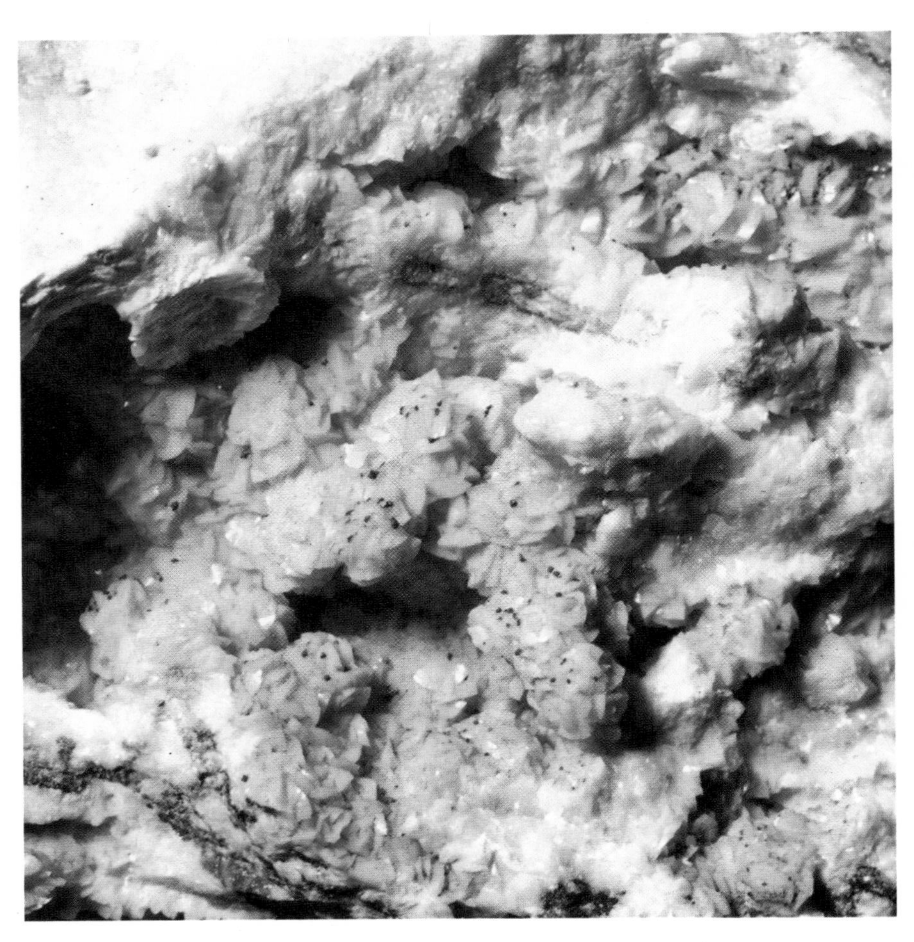

Above: Pink dolomite from Sardinia. Dolomite is a major component of limestone in some places

of calcite and many other carbonates. However, in minerals such as witherite, $BaCO_3$, where the cation is comparatively large (1.35 angstroms radius), the ions pack in such a way as to give orthorhombic symmetry. The compound $CaCO_3$ exhibits polymorphism because, although usually crystallizing in the hexagonal system as calcite, it also occurs as a less common orthorhombic mineral called aragonite. Most carbonates are comparatively soft and soluble in dilute acids.

Sulphates. Sulphur, the fourteenth most abundant element in the earth's crust, is found in nature as the pure element, as the anion S^{2-} in sulphides and as the complex anion $(SO_4)^{2-}$ in sulphates. Gypsum, $CaSO_4.2H_2O$, and its close relative anhydrite, $CaSO_4$, are by far the most common sulphate minerals. They occur in sedimentary rocks and are often associated with salt deposits because, like them, they are precipitated from evaporating bodies of highly saline water in landlocked lakes or at the margins of the sea in hot desert areas, such as the Persian Gulf. The only other relatively common sulphate mineral is baryte, $BaSO_4$, which typically occurs in veins deposited by hot circulating groundwater within the earth. Its chief uses are in the

Right: A fine group of baryte crystals, a fairly common sulphate. This specimen was part of the crystal lining of a druse or rock cavity

Right: Violet apatite from Saxony. Apatite is one of the world's main sources of phosphorus, used in fertilizers

special mud that is pumped down an oil well during drilling to carry rock fragments up to the surface, and in the manufacture of high-quality paper.

Sulphides. These minerals have S^{2-} as their anion. They crystallize only in places where the concentration of oxygen is low, so that the formation of sulphate is prevented. Most of the major metal ores are sulphides. They usually have high densities and variable hardness, mostly low. The metal and sulphur atoms link in a variety of structures, with bonding more frequently covalent than ionic. The only sulphides that you will meet frequently outside mining areas are those of iron, namely pyrite, FeS_2, and pyrrhotine, $Fe_{1-x}S$ (discussed above).

Halides. The last group of minerals that we have space to mention in this chapter contain as their anion one of the group of elements called *halogens*. These are fluorine, chlorine, bromine and iodine. The bonding in most of the halide minerals is ionic, with cubic symmetry characteristic of all the commoner species. A halide mineral that is widespread at the earth's surface is fluorite, CaF_2, a characteristic associate of

Above: Salt mounds at Trapani, Sicily. The salt comes from deposits formed by the evaporation of sea water in manmade lagoons. It consists of a mixture of halides, among which sodium chloride predominates

calcite and baryte. Other common halides, halite ($NaCl$), sylvine (KCl) and carnallite ($KMgCl_3 . 6H_2O$), are all formed by the evaporation of brine in such places as inland salt lakes. They have low densities and are very soft and soluble in water. If you want to check this, you will have to visit a salt mine; halides are dissolved by groundwater if they are near the surface of the earth and are only preserved where they are buried beneath impervious clay.

METHODS OF CHEMICAL ANALYSIS

The procedure for identifying an unknown mineral begins with a study of such physical properties as hardness, specific gravity and refractive index, along the lines indicated in the chapter in this volume called Identifying Minerals. The next step is to study its atomic structure by means of x-ray diffraction. Measuring the angles of the three most intense diffracted rays is usually sufficient to identify the mineral from standard tables, unless it is your misfortune to have found a new one!

Naming a mineral is a start, but detailed study requires complete chemical analysis. The principal current analytical techniques will be summarized in this section; they call for apparatus and reagents that are far outside the scope of the amateur.

Wet chemistry

In its simplest form this method involves dissolving the mineral and treating the resulting solution with a series of reagents. These are chosen to precipitate each constituent element in turn as part of a known compound, which may be filtered, dried and weighed. In oxygen-containing compounds the components are taken to be simple oxides, and the proportions of these by weight are calculated. In cases where the mineral is an oxygen-free type, such as a sulphide, the analysis is given in the form of separate elements.

Our modern knowledge of mineral chemistry is based on the accumulated labours of a line of skilled 'wet chemists' extending back into the last century. Their methods have been modernized and speeded up during the last few years by, for example, the use of reactions which convert each element into a brightly coloured solution. The intensity of these colours may be measured very accurately with a photoelectric

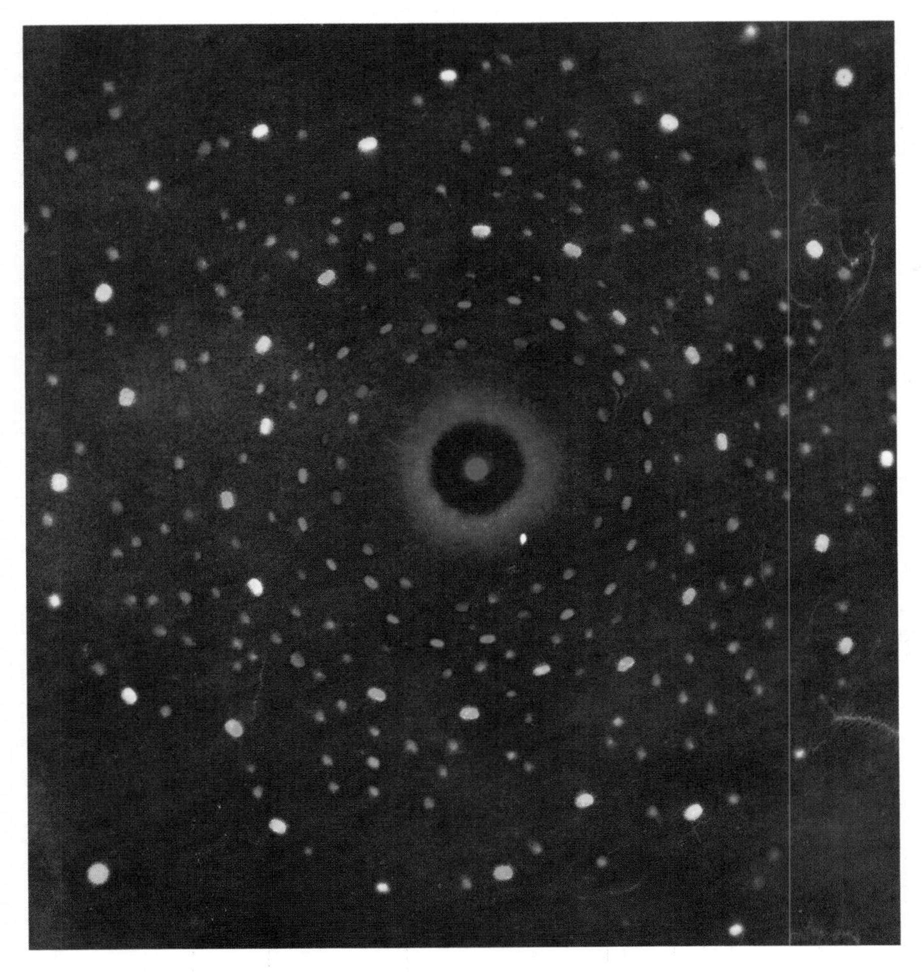

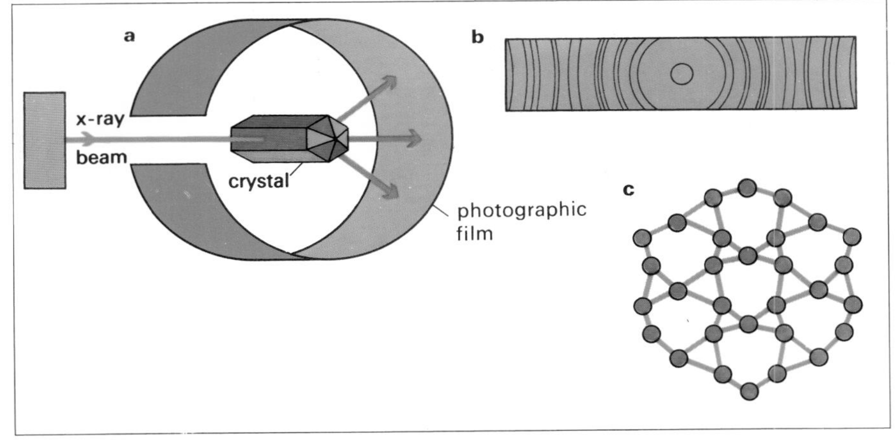

Above: X-ray diffraction. (a) x-rays are passed through a crystal; (b) rays that are diffracted (scattered) are registered on film; (c) the atomic pattern is then worked out

Top: X-ray diffraction pattern of quartz

cell. However, such methods remain too slow to produce the vast amounts of data needed for such purposes as mineral prospecting or quality control on the mineral input of industrial processes. The physical methods that we shall now consider can be automated to handle large-scale projects.

Physical methods

A good analogy for the basic principle used in most of the physical analytical methods is the beehive. If you throw a brick at a beehive, the occupants rush about in an extremely agitated manner and emit an angry buzz. Similarly, if energy is pumped into an atom, its orbiting electrons begin hopping from shell to shell, emitting electromagnetic radiation such as light or x-rays. The radiation from each element is characteristic and can be measured accurately.

The energy used in conjunction with the *optical spectrograph* is heat. A powder of the mineral is vaporized between the electrodes of a carbon arc. The light emitted is then split up into its component wavelengths and recorded on a photographic plate. For *x-ray fluorescence spectrography* the mineral is formed into a compressed powder or fused glassy disc and is irradiated by a powerful x-ray beam. The atoms of the target are stimulated to emit *secondary* x-rays, whose intensities are compared with those of the radiation from a standard sample. The chemistry of tiny crystals and the components of complex mineral intergrowths may be determined nondestructively by use of the *electron probe microanalyzer*. Its energy source is an electron beam, which may be focused to hit an area of the polished target mineral no more than 10,000 angstroms in diameter. Secondary x-rays are emitted from the impact area.

The two remaining major physical methods of analysis both involve tampering with the nuclei of atoms in the mineral. *Neutron activation* may most easily be understood by considering elements such as uranium and thorium. In nature these elements consist of mixtures of isotopes, some of which are radioactive. As these unstable atoms 'decay' one by one to stable products, they emit gamma rays, which are close relatives of x-rays. A gamma-ray detector is therefore all that is needed to measure the uranium or thorium content of a mineral. In order to tackle other elements nature must be helped a little by bombarding them with neutrons from an atomic reactor until they are partially converted to unstable isotopes, each of which emits characteristic gamma rays.

The final method for us to consider is *mass spectrometry*. Here the mineral sample is vaporized in a high vacuum and passed through the field of a powerful magnet. The product is a stream of ions, which are separated as they pass through the magnetic field in a manner analogous to the splitting of light in a prism. As the ions are segregated according to their nuclear masses, it becomes possible to measure the abundances not merely of the elements but of the individual isotopes in the mineral.

The Crystalline State

Left: Chabazite crystals (white) enclosed in pink apophyllite

Below: Atomic arrangements in an amorphous (disorderly) solid contrasted with those in a crystal, where long range order is the dominant feature

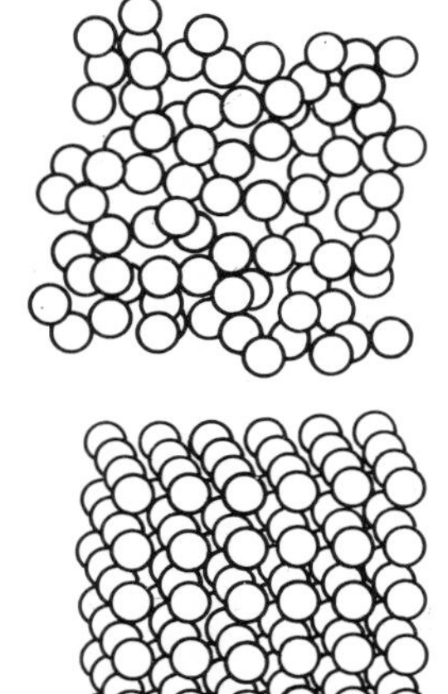

From the earliest times, even casual observers have noticed rare, striking mineral forms occurring in nature, possessing flat faces meeting in well defined edges, and having a certain degree of overall symmetry. Not all the faces of such a formation would be flat, and those that were would invariably have many flaws. Such a body was like a highly imperfect copy of some ideal geometrical shape.

Such materials sometimes occur in transparent form. The Greeks named them *krustallos*, meaning 'clear ice'. The word became *crystal*, which was applied to any mineral body of regular shape bounded by flat surfaces.

Today we know that crystals, far from consisting of ice, can be formed by almost any solid substance. They owe the regularity of their outward appearance to their atomic structure—to the way in which their atoms are arranged to make a three-dimensional pattern. In crystals this arrangement is extremely orderly in ways that will be explained in more detail below. A material that has a very orderly arrangement of atoms over distances large on the atomic scale is called *crystalline*. If for some reason regular faces are prevented from developing when a crystalline substance is formed, there may be no outward sign of the underlying atomic regularity. Most minerals are like this—orderly at the atomic level, while possessing rough and irregular outer surfaces. Well shaped crystals are simply masses of crystalline matter that have formed under conditions favourable to the production of regular faces. So mineralogy is largely concerned with the crystalline state.

The opposite of crystalline is *amorphous*, 'without form'. Amorphous materials do not have a regular atomic structure. They are rather rare among minerals. However, many artificial materials, including glass and plastic, are amorphous at the atomic level.

THE GROWTH OF CRYSTALS

Nearly all crystals form by the repetitive addition of new matter to a growing crystalline mass. Some crystals form by the solidification of a melt (a molten material), some are built up from vapours, and some from liquid solutions. The last include some of the finest crystals for the collector of minerals.

Crystals derived from melts can be seen in granite, which is widely used in facing stones on public buildings. The crystals in granite are relatively large and consist of quartz (grey, white, or colourless), feldspar (white or pink), and mica (black or white). The molten rock mass cooled slowly and atoms of the mineral constituents grouped together to form the essentially regular structure determining the final crystal shape. Some minerals crystallize before others; the first may form crystals with flat faces, but later crystals are crowded together as they grow, and their faces do not develop so well.

Common salt dissolves in water at room temperature, but quartz crystals do not. However, quartz does dissolve in water at temperatures and pressures far above the normal range, and synthetic quartz is made from such solutions. Many natural crystals grow from watery solutions, including quartz, calcite, the zeolites and some clay minerals. Crystals grown from vapour are commonly found in volcanic areas, particularly around fumaroles (vents from which volcanic gases escape). The materials condense directly from the hot gases to the solid state. The best example of vapour crystallization is native sulphur, which quite often forms crystals with well developed faces; other vapour-grown mineral crystals tend to be very small.

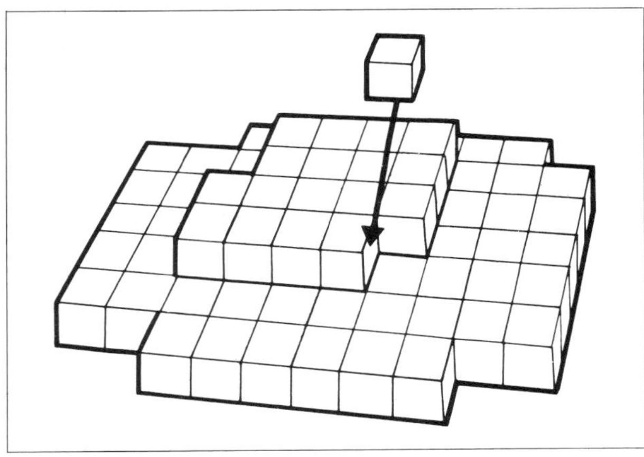

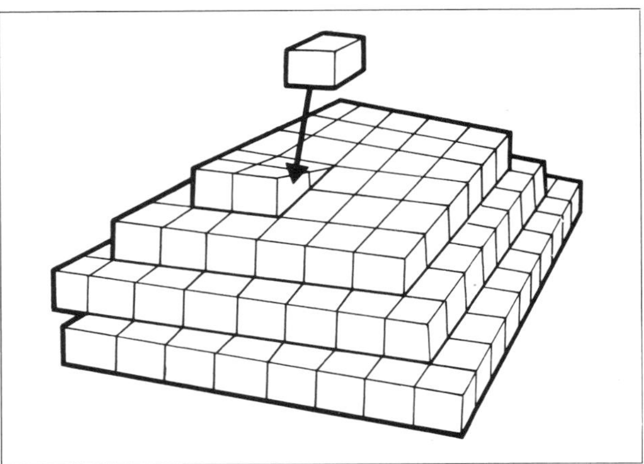

Left: An incomplete plane of atoms at the surface of a crystal. New atoms are attracted to the 'step'. Such a plane spreads until it occupies the whole surface

The growth process

We must now examine the growth process itself to see how atoms group themselves so that the product has the regularity characteristic of the crystalline state. For more than a century, there have been attempts to explain how crystals grow, and even nowadays many aspects of the problem remain unclear. However, it is now known that a crystal can begin to grow only when it has a *nucleus* of a certain size to grow around. A crystal that is smaller than a certain critical size, which varies according to the conditions, will not be able to grow. A crystal or any other solid body, such as a grain of dust, that is larger than this critical size can act as a nucleus. If there are no such bodies present in the medium to act as nuclei, crystallization will not begin until groups of atoms have come together by chance just long enough to form a minute nucleus. The chances of a nucleus arising accidentally generally increase as the temperature is reduced. Sometimes it is necessary to cool a solution or melt well below its normal crystallization temperature before nuclei appear and crystallization begins.

Crystals grow fastest when the arrival of fresh particles on the surface continually generates sites suitable for attaching further particles, which themselves reproduce the attachment

Left: Dislocation of the screw type. Again growth takes place at the 'step'. But though the step changes position, it never vanishes

Above: Synthetic tungsten crystals seen under the electron microscope. These manmade crystals show sharp edges and smooth faces. Some crystals have a hollowed-out look, where growth has taken place fastest at the edges

Left: Growth spirals on another manmade crystal. New material tends to be added to the crystal at the 'step'. As the crystal grows, the spiral spreads outward

sites, and so on. This occurs when a crystal contains a *screw dislocation*, one of several possible kinds of 'error' in atomic arrangement. This gives rise to a *growth spiral* on the face of the crystal. Such spirals are not visible to the unaided eye, but they can frequently be seen under the microscope. The dislocation extends through the body of the crystal and appears at the surface as a step. The step provides a niche to which new particles are strongly attracted. As they join the crystal, the step moves. Fresh turns are continuously generated at the centre and the spiral spreads outwards. Such spirals expand in this way until they reach the edge of the face, where the steps are lost. A spiral thus forms a very flat hump on the face.

Another mechanism does not involve dislocations. When a minute part of a fresh layer has built up by chance on a face of a crystal, the step at its edge forms a place where new particles readily attach themselves. Such a partial layer—a two-dimensional nucleus—expands until it covers the entire face. A new part-layer has to be produced by chance before growth can continue. This is therefore not so favourable to growth as the dislocation mechanism, in which new nuclei do not have to be continually produced. However, it produces crystals free from structural defects (dislocations), which is especially important in artificial crystal-growing.

It is possible for atoms of a different kind to take up places in the crystal structure during its growth. They can do this if they are similar in size and chemical properties. The substance formed when these impurities are present is described as *isomorphous* with ('having the same form as') the pure substance, for the atomic structure is the same. For instance, minerals containing calcium as a major component frequently have small amounts of the much rarer element strontium (which closely resembles calcium) present as an impurity replacing some of the calcium. Another example is ruby, in which the basic structure of aluminium oxide, which is colourless, accommodates small amounts of chromium. The chromium resembles the aluminium to some extent but causes the red colour.

Influences on growth

When new layers of atoms are deposited rapidly on a particular face, the crystal grows fast in the direction at right angles to it. Such a face tends to become smaller as more material is deposited. Imagine that the face is like the top surface of a partly built pyramid, bounded on each side by a sloping face. As the height of the pyramid increases, the size of the upper surface shrinks, finally becoming the pyramid's point. Similarly a crystal face on which material is deposited faster than on its neighbours will tend to shrink, and even be lost from the crystal if it develops into a point or edge. So in general the best developed faces of the final crystal are those that have built up most slowly.

But other factors can interfere with this development. For instance, as a copper sulphate crystal grows from a solution in water, it removes the dissolved copper sulphate from the nearby liquid and so leaves the solution there slightly weaker. The weaker solution is replenished from other parts, and this occurs more readily at edges and (especially) corners than at the centres of faces. There is therefore a tendency for edges and corners to grow preferentially. Preferential edge growth is partly responsible for 'hopper' crystals, illustrated on previous pages; preferential corner growth can similarly result in dendritic (tree-like) crystals, which are particularly common when crystals grow rapidly from solution. Dendritic crystals are, however, not of great interest to every collector.

Impurities in solutions and melts have marked effects on crystal growth. In chemistry, it has long been known that often a good method of

purifying a substance is to dissolve it in a suitable solvent and crystallize it out again. The reason is that a crystal has a natural tendency to reject impurity atoms as it grows, unless they happen to be of just the right size and chemical properties to be taken into the structure. Impurities thus tend to accumulate ahead of a growing face; they hinder new material from reaching the face and so slow its growth. This effect, called 'poisoning', may vary from one face to another, and so the final shape of the crystal may be affected by the impurities, even though they are not built into the crystal at all. Again, the impurities, such as dust particles, may be trapped and thus be incorporated purely mechanically into the crystal, without truly being a part of its structure; these are called inclusions and are

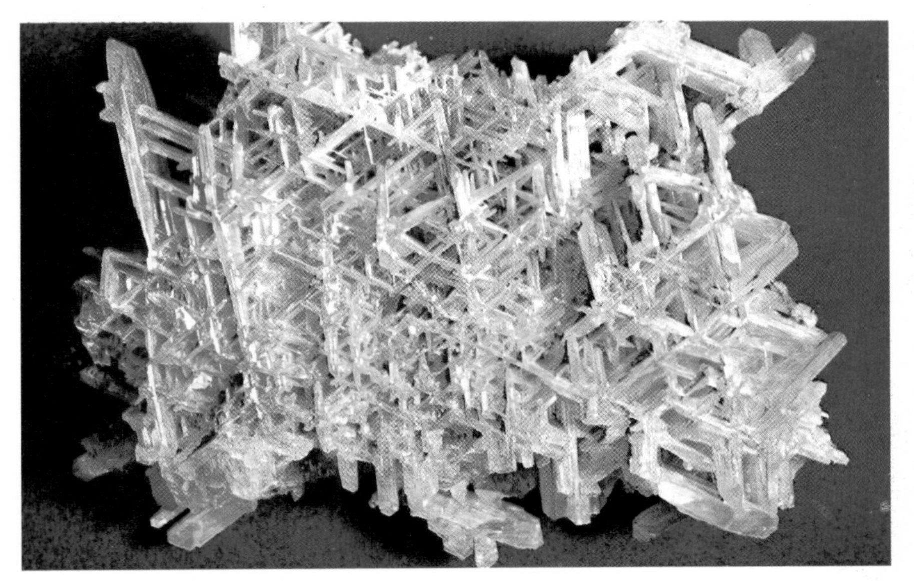

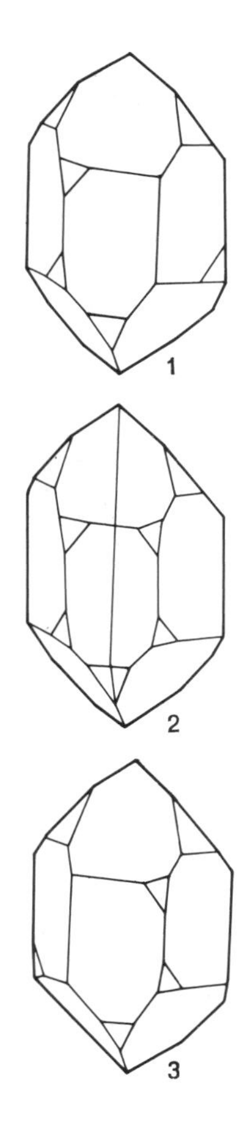

Above: Left-handed and right-handed quartz crystals (1, 3) and a twin combining both forms (2)

Left: Prismatic calcite. Inside it cleavage planes can be seen

Right: A twinned specimen of calcite. The two parts are 'mirror images' of each other

Left: Cerussite from South-West Africa, forming multiple intergrowths

Far left: A columnar specimen of heliodor, a variety of beryl

very common in natural crystals. They are discussed in more detail later in this chapter.

Some faces of a natural crystal are very smooth and unblemished, while others are quite rough. In general, the larger a face, the more accidents are likely to happen to it, since it takes longer to grow. Only in laboratory-made crystals are smooth regular faces common. Look at a natural crystal with a × 10 lens; you can immediately see pitting, grooving and other defects, all of which are due to growth processes. These imperfections may make a specimen more rather than less desirable to a collector; for example, a natural ruby will have inclusions different from those in a synthetic one.

Another growth effect is *overgrowth* of various kinds. A very attractive example of this is 'sceptre' quartz, in which the original prisms are capped by expanded sections of the same shape, which give rise to the sceptre effect. Such crystals are commonly found growing from the walls of cavities in mineral veins or the like. It may be that the tips have readier access to fresh material in the cavity than the parts closer to the wall.

The overgrowth mineral may be different from the original one; examples are quartz on calcite and rutile on hematite. Some faces of a host crystal are preferred in overgrowth. For example, hematite can colour alternate faces of a quartz crystal red because only those faces attract the overgrowth. This is called selective encrustation. Old overgrowths may themselves be covered by fresh layers of the host. In transparent crystals such past overgrowths can be seen to form so-called ghost or phantom crystals that show the size and shape of the crystal at earlier stages of its growth. This is best seen in quartz.

It is possible for different parts of a crystal to grow with different alignments. The final result is actually a mosaic of individual crystals, though it appears to be a single individual. This is responsible for the curved quartz crystals found in the Swiss Alps. Dolomite crystals may also be curved and are sometimes known as saddle crystals.

The small pits frequently seen on crystal faces —they are often prominent on beryl crystals— result from changes in the environment that cause the crystal to dissolve rather than grow further. If you place a calcite crystal in dilute hydrochloric acid for a few minutes, etch pits develop on the sites of dislocations, and these can sometimes be seen to be geometrically related to the crystal shape. For example, the pits on certain faces of beryl are likely to be hexagonal and so reflect the crystal structure's hexagonal symmetry.

Crystal habit

Another important aspect of the study of crystals is the *habit* of a crystal—that is, its general shape. Habit can be so characteristic of a particular mineral that no other feature is needed to establish its identity. Several terms are in common use to describe the habit of individual crystals and the appearance of aggregates. The ones most likely to be encountered, with their meanings, are:

acicular : needle-like (as in natrolite);
capillary : hair-like (as in some millerite);
columnar : like a column, relatively thick (as in beryl or tourmaline);
platy : (as in mica);
bladed : long and flat (as in kyanite);
tabular : flat (as in some varieties of corundum, baryte or wulfenite);
dendritic : treelike (as in some native metals, especially copper and silver);
wedge-shaped (as in gypsum);
spear-shaped (as in descloizite).

Many minerals produce *twinned* crystals. A twinned crystal looks as if it is made up of several parts, each of which is a simple crystal. The parts may be similar or identical to each other, or they may be like mirror-images of each other (like a left-hand and a right-hand glove). These portions may look as if they have been joined together, in which case they make up a *contact* twin, or they may seem to penetrate each other, in which case they are described as *interpenetrant* or *penetration* twins. Though they are hard to describe, twinned crystals are often easy to recognize; they may have faces meeting in re-entrant (V-like) angles, and may show swallowtail or butterfly shapes.

Twinning can be repeated many times in a crystal and a complex shape then results. The product is described as a *multiple* twin. If the individual portions are very thin, a lamellar appearance (resembling the pages of a book) can result. This is common in plagioclase feldspars.

Inclusions and 'false forms'

Environmental changes during the crystal's formation can leave traces in the form of *inclusions*. These fall into three basic types: pre-existing, or *protogenetic*; contemporary or *syngenetic*; and post-contemporary or *epigenetic*. These inclusions are particularly valuable in the study of gemstones, since they vary not only from species to species but from area to area, especially in the case of emerald. Inclusions help to distinguish gems from synthetic minerals.

Inclusions may be formed early and then be covered by their host; for example, they may be present on the walls of cavities in which later crystals form. Rutile can occur in quartz in this way. The lack of any definite orientation relative to the quartz shows that the rutile was formed first. Quartz commonly forms late and envelops other species; tourmaline and pyrite are further examples. Syngenetic inclusions generally consist of bubbles of liquid or gas. Sometimes solids, liquids, and gases all occur together in inclusions.

The behaviour of inclusions when the crystals are heated can tell the scientist a great deal about their formation. The sequence of events as inclusions melt, dissolve or burst provides a kind of reverse motion picture of the process of crystallization, though one needing a careful interpretation.

Sometimes the host crystal and its inclusions have a definite mutual orientation, as with the tiny plates of green fuchsite mica that occur in quartz and give a spangled effect, owing to reflection from the numerous parallel faces. This glitter is called *aventurescence*. It occurs also in the sunstone variety of feldspar, in which the plates are of hematite or goethite. A different included crystal pattern gives the *cat's-eye* effect (chatoyancy); in this case the inclusions are tiny tubular or needle-shaped crystals, which reflect light as does cotton on a reel, giving rise to a band of light at right angles to their length. Hence a band of light appears to cross a polished specimen

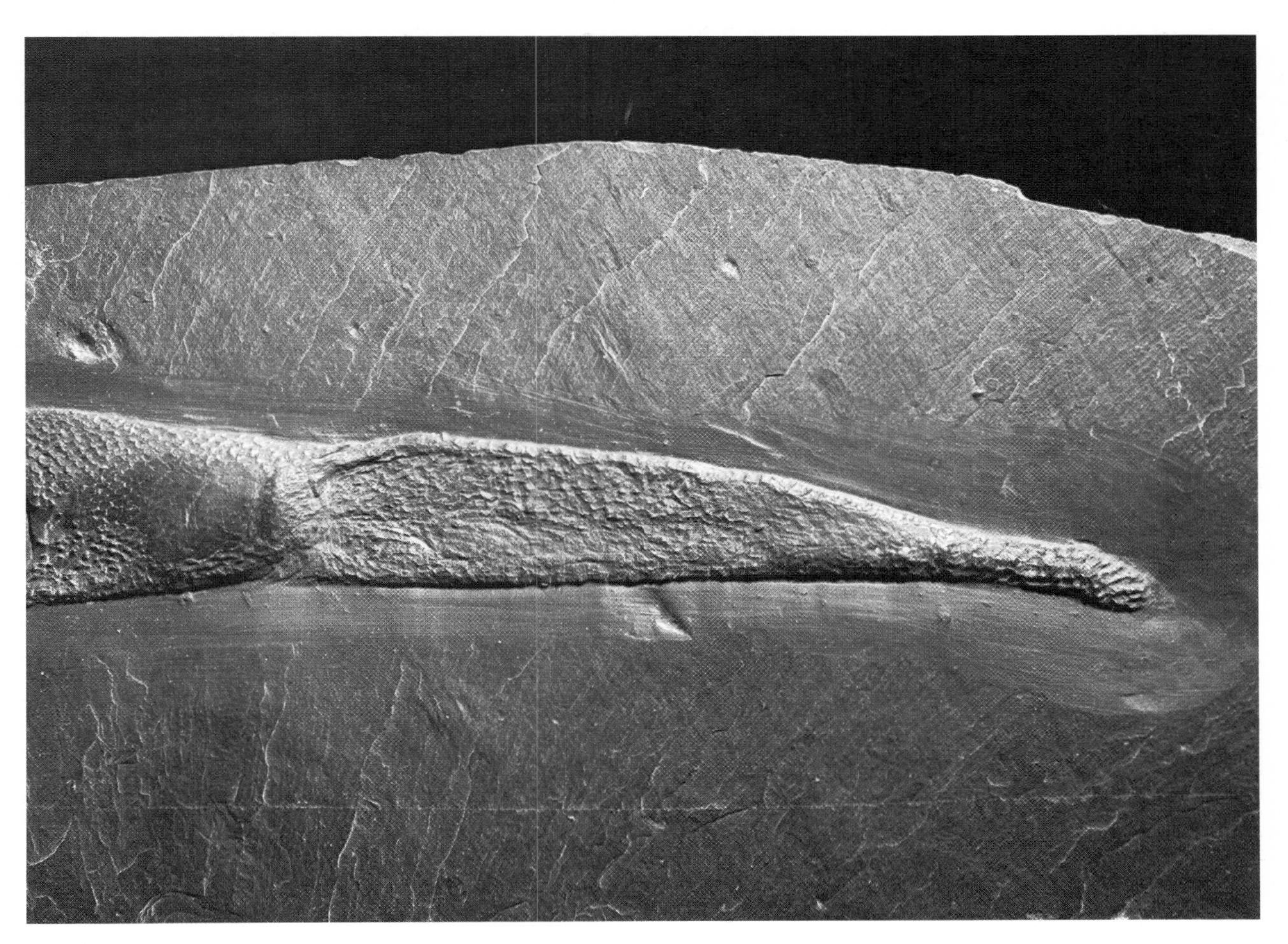

Right: Fossil sponge, almost 400 million years old. Fossils are mineral but preserve the forms of once-living creatures. Some form as moulds or casts of the creature, others consist of minerals that replace its tissues. They are examples of pseudomorphism

Left: Penetration twins of calcite from Derbyshire

as the eye is moved. Chatoyancy can occur in various minerals but is most common in quartz, tourmaline and chrysoberyl, whose chatoyant variety is called cymophane. *Asterism* has a cause similar to the cat's-eye effect; here crystals at right angles to an axis through the crystal show a star-like pattern of reflected light. Such crystals are formed by a process known as *exsolution*; the inclusion material remains dissolved in the host crystal at high temperatures, forming an isomorphous material, but it crystallizes out, or exsolves, to form long thin crystals in particular directions as the temperature falls. Asterism is particularly prominent in corundum (ruby and sapphire) and in quartz. Artificial star-stones have much sharper stars because the amount of the included material, such as rutile in corundum, can be controlled to give the best results.

One might think that natural crystals, since they have already existed for a long while, will not alter. In fact they do; for example, realgar crystals will turn into orpiment in time if exposed to air and strong light. Other crystals will take up water from the air, while some are attacked by chemicals present in polluted atmospheres. Sometimes a crystal within a rock will alter in such a way that its chemical composition and structure are completely changed while its shape remains the same. Such crystals are called *pseudomorphs* (false forms) and are particularly common in varieties of quartz. Aragonite may replace copper, clay minerals may replace beryl or feldspar, and chlorite may replace garnet.

Another very common example of pseudomorphism is the replacement, often by silica, of fossil material such as shells. Opalized shell, a particular kind of pseudomorph, is very beautiful and prized by collectors. Pyritized fossils are also very common. When describing examples of pseudomorphism, mineralogists speak of 'opal pseudomorphous after wood', or 'quartz pseudomorphous after fluorite'.

Many minerals occur not as individual crystals but as masses with no characteristic external form. They may be of one mineral species or of several. The appearance (for example, radiating or fanlike) will often serve, with the colour, to identify them. These masses, and groups of individually recognizable crystals, are known as *aggregates*. Some descriptive terms that refer to the external appearance of aggregates are:

botryoidal: like a bunch of grapes;
reniform: kidney-shaped;

Left: Wavellite from Devonshire. Radiating fibres of the mineral form globular masses

mamillary : breast-shaped;
spherulitic or *globular :* more or less spherical;
fibrous : consisting of long thin crystals;
stalactitic : resembling stalactites, the growths hanging down in caverns;
foliated or *lamellar :* consisting of sheets that easily split apart;
plumose : resembling plumes;
divergent or *radiated :* spreading from a single point like a fan;
stellate : divergent, forming a star;
reticulated : consisting of thin crystals in a lattice formation;
dendritic : resembling fern branches;
arborescent : resembling the branches of a tree.

A mineral's surface often has a characteristic appearance called *lustre,* which is due to the mineral's individual way of reflecting light. It is another feature for which mineralogists have developed an elaborate vocabulary. Some of the terms used to describe lustre are: *metallic ; sub-metallic ; vitreous* (having the lustre of broken glass); *resinous ; pearly ; silky ;* and *adamantine* (having the lustre of diamond). Still other terms refer to the intensity of the lustre. Minerals that have no lustre at all are described as *dull.*

Crystal cleavage

The orderly pattern of the atoms in a crystal influences many other properties apart from the external appearance of individual crystals or of aggregates. Many crystals show *cleavage*—that is, they split readily along certain directions. The direction of the cleavage is always parallel to a possible crystal face. Cleavage planes are dependent on the atomic structure, and they pass between sheets of atoms in well-defined directions. The ease of cleavage and its effects vary from one mineral to another. The species in

Below: A piece of halite, or rock salt, that has cleaved several times. Halite tends to grow in the form of cubes, and the faces exposed by cleavage are parallel to cube faces

Above: Globular masses of pyrite, an iron sulphide

which cleavage gives exceptionally smooth surfaces are said to show *eminent* cleavage; other terms are *distinct* and *poor*. Cleavage may also be said to be *easy* or *difficult*. A diamond has quite easy cleavage despite its great hardness.

One of the first things the collector does on finding a sample of a mineral is to study the signs of naturally occurring cleavage on its faces. The directions in which a crystal tends to cleave are clues to its atomic arrangement. This topic is discussed more fully in connection with crystal symmetry later in this chapter.

When a mineral shows cleavage, it appears in all specimens. Quite distinct from it is *parting*, which may appear in some specimens of a mineral, but not all. A further difference is that there are only a limited number of planes of parting in any specimen. If a crystal cleaves at all, however, it can in principle be cleaved along an unlimited number of parallel planes indefinitely close to each other.

Parting planes arise because of some peculiarity in a specimen's development. For example, the planes along which the components of a contact twin meet are often planes of weakness, and parting can take place along them.

When a mineral specimen's surface is broken it may not cleave or part along some definite plane but simply *fracture* to leave an irregular surface exposed. Even this kind of breakage can be characteristic of the mineral. In quartz, flint and some glasses, fracture is typically *conchoidal*—that is, whorled like a spiral seashell. Other kinds of fracture are described as *even, uneven* (when the exposed surface has a limited degree of roughness), *hackly* (like the surface of broken iron) or *splintery*.

Crystals sometimes have distinctive electrical properties. Tourmaline crystals display *pyroelectricity*; they become electrically charged if they are subjected to changes in temperature. Specimens of tourmaline displayed in shop windows often attract dust from the air when they are heated by sunshine. Quartz crystals show *piezoelectricity*; pressure produces small electric voltages across them. This has been of great value in electronics and has been one of the main reasons for the extensive production of synthetic quartz crystals.

There are many more properties distinctive of crystals but they are outside the scope of this book. Those discussed above are the most useful to the collector and lapidary. The effects of light in crystals will be discussed later in this chapter.

Above: Globular masses of pyrite, an iron sulphide

which cleavage gives exceptionally smooth surfaces are said to show *eminent* cleavage; other terms are *distinct* and *poor*. Cleavage may also be said to be *easy* or *difficult*. A diamond has quite easy cleavage despite its great hardness.

One of the first things the collector does on finding a sample of a mineral is to study the signs of naturally occurring cleavage on its faces. The directions in which a crystal tends to cleave are clues to its atomic arrangement. This topic is discussed more fully in connection with crystal symmetry later in this chapter.

When a mineral shows cleavage, it appears in all specimens. Quite distinct from it is *parting*, which may appear in some specimens of a mineral, but not all. A further difference is that there are only a limited number of planes of parting in any specimen. If a crystal cleaves at all, however, it can in principle be cleaved along an unlimited number of parallel planes indefinitely close to each other.

Parting planes arise because of some peculiarity in a specimen's development. For example, the planes along which the components of a contact twin meet are often planes of weakness, and parting can take place along them.

When a mineral specimen's surface is broken it may not cleave or part along some definite plane but simply *fracture* to leave an irregular surface exposed. Even this kind of breakage can be characteristic of the mineral. In quartz, flint and some glasses, fracture is typically *conchoidal*—that is, whorled like a spiral seashell. Other kinds of fracture are described as *even*, *uneven* (when the exposed surface has a limited degree of roughness), *hackly* (like the surface of broken iron) or *splintery*.

Crystals sometimes have distinctive electrical properties. Tourmaline crystals display *pyroelectricity*; they become electrically charged if they are subjected to changes in temperature. Specimens of tourmaline displayed in shop windows often attract dust from the air when they are heated by sunshine. Quartz crystals show *piezoelectricity*; pressure produces small electric voltages across them. This has been of great value in electronics and has been one of the main reasons for the extensive production of synthetic quartz crystals.

There are many more properties distinctive of crystals but they are outside the scope of this book. Those discussed above are the most useful to the collector and lapidary. The effects of light in crystals will be discussed later in this chapter.

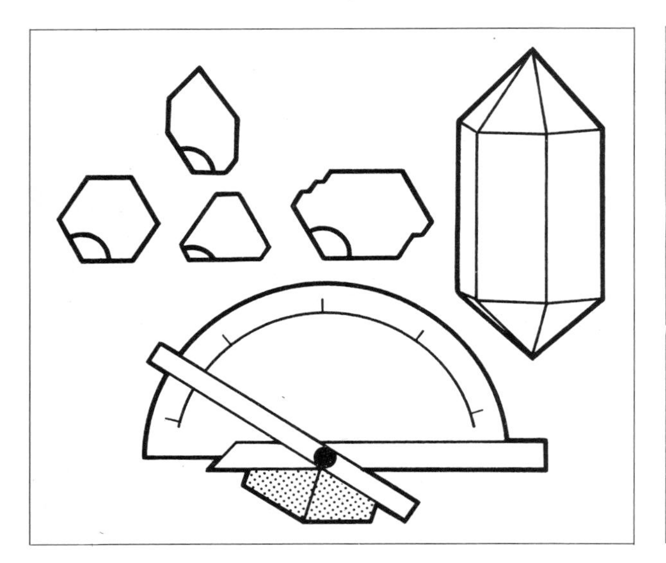

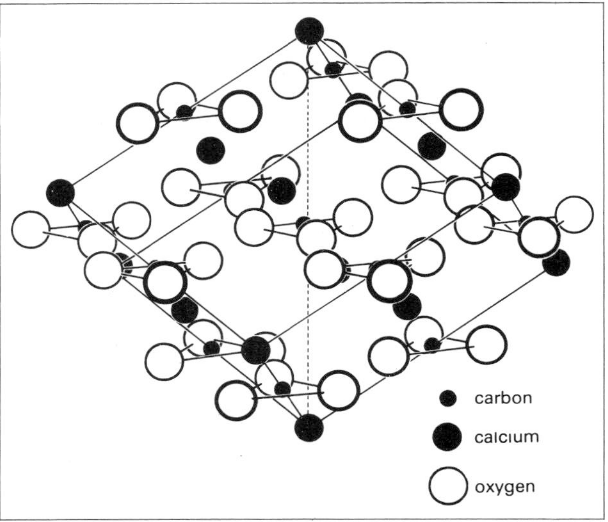

Left: Unit cell of calcite. This is the crystal's building block and is itself made up of atoms of carbon, calcium and oxygen

Far left: Constancy of angles. These cross-sections of quartz crystals vary in shape, but corresponding angles are the same. A simple goniometer is shown with them

CRYSTAL STRUCTURE

We now have some idea of the processes underlying the growth and shaping of crystals, and we have seen that they consist of atoms built up in a regular array. However, this knowledge was gained only late in the study of crystals. The earliest achievements in crystallography were in describing external appearance.

One of the early discoveries on crystal shape was published by the Danish scientist Steno in 1669. He cut cross-sections from quartz crystals of various shapes. These sections, too, were of varying shapes, but in every case he found that the angles between corresponding sides (representing faces of the whole crystal) were the same. In 1772–82 Romé de l'Isle published similar findings based on further measurements. He formulated the Law of Constancy of Angles, which states that the angle between corresponding faces has a constant value for all crystals of a given substance.

We should be clear what is meant when we speak of the angle between two faces; it is defined as the angle between lines perpendicular to the faces. An instrument called a *contact goniometer* provides the simplest way of measuring this angle; the amateur can make one for himself from a piece of wood and a protractor.

The Abbé Haüy pursued his studies of crystals in the late eighteenth and early nineteenth centuries. It is said that one day he dropped a calcite specimen belonging to a friend, and it shattered. He noticed that the fragments were all rhombohedra—that is, their surfaces were all flat parallel-sided faces. When such a piece was further broken, the new pieces produced were also rhombohedra. This led him to the idea that all crystals were made of tiny identical 'building blocks' that stacked together to make the whole crystal.

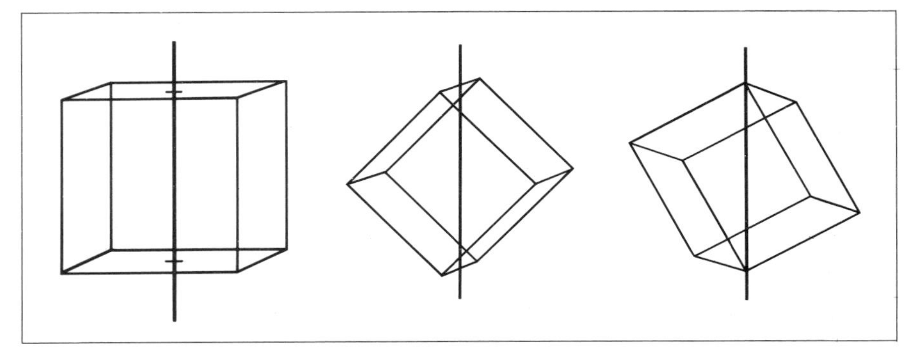

The idea of a basic building block is still retained today, but it is not a tiny solid body as Haüy thought. Now called a *unit cell,* it is the basic unit of the crystal's atomic pattern. It is repeated endlessly through the body of the crystal rather as a wallpaper pattern is repeated regularly over the paper's whole area.

In 1912 Max von Laue found that a narrow beam of x-rays is split into several beams when it is passed through a crystal. He ascribed this to *diffraction.* Diffraction of visible light can be seen when a distant street lamp is viewed through a nylon umbrella. Several spots of light are seen surrounding the lamp; they are due to light rays diffracted by the regularly spaced nylon threads. The spots are separated by distances governed by the fineness of the mesh. A very fine mesh causes the spots of light to be widely spaced, while a coarser one makes them appear closer together. In a similar way, closely spaced sheets of atoms in a crystal diffract x-rays at larger angles than do more widely spaced ones. The diffracted x-rays strike a photographic film, and the distances between the spots enable scientists to compute the atomic positions.

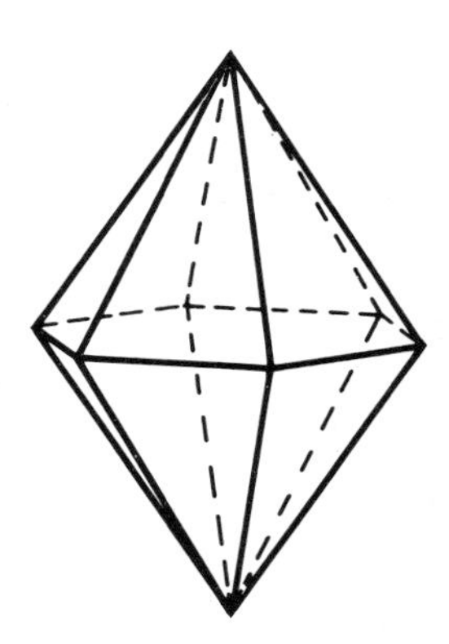

Above centre: Some axes of symmetry in a cube. From the left they are: fourfold; twofold; and threefold

Above: Hexagonal symmetry. This body has sixfold symmetry about its long axis

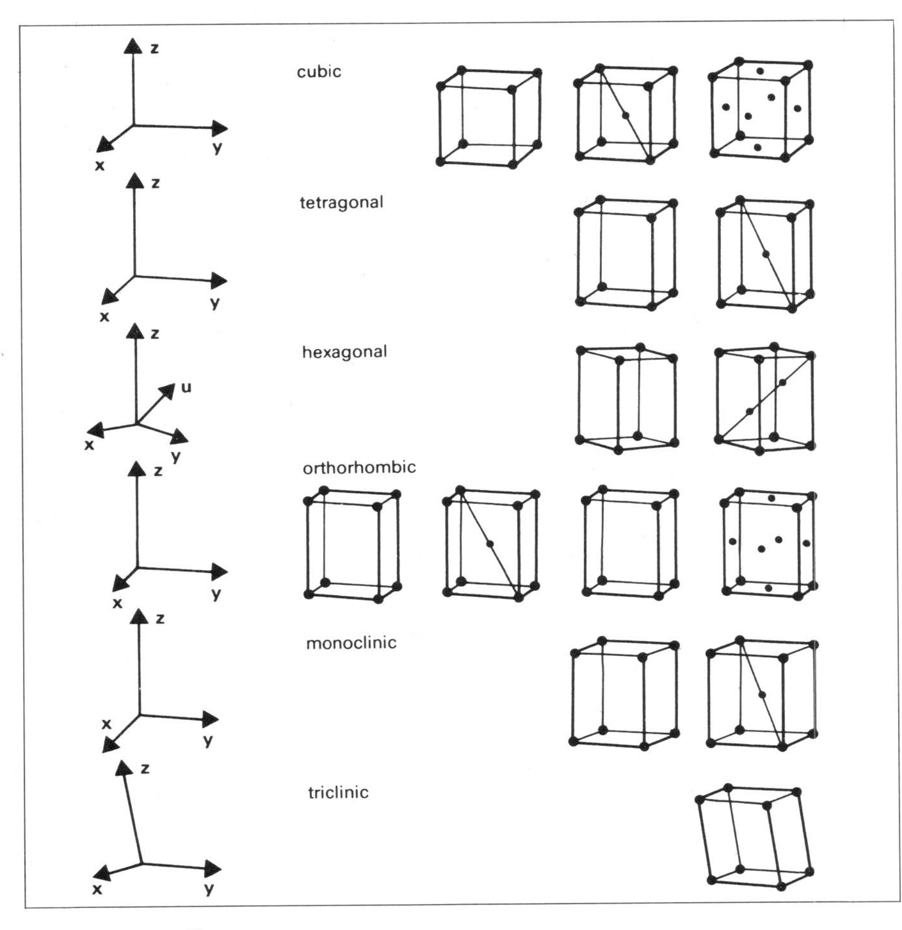

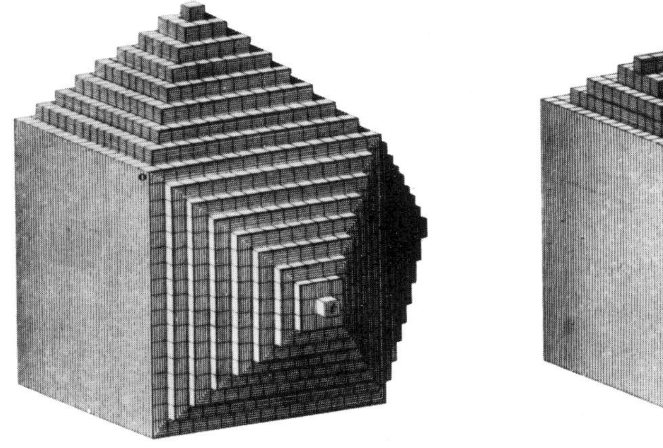

Above: Haüy's idea of crystal structure. Here two crystals are built up from cubes regularly stacked together

Top: The six crystal systems. They include 14 possible 'lattices' or atomic arrangements. Axes of reference are chosen according to the symmetry of each system

The idea of symmetry

The most important concept required to describe both crystal shapes and atomic arrangements is *symmetry*. In the everyday world symmetry is shown by many manmade things. A four-bladed propeller displays *rotational symmetry*. If it is rotated, there are four positions in each revolution in which it presents exactly the same appearance as it did in its starting position. It is said to possess *fourfold* symmetry. A cube likewise has fourfold symmetry about an axis passing at right angles through the centre of any of its faces. An ordinary brick, on the other hand, has only twofold symmetry: if it is rotated about an axis passing at right angles through the centre of any face, it presents the same appearance only twice in each revolution.

Another kind of symmetry is *reflection* symmetry. A left and a right hand have the same shape, in one sense; but they will not look the same, no matter what position they are turned into. One is a 'mirror-image' of the other. The same relationship holds between a left and a right foot, and between the left and right sides of a human face.

Ideas of symmetry are important in describing the shapes of crystals and also the arrangement of atoms in the unit cells. Natural crystals usually display highly imperfect symmetry because some of their faces have developed more strongly than others and there are always defects on their surfaces. But their atomic pattern is, relatively, much closer to being perfect.

One kind of idealized beryl crystal has a hexagonal cross-section, and has sixfold rotational symmetry about an axis. In addition, if it is imagined to be divided by a plane joining directly opposite edges, the two parts so obtained are mirror-images of each other. Hence the crystal has reflection symmetry about these three imaginary planes (among others).

There is another kind of symmetry that is important in crystallography. Sometimes a crystal has a *centre of symmetry*. In this case, every edge or face has a corresponding parallel edge or face on the far side of the crystal. A cube has a centre of symmetry (its geometric centre); a pyramid does not.

The crystal systems

Crystallographers assign crystals to six *systems* in accordance with their symmetries. (These are further divided into a total of thirty-two classes, but we shall not consider these here.) The crystals in a given system can have a wide variety of shapes, but they all have certain symmetry properties in common. For example, any member of the tetragonal system has an axis of fourfold rotational symmetry. There may be other axes of symmetry, planes of reflection symmetry and a centre, but a fourfold axis is always present. (It is also possible for some crystals in other systems to have fourfold rotational symmetry.) On the other hand, crystals in the least symmetric system, the triclinic, have no symmetry axes.

The systems are most easily discussed in terms of their *axes of reference*. These are three (or, in the 'hexagonal' system, four) imaginary

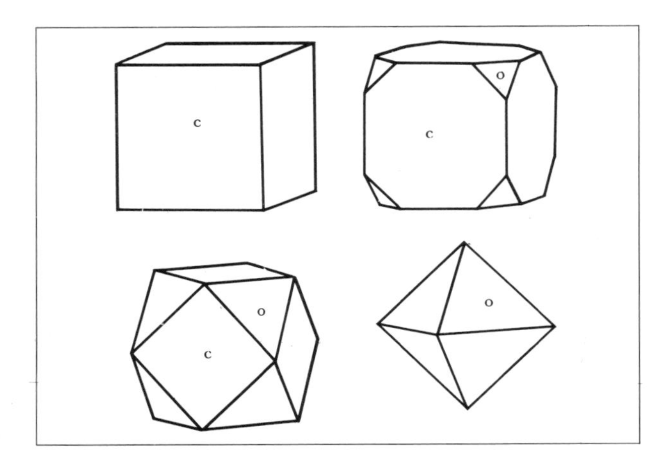

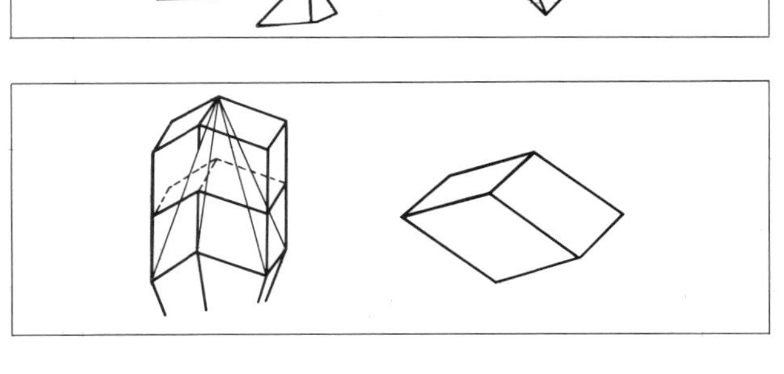

Left: Crystal cleavage. Here a cube becomes an octahedron (which is like a double pyramid) by repeated cleavages

Left: Cleavage of calcite. Whatever its original shape, a calcite crystal (hexagonal system) readily cleaves into 'rhomb' shapes

Far left: Forms in the cubic system. Octahedra and cubes occur alone or combined

straight lines parallel to possible or actual crystal edges. They are not necessarily axes of symmetry, though whenever possible they are. Ultimately the choice of lengths and directions of the axes is related to the lengths of the sides of the crystal's unit cell, as the accompanying diagram shows. The axes are labelled x, y, z; in the most symmetrical system, the cubic, all three are equal and at right angles; in the least symmetrical, the triclinic, none of the angles is a right angle, and both they and the lengths of the axes may very well all be unequal.

One useful way of approaching the crystal systems is to consider the *forms* that are found in each one. The term 'form' is used in a technical sense in crystallography. It is a basic geometrical shape that can appear in a crystal, and it has all the symmetries of the relevant system. For example, in the cubic system, one form is the cube; another is the octahedron, the eight faces of which are equilateral triangles. Crystals can occur as simple cubes or octahedra—diamond crystals are frequently octahedra—and combinations of the two can also occur. The diagram here shows a sequence of shapes that are combinations of the cube and the octahedron, with each dominating to varying degrees. The shape arising from a particular combination is the individual crystal's habit.

Some forms are *open*; that is, they do not enclose a volume of space, and so can only occur in combination with other forms. For example, in a crystal with fourfold rotational symmetry, one possible form is a four-faced *prism*, which is like a box whose top and bottom have been removed. Prisms can also be three- or six-faced. Clearly they are open forms. In crystals the ends are closed by other faces—for example, by pyramid-shaped caps. The terms prismatic, tabular and pyramidal are applied to combinations of a prism with other forms.

Another important open form is the *pinacoid*, which simply consists of a pair of parallel faces. Pinacoids occur whenever there is a single plane of reflection symmetry or a centre of symmetry.

Sometimes a crystal's cleavage reveals a form. Galena, for example, easily cleaves into cubes. Calcite easily cleaves into rhombohedra, as Haüy found. These cleavages are described as cubic and rhombohedral respectively. Other kinds of cleavage include prismatic, octahedral and so on. Cleavage is often parallel only to certain faces of a form, not all. Many substances have no good cleavage at all. Substances cleave well when the chemical bonds between certain sheets of atoms are weaker than those within the sheets. The micas, which are easily cleaved into thin sheets, are an extreme case. In many substances the chemical bonds do not show an overall direction in this way, and they do not cleave.

We now proceed to discuss the crystal systems and the forms that are found in them.

Systems and forms

The *cubic* system (all axes perpendicular and equal) has fifteen forms in all, some of which are rarely encountered; those we have mentioned, the cube and octahedron, are the commonest. Some more complex ones are the dodecahedron, with twelve faces, and the icositetrahedron, with twenty-four faces.

Crystals of the cubic system often have a generally globular appearance and lack elongation. This is especially noticeable in the garnets. When the forms occur singly, recognition is aided by the shapes of their faces: the cube has square faces, the octahedron equilateral triangles and the dodecahedron four-sided (diamond-shaped) faces; the icositetrahedron's faces are also four-sided.

Minerals often forming cubes include pyrite

Below: Crystal forms: top, a 24-faced body or icositetrahedron (cubic system); centre, a four-sided prism (tetragonal system); bottom, a hexagonal bipyramid (hexagonal system)

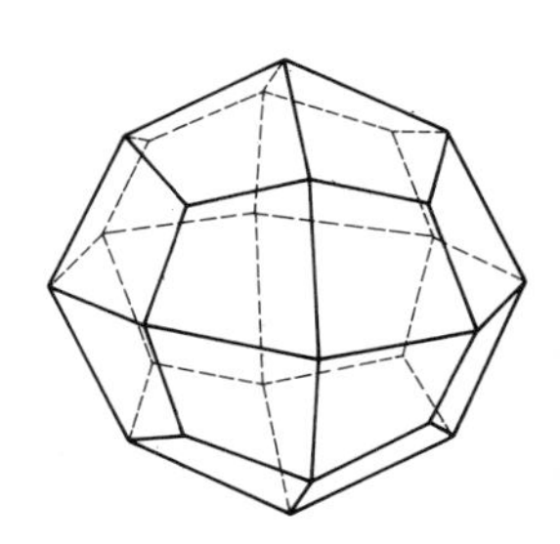

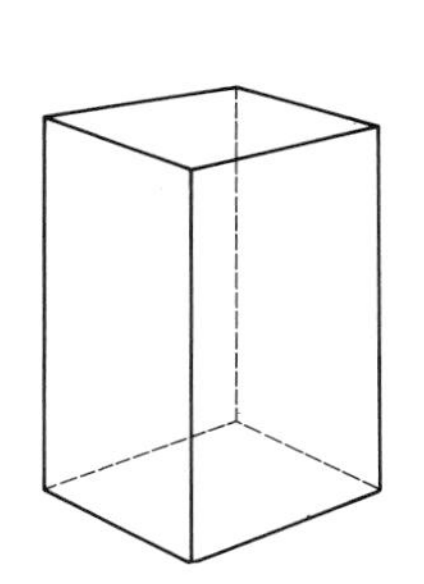

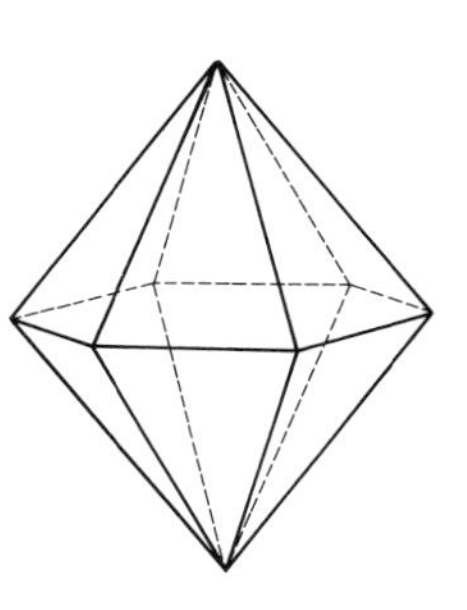

Right: Analcime crystals. This mineral crystallizes in the cubic system. Here it has an icositetrahedral or 24-faced form

Far right: Wulfenite. This specimen is tabular—it has a flattened shape. Wulfenite belongs to the tetragonal system

Bottom: Open forms. Prisms have faces meeting in parallel edges. The pinacoid is a pair of parallel faces. Open forms cannot appear alone

Below: Types of habit. A bipyramid, prism and pinacoid can combine to give (1) a prismatic, (2) a pyramidal or (3) a tabular habit

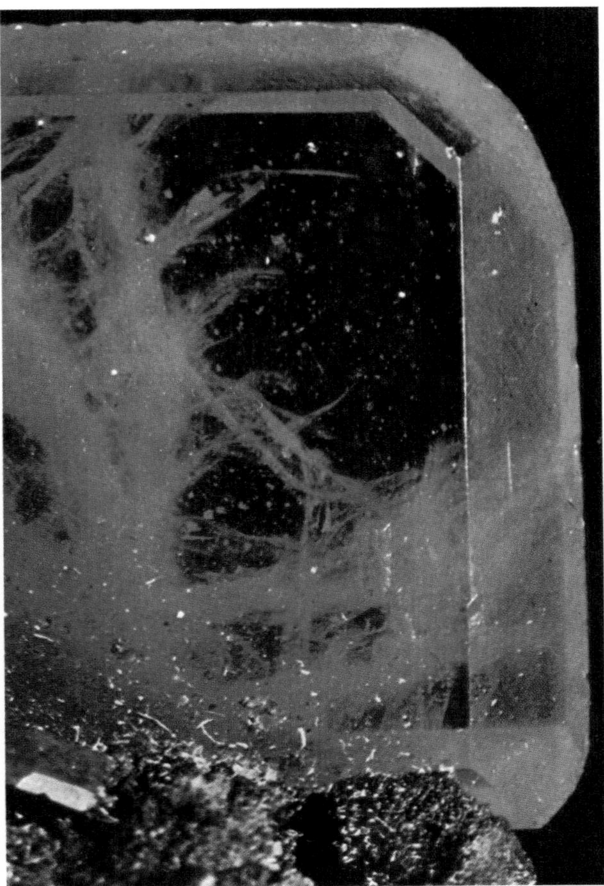

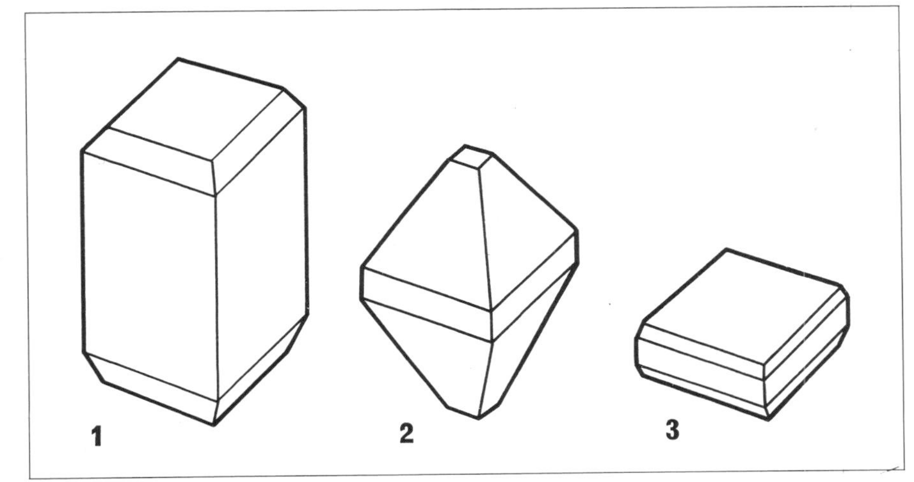

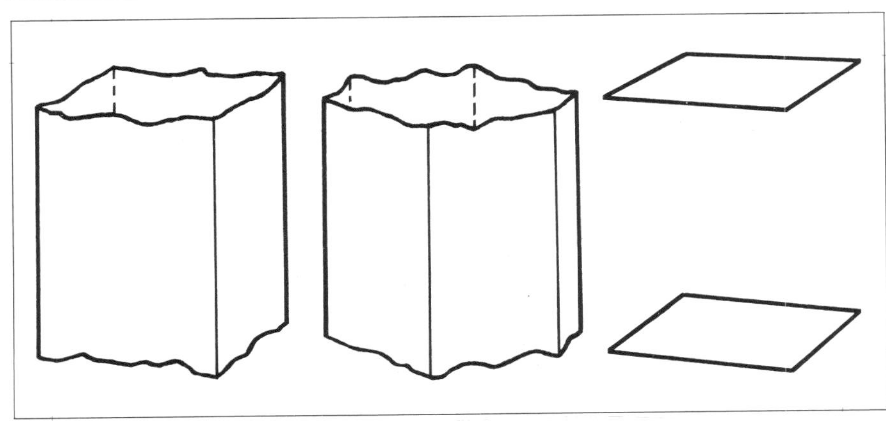

and galena; spinel and diamond form octahedra; garnet and cuprite form dodecahedra; and garnets form icositetrahedra. The mineral tetrahedrite is named from its tendency to form tetrahedra (which have four identical faces that are equilateral triangles). Minerals in the cubic system showing pronounced cleavage include galena (cubic), diamond and fluorite (octahedral), and sphalerite (dodecahedral).

The *tetragonal* system has mutually perpendicular axes of reference, but with only two of the same length. The forms occurring in it include the bipyramid (resembling two pyramids joined base to base) and the prism.

Idocrase crystallizes in the tetragonal system, forming prisms of square cross-section; zircon forms similar prisms with pyramidal ends. Scheelite often forms bipyramids and wulfenite often forms square plates. Apophyllite can crystallize with a pyramidal or a tabular habit, and it has a perfect cleavage parallel to the basal pinacoid.

Although only three axes are needed in any crystal system, it is convenient to define four in the *hexagonal* system. Three lie in a plane at angles of 120° to each other; the fourth is at right angles to the others. Some crystallographers distinguish part of the system as a separate 'trigonal' system, which has basically threefold symmetry rather than sixfold. Characteristic forms of this part of the hexagonal system include the rhombohedron (as found in calcite) and the trigonal prism, which has three faces.

In the rest of the hexagonal system we find prisms and pinacoids, as in beryl. Hexagonal bipyramids are characteristic of some sapphire crystals. These are composed of two pyramid-like shapes (each with six sloping faces) joined base-to-base. Hexagonal pyramids also occur.

In the hexagonal system, the collector should look for signs of threefold or sixfold symmetry. Rhombohedra suggest calcite or dolomite; both of these have rhombohedral cleavages. Hexagonal prisms may indicate quartz, beryl, or apatite, while trigonal prisms (having a triangular rather than a hexagonal cross section) suggest tourmaline, which also has striations parallel to the length.

In the *orthorhombic* system axes are defined that are at right angles but are unequal. A notable form here is the pinacoid; a bipyramid also occurs. The pinacoid frequently forms the base of a prismatic crystal and is then called the basal pinacoid.

Topaz crystallizes in the orthorhombic system and forms prisms that often show cleavage at right angles to their length. Natrolite forms long thin fibre-like prisms. Baryte, on the other hand, has a tabular (flattened) or prismatic habit; it, too, cleaves parallel to the basal pinacoid. Chrysoberyl is frequently twinned, and this gives it a pseudohexagonal near-symmetry.

In the *monoclinic* system, the three crystallographic axes are no longer all at right angles; one is at 90° to the plane of the other two, which are not at right angles to each other. Forms in this system include prisms and pinacoids.

The mica minerals, which crystallize in this system, form platy crystals, while gypsum forms tabular shapes or long prisms. Pyroxenes and amphiboles may form shorter prisms, and they both show prismatic cleavages. The two groups can be distinguished by their cleavage angles; in pyroxenes the cleavage directions are separated by about 90°, while in amphiboles they are separated by about 124°.

In the *triclinic* system, which has the least symmetry, none of the axes is at right angles to any other. The crystals show a distinct lack of

Above: Mimetite, which crystallizes in the hexagonal system. This specimen combines a hexagonal prism, a hexagonal bipyramid (sloping faces) and a pinacoid (top and bottom faces)

Above left: Celestine (orthorhombic system)

Right: Two forms in the orthorhombic system: 1, a bipyramid; 2, a prism

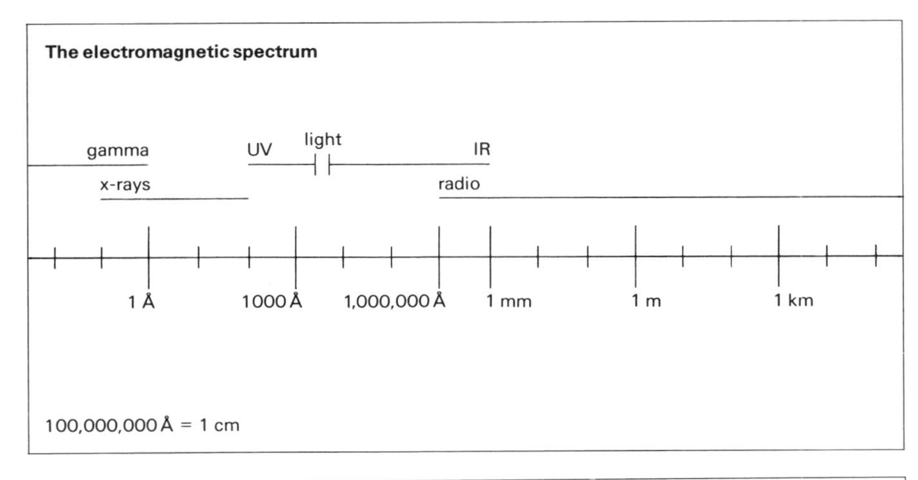

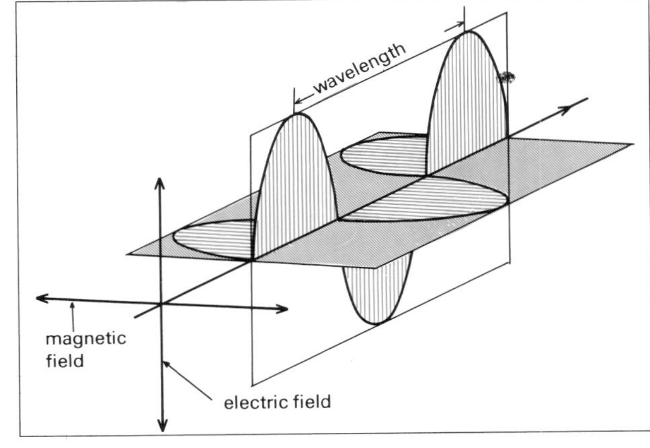

Above: The EM (electromagnetic) spectrum. Wavelengths range from less than a million-millionth of a metre (gamma rays) to many kilometres (radio)

Right: Polarized electromagnetic wave. It consists of electric and magnetic fields varying in intensity

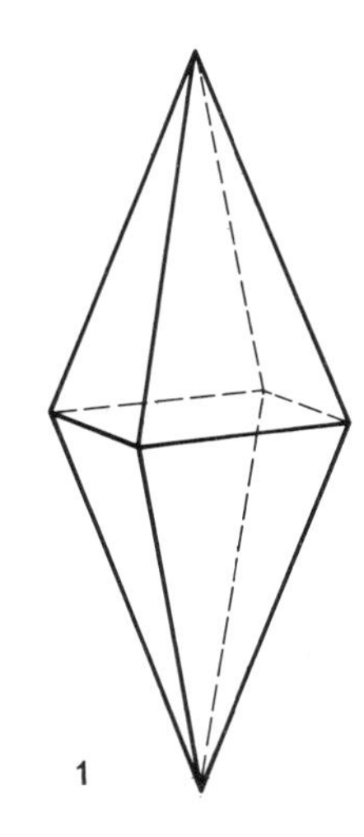
1

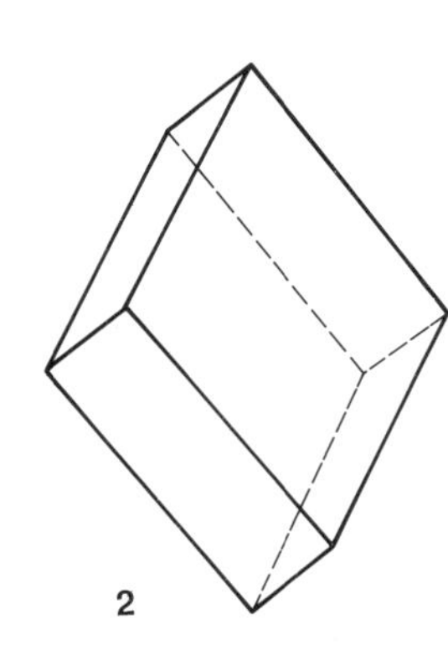
2

symmetry. Triclinic crystals show pinacoidal forms, however.

The triclinic system includes crystals that have a platy habit, such as axinite, sphene and kyanite, while the microcline variety of feldspar is interesting in that it is only just triclinic and its crystals may closely resemble crystals of the monoclinic system.

CRYSTALS AND LIGHT

The effects of crystals on light are important to the lapidary who wishes to turn them to advantage when fashioning gems, and to the gemmologist or mineralogist who wishes to identify a transparent specimen. We must first discuss the nature of light.

Light consists of electromagnetic (EM) waves —rapidly varying electric and magnetic fields whose variations travel through space much as sound waves spread outwards through air or as ripples spread out across the surface of water. Like other forms of wave motion, EM waves have a definite *wavelength*. The wavelength of ripples on water is the distance between one peak and the next. The wavelength of EM waves is the distance from one maximum of the EM field to the next. EM wavelengths range from many miles, in the case of some radio waves, to less than a billionth of an inch, in the case of 'hard' x-rays. The wavelengths of light are measured in a convenient unit called the angstrom (symbol: Å), which is a hundred-millionth of a centimetre (about a two-hundred-millionth of an inch). Red light has the longest wavelength of the visible colours, 7,000 Å. Blue light has the shortest wavelength, about half this.

White light is a mixture of all colours of light. Whenever it passes through a substance or is reflected from it, some wavelengths may be weakened compared with others. The reflected or transmitted light is accordingly coloured. It is this that is responsible for the colours of the objects we see around us.

A mineral will usually contain several elements. If a major element in the mineral (for example, copper in turquoise) absorbs some colour of light substantially, the mineral is said to be *idiochromatic* (self-coloured). If the absorption is due to an element present as an impurity, as with chromium in ruby, then the mineral is said to be *allochromatic*.

The subject of colour in minerals is too complex to be treated in detail here, but it can be said that many colour effects are due to a limited number of metal impurities. The following are the effects of certain important elements on the colours of gem materials:

titanium: acts with iron to give the blue in sapphire;
vanadium: thought to give colour in blue zoisite, and also some green in emerald and other green beryls;
chromium: gives the red in ruby and the green in emerald;
manganese: gives pink, red, and orange, as in spessartine garnet and rhodonite;
iron: gives red, green, yellow, and blue, usually less bright than the colours induced by chromium, in several gemstones including yellow sapphire and blue spinel;
cobalt: colours some glasses and synthetic stones blue;
nickel: gives the bright green of chrysoprase;
copper: gives the blue-green of turquoise.

Sometimes, however, colour arises from other causes. For example, the purple colour of fluorite is due to structural damage. Sometimes colour can be altered by heating, and many gem materials such as aquamarines and zircons are heated to give them more desirable colours before being sold. The fashions in colour change with the years, and sea-green aquamarines, once regarded as desirable, are now replaced in public esteem by those of a bluer colour. Various types of radiation have been used to alter the colour of diamonds.

Refraction and interference

Crystals affect light passing through in a fairly complex fashion. The light is slowed down and deflected; in certain materials it is split into two separate beams. The deflection is called *refraction* and is responsible for the bent appearance of a pencil placed in water. Light travels through air or a vacuum at 186,000 miles (300,000 km) per second. Its speed changes whenever it goes from one medium to another, and it is this that is responsible for refraction.

In non-crystalline materials like glass and in crystals of the cubic system only one refracted ray is formed. As light enters the medium it is slowed down; if it meets the interface at right angles it travels on without deviation but otherwise its path is bent towards the perpendicular to the interface. When light leaves the medium and enters the air, it speeds up; unless it meets the interface at right angles it is bent away from the perpendicular.

There is a simple relationship between the arriving or 'incident' ray and the refracted ray in a given medium: the sine of the angle of incidence has a constant ratio to the sine of the the angle of refraction. (Both angles are measured between the light ray and the perpendicular.) The ratio is called the *refractive index*; it can be measured with an instrument called a refractometer.

In crystals of systems other than the cubic, an additional refracted ray is usually formed. It is called the *extraordinary* ray and is generally close to the *ordinary* refracted ray, which behaves in the manner described above. In some materials, such as synthetic rutile, the two rays diverge quite strongly, however, giving impressive optical effects. The numerical difference between the refractive indices for the two refracted rays is called the *birefringence* of the material.

A specimen can show colours in other ways than selective absorption. One is called *dispersion* and occurs when white light is split into its colours by a transparent medium such as a gemstone. The various component colours are refracted by different amounts on entering and leaving the stone, the red component least and the violet most; or, to put it another way, the refractive index is dependent on the colour.

The other cause of colour is called *interference* and can be seen in mother-of-pearl, oil films on water, soap bubbles, and so on. Whenever light falls on a thin transparent film, like a soap bubble, some light is reflected from the first surface and the rest travels on and is partly reflected from the other surface. Hence the light that travels back towards the observer consists of two reflected rays, which overlap. If the wave peaks and troughs of the two rays coincide, the light is intensified; but if they are out of step (the technical term is *out of phase*), the peaks and troughs cancel out and no light is seen. The result depends on the thickness of the film and the wavelength of the light. Hence some colours will be intensified in the reflected

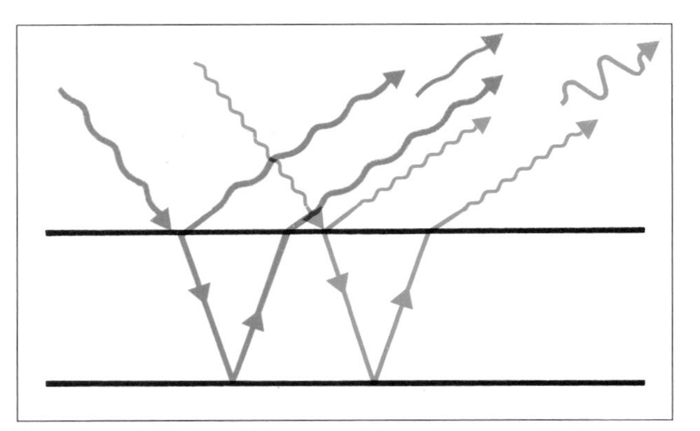

Far left: Refraction and reflection. Light entering a transparent medium from air is bent towards the perpendicular (top). When leaving the medium it is bent away from the perpendicular. But if the light meets the interface at too great an angle it is reflected back into the medium (bottom)

Left: Interference. Light passing into a thin transparent film is reflected from both surfaces. Here the blue (short wave) rays are in step and reinforce each other; the red (long wave) beams are out of step and cancel each other

Right: Fluorescence in ultraviolet light. The specimen includes hyalite, which is hard to distinguish in ordinary light (near right). Under UV light, however, it glows bright green (far right)

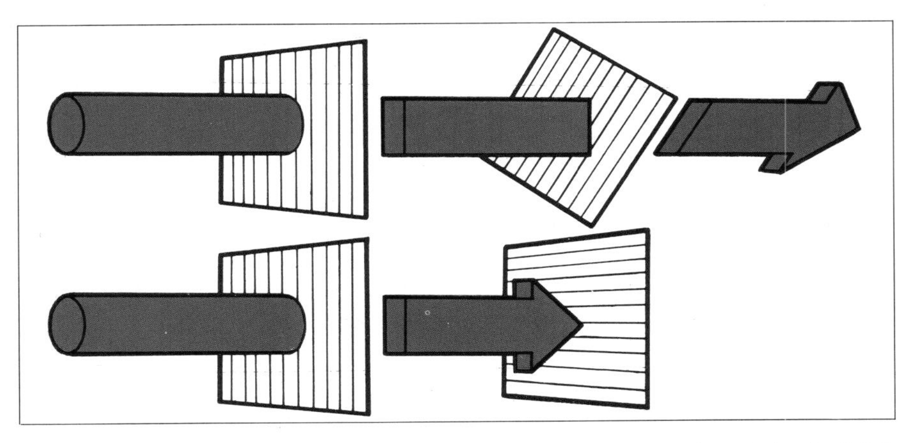

Above: Polarization. Light passing through Polaroid becomes 'polarized'—its vibrations lie in only one plane. Only part of this light passes through a second Polaroid oriented at an angle to the first, and its direction of polarization changes. Filters at right angles completely block the light

light, while others will be extinguished. This gives rise to the iridescent colours seen in interference. They arise in minerals when there are many thin layers present, as sometimes in mica. They can also be caused by the thin gaps around cracks or other flaws.

Polarization

Ordinary light consists of components that vibrate in all directions at right angles to the direction of travel; such light is called *unpolarized*. This light can be split into two rays by a suitable material, each ray vibrating in one direction only; it is then said to be *plane polarized*. Polaroid is a synthetic material that produces polarized light by selective absorption. A piece of Polaroid transmits light vibrating in a certain direction but blocks light vibrating at right angles to that direction. No light passes if two pieces of Polaroid are placed with their transmitting directions mutually at 90°. This is called the crossed position.

The two refracted rays formed in a birefringent crystal are in general polarized in different directions. They can be differently affected by their passage through the crystal, with some wavelengths being selectively absorbed in one ray but not the other. The result is that birefringent crystals can display different colours when viewed in different directions. The colour difference is especially marked when the crystal is viewed through a Polaroid filter. The effect is called *pleochroism*. Crystals in some systems show three different colours, others show only two (when the effect should strictly be called *dichroism*).

Luminescence

When some minerals are illuminated with ultraviolet light, they will begin to glow, and often will continue to do so after the ultraviolet light has been turned off. They are described as *fluorescing*. A similar effect occurs in other minerals when visible light is used as the stimulus; in this case the effect is called *photoluminescence*.

These phenomena are best considered in terms of the *quantum* theory, which regards light as made up of 'packets' of energy called quanta. Short-wavelength light occurs in high-energy quanta, while long-wavelength light occurs in lower-energy quanta. Ultraviolet light quanta have higher energies than visible light quanta because they have shorter wavelengths. When an ultraviolet quantum is absorbed by the mineral it gives up its energy to an atom or other particle, which is described as being in an 'excited' state. After a delay the particle gives out some of the energy in the form of a relatively low-energy quantum of visible light.

Ultraviolet lamps are readily obtainable by the amateur collector, and he can use them in identifying many minerals that fluoresce in characteristic colours.

Geology for the Collector

Left: Grand Canyon, Arizona. A mile-deep cross-section of ancient sedimentary rocks has been exposed by the erosion of the Colorado River. The rocks were originally deposited as sediments on the floors of ancient seas. Then the land was raised by geological forces and the river cut into it

This chapter aims to throw light on the geological relationships to be found among rocks and minerals, and to give some brief practical hints on field work. The term 'field work' refers to the outdoor activities associated with the collecting and observation of minerals. For the professional geologist and mineralogist these may include mapping, sketching, photography and other scientific techniques that have proved useful in locating and understanding mineral deposits. For the amateur collector, however, only the simplest tools are needed for the successful pursuit of his interest in minerals. Before detailing these, two very important points must be stressed, which can be summed up by the words 'conservation' and 'safety': preservation of the environment and preservation of oneself.

FIELD WORK

The mineral specimens in which the collector is interested, whether because they are fine crystal aggregates or represent rare species, are only occasionally to be found, and often at localities of very limited extent. This means that the better localities, when they become known, have a magnetic attraction for the enthusiast and sometimes for student parties, and hence are very likely to be badly damaged. The damage that can be done by one sturdy wielder of a two-pound geological hammer can be considerable, while the havoc caused by a party of thirty competing students can be catastrophic. If the locality is, for example, a working quarry, this may be of little consequence, but if it consists of a few narrow veins, or rock layers or horizons, the collector must have some sense of the value of the site. Further, broken rock and mineral fragments scattered over grazing land can be a hazard to livestock, to say nothing of the justified anger of the farmer who finds his banks eroded or holes torn in his fields. The mineral collector should also remember that indiscriminate hammering does a great disservice to his fellow enthusiasts. At many localities no hammering is needed at all, for there is ample loose material already available, provided by the efforts of previous visitors and the normal weathering processes of nature.

A number of geological sites in many countries are now protected because of their scientific importance. It must also be borne in mind that many mineral-collecting localities are on private land, so that permission must be obtained to visit them; no attempt should be made to enter working quarries without prior permission.

Safety

Although mineral collecting, like many other vigorous outdoor pursuits, is a very healthy pastime, a little attention must be directed towards self-preservation. If it is intended to visit remote or mountainous country, then it is obviously not advisable to go unaccompanied. However many there are in the party, it is always sensible to tell someone not travelling with it the intended destination, approximate route, and time of return.

A hazard peculiar to geological and mineralogical field work results from undisciplined hammering. Sharp mineral and rock fragments can be projected at considerable speed by hammering, putting both the collector and nearby colleagues in jeopardy. Eyes are particularly vulnerable to damage from such projectiles and precautions should be taken. Care should also be taken when hammering at the base of quarry faces and cliffs not to dislodge loose boulders from above; quarries are particularly dangerous in this respect, because of the inherent instability of the near-vertical working faces. In many countries, including Britain, it is now obligatory for protective headwear to be worn when visiting working quarries.

Clothing

There are perhaps only two things to be remembered when clothing oneself for field work; the first is the necessity of stout boots or shoes, which are required for rough walking and for work in such places as quarries and mine dumps. The second requirement is for old, expendable clothes. Mineral-collecting invariably involves a certain amount of scrambling over and among rocks, so that one must be prepared for the occasional layer of dust or mud. Some enthusiasts sport trousers and jackets with many and diverse pockets for the various pieces of small equipment useful to the mineral collector, but generally most collectors simply dress according to the dictates of the climate.

Collecting equipment

The one essential piece of equipment for any mineral collector is a geological hammer. These hammers usually have a head that is square on one side and chisel-edged on the other; but sometimes the chisel edge is replaced by a pointed pick. Weights vary from about half a pound to two and a half and occasionally four pounds. The size used depends partly on the work to be done, but more often than not simply

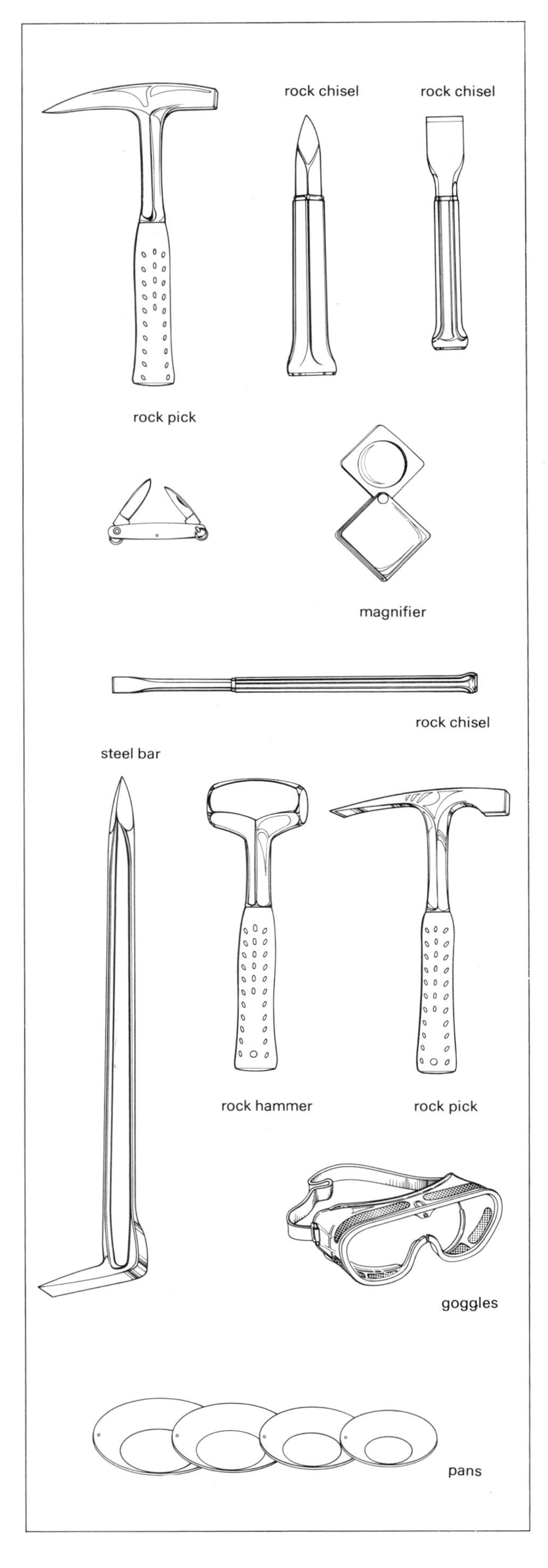

Tools for the well equipped mineral collector. The essential implement is a rock pick. It is made of hard steel and combines a point or chisel-blade with a hammer head. Large specimens can often be broken off with a heavy hammer. If the rock is tough it may be necessary to first use a chisel blade or steel bar. It is essential to wear safety goggles during any kind of hammering. Mineral specimens of differing densities occurring in stream gravels can be separated by panning. Once collected, specimens can be examined with a small hand-magnifier. Preliminary hardness tests can be carried out with the aid of a penknife blade and coins, while the streak can be obtained by rubbing the mineral on an unglazed ceramic tile

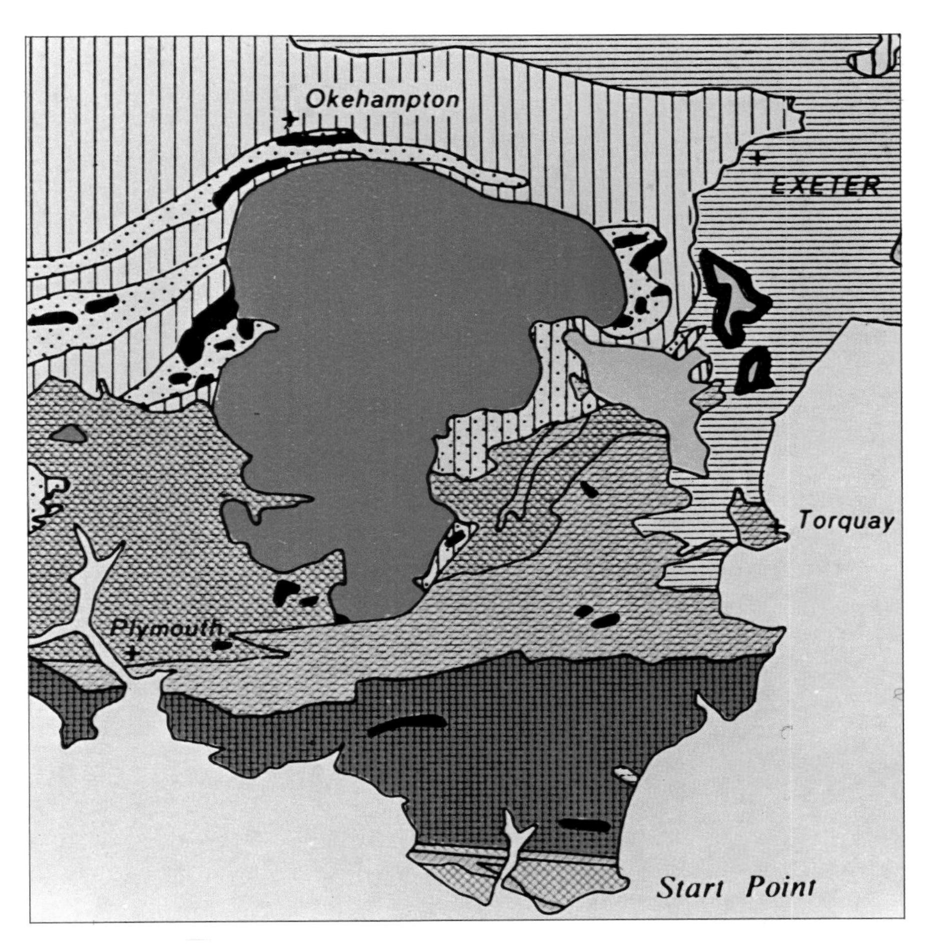

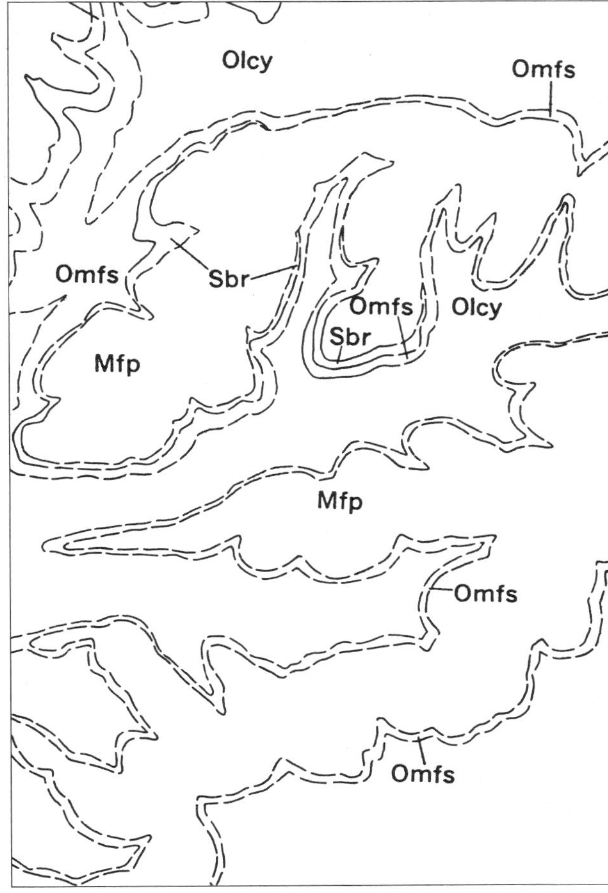

Geological maps can help to guide the collector. They vary in style from country to country. Above: Southwestern England, a traditional mining area. Colours show rock types. Black is dolerite, basalt or greenstone, red is granite, and so on. Top right: Geological map of part of Tennessee. Letters indicate rock types. Thus, Mfp consists mainly of cherts and shale, Sbr is limestone. Contour lines are included

on the predilections of the user. Occasionally large sledge hammers are called for. The handles must be of some tough, straight-grained wood such as hickory, but hammers may be bought that are fashioned from a single piece of steel; these are favoured in North America.

Steel chisels are nearly as useful to the mineral collector as hammers. A chisel with a half-inch or one-inch wide cutting edge is perhaps the most useful, but smaller sizes are sometimes called for in more delicate work such as removing small crystals.

Another most useful piece of equipment is a pocket knife. This may be used for cleaning around specimens that are still attached to the rock, particularly if this is relatively soft, and for digging or prising out small crystals and nodules. Hardness is one of the most useful physical properties for distinguishing minerals, and is readily determined in the field—for example, to distinguish quartz from calcite, or pyrite from chalcopyrite. For this purpose a good steel knife-blade is necessary.

For the study of fine-grained rocks or small crystalline masses, which will certainly constitute most of the material found in the field and elsewhere, a lens is essential. Lenses with a magnification of eight to twelve times are most satisfactory, and can be obtained with protective handles into which they retract. They are quite small, rarely more than two inches across, so that they are easily carried in the pocket.

Collectors' guides

An important consideration for any collector is to know where he can obtain specimens. Short of buying his collection—and there are now many shops, in most countries, that sell minerals—he needs to know where he can get information about the location and nature of collecting sites. There is no point whatsoever in the collector going out into the countryside at random looking for minerals, because material worth collecting is very restricted in its occurrence, and is unlikely to be encountered by accident. Instead, collecting expeditions must be directed to known collecting localities, and some preparatory work should be done searching the literature to discover the form of the deposit and the minerals to be found in it. The appearance of unfamiliar minerals can be learned in advance, to help with their recognition, and hence discovery.

The problem for the beginner, of course, is how to find out where the best localities are in his

Left: Dodecahedral crystals of almandine garnet embedded in schist. By carefully removing the surrounding rock the collector can often obtain perfect crystals from material of this type

neighbourhood. Perhaps the most useful action he can take in order to get himself started is to try to discover if there is a mineral club, 'rockhound' group or geological society in his area. Such societies and groups are very numerous in some countries, and range from the bodies dominated by professionals, such as the American Mineralogical Society and the Mineralogical Society of Great Britain, to local groups of enthusiasts. For the collector the local group is the more likely to be useful because it will put him in contact with fellow enthusiasts who are likely to know the best collecting localities in the area, and it will probably organize club or society field trips to just such places. There will certainly be talks and lectures, there may be a journal or newsletter, and there may be a society library. The reader who finds such a group and joins it will probably need little further help, but the collector who has to pursue his interest in isolation must have recourse to the literature.

The vast body of information that comprises the mineralogical and geological literature is contained in books and pamphlets, and geological maps. Most geological surveys publish memoirs, reports and bulletins describing the geology of particular areas, and such publications usually include descriptions of mineral occurrences, mining areas and so on. These, together with papers in research journals usually constitute the most detailed knowledge available. However, most mineral collectors need guidance that is more accessible and less academic, and to this end many geological surveys, including the American, Canadian, and British, publish material of more general interest. The Institute of Geological Sciences in London, for instance, publish eighteen guides in a series called *British Regional Geology*, which cover the whole of Britain. In the United States there are the numerous publications of the US Geological Survey and many of the State Geological Surveys put out their own publications. By enquiring at his local library or museum the reader should be able to discover what is available for his own area or the areas in which he is interested. Apart from essentially academic journals there are a number that cater solely for the amateur collector. Some of the better known ones are listed in the bibliography at the end of this book.

Most collectors must have need from time to time to consult geological maps. These maps, usually printed in colour, indicate the areal distribution of different rock types, and the larger-scale maps will distinguish most of the rock types referred to in this chapter, so that their utility for the mineral collector is obvious. Particularly on the larger-scale maps, features such as dykes and sills, and sometimes pegmatites, mineralized veins and contact-metamorphic aureoles are shown, so that such maps together with a mineralogical guide are an extremely useful tool in the armoury of the collector.

Because geological maps are published by the geological surveys of all countries, often on a number of different scales, it is not possible to detail specific sheets here. Instead enquiries should be made as to what maps are available and where they can be bought and consulted.

Specimen collecting

Minerals that are interesting enough to be worth collecting occur in a broad spectrum of geological environments. It is not practicable, therefore, to present a comprehensive account of mineral collecting, but a few general hints may be of use. Beyond these the collector must rely on his own judgement and experience.

Many of the better-crystallized minerals are found projecting into cavities into which they originally grew. In these circumstances, it is

Right: Collectors working on the spoil tip of a mine. Quarries and mine workings remain a fruitful source for the amateur collector

Far right: Panning for gold in a small stream. The technique of separating denser minerals from lighter ones by swirling them in water is easily learned. Ore minerals are generally denser than the common rock-forming minerals

generally better not to try to remove the crystals in isolation, but to attempt to free a piece of the matrix to which the crystals are attached. If successfully removed in this way, the matrix will provide an easy way of handling the specimen and a natural way to lay it down without the risk of damage to the crystals. You will see that many of the mineral specimens displayed in museums comprise groups of crystals sitting on matrix, and often these look more attractive than solitary crystals. Apart from ease of handling, the matrix gives strength to what might be very fragile material. Some minerals can hardly be separated from their matrix because they comprise very fine-grained encrustations, fine powderings, or fillings in narrow cracks.

When removing specimens with matrix the way the rock is likely to break should be judged carefully. Most rocks split or fracture in certain preferred directions, determined perhaps by bedding, schistosity, or jointing, so this must be taken into account. A geological hammer is a most versatile tool for detaching chunks of rock, but when fine crystals have been found that would be at risk from a misdirected hammer blow, or when working space is confined, or again when rock is proving particularly resistant, recourse must be made to a steel chisel. It is usually possible to find a convenient joint (crack) in which the chisel can be started, and few rocks can withstand the force that can be applied when it is well driven home.

When you are collecting from such rocks as coarse pegmatites, it is important to note that fragments are usually bounded by the cleavage surfaces of the very large crystal constituents, so that this must be taken into consideration when freeing specimens. A mineral which must be treated most carefully because of its ready cleavage is calcite. This is a common mineral and often found as fine crystals in a wide variety of forms; but if a good crystal is found do not attack it directly with a chisel at its base, for it will certainly cleave, and may even end up as a collection of calcite rhombs. Instead, try to free the crystal with some of the rock to which it is attached; it may prove possible to separate them later.

Some interesting minerals may be found in soft sediments as concretions, in nodules, or as single crystals, or be isolated by the weathering of their host rocks—as for instance agates in weathered basaltic lavas. In these cases the problem is merely to find them.

Labelling and note taking

Although minerals are well worth collecting in their own right, a well organized collection calls for more documentation than the mere identification of its constituent specimens. This may not be possible if specimens are obtained from collectors or dealers who do not take much care in such matters, but for specimens collected by oneself every care should be taken to record the fullest details. To this end a notebook should be used in the field for recording the locality as precisely as possible, preferably with a map reference, together with notes, possibly illustrated by sketches, of the field relationships of the collected material. These notes will form the basis for labels to be made out later to be kept with the specimens, and for entries in a master catalogue of the collection. Some sort of numbering system must also be adopted in the field so that individual specimens can be identified with the notebook entries. These numbers will later be superseded by whatever numbering system has been adopted for the collection as a whole.

A wide variety of methods can conveniently be used in the field for numbering specimens. For instance, a number can be inked onto the paper in which a specimen is wrapped. This method has the disadvantage that the number can be obscured by rubbing or tearing of the paper during transport. It is perhaps better to write the number on a slip of paper which is then wrapped in with the specimen, though loose pieces of paper tend to get mislaid. Probably the most satisfactory methods are either to write directly onto the specimen with a felt-tipped pen, provided there is some matrix that will not be spoiled in this way, or to write on a piece of adhesive linen tape, which is then stuck onto the specimen. Such tape is easily removed when the specimen has been permanently numbered.

Carrying specimens

The transport of mineral specimens presents two principal problems: firstly they can be heavy, and secondly they can be easily damaged. The majority of collectors favour some sort of canvas bag for carrying their specimens in the field. This may be a simple shoulder bag, but for heavy loads a rucksack carried high on the back is much more efficient and comfortable. However, the relative softness of such bags makes it essential that specimens are carefully packed.

Although many mineral specimens may be satisfactorily protected from breakage, scratching and general abrasion during transport by wrapping them firmly in newspaper, others, particularly well crystallized specimens, present more of a problem. If the crystals are attached to a good piece of matrix and are themselves robust, then a stout paper or linen bag or plenty of newspaper will probably suffice, so long as other specimens are not carried on top of them. More delicate material should be carried in a tin or wooden box to prevent crushing, and should be well packed in with suitable material.

ROCKS AND THEIR MINERALS

Mineral specimens that are displayed as single crystals or aggregates of crystals were originally found adhering to some matrix. Matrix material, which is of lesser interest to the mineral collector, is of great importance to the petrologist. Petrology is the science of rocks and is concerned with their mineralogical and chemical constitution and other features that will throw light on how they were formed. The material referred to by the petrologist as *rock*, which is simply an aggregate of minerals, is the same stuff that the mineralogist calls matrix. Whereas the mineralogist tends to look at particular mineral species for their own sake, and prefers large, well formed or rare specimens if he can obtain them, the petrologist is interested in the properties of minerals in aggregate, and tends also to be more interested in the few mineral species that constitute the bulk of the commoner rocks.

Although the petrologist evidently must call upon the science of mineralogy to help him to understand the minerals in his rocks, the mineralogist must also rely to some extent on petrology to tell him where he will find his minerals. The reason for this is that the various mineral species are not randomly distributed through the Earth; particular minerals tend to be associated with only one or a few kinds of rock. Therefore, the search for particular minerals must be directed to the right rocks. To take a simple case, the mineral collector who searches over a mass of granite for the minerals nepheline, chromite or halite will be disappointed, but he might find various forms of quartz, particular feldspars or micas, or perhaps beryl, cassiterite, or tourmaline.

The rest of this chapter is intended, therefore, to introduce the mineral collector to the various kinds of rocks that constitute the crust of the Earth, the geological environments in which they are found, their classification, and particularly their mineralogy. And it is hoped that this will not only help to make mineral-collecting a more

Right: A lake of incandescent lava in the crater of Kilauea Volcano, Hawaii. Fountains of lava are thrown into the air by the release of dissolved gases. Lava such as this solidifies to form the igneous volcanic rock called basalt. Below: The crust has darkened in patches as it temporarily cools. Currents in the underlying liquid sometimes break up the crust, revealing the glowing lava

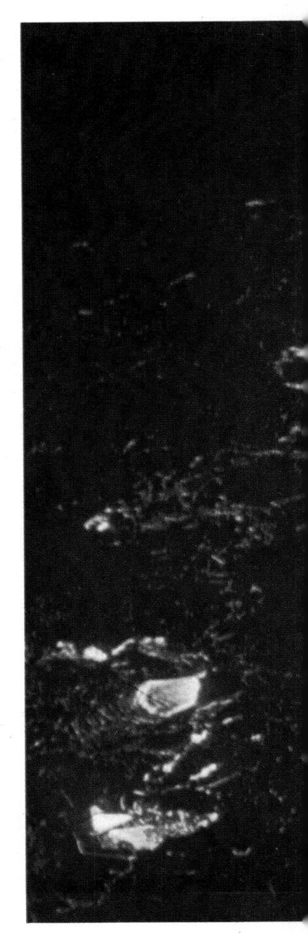

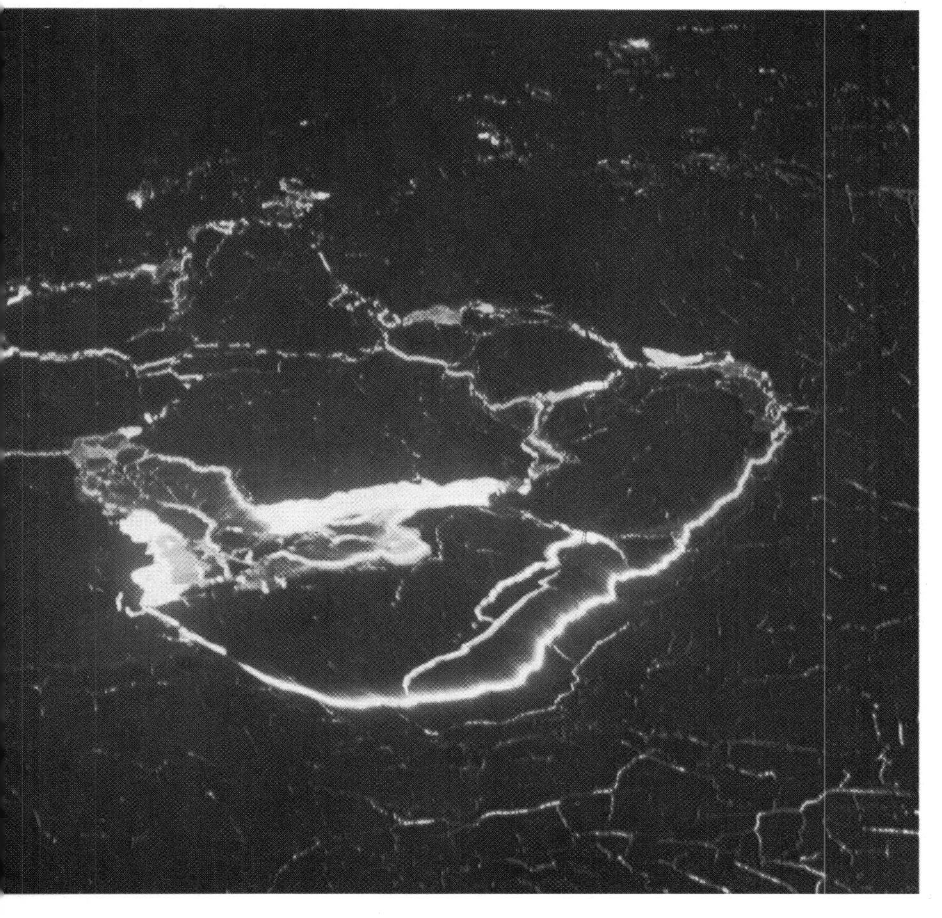

scientific and successful pursuit, but also will add to the intrinsic interest of minerals by placing them in their natural contexts and explaining something of why they occur in the places where they are found.

Rocks can be divided into three fundamental categories: igneous, metamorphic and sedimentary rocks. The *igneous* rocks are formed by the solidification of molten material; *metamorphic* rocks are formed by the alteration of igneous and sedimentary rocks by heat and pressure; while *sedimentary* rocks are produced by the accumulation of material at the Earth's surface.

IGNEOUS ROCKS

Igneous rocks are formed from the solidification of molten rock material, which is called *magma*. Magma is generated by melting deep in the Earth, sometimes at depths in excess of 100 kilometres. When magma migrates upwards towards the surface of the Earth it may be injected in masses of varying shape and size among the near-surface rocks, forming igneous *intrusions*, or it may reach the surface and pour out over it as *lava*.

Igneous intrusions are conveniently divided into major and minor intrusions. Of major

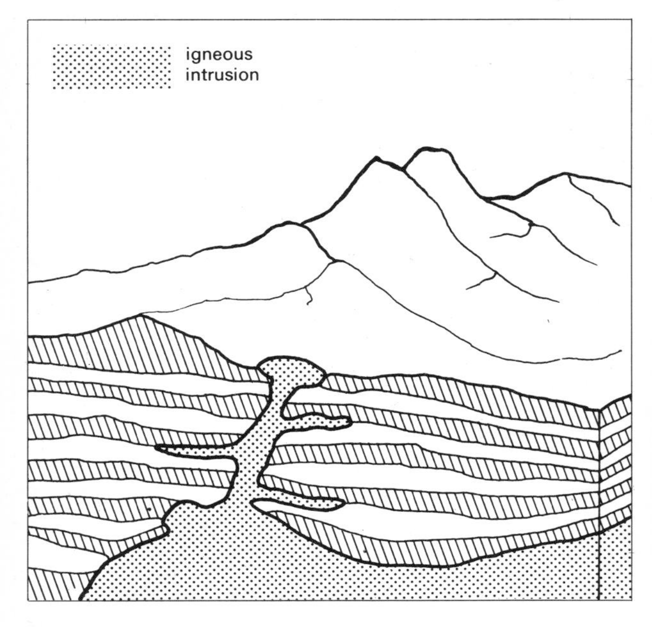

intrusions, the largest are *batholiths*, which are very extensive bodies usually covering hundreds or thousands of square kilometres. They cut across the country rocks (the rocks into which the intrusion is injected), and show no signs of having a floor. Intrusions similar in shape but smaller, covering a few square kilometres or tens of square kilometres, are called *stocks*. Pipe-like cylindrical intrusions called *plugs* are usually circular or oval in plan, and often fill the feeder channels for volcanoes; they are rarely more than one or two kilometres in diameter.

The principal types of minor intrusions are *dykes* and *sills*. Dykes are sheet-like intrusions that are vertical or nearly vertical in attitude, and cut sharply across any structures in the country rocks. They range from a few centimetres to hundreds of metres in width, and may be short or many kilometres long. Sills are similar in form to dykes, but are horizontal or near-horizontal in attitude, and lie parallel to any layering in the rocks they cut.

Because of their large size major intrusions cool down rather slowly, allowing their constituent minerals to form big crystals. The coarse-grained rocks that this gives rise to are called *plutonic* rocks, and contrast with the medium- and finer-grained rocks of the more rapidly cooled minor intrusions.

Igneous rocks formed from material poured out over the surface of the Earth are known as *extrusive* rocks, as opposed to the *intrusive* rocks of igneous intrusions. Lavas may spread widely, forming lava plains or plateaux, or they may accumulate around the pipe or fissure from which they were extruded, building up a volcano. All

Mineralogy of igneous rocks

principal rock-forming minerals		accessory minerals
leucocratic minerals	quartz	magnetite ilmenite
feldspars	orthoclase, microcline, sanidine alkali feldspar series plagioclase feldspar series	pyrite pyrrhotine apatite corundum
feldspathoids	nepheline, leucite, sodalite, cancrinite	sphene zircon fluorite
mafic minerals	olivine	chromite zeolites
pyroxenes	augite, diopside, hypersthene, aegirine	tourmaline
amphiboles	hornblende, riebeckite	
micas	biotite, muscovite	

magmas contain a greater or lesser amount of dissolved gas, and the reduction of pressure as the magma comes close to the surface may be sufficient to allow the gas to come out of solution. It may simply form bubbles in the lava, which become cavities called *vesicles* in the rock that is formed, but it may be so abundant that it violently disrupts the lava into pieces that are thrown out onto the surrounding land surface. Rocks formed from lava fragments are known as *pyroclastic* rocks and range from those containing large fragments many metres across, which are called volcanic *bombs*, to fine-grained volcanic *ash*. Most volcanoes are built up of a mixture of lava flows and pyroclastic rocks laid down in the course of thousands of years.

Because extrusive rocks are cooled very rapidly, unless they form very thick lava flows, their constituent minerals have little time to

Right: A thick sequence of pyroclastic rocks thrown out by past volcanic activity, forming a cliff on the island of Ischia. About halfway up, the horizontal beds of ashes can be seen to be partly truncated by the overlying beds, indicating two distinct periods of volcanic eruption

Far left: Igneous intrusions. The large body of igneous rock is called a batholith. The sheets of rock that extend from it are called dykes when they cut across the layers of surrounding 'country' rock and sills when they lie parallel to them

Left: A fine-grained igneous dyke cutting through coarser-grained granitic rock in Sardinia. Note the nearly parallel edges of the dyke

crystallize, so that they are fine-grained; indeed some are chilled so rapidly that no crystallization takes place at all, and the result is a glass, which is called *obsidian*. However, the magma often starts to crystallize even before it is erupted. The usual result is that the rock contains some larger crystals in a finer-grained matrix; the large crystals are called *phenocrysts* and the whole rock is referred to as a *porphyry*. Sills and dykes are also often composed of porphyry.

Mineralogical classification of igneous rocks

It is essential for the mineral collector to know something about the classification of rocks. The reason is that rocks are classified principally according to the main minerals of which they are composed. Igneous rocks are made up overwhelmingly of just seven minerals or mineral groups, which are all silicates: quartz, feldspars, feldspathoids, pyroxenes, amphiboles, micas and olivine. There are other minerals present in most rocks, but usually in minor amounts, and they are generally referred to as 'accessory minerals'. The commonest are probably magnetite, ilmenite, apatite and sphene. Some rocks consist of only a single mineral; dunite, for instance, is a rock composed wholly of olivine, and pyroxenite consists wholly of pyroxene. However, most igneous rocks include three or four different minerals, though some minerals never, or only very rarely, occur together in the same rock. Specimens of quartz, for example, are rarely found in rocks containing olivine or the feldspathoids.

Apart from quartz the minerals listed above as comprising the principal constituents of igneous rocks are in fact mineral groups consisting of several different mineral species. The group of minerals called the amphiboles contains dozens of different minerals, although they do not all occur in igneous rocks. The feldspars, which are the most abundant minerals of igneous rocks, averaging about sixty per cent by volume, similarly comprise a number of distinct species. The main groups found in igneous rocks together with the commoner individual species are given in the table on the preceding page.

The classification of igneous rocks is based on the nature and proportions of the minerals occurring in them, and of these minerals the feldspars play a particularly important role. It is therefore necessary at this point briefly to explain something of the organization of this mineral group. The feldspars are aluminium silicates of calcium, sodium and potassium. Pure calcium feldspar is called anorthite, pure sodium feldspar albite, and pure potassium feldspar can be orthoclase, microcline, or sanidine depending on the atomic structure. There is a complete series of feldspars between anorthite and albite; that is, feldspars occur which contain all possible proportions of calcium and sodium. The minerals of this series are called the plagioclase feldspars. The plagioclases are grouped into six divisions depending on the proportions of calcium and sodium present. For purposes of classifying igneous rocks the plagioclase series is an important one. Rocks containing plagioclase which has more sodium than calcium are usually differentiated from those in which the reverse holds, and some types depend on the even finer distinctions of their plagioclase compositions.

There is a second important series of feldspars known as the alkali feldspar series. This ranges from sodium feldspar, albite, to potassium feldspar—orthoclase, microcline or sanidine. Some rocks are named according to whether their feldspar is mainly a potassium or a sodium type.

Other mineralogical features which are then considered in classification are whether quartz or feldspathoid are present, and the amount and nature of what are collectively called the dark or *mafic* minerals, that is pyroxene, amphibole, olivine and biotite mica. These criteria are summarized on the following page in a table that gives a basic classification of igneous rocks. It is to be noted that the rocks are divided into a coarser-grained series, which comprises essentially the plutonic rocks of large intrusions, and a medium- to fine-grained series, which comprises the rocks of minor intrusions and the extrusive rocks. The proportions of mafic minerals and light, or *leucocratic*, minerals (quartz, the feldspars and the feldspathoids) vary through the two series. In the series granite, syenite, diorite, gabbro, peridotite the rocks gradually change from having very little mafic mineral to rocks comprising mafic minerals only, so that granite and syenite are light-coloured, diorite and gabbro are dark, and peridotite is black or dark green in colour.

If it is intended to collect any of the essential rock-forming minerals listed on page 50, then attention should be directed to the appropriate rocks tabulated on page 53, though it must be realized that these minerals can be found in other environments, to be described below. From the brief descriptions of the rocks given here it will probably not be possible to recognize them without further study and practice, but these rock types are usually distinguished on geological maps and in the literature, so they are readily to be found.

Most igneous rocks, whether coarse- or fine-grained, are fairly homogeneous. The minerals present may be recognized in hand specimen—for instance feldspar, quartz and perhaps mica in a piece of granite—yet consist of relatively small, intergrown crystals, not usually considered good enough for a mineral collection. There are, however, often situations where the minerals of the rocks, sometimes including the accessory minerals, are either segregated into minerals of one species, or form bigger or better developed crystals; some of these situations are now described.

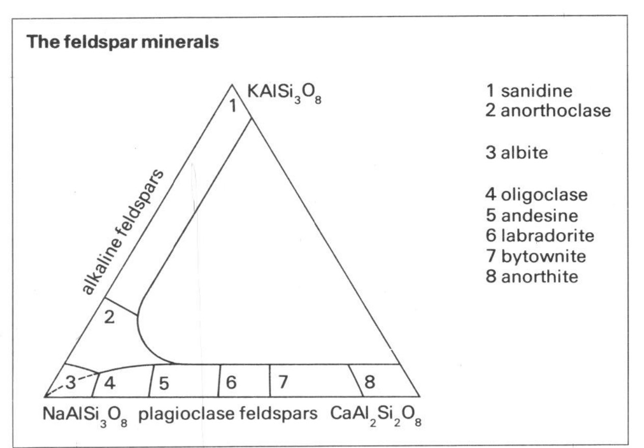

Left: The feldspars. In the plagioclase series the proportions of sodium and calcium vary from anorthite (pure calcium feldspar) to albite (pure sodium feldspar). The alkali feldspars run from albite to a pure potassium feldspar, sanidine microline or orthoclase

Above: Hexagonal white crystals of nepheline on finely grained lava

Pegmatitic segregations. Sometimes among plutonic rocks patches of rock are found in which the minerals are much coarser than the average. These patches may be small or large, and are often lens-shaped or vein-like in form. Such coarse textures are referred to as *pegmatitic*. One can have pegmatitic granite, gabbro, nepheline syenite and so on. The minerals are the same as those of the rock as a whole, though sometimes in different proportions. These local pegmatitic segregations are not to be confused with the discrete and distinctive pegmatite bodies that are described below.

Right: Mineral classification of igneous rocks. 'Plagio.' means 'plagioclase'; 'Na > Ca' refers to plagioclases containing more sodium than calcium; 'Ca > Na' means the reverse

Mineral layering. Some magmas, particularly those having the composition of gabbro or peridotite, have relatively low viscosities, so that when they start to crystallize the early-formed crystals, which tend to be denser than the magma, sink to the bottom of the magma chamber, where they accumulate. Because of the resulting change in composition of the remaining magma, or because of changes in the physical conditions, the mineral species crystallizing may change, and hence a layer of different composition will start to accumulate on the first layer. Such changes may be frequent, and commonly result in alternating layers of two or perhaps three minerals. Individual layers may be less than a centimetre to more than a metre in thickness, and may contain one or several mineral species. Minerals that occur in such layers and which may be obtained as pure monomineralic rock include plagioclase, olivine, pyroxene, magnetite and chromite. Many economically valuable chromium deposits, for instance, occur as such rocks, and the platinum deposits of the famous Merensky Reef in South Africa occur in accumulations of this kind.

Below: A fine crystal of leucite of icositetrahedral form, from Parco Chigi, Ariccia, Rome

Phenocrysts. As explained earlier, the term 'phenocryst' is applied to large crystals set in a finer-grained matrix. This texture is common amongst extrusive igneous rocks and in minor intrusions, but may occur in plutonic rocks also. Because phenocrysts are crystals that started to grow early on in the cooling history of a magma, they were usually able to form well-shaped crystals, because they were not confined by other close-packed crystals. Phenocrysts are therefore often good mineral specimens. Some of the commoner minerals that form phenocrysts, with the rocks in which they are to be found, are these:

sanidine in trachyte;
orthoclase in granite;
orthoclase, quartz in rhyolite;
nepheline, leucite, aegirine in phonolite;
olivine, augite, plagioclase in basalt;
plagioclase, hypersthene, hornblende in andesite.

Pegmatites

Towards the end of the crystallization of magma in a major intrusion, particularly if the magma has the composition of granite or nepheline syenite, a residual melt is concentrated that is very rich in water, chlorine, boron, and other substances that are taken up to only a small degree by the minerals that have crystallized so far. The melt increasingly becomes a watery solution as crystallization proceeds, and this solution has a relatively low viscosity. Because of this fluidity these residual liquids can penetrate fissures and cracks in the already solidified parental igneous mass, and perhaps also the surrounding country rocks. In these fissures they crystallize as bodies known as *pegmatites* or pegmatite veins. The fluidity of the melt facilitates the growth of large crystals, so that pegmatites are characterized by their coarse grain size. However, it is not only the large size of the

Classification of igneous rocks

rock name			feldspars		feldspathoid/ quartz content	dark minerals	dark mineral content	general rock type
coarse-grained	medium-grained	fine-grained						
IJOLITE		NEPHELINITE	none		feldspathoid	alkali pyroxene	important	alkaline
NEPHELINE SYENITE		PHONOLITE	alkali feldspars dominant					
GRANITE		RHYOLITE OBSIDIAN			more than 10% quartz	biotite and hornblende	minor	acid
SYENITE		TRACHYTE			not important		important	intermediate
DIORITE		ANDESITE	plagio.* dominant	Na > Ca*	quartz and feldspathoid absent			
GABBRO	DOLERITE	BASALT		Ca > Na*		pyroxene and olivine		basic
PERIDOTITE DUNITE PYROXENITE SERPENTINITE			little or no feldspar				dominant	ultrabasic

Left: A group of well formed mineral crystals from Elba. They include a fine specimen of 'Negro's head' tourmaline, white orthoclase feldspar and colourless quartz

crystals of pegmatites that makes them a great attraction to the mineral collector but also the wide range of rare minerals that are often to be found in them.

Another feature of pegmatites that enhances their importance as a source of fine mineral specimens is the fact that *druses* or *drusy cavities* are relatively common in them. Druses are irregular cavities in the pegmatite that are lined with crystals, which grow inwards from the walls with little interference from surrounding crystals. This unrestricted growth accordingly allows them to develop simply according to the dictates of their atomic arrangement, so that perfectly formed crystals result, usually as crusts lining the cavity.

Pegmatites are essentially dyke-like bodies, but because they tend to be irregular in form they are perhaps best thought of as veins. They vary in thickness from a few centimetres to tens of metres. In length they are likewise variable, but only rarely do they reach as much as one or two kilometres. They commonly occur as clusters or 'swarms' around the outer margins of granite stocks, or in the adjacent country rocks.

Pegmatite minerals, although by definition coarse-grained, may vary in grain-size from species to species, and the pegmatite may be zoned in distinct layers varying both in mineralogy and in coarseness of grain. A common feature is for prismatic crystals to grow perpendicular to the length of the vein, either from the walls, or across individual zones. Some pegmatites are noteworthy for the giant size of the crystals within them, and such crystals as the following have been recorded: quartz crystals 5.5 metres long by 2.5 metres in diameter; orthoclase crystals 10 metres by 10 metres and weighing up

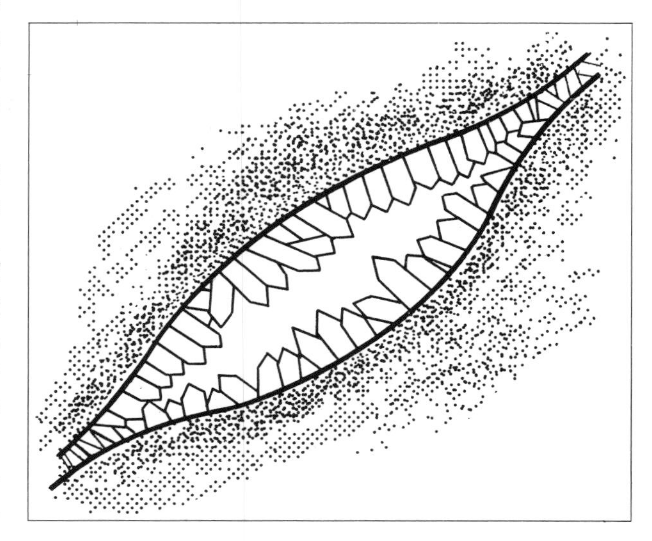

Left: A crystal-lined cavity or druse. Because the crystals in such cavities are relatively uncrowded they are often very well formed, and can provide the collector with fine specimens

Right: Crystals of green muscovite. Muscovite is a common pegmatite mineral

Below right: 'Graphic' granite, with quartz forming strange shapes like ancient script

to one hundred tons; beryl crystals 6 metres or more long and weighing up to two hundred tons; tourmalines up to 3 metres long; micas with surface areas of 7 square metres; and so on.

The elements that are concentrated in the late residual fluids of granite and nepheline syenite magmas are essentially of two kinds. Firstly, there are those major elements in the magma that form ingredients that melt at the lowest temperatures, and so are concentrated in the pegmatite stage. These are principally silicon, aluminium, sodium and potassium. Secondly, there is a group of relatively rare elements that were originally thinly dispersed through the magma, but become concentrated in the late fluids. The reason for their late incorporation into rock is that they have ions that are either too small or too large to be taken into the atomic structures of the principal rock-forming minerals. Typical elements concentrated in this way are boron, phosphorus, fluorine, chlorine, sulphur, lithium, beryllium, rubidium, caesium, molybdenum and the rare-earth elements. There are some differences in the elements concentrated in granite pegmatites and nepheline syenite pegmatites, which reflect differences in the original magmas and lead to different sorts of minerals.

Apart from the principal mineralogical differences between granite and nepheline syenite pegmatites, different suites of minerals occur in different examples within each type, and this applies particularly to granite pegmatites. These differences are partly due to differences in the original magmas from which they were derived; but probably of greater importance is the possibility that different pegmatites may represent different evolutionary stages, with different temperatures and pressures at which crystallization occurred. In the USSR in particular a great deal of research has been devoted to investigating the evolutionary trends of pegmatites and to the chemical and physical constraints that determine these. The reason for this effort is that pegmatites are a very important source of valuable mineral deposits, particularly ores of tungsten, tin, niobium, tantalum, titanium, beryllium and lithium, as well as many non-metallic industrial minerals, such as mica and feldspar, and a wide range of gemstones.

The basic mineralogy of pegmatites is simple, comprising quartz, alkali feldspar, plagioclase (sodic varieties, mainly oligoclase grading into albite) and muscovite mica; though in nepheline syenite pegmatites the quartz is replaced by one or more feldspathoid minerals, of which nepheline is the most common. A characteristic feature of many granite pegmatites is a complex intergrowth of quartz and orthoclase feldspar known as 'graphic granite'.

Of the rarer minerals the following is a selection of those to be found in granite pegmatites: beryl (emerald, aquamarine, etc), chrysoberyl, columbite-tantalite, apatite, allanite, monazite, xenotime, fergusonite, euxenite, thalenite, gadolinite, uraninite, tourmaline, spodumene, petalite, amblygonite, lepidolite, topaz, danburite, microlite and pollucite.

Minerals commonly occurring in nepheline syenite pegmatites include aegirine, eudialyte, natrolite, analcime, chabazite, gibbsite, arfvedsonite and melanite garnet.

Hydrothermal deposits

In the previous section the formation of pegmatites as a late-stage product of magma crystallization was explained, and it was pointed out that the pegmatitic fluid can be considered to consist of a mixture of residual magma plus certain volatile constituents, including water. The final consolidation of the pegmatite will leave a residual fraction consisting of a water-rich fluid, which is essentially a hydrous solution rather than a magma. Such hydrous fluid phases are produced particularly often in magmas with compositions ranging from granite to diorite during the final stages of crystallization. These fluids force their way into the country rocks and there deposit the various elements that they hold in solution. The resulting rocks are called hydrothermal deposits, and are perhaps the most numerous and diverse of all economic mineral deposits; though nowadays, when it is technically possible to exploit economically ores of much lower grade, the hydrothermal deposits are not so pre-eminent as they formerly were. However, from them still comes a significant proportion of the world's gold, silver, zinc, lead, copper, tungsten, tin, mercury, molybdenum, antimony and most of the minor metals, and many non-metallic minerals, including a broad range of sulphides, selenides, tellurides, arsenides, silicates, oxides, carbonates, sulphates and others. It is because of this diversity that the spoil tips of mines working such deposits have a considerable attraction for the mineral collector.

The minerals of hydrothermal deposits are often divided into two groups, according to their supposed economic importance. The sulphides are typical of the 'economic' minerals and are probably the most important. The following sulphides are some of those commonly found: pyrite, chalcopyrite, sphalerite, galena, bornite, chalcocite, marcasite, tetrahedrite, arsenopyrite and stibnite. The second group of minerals, formerly considered to have little or no commercial value, are called *gangue* minerals, and include quartz, calcite, baryte, siderite, dolomite, rhodochrosite and fluorite. However, the term gangue mineral is now something of a misnomer as many now find an important place amongst the raw materials of industry.

The temperature of hydrothermal solutions can be inferred with some precision from the mineralogy of the resulting deposits. Thus, certain minerals alter from one species to another at definite temperatures, and some minerals have polymorphs that are stable over different temperature ranges; yet other minerals form solid solutions that separate into two or more phases below particular temperatures. From evidence of this kind hydrothermal deposits can be shown to have been formed in some range of temperatures. Lindgren has categorized hydrothermal deposits as high-, medium- and low-temperature, which are called *hypothermal, mesothermal* and *epithermal*. Hypothermal deposits are formed at about 300–500°C; mesothermal deposits at about 200–300°C; and epithermal deposits at about 50–200°C. These types necessarily grade into one another, but many deposits can be assigned to one of these groups from their mineralogy and geological environment. The hypothermal deposits tend to lie close to the magmatic source, and mesothermal and epithermal deposits farther away. Many hydrothermal deposits are accordingly zoned, with different elements, and therefore suites of minerals, concentrated at different distances from the igneous intrusion. This is expressed in the accompanying table, due to G Berg, which shows the relationship between the elements of hydrothermal deposits and proximity to the parental intrusion.

This is expressed mineralogically in the following way:

Hypothermal deposits may contain gold, gold tellurides, magnetite, ilmenite, cassiterite, scheelite, wolframite, molybdenite, garnet, mica, apatite, tourmaline, topaz, etc. The principal gangue mineral is quartz.
Mesothermal deposits may include gold, galena, chalcopyrite, pyrite, sphalerite, bornite, arsenopyrite, enargite, tetrahedrite, etc. The principal gangue minerals are quartz, calcite, siderite, dolomite and baryte.
Epithermal deposits often include pyrite, marcasite, cinnabar, stibnite, native silver, pyrargyrite and proustite. Gangue minerals include quartz, often in the form of chalcedony and opal, fluorite, adularia and baryte.

These lists are by no means comprehensive, but they give some idea of the range of minerals occurring and the overlap across the groups. However, it must be appreciated that the full range of minerals is rarely found in one deposit,

Zoning of elements in hydrothermal deposits

higher temperature			
(closer to intrusion)	tungsten	tin	molybdenum
	uranium	bismuth	
	gold	arsenic	
	copper		
	silver	zinc	
	lead		
lower temperature	antimony		
(farther from intrusion)	mercury		

Above: A specimen of blende, of the variety known as marmatite, together with pyrite (yellow) and quartz (colourless). This is a typical mineral assemblage of some hydrothermal deposits

and certain minerals tend to occur together, forming characteristic associations.

So far no mention has been made of the form that hydrothermal deposits take, but this is often very distinctive, and a most interesting aspect of these rocks. When hydrothermal solutions are expressed from the mass of consolidating magma they must find channels and spaces in the rocks through which they can pass. The most common form of passageway for the solutions is fissures produced by various geological processes, and the new minerals deposited in them give rise to mineral *veins*. Because veins are such a common form for hydrothermal deposits the whole range of these rocks is often referred to as the *vein association*.

One of the most common types of fissure exploited by hydrothermal solutions is the fault plane, so that many hydrothermal deposits are associated with fault systems. Joints and bedding planes, the pore spaces between mineral grains, volcanic pipes filled with only partly consolidated pyroclastic rocks, vesicles in lavas, and even solution caves in limestone have also provided some of the routes and depositories for hydrothermal solutions.

Hydrothermal veins vary widely in their mineralogy, form, size and geological environment. Some veins consist of only one or two minerals; for instance, almost monomineralic stibnite veins occur widely, and veins of galena and sphalerite are commonly deposited in limestones. In contrast, some veins contain assemblages of perhaps thirty or more minerals, usually deposited in a sequence, which is sometimes repeated. The minerals have been built up layer by layer, gradually filling up a cavity, and often there is a final unfilled space or vug into which well developed crystals project. Because minerals of the same type are deposited simultaneously on both walls, veins are often symmetrical with the same mineral sequence repeated on the two sides.

Usually the rocks on each side of a vein have been affected to some degree by the hydrothermal process, and the alteration usually produces a very obvious colour change in these zones. Although new minerals such as white mica and orthoclase are developed in the alteration zone, they are rarely in a form that attracts the mineral collector.

Although veins are probably the most characteristic feature of hydrothermal deposits, in fact the greater part of these deposits owe their origin to *metasomatic replacement*, and this process is always the dominant one in the high- and medium-temperature deposits. Replacement is a two-part process, involving the taking into solution of the original rock minerals and the deposition of new ones in their place. The process can be a very delicate one, so that the shapes of original crystals or structures may be duplicated exactly by the replacing minerals. However, the overall replacement may be on a huge scale, and some enormous mineral deposits have been formed in this way.

Alteration of hydrothermal deposits

At the surface of the Earth and to varying depths beneath it rocks are affected by *weathering* processes. These involve a combination of physical and chemical changes in which rocks and their constituent minerals are broken down to varying degrees and new minerals and rocks are formed. Hydrothermal deposits are highly susceptible to attack because they contain a high proportion of sulphide minerals, which are particularly unstable under near-surface conditions. The main agents of chemical decomposition are water, carbon dioxide, which dissolves in water to produce a powerful solvent, and oxygen, which

Below: An oddly shaped but superb specimen of azurite growing on green layered malachite. Layering is often found in minerals in hydrothermal deposits that have been altered by weathering

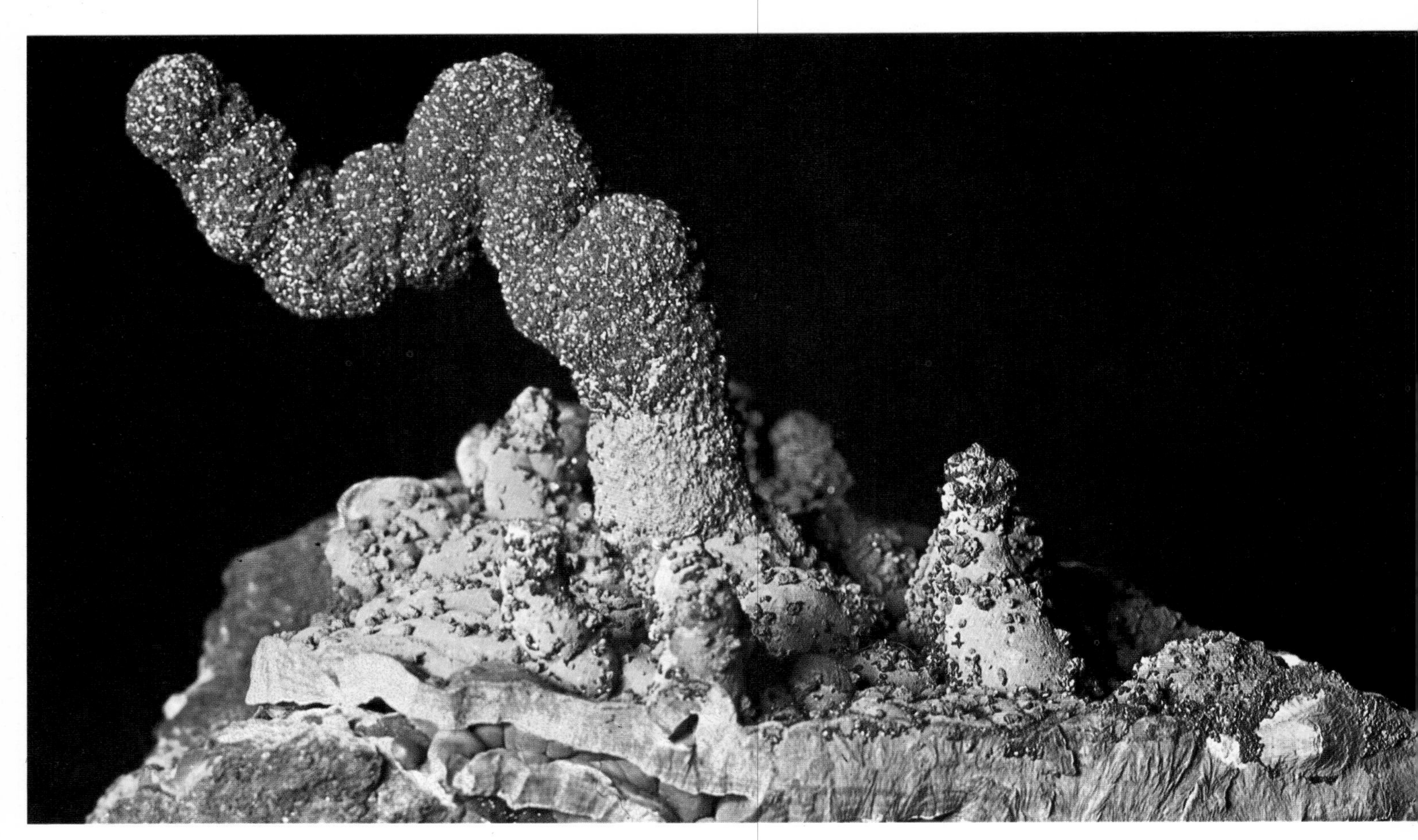

allows oxidation. The oxidation of sulphide minerals produces sulphuric acid and sulphates, both of which are effective decomposing agents.

The total effect of chemical activity is to take into solution many of the components of the rock and to carry them down to lower levels. Thus a hydrothermal deposit is leached of its sulphides, and near the surface may become just a cavernous rock rich in silica and iron oxides and known as *gossan* or 'iron hat'. The weathering and leaching process may extend downwards for several hundred feet, but stops at the water table, below which quite different chemical conditions prevail. Below the water table there is an oxygen deficiency and because of this the metals carried downwards in solution are redeposited as sulphides, rather than, for example, sulphates or carbonates. At greater depths still the primary hydrothermal deposit will be reached, in the form of veins or massive replacement deposits. It is apparent that there are effectively three zones, an upper zone of oxidation above the water table, a zone in which sulphide minerals are deposited, just below the water table, and the unaffected primary rocks. This sequence is shown in the diagram here.

Below: Wulfenite, often found in the oxidized zones of hydrothermal deposits

Right: Weathering of ore-bearing rocks. Near the surface ground water, oxygen and carbon dioxide tend to leach out minerals. Often this leaves an iron-rich 'gossan'. Oxidized minerals are redeposited at lower levels. Below the water table, however, dissolved minerals are deposited as sulphides. Still deeper is the original primary ore

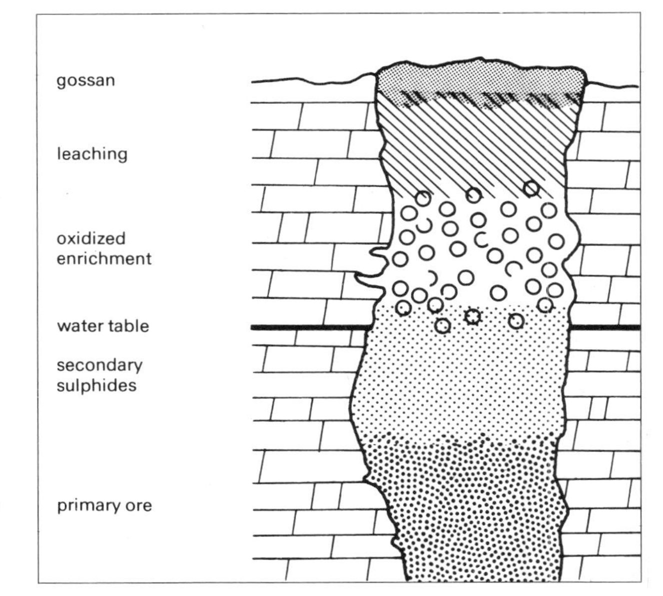

It will be seen from the figure that the oxidized zone is composed of two parts; towards the top is the zone of leaching from which many soluble minerals have been removed. As the cold leaching solutions trickle down through the zone of oxidation a greater part, or all, of their dissolved metals may be precipitated to form oxidized-ore deposits. The net effect of this process is that a high proportion of the metals originally distributed through the whole of the zone of oxidation are now concentrated towards the bottom of the zone. In this way many very rich ore deposits have been formed. Of great interest from the present point of view is the fact that a wide spectrum of often spectacular minerals can be produced in this way. For instance, many of the best examples of the carbonates, silicates, sulphates and oxides of copper and zinc are to be found in this environment, as well as the oxides of cobalt, antimony, molybdenum and bismuth, and the chlorides, iodides and bromides of silver. Among the very many individual minerals found are malachite, azurite, cuprite, chrysocolla, chalcanthite, smithsonite, cerussite, hemimorphite, anglesite, wulfenite and embolite.

The dissolved metals that are not precipitated in the zone of oxidation trickle down to levels beneath the water table, from which oxygen is excluded. There they are deposited as secondary sulphide minerals. As a result, there is an enrichment in sulphide minerals in this zone, so that rich sulphide deposits are made richer and poorer deposits are so enhanced that they may become economically valuable. With time erosion will gradually lower the land surface, together with the water table and the zone of oxidation, so that the secondary sulphide deposit will again be oxidized and its metals carried to lower levels. By repetition of this process the primary ore deposit may be enriched by a factor of up to ten. Many important copper deposits, for instance, have been formed in this way.

Minerals such as pyrite, chalcopyrite, covellite, bornite and chalcocite are deposited in the zone of secondary sulphide enrichment, but since these minerals are the same as the primary sulphides, they are not diagnostic of this zone, and the fact that secondary enrichment has taken place is indicated by other features.

Fumaroles

Among the few places in nature where minerals can actually be seen growing are volcanic *fumaroles*. These are localities, often found in large numbers in volcanic regions, where volcanic gases escape at the surface. The gases were originally dissolved in the magma but have been released into the surrounding rocks and may find their way to the surface. Substances present in the gases, or formed by interaction among them, are deposited around the fumaroles. Fumarolic mineral deposits are usually short-lived, either because they are overwhelmed by further volcanic activity, or because many of the minerals are unstable, many being soluble in water so that they are destroyed by the first heavy rain.

The minerals formed in fumarolic deposits are numerous, and around Vesuvius alone more than fifty have been identified. Sulphur is one of the most common minerals of this environment and together with alunite and agate has been worked commercially. Others include chlorides, particularly ammonium chloride, fluorides, sulphates, sulphides, tellurides and arsenides; amongst the species found are hematite, magnetite, realgar, pyrite, molybdenite and covellite.

A particular manifestation of mineral deposition associated with volcanic activity that is of great interest to the mineral collector is the formation of amygdales and geodes. Amygdales are mineral aggregates deposited in gas cavities in lavas. They are commonly almond-shaped or pipe-like in form and a few centimetres across, but can be very much bigger. The minerals are deposited from the walls inwards, and if they fill the gas cavity only partly the inner crystals may be beautifully developed. The zeolites are particularly characteristic of this environment, and the delicate aggregates of fibrous zeolite to be seen in many mineral collections were formed in this way. Quartz, chalcedony, calcite and other minerals are also found in amygdales.

Geodes are a particular type of amygdale and a great attraction for the mineral collector. They are generally filled with silica minerals and may vary from one or two centimetres to tens of centimetres in diameter; they may be completely or only partly filled. They are generally lined with agate which is finely banded parallel to the margin. The agate may fill the whole geode, or the central portion may be filled with quartz (sometimes the purple variety, amethyst) and there may or may not be a central cavity. Occasionally an outer layer of agate terminates against a central area of onyx in which the mineral bands are parallel and horizontal. Occasionally some fluid, from which the agate and quartz were probably deposited, remains trapped inside, and can be heard slapping about as the geode is shaken; such geodes are known as *enhydros*.

Above: Spectacular terraces of tufa, a form of calcite, deposited from hot spring waters at Pamukkale, Turkey

Above right: Many minerals are deposited from volcanic gases. Here yellow sulphur coats rubble near a fumarole

Right: Geode. The finely banded outer layers are agate, while the inward growing crystals are amethyst

Hot-spring deposits

One further manifestation of the activity of magmatic hydrothermal solutions appears in hot springs. The water in such springs is invariably a very dilute solution because many of the elements originally dissolved in it have been deposited elsewhere, and because it has been diluted by groundwater. However, minerals are deposited by hot springs. Most commonly these are siliceous sinter or geyserite (opaline silica), calcite (usually as a banded, often cavernous rock called *tufa*), iron oxides in the form of ochre, and manganese wad. Hot-spring waters also often

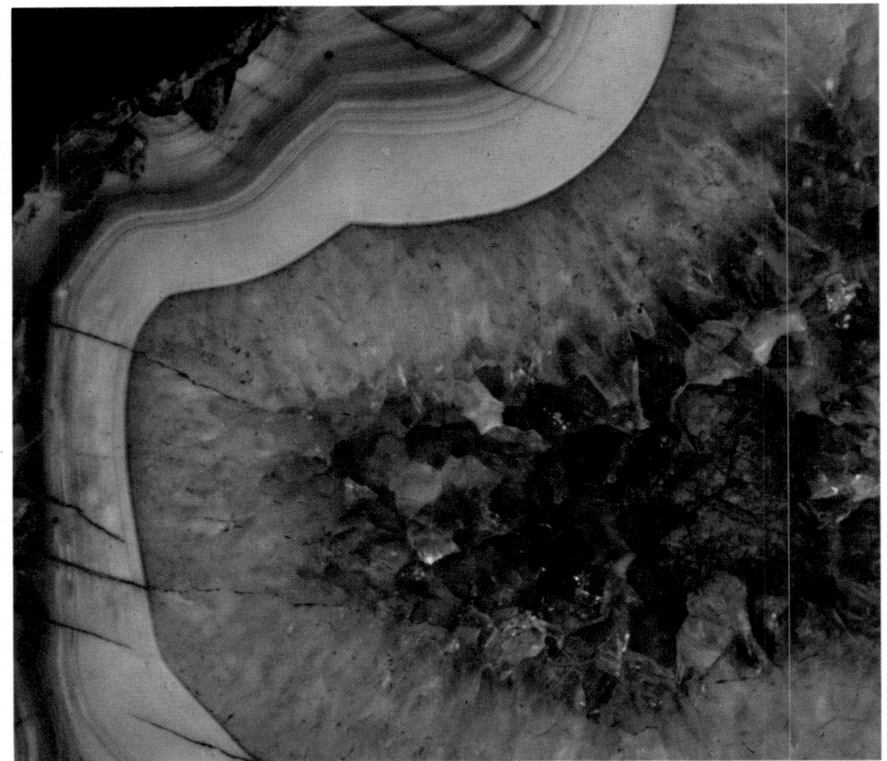

carry traces of many metallic and non-metallic elements in solution, and these are deposited, usually in the form of sulphides. One hot-spring deposit in California has been worked commercially for mercury.

METAMORPHIC ROCKS

The term 'metamorphism' is used to denote those processes working beneath the surface of the Earth by which sedimentary and igneous rocks are mineralogically, texturally, and structurally changed to produce the metamorphic rocks. The changes are brought about by the action of heat, pressure and, sometimes, by the addition and subtraction of material. Although many of the minerals occurring in metamorphic rocks are to be found in igneous rocks, and sometimes in sedimentary rocks also, many minerals are only produced in the metamorphic environment.

Except for some metamorphic rocks that have been heated to such elevated temperatures that they have been partially or wholly melted, and so pass into an igneous state, metamorphic minerals must grow in a solid environment. This means that the boundaries of growing crystals must find some mutual accommodation with the boundaries of the surrounding crystals. Because of this, minerals such as quartz, feldspar, calcite and dolomite rarely develop good crystal forms in metamorphic rocks. On the other hand certain minerals tend to establish their own shapes at the expense of their neighbours to produce perfectly formed individual crystals. Such minerals tend to be relatively sparsely distributed through the rock, but grow to above-average dimensions, often of a centimetre or more in diameter; such large crystals are known as *porphyroblasts* and are the metamorphic equivalent of the phenocrysts of igneous rocks. Typical porphyroblast minerals are garnet, staurolite, andalusite, kyanite and epidote. Before discussing in more detail the mineralogy of metamorphic rocks some space must be devoted to outlining something of the field relationships, appearance and nomenclature of these rocks.

Field Relationships

Two types of metamorphism are generally distinguished, *regional* and *contact*.

From time to time linear belts of the Earth's surface become geologically very active, with the deposition of great thicknesses of sediments, much igneous activity, and eventually the elevation of the accumulated rocks into a chain of

Far left: Garnet-mica schist, a metamorphic rock. The micas give the rock a finely layered appearance, while the garnets form large embedded crystals

Left: Gneiss, another metamorphic rock. The foliation or banding of light and dark minerals is here very noticeable

Below left: Phyllite, a metamorphic rock with a sheen due to myriads of parallel crystals of chlorite and/or muscovite mica

mountains. Subsequent removal of much of the elevated material by the forces of erosion reveals that the underlying rocks have been metamorphosed during the mountain-building process. Such rocks are referred to as regional-metamorphic rocks, and they may cover thousands of square kilometres. Other large areas of metamorphic rocks may have been formed by other processes.

The contact-metamorphic rocks are much more restricted in area, being confined to the immediate vicinity of igneous intrusions. When magma, which is usually in the temperature range 700–1200°C, is injected into the crust to form intrusions, it heats the surrounding rocks, and in so doing metamorphoses them. The resulting *metamorphic aureole*, as it is known, may be several kilometres wide around a large intrusion, or only a few centimetres thick around a small one, such as a narrow dyke. In larger aureoles in particular the degree of metamorphism usually becomes greater closer to the intrusion, and sometimes the country rocks immediately adjacent are melted. In some contact aureoles there is evidence that the alteration was not solely a matter of heating, but that there was addition to, and possibly subtraction from, the rocks close to the intrusion. This phenomenon is known as *contact metasomatism* and sometimes leads to the production of rather exotic rocks known as *skarns*.

The degree to which a rock has been metamorphosed is called the *grade* of metamorphism, and may be low, medium or high. As metamorphism progresses new textures and structures develop in the rocks, but many of the primary features of the original sedimentary or igneous rock may be retained. Thus it is sometimes possible to recognize fossils and various types of bedding structures in metamorphosed sedimentary rocks. Even after high-grade metamorphism it is usually quite easy to distinguish the original stratigraphy, so that original shale, sandstone and limestone horizons can be recognized. Similarly, features such as pillow lavas, amygdales and dykes are often still recognizable.

The new textures that develop with metamorphism are distinctive and to a great degree reflect the pressure or stress to which the rock was subjected. Regionally metamorphosed rocks were usually not only deeply buried but subjected to

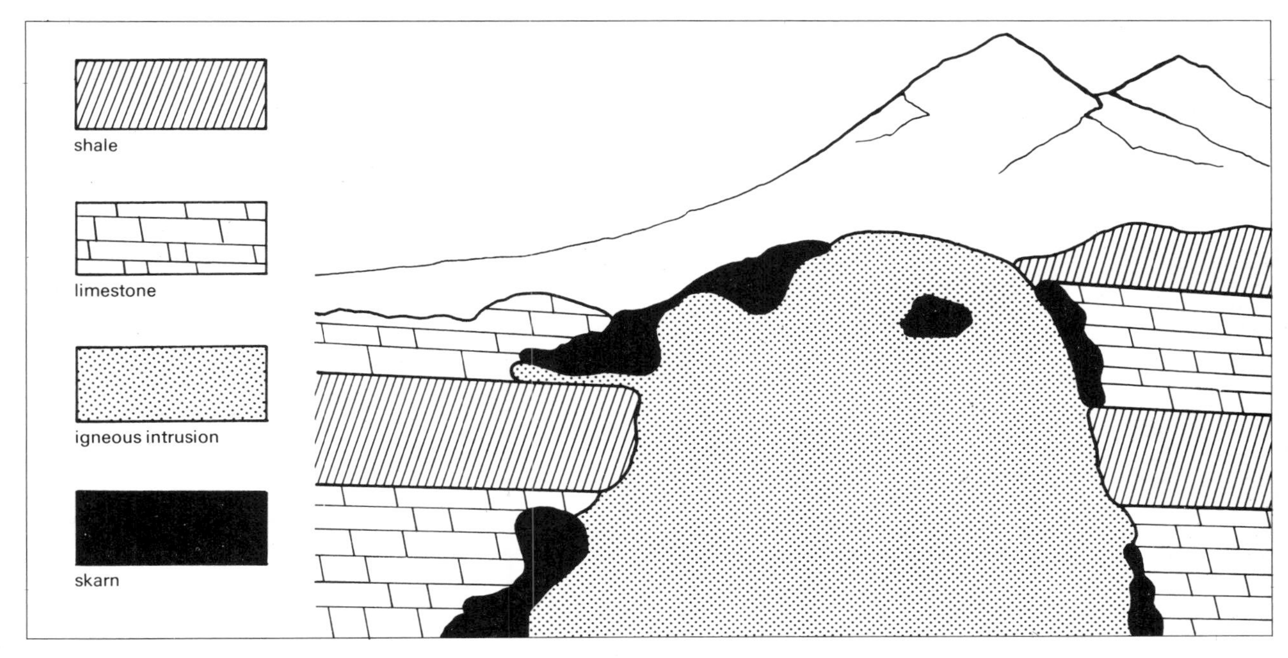

Right: Some features of igneous intrusions. Certain neighbouring rocks, especially limestones, are metasomatized (changed by hot fluids from the intrusion), forming rocks called 'skarns'. Fragments of country rock left surrounded by the intrusion are called 'xenoliths' and are usually metasomatized

Below right: Highly deformed schist layers at Hammerfest in Norway

stresses associated with the squeezing and folding of the rocks during mountain-building. The resulting rocks are generally foliated, the commonest being *schists* and *gneisses*. The foliation is the result of the crystallization of the constituent minerals under pressure. Further evidence for the reality of these pressures is afforded by the folds that are common in metamorphic rocks. These may be tiny crenulations a centimetre or less across or larger structures on the scale of tens, hundreds or thousands of metres.

Contact-metamorphosed rocks are generated by high temperatures but low pressures. This produces a texture known as *hornfelsic*, in which there is no preferred orientation of the minerals and the rock has an even, granular texture.

Naming metamorphic rocks

The nomenclature of metamorphic rocks is relatively simple, and based on structural and mineralogical criteria. The principal name is determined by the structure, and occasionally the mineralogy, while qualifying prefixes indicate the principal minerals present. The main terms based on structure follow:

Slate is a fine-grained, homogeneous rock with a single prominent cleavage, which is not related to bedding.

Phyllite is a fine-grained rock, though somewhat coarser than slate, with a lustrous greenish or silvery sheen on the well developed cleavage surfaces.

Schist is a coarser-grained rock with a marked lamination or foliation (commonly called *schistosity*) defined by platy or elongate minerals (usually micas).

Gneiss is a coarse-grained rock with a marked structure of darker and lighter layers; it is rich in quartz and feldspar.

Hornfels is a fine-grained, even-textured rock with no cleavage or schistosity, though porphyroblasts may be present; produced by contact metamorphism.

The following principal names are based on mineralogy:

Quartzite is a rock composed mainly of recrystallized quartz.

Amphibolite consists essentially of hornblende with a variable proportion of plagioclase feldspar.

Marble is a coarse- to medium-grained rock of recrystallized calcite or dolomite.

Skarn is a rock rich in calcium, magnesium and iron silicate minerals, formed by contact metasomatism of limestone and dolomite by plutons of granitic to intermediate composition.

The principal names may be used alone or made more precise by mineralogical prefixes; for example: garnet-biotite schist, sillimanite gneiss, tremolite marble, hedenbergite-sphalerite skarn, and so on.

Contact-metamorphic rocks

The nature of the minerals that crystallize in contact aureoles is determined by two principal factors: the composition of the metamorphosed rock and the degree or grade of the metamorphism. The assemblage of minerals in the outer part of an aureole, where the grade is low, may be quite different from the assemblage close to the intrusion where higher temperatures prevailed. This means that the mineral collector should be aware of the relationship between metamorphic grade and mineralogy. However, since the nature of metamorphic minerals also varies according to the type of rock that has been metamorphosed, many different assemblages are encountered, and it is beyond the scope of this account to list them in detail. Further, many contact-metamorphic rocks are rather fine-grained, so that it is necessary to have recourse to a microscope to determine the minerals present, and few mineral-collectors are interested in material they cannot study with the naked eye or with a hand lens. Fortunately, many minerals do form porphyroblasts, which often develop good crystals. In Table 4 some of the commoner contact-metamorphic minerals are listed according to the nature of the parental rock. Some of these minerals, such as andalusite, commonly form good porphyroblastic crystals, but with others crystal forms are more elusive and identification proportionately more difficult. The term 'impure limestone' indicates that there is some admixture of shale or sandstone or both amongst the calcium and magnesium carbonate (calcite and dolomite), so that chemical components are present for the production of a wide range of minerals.

Contact-metasomatic rocks

In the contact-metamorphic rocks just described the minerals present were determined by the chemistry of the metamorphosed rock, the only contribution made by the igneous intrusion being heat and perhaps some water. In some aureoles, however, there is evidence that the igneous intrusion played a more active role in the form of volatile constituents that streamed outwards into the country rocks, facilitating the metamorphic process and carrying with them various chemical elements, which were combined in the new minerals being formed. Apart from water the principal volatiles appear to be boron, fluorine and chlorine. Not all magmas appear to be capable of promoting metasomatism, and these effects are mainly associated with nepheline syenites and intrusions ranging in composition from granite to diorite. The nature of the country rocks is also a potent factor; the metasomatic rocks known as skarns are confined mainly to limestones and dolomites.

When boron has been introduced by metasomatism into rocks such as shales and slates, distinctive rocks commonly result. They are rich in tourmaline, often concentrated in clusters of well developed prismatic crystals or in veins with quartz. On the other hand, impure limestone and, sometimes, basic igneous rocks react with boron-rich volatiles to produce minerals such as datolite and axinite, which is often associated with andradite and hedenbergite. The presence of fluorine in the metasomatizing fluids gives

Metamorphic minerals and their parent rocks

metamorphosed rock	minerals formed by contact metamorphism
shale	biotite, cordierite, andulusite (chiastolite), garnet, hornblende, sillimanite, feldspar, brookite, anatase, hypersthene
impure limestone	wollastonite, grossular, idocrase zoisite, clinozoisite, anorthite, sphene
impure magnesian limestone (dolomite)	periclase, brucite, forsterite, serpentine minerals, diopside, tremolite, spinel grossular, idocrase, wollastonite, melilite, monticellite, merwinite, larnite,
basic igneous rock (dolerite)	chlorite, biotite, garnet, hornblende, feldspar, augite

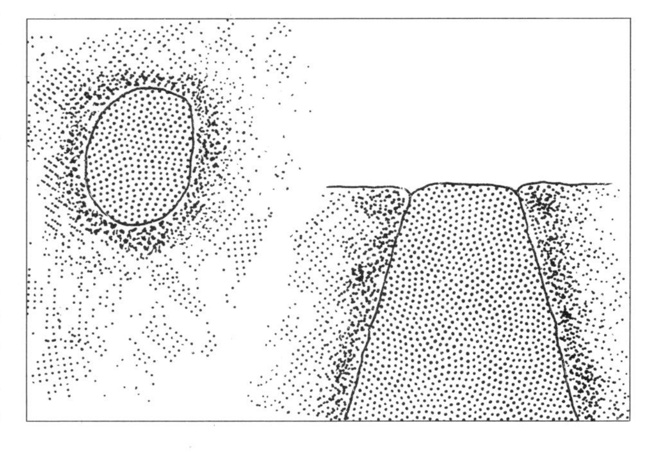

Left: The zone of altered rocks around an igneous intrusion. The changes are due to heat and chemically active fluids from the intrusion. The area of altered rocks is called the metamorphic aureole of the intrusion

Above: Pale green prisms of diopside lie among crusts of hessonite garnet in a rock type typically produced by the contact metamorphism of impure limestones. This specimen is from Val d'Ala, Piedmont, Italy

rise, in impure limestones and dolomites, to minerals such as lepidolite, muscovite, phlogopite, chondrodite, humite, clinohumite, fluorapatite and cuspidine, while chlorine causes the development of scapolite, particularly the marialite variety. The metasomatic rocks associated with nepheline syenites are characterized by minerals rich in alkalis; of these, aegirine and aegirine-augite, alkali-rich amphiboles and alkali feldspars are the most distinctive.

Of all contact-metasomatic rocks, those distinguished as skarns are probably the most interesting from the point of view of the range of minerals to be found in them. The production of these rocks has involved not only the migration of the volatile elements outlined above, but also the movement of sulphur, silicon, potassium, sodium, magnesium and a wide spectrum of metallic elements, including iron, zinc, copper, lead, tin, etc. The presence of these metals, usually in the form of oxides and sulphides, has led to many skarn deposits being exploited as ores, though they rarely form deposits of very large size and are not of such great economic importance as hydrothermal deposits.

Skarn deposits occur immediately adjacent to igneous intrusions and are concentrated in limestones and dolomites in contact with the intrusion, as shown on page 63, for it appears that limestones are susceptible to attack by, and combination with, the emanations from the intrusion. If the limestone or dolomite is somewhat impure, that is, if it contains a small proportion of shale and/or sandy material, then it is even more prone to alteration. Other sedimentary rocks, such as shales, may be affected to some degree but very rarely to the same extent.

The structure of the sedimentary rocks has a strong controlling influence over the form of the skarn body. Bedding planes, particularly if they dip inwards towards the intrusion, provide ready access for metasomatizing fluids. Similarly, faults may provide routes along which the fluids are channelled. Particularly susceptible to alteration are roof pendants, masses of rocks such as limestone extending downwards into the intrusion from the roof. Commonly they become detached and form large *xenoliths*, rafts that 'float' in the magma. Often the skarns are developed in particular horizons and not others, though the chemical or physical differences that determined this may not be readily apparent.

The most characteristic minerals of skarn deposits are probably iron-rich garnet (andradite) and pyroxene (hedenbergite), but many other silicate minerals are commonly present, including grossular garnet, wollastonite, diopside, hastingsite, etc.

'Metallic' minerals—that is, roughly, those with a metallic lustre—may or may not be present. When they are, they may include magnetite, hematite, chalcopyrite, pyrite, bornite, pyrrhotite, etc.

Regional-metamorphic rocks

To describe the regional-metamorphic rocks and their constituent minerals in the limited space here available is an almost impossible task, because these rocks represent, not only the metamorphosed equivalents of nearly all kinds of sedimentary and igneous rocks, but the resulting diverse mineralogical assemblages produced over a wide range of temperatures and pressures. Thus a rock that was originally a shale can be represented by half a dozen different assemblages of minerals, depending on the degree to which it has been metamorphosed. There are three principal factors that determine the mineralogy of regionally metamorphosed rocks: the initial composition of the rock (for instance, whether it was a shale, limestone, basalt, etc); the pressures and temperatures to which it has been subjected; and the type of metamorphism.

One of the most abundant rock types that may be metamorphosed is shale. With increasing metamorphism it forms firstly slates, then phyllites, and then schists. It is found that at successively higher grades chlorite, biotite, almandine garnet, staurolite, kyanite and sillimanite appear. These are known as *index* minerals. Minerals of lower grade persist into higher grades, but the first appearance of an index mineral marks the beginning of a grade. It must be appreciated that the index mineral is not the sole component of a rock, but only part of an assemblage of minerals; see the table below, which shows the principal minerals found in metamorphosed shales from a typical area of the Highlands of Scotland.

If now rocks other than shale are considered, it is found that certain differences of mineralogy occur. Pure quartz sandstone merely recrystallizes to form quartzite, and pure limestones recrystallize to form marble, but the minerals present in basic igneous rocks that have been metamorphosed may include chlorite, albite, epidote, sphene, calcite, hornblende, biotite and almandine garnet; a metamorphosed impure non-dolomitic limestone can include zoisite, grossular garnet, idocrase, wollastonite, diopside, anorthite, hornblende, and scapolite; metamorphosed dolomitic limestones may have forsterite, serpentine minerals, spinel, diopside, tremolite-actinolite, hornblende, phlogopite, chondrodite and others. Metamorphosed rocks which were originally peridotites commonly occur in regional metamorphic terrains, but they are now rich in serpentine minerals, talc, anthophyllite and tremolite, rather than olivine.

The above examples serve to illustrate some of the metamorphic minerals determined by initial rock type and metamorphic grade, but there remains the third factor referred to earlier, the pattern, or type, of metamorphism. There are a number of these types, which are apparently related to the pressures and temperatures prevailing at the time of metamorphism. One is characterized by mineral assemblages including andalusite and cordierite (already mentioned as being common in some contact-metamorphic rocks); a very different one includes the minerals jadeite, glaucophane, omphacite, pumpellyite, stilpnomelane and lawsonite. It is not possible to give a comprehensive list of these variants, but these two examples should give some idea of the complexities encountered in these rocks.

From the point of view of the mineral-collector the foregoing account is intended to illustrate something of the range of minerals to be found in these rocks, and something of the patterns of mineral distribution. This should afford some help in looking for particular species.

Right: Prismatic actinolite, a metamorphic mineral

Index minerals

index mineral	mineral assemblage
chlorite	muscovite — chlorite — quartz
biotite	biotite — muscovite — chlorite — quartz
almandine	biotite — muscovite — almandine — quartz
staurolite	biotite — muscovite — staurolite — almandine — quartz
kyanite	biotite — muscovite — almandine — kyanite — quartz
sillimanite	biotite — muscovite — almandine — sillimanite — quartz

Many of the minerals mentioned above occur in particular rocks in small quantities or in relatively small grains, so that they can be readily identified only with the microscope. But fortunately for the mineral-collector, many metamorphic minerals commonly form large, well shaped porphyroblastic crystals, and many metamorphic rocks, especially at higher metamorphic grades, are very coarse-grained, so that the major mineral constituents are readily identifiable. Kyanite or staurolite crystals several centimetres long and garnets tens of centimetres across are not unusual. Indeed, garnet schists with easily removable garnet crystals a centimetre or so in diameter are relatively common. The difficulty with these rocks from the point of view of the mineral collector is that they are not confined to small, well defined areas like the contact metamorphic rocks and skarns, but may occur over hundreds or thousands of square kilometres, so that it is necessary to find references to good mineral localities by searching the literature.

SEDIMENTARY ROCKS

Sedimentary rocks are formed by the accumulation on the surface of the Earth of sediment, comprising the weathering products of rocks and detritus of organic origin. Because sedimentary rocks are formed at the surface they are not the product of high temperatures or pressures, as are the igneous and metamorphic rocks. But they are subjected to processes known as *lithification* and *diagenesis* in which the soft, unconsolidated sediment becomes hardened into a rock that may be just as tough and intractable as any igneous or metamorphic rock.

Field relationships

The weathering processes by which even the toughest rocks are broken down involve complex chemical reactions with ground water. Some minerals are little changed by weathering; quartz is the most outstanding example of these. The feldspars are less resistant, while mafic minerals such as the pyroxenes, micas and amphiboles are very susceptible to decay. Breakdown involves the production of new minerals, many of which are rich in water, and of which the clay minerals are the most abundant. Some elements, particularly sodium, potassium, calcium and magnesium, tend to go into solution. These three—the unaltered minerals, the newly formed minerals and the elements in solution—provide the raw material for the formation of sedimentary rocks.

This material is transported, principally through the agency of wind, running water and ice, to the place of accumulation, which is usually the sea.

The sedimentary rocks are often characterized in the field by the presence of fossils, and by layering, which is called *bedding*. As sediment is washed into the sea or a lake it builds up layer upon layer. Variations in the speed of deposition and in the grain size, texture and mineralogy of the sediment accentuate the layering. Individual layers, or *strata*, can often be followed for considerable distances, but sometimes they are disrupted in various ways, giving rise to various types of bedding. Along with bedding and fossils there are numerous other features that can be observed in sedimentary rocks, which are all products of the mechanical processes operating when the rocks were laid down. Structures such as ripple marks, mud cracks and rain prints are relatively common. These features are an invaluable guide to the geologist trying to reconstruct such features as the geography and climate that prevailed at the time the rocks were being formed.

Classification

Numerous schemes of classification have been proposed for the sedimentary rocks, but the relatively simple one outlined below is quite adequate. The rocks are first divided into three groups according to their mode of origin. The first comprises those rocks that were transported as solid particles, and so includes minerals, such as quartz, that were not altered by the weathering process, together with the secondary minerals, such as the clays, which were produced during weathering. This group of rocks, which can be

Classification of sedimentary rocks

mechanical origin		chemical origin		organic origin	
coarse	conglomerate, breccia	calcareous	calcareous mudstone, limestones, dolomite, travertine, tufa	calcareous	organic limestone
medium	sandstone, quartzite, grit			carbonaceous	coal
fine	siltstone, mudstone, shale	siliceous	flint, chert	phosphatic	phosphate rock (some types)
		saline	rock salt, gypsum rock and other evaporites		
		ferruginous	ironstone		
		phosphatic	phosphate rock		

Left: Pebble banks exposed in a river at low water. Rivers are an important mechanism by which sediment is transported and finally deposited. The pebbles seen here might eventually become consolidated to form the sedimentary rock known as conglomerate

Below left: Sedimentary rock layers or 'beds'. Geological upheavals have tilted the beds from their original horizontal position. The different degrees of weathering, or erosion, of the shale (dark) and limestone (light) are clearly visible

said to have a mechanical origin, is further subdivided according to the size of the constituent grains. The second group includes the rocks formed from the elements that are taken into solution during the weathering process. These are carried by rivers into the sea and into lakes, where they are eventually precipitated. This group is subdivided according to the chemical composition of the precipitated rock, and can be said to be of chemical origin. The third group consists of those rocks that have an organic origin, comprising as they do the accumulated remains of animals and plants. This group is also subdivided according to the chemical nature of the rock.

Rocks of mechanical origin

The table shows that these rocks include the coarse conglomerates and breccias, which contain pebbles and boulders that are rounded and angular respectively; the medium-grained sandstones; and the fine-grained shales and mudstones. The finer-grained rocks are composed predominantly of clay minerals. This is a large group of hydrated aluminium silicates, which are invariably extremely fine-grained and require special techniques to identify and investigate them. For this reason, and because of their unprepossessing appearance, the clays hardly figure in most mineral collections, and will not be considered further here.

Apart from the clays and other secondary minerals the sedimentary rocks of mechanical origin comprise minerals derived from primary igneous, metamorphic and other sedimentary rocks. Because these minerals were transported they are invariably broken, rounded and generally abraded, so that they rarely show good crystal shapes. However, certain minerals that are resistant to the weathering process, and which may be relatively rare in the primary rocks, can become concentrated. The *alluvial deposits* or *placers* are formed in this way. Minerals that can be concentrated in such deposits include topaz, garnet, zircon, tantalite, columbite, rutile, andalusite, monazite, magnetite, ilmenite, cassiterite, gold, the platinum minerals and diamond. Many such deposits are exploited commercially, but usually the individual mineral grains are rounded and unattractive. However, certain very hard minerals such as diamond and corundum may retain their crystal shapes during transport and be found in placers as slightly rounded but well formed crystals. For instance, certain placer deposits in Sri Lanka are the world's principal source of most of the near-gem quality corundum, including sapphire and ruby. Gold, although very soft, tends to retain its particle size during transport, presumably because of its lack of cleavage, and is a relatively common mineral in placer deposits, though in small quantities.

Rocks of chemical origin

For the collector of minerals these are undoubtedly the most interesting and attractive of the sedimentary rocks, as they include a wide range of minerals that do not occur in other kinds of rock. The chemically derived sediments can be divided conveniently into two groups: the first comprises those rocks formed directly by primary precipitation from water, mainly sea and lake water; the second comprises the volumetrically insignificant but mineralogically interesting rocks formed by chemical reactions within the sediment, promoted principally by the agency of percolating solutions.

The rocks formed by primary precipitation are a very varied and extensive group, which includes calcium and magnesium limestones, iron-rich sediments, pure silica rocks, phosphorus-rich deposits, and a very varied group known as the evaporites, which consist principally of chlorides, sulphates, carbonates and rare nitrate minerals.

A good proportion of the world's limestone rocks have been formed either by direct precipitation from sea water, or by precipitation induced by the intervention of living organisms. The principal minerals involved are calcite and aragonite, but limestones composed of dolomite are also common, though more often than not the dolomite appears to have been formed by replacement of primary calcite. Most limestones are rather fine-grained, compact rocks, but occasionally well crystallized masses can be found in veins or cavities where it was probably deposited after consolidation of the rock as a whole. Locally limestones are deposited from spring waters as the spongy-textured rock known as tufa, while in limestone caverns a dense, usually banded limestone called travertine may form, often associated with stalactites and stalagmites.

The sediments produced by the precipitation of iron from solution are known as ironstones or iron formations, and constitute by far the most important source of iron ore. As with the limestones these rocks rarely provide spectacular mineral specimens, but a very wide range of iron-bearing minerals is found in them. Not all of these are primary precipitates, many being

Above: A magnificent stalactitic grotto at Castellana in Italy. Ground water trickling through the roof deposits limestone, which grows into the dangling stalactites. Water dripping from each of these builds up a corresponding stalagmite on the cave floor

produced by later processes. Hematite and magnetite are the commonest minerals but goethite and limonite, siderite, chamosite, greenalite, glauconite, pyrite, marcasite, pyrrhotine, thuringite, and stilpnomelane are also relatively abundant.

Deposition of silica gives rise to the rock known as chert, which is composed of chalcedony, opal, and very finely crystalline quartz. Flint has the same composition as chert, but the term is generally applied to nodular masses rather than to continuously bedded deposits. Deposition of phosphorus produces the rocks known as phosphorite or phosphate rock. These are mineralogically rather complicated, dozens of different phosphate minerals having been identified in them, but the principal ones are various types of apatite, including francolite and dahllite, which are usually referred to by the name collophane.

The rocks grouped under the name evaporites are formed by precipitation from concentrated brines, either in seas isolated from the open oceans, or in continental saline lakes. These rocks are of considerable economic importance, providing much raw material for industry and, incidentally, most of our table salt. The major evaporite deposits are marine in origin and the minerals comprising them are predominantly chlorides, sulphates and carbonates of calcium, sodium, potassium and magnesium. The most common mineral species are gypsum, anhydrite, halite (rock salt), sylvine, carnallite, langbeinite, kainite, epsomite and kieserite. Rarer minerals include bischofite, glauberite, picromerite and rinneite. Not all these minerals are primary precipitates.

The non-marine evaporites are more restricted in occurrence, and form predominantly in salt lakes. Although of limited extent they supply all the naturally occurring nitrate minerals, most of the world's boron compounds and iodine, and various other salts. The only significant deposits of nitrate minerals are in northern Chile, where the principal mineral is nitratine, also known as Chile saltpetre or soda nitre. Some of the minerals of marine evaporite deposits detailed above also occur in non-marine evaporites. Apart from these, the following may be found: borax, blödite, hexahydrite, colemanite, ulexite, thenardite, mirabilite, probertite, hydroboracite, danburite and others.

Chemical reactions within the sediments after deposition, and some movement of material in solution, commonly concentrates certain minerals into masses of varying shape known as *nodules*. Among the more common ones are pyrite nodules, often to be found in black shales and sometimes in limestones, and masses of marcasite, often showing fine twinned crystals as typical spear-shaped forms. Other common types of nodule are made of flint and phosphate minerals.

Concretions are spherical to ellipsoidal masses to be found mostly in sandstones and shales, formed by the cementing of the rock by local concentrations of silica, iron oxides, carbonates, etc. Occasionally well formed crystals are to be found inside them; one of the world's major sources of strontium, the mineral celestine, is found in such concretions. Some minerals, when they crystallize in loose sediment, produce well formed crystals known as sand crystals. In low-rainfall areas baryte and gypsum crystals often grow in this way to form concretions called desert roses.

Above: A fine group of interpenetrating crystals of halite, which is simply common salt. This mineral is one of the 'evaporites' which are deposited when salt water dries up

Right: A desert rose, formed by the crystallization of baryte in sand. The baryte has incorporated sand grains into its structure while developing its own characteristic crystalline form

Rocks of organic origin

Many limestones are built of the accumulated calcareous skeletons of various organisms. Although of great interest to palaeontologists they have little attraction for the mineral collector, unless cut by mineral veins. However, in certain fossiliferous rocks, particularly some shales, the fossils are replaced by minerals such as pyrite, which makes them extremely attractive mineral specimens. Certain recent bone deposits are a source of the mineral vivianite, and fossilized wood may be replaced by opal, or occur in the hard black form known as jet. Overall, however, the rocks of organic origin are not a bountiful source of material for the collector of minerals.

Minerals Valuable to Man

Left: A huge open-pit copper working near Salt Lake City, Utah. As man seeks to exploit ores of ever lower quality, large areas of great natural beauty may be destroyed by surface workings like this

Since his earliest days, man has used the minerals and rocks of the earth to his advantage, whether as tools and weapons, building materials, or personal ornaments. One of the first forms of trading was concerned with the mutual exchange of useful rocks or minerals, and certainly the selection of suitable rocks as material for tools set man on the path to civilization.

It is known that early man established a trade in obsidian centred on the Mediterranean, and that flint, a similarly tough and easily worked material, was a prized mineral resource of neolithic man. Extensive flint mines have been found in such areas as the chalk downs of southern and eastern England. Modern man still uses rocks and minerals for the same basic reasons, but the winning of minerals is now of foremost importance in maintaining technological civilization; indeed our industrial society depends heavily on mining and the practical use of the materials obtained. Enormous investment by companies or countries is necessary to locate, establish and work mineral deposits. These may be of metalliferous (metal-bearing) minerals, valuable for their metal content; or they may be of non-metalliferous minerals such as asbestos, worked for their own valuable properties.

ECONOMIC MINERALS

Economic mineral deposits can be defined as those that are of use to man and that can be worked at a profit. Their importance is therefore dependent on the whims of world metal or commodity markets, so it may be economic to mine a deposit one year but not the next.

Abundance and profitability

By chemically analyzing samples of rocks from all parts of the world, geochemists have shown that many of the elements upon which technology relies are present in only extremely small amounts in the earth's crust as a whole, often as low as one part in a million. In order to extract such elements economically, areas must be found where they have already been concentrated by geological processes to many times the average figure. For instance, in the Witwatersrand goldfield of South Africa, gold exists in a concentration some 1,750 times the crustal average. Similarly the rock mined in the rich copper, lead and zinc deposits of South-West Africa has concentration factors of a thousand times for copper and zinc, and up to ten thousand for lead. Even a low-grade copper ore containing 0.8 per cent of the metal by weight has about two

Abundance of some of the elements

% (by weight) of the earth's crust

Element	%	Element	%
oxygen	46.60	sulphur	0.05
silicon	27.72	chromium	0.02
aluminium	8.13	nickel	0.008
iron	5.00	zinc	0.0065
calcium	3.63	copper	0.0045
sodium	2.83	lead	0.0015
potassium	2.59	tin	0.0003
magnesium	2.09	silver	0.00001
titanium	0.44	platinum	0.0000005
manganese	0.10	gold	0.0000005

hundred times the average amount found in rocks. Of the ninety-two naturally occurring elements, eight constitute nearly 99 per cent by weight of the earth's crust. The percentage abundances of some of the elements are shown in the accompanying table.

Some elements are difficult to extract because, although present in the crust in considerable amounts, they are systematically dispersed through common minerals and never occur in any great concentration. Examples of these dispersed elements are rubidium, which is dispersed in potassium minerals, and gallium, which is dispersed in aluminium minerals.

A distinction must be drawn, however, between the abundance of an element and its availability, which also depends on the element's chemical properties. The element's chemistry dictates whether it readily occurs in compounds, such as oxides, silicates or sulphides, or occurs native (that is, uncombined with other elements), like gold, or as a gas, such as argon. Elements of quite high abundance may not occur in easily worked ore minerals. Perhaps the best example is the important metal aluminium. It is the third most abundant element in the earth's crust and a major constituent of many common minerals, including feldspar, which is found in virtually all igneous rocks and in sedimentary clays and shales. Nevertheless it has only one *ore mineral*—that is, a mineral from which it can be profitably extracted. This is bauxite, which is a mixture of fine-grained hydrous aluminium oxides, formed by the weathering of rocks with a high alumina content in a tropical or subtropical climate. To extract the aluminium from the ore it is necessary to use an electrolytic process, which requires a large amount of electricity. Cheap power is therefore essential to the process of extraction. Similarly, magnesium is extracted from sea

Above: A cleft remaining in country rock after the extraction of a pegmatitic vein. The coarse texture of the residual pegmatitic pillar is clearly visible

water, in which it is present in low abundance, even though there are vast deposits of the mineral olivine, which contains up to 30 per cent magnesium, but which is difficult to process. In general, it is easier to separate elements from oxide and sulphide ores than from silicates.

So, to be of economic value, elements important to man must not only occur in local concentrations, but must also be capable of being readily extracted from the ore. Examples of the concentration process will be considered later.

Metal-bearing ores

Economic metalliferous mineral deposits—ore deposits—consist of a mixture of the desired mineral (the ore) with unwanted minerals (the *gangue* minerals). The metal content of the ore is termed the *tenor*. Owing to the changing demands of industry for raw materials, a mineral which was once rejected as a gangue mineral may now be of sufficient economic value for old mine waste tips to be reworked for it. A good example is the mineral fluorite, which is associated with the lead and zinc ores in the Pennine ore field of northeast England. Mining for galena (lead sulphide) and blende (zinc sulphide) ceased when most of the deposits were worked

Above: Stockpiling crushed iron ore in the Hamersley Range of northwest Australia. It is estimated that this ore deposit contains 600 million tons of high grade hematite

out or became uneconomic. Extensive veins of fluorite and old tips containing concentrations of fluorite are now being reworked for this mineral, which has become important in the chemical and steel industries. Lead and zinc minerals are recovered as secondary products.

Crystallization of magmas

One of the natural processes that is responsible for raising the concentrations of elements above the average is the crystallizing of magma (molten rock) as it cools and solidifies. Detailed discussion of how a magma crystallizes is given in another chapter, but a few examples of the formation of economically important minerals will be given here.

Concentration of minerals may take place in several ways and at different times during the crystallization of a magma. Heavy minerals crystallizing early may accumulate by sinking under gravity to the bottom of the magma chamber to form *magmatic segregations*. This usually takes place in the temperature range 1,500°C to 700°C. One of the best known economic deposits of this type is the rich chromite and magnetite deposit of the Bushveld igneous complex of South Africa.

Pegmatite ores

As a magma crystallizes, volatiles—materials of low boiling-point—are not taken up by the crystals that form early. They become more concentrated in the remaining fluid material, together with elements originally dispersed throughout the magma that, by reason of their atomic structure, do not enter readily into mineral structures. These liquids may be forced into cracks and fissures in the parent igneous rock or in the surrounding 'country' rock, and crystallize as high-temperature *pegmatite* bodies. Pegmatites characteristically consist of large crystals. In granite pegmatites, quartz, feldspar and mica, which make up the bulk of ordinary granites, form crystals large enough to be of economic interest. Mica crystals from Ontario have been recorded up to 14 feet in diameter and weighing up to 90 tons. In the famous pegmatites of the Black Hills of South Dakota spodumene crystals 40 to 50 feet long occur, and a single beryl crystal weighing 100 tons has been reported. The mineral collector finds that pegmatites are the source of many rare minerals, often well crystallized and including gem species such as topaz, beryl and spodumene.

Lithium is an important element in many pegmatites and is becoming increasingly important in modern technology. In the Bikita pegmatite of Rhodesia, the concentration of lithium is extremely high; hundreds of tons of spodumene and petalite (lithium aluminium silicates) and lepidolite (lithium mica) are present. Pollucite, a rare mineral containing caesium, is also found as an associated mineral.

Beryllium is a light metal that has the structural properties of aluminium, and has many uses in the 'space-age' technology of space vehicles, missiles, and aircraft. It is obtained from the mineral beryl, which is often found in granite pegmatites. Although the present world production of beryllium for industrial purposes is small, new applications for the metal are constantly being found.

Hydrothermal deposits

The processes involved in hydrothermal ore deposition were described in the last chapter. Hydrothermal deposits are laid down from the high-temperature watery solutions left over following the consolidation of magma. These watery fluids force their way into the adjacent country rocks and there form the hydrothermal deposits.

The area of southwest England comprising

the counties of Devon and Cornwall has a long history of mining. It is a classic area both in the development of mining techniques and in the understanding of hydrothermal ore deposition. The formation and distribution of the minerals of the region are believed to be related to the intrusion of granites there 270 to 280 million years ago. If this is so, the fluids responsible for forming the minerals were a late-stage product of the crystallizing granite, formed as pockets of volatile-rich material, which travelled upwards under pressure through fissures in the country rock. The metallic, ore-forming elements existing there before the granite formed, ments existing there before the granite intruded, or they may have come from the fluids associated with the granite.

In southwest England both the ore and the gangue minerals are distributed in a series of concentric belts or zones. These mineral zones can be related, both laterally and in depth, to Lindgren's three subdivisions of hydrothermal deposits. The ore minerals crystallizing at the highest temperatures—that is, in the 'hypothermal' zone—are cassiterite, wolframite, scheelite and arsenopyrite. They are associated with gangue minerals such as tourmaline, fluorite, chlorite and hematite. At slightly lower temperatures a range of copper minerals will crystallize in the hypothermal zone. At still lower temperatures, in the 'mesothermal' zone, niccolite, smaltite, argentite, galena and sphalerite are formed, associated with the gangue minerals fluorite, baryte, dolomite and calcite. In the lowest-temperature 'epithermal' zone, iron minerals such as hematite and siderite are common, along with stibnite, bournonite and tetrahedrite.

In much of this mining field the upper layers consist of *gossan* or 'iron hat', a mixture of gangue minerals with iron and manganese oxides, from which soluble minerals have been dissolved. These have been washed down and redeposited in lower enriched layers containing malachite, cuprite and azurite. This in turn overlies a layer

Composition of sea water

% (by weight) of sea water

%	Constituent	%	Constituent
96.24	water	0.14	calcium sulphate
2.94	sodium chloride	0.06	sodium bromide
0.32	magnesium chloride	0.05	potassium chloride
0.15	magnesium sulphate	0.01	calcium carbonate

Below: The derelict engine house of an old tin mine, Wheal Coates, Cornwall. The mine was active in the nineteenth century and worked some lodes that extended beneath the sea

of 'secondary sulphide enrichment' containing such minerals as bornite and chalcocite.

Sedimentary concentration of minerals

The mineral deposits considered in the previous sections were all formed within the earth's crust, often at considerable depth, but now outcrop at or near the surface as a result of geological uplift and erosion. Another group of mineral deposits are formed by surface processes and contain minerals concentrated by the sedimentary cycle —weathering, transport and sedimentation. Within this group are evaporite deposits, sedimentary rocks and residual deposits.

Residual deposits. Residual mineral ores result from the mechanical and/or chemical breakdown of a rock, which allows separation and concentration of the economically important minerals during subsequent sedimentary processes. In this way residual ore deposits of tin and gold have been formed. Gemstones are often worked from residual weathered material overlying unweathered parent rocks. An important group of iron ores are of residual origin: primary iron ores such as pyrite and siderite may be converted by weathering into more easily processed hydrated iron oxides in residual deposits like those of Bilbao in Spain, or some of the Cuban iron-ore deposits.

Impure iron oxides, such as limonite, goethite or hematite, may be formed by weathering from primary iron minerals, and deposited within sedimentary rocks, replacing the material originally present, which is carried away in solution. The Lake Superior iron ores of North America were originally sedimentary deposits of Precambrian seas, which were later leached of the dominant silica so that almost pure iron oxide (hematite) was left. These enormous deposits are worked by open-pit methods.

Sedimentary rocks. Many sedimentary rocks are of great industrial importance. Clay is used in the manufacture of bricks and tiles; sandstones, especially those that are crumbly and of even grain size, are used in concrete and mortar; pure quartz sands are used in glass-making; and foundry sands containing a small percentage of clay are used to make moulds for casting. Gravels are now dug in vast quantities as aggregates for concrete. As a result, good-quality inland deposits are becoming scarce, and the gravels of the continental shelf such as those off Britain's coast are now being worked extensively. Many types of sedimentary rocks are employed as building stone or ornamental facing stone and, of these, sandstones and limestones and their metamorphic equivalents (quartzites and marbles) are the most popular.

Evaporites. Perhaps the most valuable of the sedimentary rocks are the evaporites of saline landlocked seas and lakes. As water evaporates from an enclosed body of sea water, the dissolved salts become more and more concentrated and eventually begin to crystallize out of solution. Experiments in the laboratory on sea water indicate that the least soluble salts are deposited first, the more soluble ones later. Calcium carbonate crystallizes out first, followed by gypsum and anhydrite (both calcium sulphate), then halite (sodium chloride) and finally salts of potassium and magnesium. However, natural deposits very rarely contain a complete sequence of this type; instead, extensive deposits often consist of one of these minerals unaccompanied by others, and there are deposits in which the expected order is reversed.

Evaporite deposits were formed on several occasions in the geological past, but the most important halite, gypsum and anhydrite deposits of the world are Silurian (395 to 435 million years old) or Permian (225 to 280 million years

Below: Mechanical excavator quarrying ragstone, a sedimentary rock used in the building industry

old). Man is able today to obtain some of the salts he requires by direct evaporation from sea water—30 per cent of the world's halite is produced from salt-pans—but the reserves of past evaporite deposits still remain the main source.

Common salt is of critical importance to man in that it is an essential part of his diet. Since earliest times, therefore, trading in salt has been important. Indeed, our word 'salary' comes from the Roman term 'salarium' or 'salt money' which was paid to every Roman soldier. Salt deposits are often overlain by barren rocks. If, as is often the case, the salt deposit is too deep for surface working, a common method of extraction is to pump water through the salt-bearing layers, and then pump the resulting brine to the surface, where it is evaporated to produce halite. Salt is also mined by conventional methods.

Gypsum beds are as widespread as halite and the mineral is mined in vast quantities in many countries. It is mainly used in the building industry as plaster or wall board. Anhydrite is formed under similar conditions to gypsum but has fewer uses. Anhydrite from the Permian evaporite deposits of northeast England is used in the manufacture of sulphuric acid. Sylvine is a major source of the element potassium, an important constituent of fertilizers.

Other important evaporite deposits are found in areas where rivers and streams flow into enclosed inland lakes. Streams that flow across recent volcanic rocks may carry in solution some rarer elements derived from the volcanic material, as well as the more usual soluble products of rock weathering. Boron, which is often found associated with volcanoes or hot springs, tends to be abundant in such streams. These streams drain into lakes and after long periods of continuous evaporation large deposits of evaporites rich in borate minerals are built up. Famous deposits of this type have been discovered in California and Nevada, associated with the enclosed depressions to the east of the Sierra Nevada. Boron could again be described as a 'space-age' element since it is required for some high-energy fuels used in missile and aircraft propulsion. The synthetic compound boron carbide is harder than any natural substance except diamond.

The fossil fuels, oil and coal, are formed from organic debris by sedimentary processes. The deposits of these two fossil fuels are now of immense importance to man since present technology depends on them.

Above: The salt works at Salin-de-Giraud on the south coast of France. In the background the shallow artificial lagoons where sea water is evaporated by wind and sun are visible

Important copper minerals

% Cu (by weight)		chemical composition
native copper	100.0	Cu
cuprite	88.8	Cu_2O
chalcocite	79.85	Cu_2S
bornite	63.3	Cu_5FeS_4
malachite	57.3	$Cu_2CO_3(OH)_2$
enargite	48.3	Cu_3AsS_4
chalcopyrite	34.5	$CuFeS_2$

PROSPECTING

A re-distribution of elements takes place at the surface of the earth by the processes of weathering, transport and sedimentation. Native metals such as gold, or ores such as cassiterite, may be concentrated in the course of being transported from the weathered rocks of their origin and subsequently deposited in river beds or on the shallow continental shelf. These are called *placer* deposits, and are often more profitable to exploit than the original ores from which they derive. It is such deposits that the miners who panned for gold or silver, like those in the Yukon at the turn of the century, were seeking. An experienced miner learned which spots along a river—such as below rapids and waterfalls—were the most likely to yield good placer deposits.

Most prospecting, however, is not concerned with placer deposits. The minerals that the prospector is likely to encounter are secondary minerals. These are formed by the action of water on exposed primary ores—that is, ores deposited by hydrothermal solutions or formed by the crystallization of a magma. The water dissolves some components of the primary ore, percolates downwards through the rocks and redeposits them as secondary minerals.

Above: Blasting the terraced quarry faces of one of the world's largest copper workings, at Chuquicamata, Chile

At the start of many mining operations, secondary minerals make up the bulk of the ore from which the metals are obtained. Such surface reactions are shallow, however, and the secondary minerals give way at greater depths to the primary ore in which most of the great metal mines of the world operate. Unfortunately most of the deposits that appear at the surface have now been located, and many worked out. As known mines are worked out, prospectors are obliged to go into areas previously ignored because the greater part of the bedrock is covered by unwanted material (sometimes of glacial origin), or because no natural outcrop of ore has been found. It is important therefore to develop techniques that give information about structure, or hint at ore deposition, at some considerable depth. Conventional prospecting, which involves the collection of samples and the regional mapping of the geology of a likely area, remains indispensable to a scientifically based mineral exploration programme, but the more advanced geophysical and geochemical methods are becoming increasingly important.

EXTRACTION

As most high-grade ores have been located, and in many cases worked out, man has developed techniques of working bodies of lower-grade ore, in sufficiently large quantities (up to 100,000 tons a day) for the contained metalliferous ore minerals to be extracted economically. The extraction of copper ores during the last hundred years provides a good example of this trend. During most of the nineteenth century, Devon and Cornwall in southwest England were among the world's largest copper-producing areas, and were worked for the secondary enriched ores and the deeper primary sulphides, all of high grade. At the beginning of the nineteenth century, however, the Parys Mountain mines in Anglesey, Wales, were a major competitor to the Cornish mines, working ores containing about 2.5 per cent copper. In 1907 at Bingham Canyon, Utah, surface quarrying of a low-grade ore body of about 1 per cent copper began on a large scale. This venture proved that economic open-pit quarrying was possible. Between 1941 and 1972 the average grade worked in these large open pits fell to 0.6 per cent. These enormous low-grade deposits are known as 'porphyry' copper deposits, and they are now known to occur around the margins of the Pacific Ocean. They are associated with the edges of the crustal 'plate' that underlies the Pacific. It may well be that with a better understanding of this type of geological environment, similar, more ancient 'porphyry' copper ores may be discovered on the sites of ancient 'plate' margins no longer geologically active. Copper deposits have been recently discovered around the Pacific in Borneo and in the Philippines, and the vast deposits on Bougainville Island in the Solomon Island chain are now being extracted. The normal copper content of oceanic and continental basaltic rocks is about 0.006 per cent but large areas of continental basalts are known where the copper content is concentrated to 0.1 per cent, and these areas, with greater efficiency of extraction, may provide the next source of copper ore.

There is a tendency towards open-pit quarrying of all deposits, whether metallic or non-metallic, if at all possible. Deep mining is an expensive operation, but for some metals it will have to continue, and mining techniques must become increasingly efficient as known surface deposits are worked out.

Processing of metalliferous ores, after extraction by conventional mining or quarrying techniques, first involves the separation of ore minerals from gangue minerals and rock. Most mining companies keep a reserve of unprocessed ore so that their processing plants can be kept in continuous operation. A very simplified pattern of the progress of ore in a general processing mill is shown here.

The ore mineral in the quarried or mined rock may be scattered through it as small crystal groups associated with other minerals, so the first process is to free the ore minerals from the rest of the rock. The unprocessed rock must therefore be finely ground so that its minerals are broken away from each other and no piece is a mixture of minerals. In most mills a continuous process occurs in crushing, grinding, sizing and sorting, often repeated many times before actual separation of the mineral particles begins.

The separation of minerals is based on their physical or chemical properties. Perhaps the most commonly used physical property is that of specific gravity. Ore minerals are on the whole much denser than gangue minerals. They can be separated in several ways.

Panning is a simple method of density separation in which light (gangue) minerals are washed away by running water leaving the heavy ore behind. The *dense-medium* separation technique is also used. In this method a mixture of minerals is put into a liquid whose density is carefully chosen so that some minerals will float and others sink, depending on their density. Unwanted material is then easily removed to obtain an ore concentrate.

Today low-grade ore is concentrated on a large scale using the process called *flotation*. A watery mixture of finely ground rock is run into tanks filled with water. Small amounts of certain chemicals are added, which are varied according to the mineral required. The water is agitated, and air bubbles are formed. Because of the

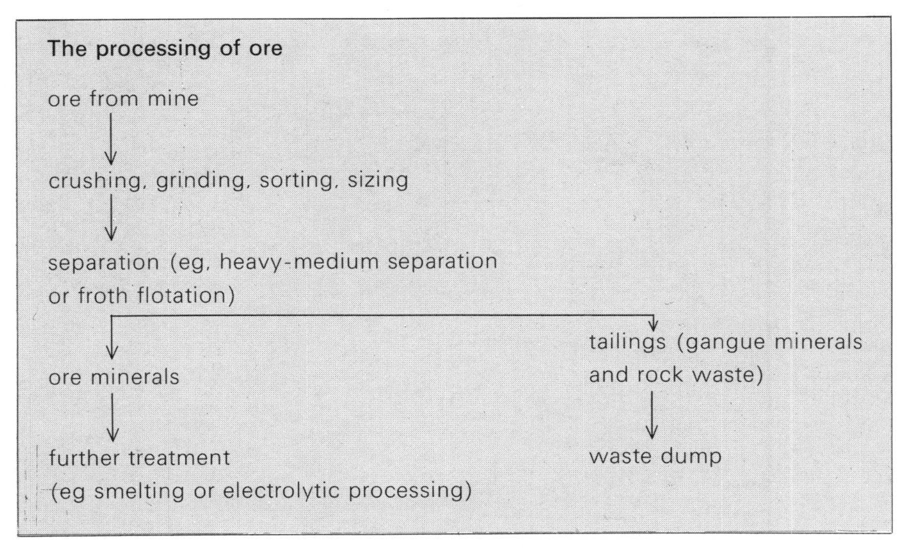

Left: A Swedish open-pit iron ore quarry and processing plant. To the right of the picture, tips of crushed gangue minerals and country rock, originally combined with the ore, can be seen

Important iron minerals

	% Fe (by weight)	chemical composition
magnetite	72.4	Fe_3O_4
hematite	70.0	Fe_2O_3
goethite	62.9	FeO.OH
siderite	48.2	$FeCO_3$
pyrite	46.6	FeS_2
ilmenite	36.8	$FeTiO_2$

additives, the air bubbles cling to some minerals but not others. Mineral particles buoyed up by the air rise to the top and are removed. Finally refined ore is sent to other processing plants, either for smelting or for chemical or electrical purification.

MINERAL RESERVES

Man is now using greater amounts of both metalliferous and non-metalliferous deposits than ever before. How long will they last? It is certainly in the exploitation of the metal-bearing deposits that the future is least secure. There are still large reserves of bulk industrial rocks and building stone, though environmental factors associated with their extraction, such as conservation and pollution, may become more important in the future.

It was stated above that most of the easily obtained mineral deposits have now been located. The emphasis in prospecting today is on instrumental methods with two broad aims in mind: discovering exploitable concentrations below the surface in areas with no surface outcrop, and discovering bodies of low-grade ore (perhaps one per cent or less) sufficiently large to make open-pit methods profitable. Unfortunately large areas of natural scenic beauty where low-grade deposits are known to occur may be used for this type of work.

Along with the development of low-grade ore extraction, the techniques of refining the ore must become increasingly efficient. Separating techniques and electrolytic processes must be further developed to the point, ultimately, where the tiny proportion of metal present in ordinary rocks or even sea water can be extracted.

About a thousand million tons of iron ore is required per annum. Attention is now being paid to lower-grade ('taconite') iron ores, containing 25 to 30 per cent iron. Hundreds of square miles of taconite deposits are known in North America alone and although processing is complicated and costly they seem an assured source of iron for the future.

Most of our presently workable mineral deposits are still found on continents or continental shelves, but there is no doubt that the ocean basins represent a vast reservoir of mineral wealth. Deep-sea investigations have found enormous numbers of manganese nodules covering large areas of the world's ocean floors, especially in the Pacific. These nodules are under great depths of water and there are difficult engineering problems to be solved before they can be recovered economically. But recent work has shown them to be an important potential source not only of manganese, but also of copper and nickel. Other useful metals, including cobalt, lead and vanadium, are also present, and could be recovered as by-products of processing for copper and nickel. Zinc has also been reported in relatively high concentrations in some types of manganese nodule.

In the 1960s oceanographic research vessels in the Red Sea established that there were three deep basins in its floor, which were rich in hot brines with extraordinarily high concentrations of metallic elements. These brines contain concentrations of various trace metals roughly 1,000 times greater than their concentrations in normal sea water; some have up to 5,000 times more iron, 25,000 times more manganese and 30,000 times more lead than normal. It is difficult to put an economic value on the metals accumulated in the hot brine deeps, but the amount of gold, silver, copper and zinc in the associated sediments is of potentially important proportions. Further oceanographic research will surely uncover more areas of metalliferous deposition in the world's seas.

Sea water is highly variable in composition

from one ocean to another. The proportions of various components shown in the table on page 76 are averages only. There it can be seen that sea water contains about 3.7 per cent dissolved salts. As man uses up the evaporite deposits laid down in past geological ages, the sea will become increasingly important as a source of the major elements contained in it. Altogether, about 70 elements have been identified in sea water. Others are certainly present, but in such small amounts that we cannot detect them by the analytical methods now available.

Not only must man endeavour to make his prospecting instruments and extraction methods more and more efficient to obtain greater tonnages of ore from the rock, but he must also develop more efficient techniques for recycling. At present, except for iron and possibly lead, the quantities recovered by recycling techniques are very low. The greater use of artificial substitutes, continually being developed, may also offset the shortage of some materials in the future. For example, artificially-made industrial diamonds, although small compared with most natural stones, are now widely used in industrial cutting and grinding processes.

Assuming that the environmental effects of open-cast metalliferous quarries are acceptable, and that the engineering and political problems associated with the recovery of ore from the sea floor are resolved, it will be possible to meet the world's demand for a considerable time by mining yet lower-grade deposits of minerals containing such elements as copper, molybdenum and iron, and possibly nickel, tin and tungsten. For other metals the future is not so hopeful, since known workable concentrations of, for example, lead, mercury, zinc and silver are limited. Unless new discoveries are made, sources of these minerals can only provide perhaps 25 to 50 years supply, and technology must be applied to produce synthetic alternatives.

MINES AND THE MINERAL COLLECTOR

To the collector, the commercial extraction of economically valuable minerals is of prime importance, for most good mineral specimens are obtained from the outcrops of rock thus revealed. Every mine or quarry, as well as road cutting and construction areas, is a possible site for the collection of a variety of interesting mineral specimens.

The best sites are probably those directly associated with the mining or quarrying of metalliferous ores. Both ore and gangue minerals are of commercial value in the international mineral specimen market. Some examples are the fine ore groups of tetrahedrite and bournonite ('cog-wheel' ore) from Cornwall and the magnificent groups of pyrite, arsenopyrite, apatite and wolframite from the Panasqueira mine in Portugal. Some of the most prized gangue minerals have come from past mining activity in the British Isles. Especially important are the calcites from Cumberland in northwest England, which are often found associated with 'kidney-ore' hematite, and specimens are still in great demand. Baryte was also found here, in a variety of colours that are all of interest to the collector. Fluorite, a gangue mineral in the lead mines of Weardale, County Durham, was frequently found as fine green and purple crystals and again can command high prices on the international market.

Of course, very good specimens of any kind have always been quite rare. A century ago, however, there was a greater chance of specimens being preserved at the mine, since much of the mining then was done by hand without heavy machinery. Often small veins of fine crystalline material would not be touched in this old method of working. In addition, the subsequent separation of the mined ore could be haphazard and good specimens of ore minerals as well as gangue, were sometimes dumped in the tips undamaged. Good specimens can still be collected from these old mine tips.

Today, the 'tailings' from major mine projects consists of intensively processed gangue minerals

Left: Mine installations and tips from the pyrite mines of Niccioleta, Italy. Despite crushing during the ore-extraction process, good specimens of ore minerals can still be collected from such old waste tips

Right: A mine and its spoil tip. The gangue minerals found in zone (a) correspond to the ore vein (a). Zone (b) gangue minerals correspond to the ore vein (b), which was exploited later than (a)

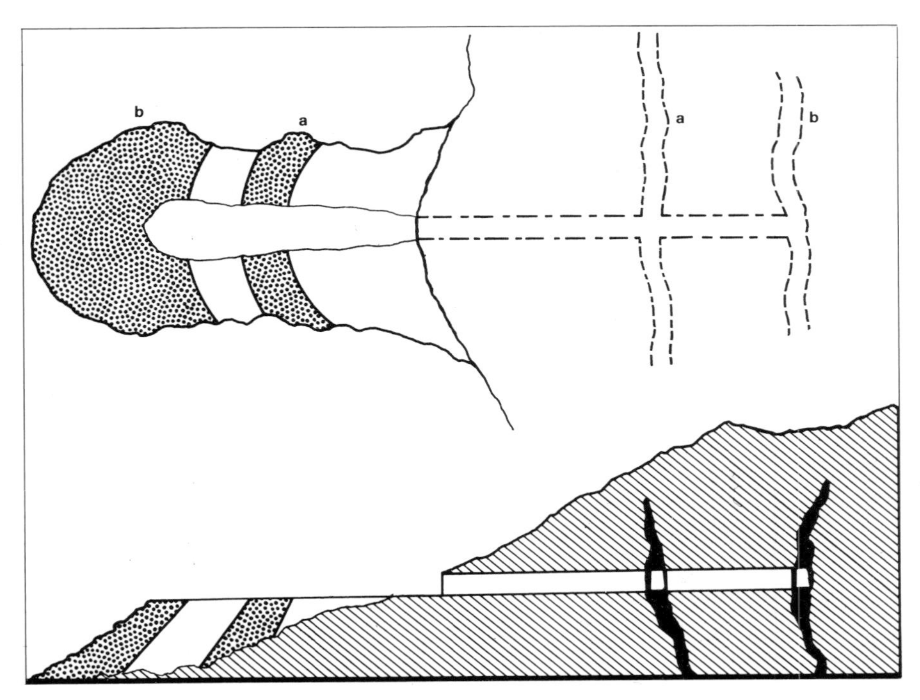

and rock waste so finely ground that they present only a powder of fine grains, or a watery slurry, to the collector. More good specimens go to the crushing plant than are recovered by the mineral collector. However, because of the increasing demand for fine specimens, some mining companies now permit their workers to collect material, so that a small but continuous supply of good specimens is produced from the major mining areas of the world.

In some old metalliferous mining areas, further specimens of interest to the mineral collector can be found associated with the main ore and gangue minerals; these are usually secondary minerals. The classic mining area of Cornwall has produced not only a large number of mineral species, but also a high proportion of exceptionally well-crystallized display specimens. These include bournonite from the Herodsfoot mine, chalcocite from the Redruth mines, and cuprite from the Phoenix mine, which are some of the world's finest specimens. The gangue mineral calcite from Wheal Wrey, St. Ive, forms crystal groups of unusual long twinned prisms. The usual secondary copper minerals, such as the carbonates malachite and azurite, are relatively scarce in Cornwall, but a variety of rarer secondary species may occur instead. Wheal Gorland and other copper mines near St. Day produced some of the finest specimens of olivenite, liroconite, clinoclase, chalcophyllite and pharmacosiderite. Southeastern Arizona and southwestern New Mexico are areas that also possess many small mines. They are prolific sources of distinctive minerals, such as the copper species ajoite and shattuckite.

In other parts of the world the amateur prospector or collector may still find new specimen localities. They are often associated with small pegmatites and other minor granitic intrusions that were first investigated for their potential economic value, but proved to have little. These sites often yield specimens of great beauty, such as the green and pink tourmalines from San Diego County, California, or they may contain large numbers of different mineral species. The Varutrask pegmatite of Sweden and the Hagendorf pegmatite of Germany both yield numerous different minerals, many of them minor phosphate minerals of interest to the mineralogist and collector, but without economic value. In the British Isles, the Meldon aplite (a fine-grained, light-coloured rock chemically similar to granite) in Devon was a past source of axinite, rubellite and petalite, but, like many British localities, now produces few fine specimens.

Quarry faces, and sometimes natural cliff sections, can provide mineral specimens, especially in areas of past tectonic disturbance. Joints and faults are often associated with the development of good quartz or calcite crystals. Minor mineralization can also occur along discontinuities in the rock sequence. Major mineralization is frequently associated with cracks or dislocations in rocks and, as blasting or cliff falls remove faces, new and richer deposits may be found. Rare lead oxychloride minerals have recently been discovered in manganese veins in the Mendips, southwest England, which follow former fault lines. Road cuttings and construction sites may also reveal pockets of mineralization in igneous or metamorphic terrains, or perhaps in sedimentary rocks like clays, in which gypsum crystals with perfect monoclinic crystal forms are often found.

All such sites of man's economic activity afford opportunities for the mineral collector. The keen enthusiast can take the opportunity to build up a specialized collection. Just as some collections are based on mineral properties like fluorescence or radioactivity, or represent the geology of a particular region, so a collection can be made up of specimens from a particular mine or quarry. The scientific interest and value of such a collection can be considerably enhanced by its theme. And the less scientifically inclined collector, too, can hope to find unusual and attractive specimens in old tips and cuttings.

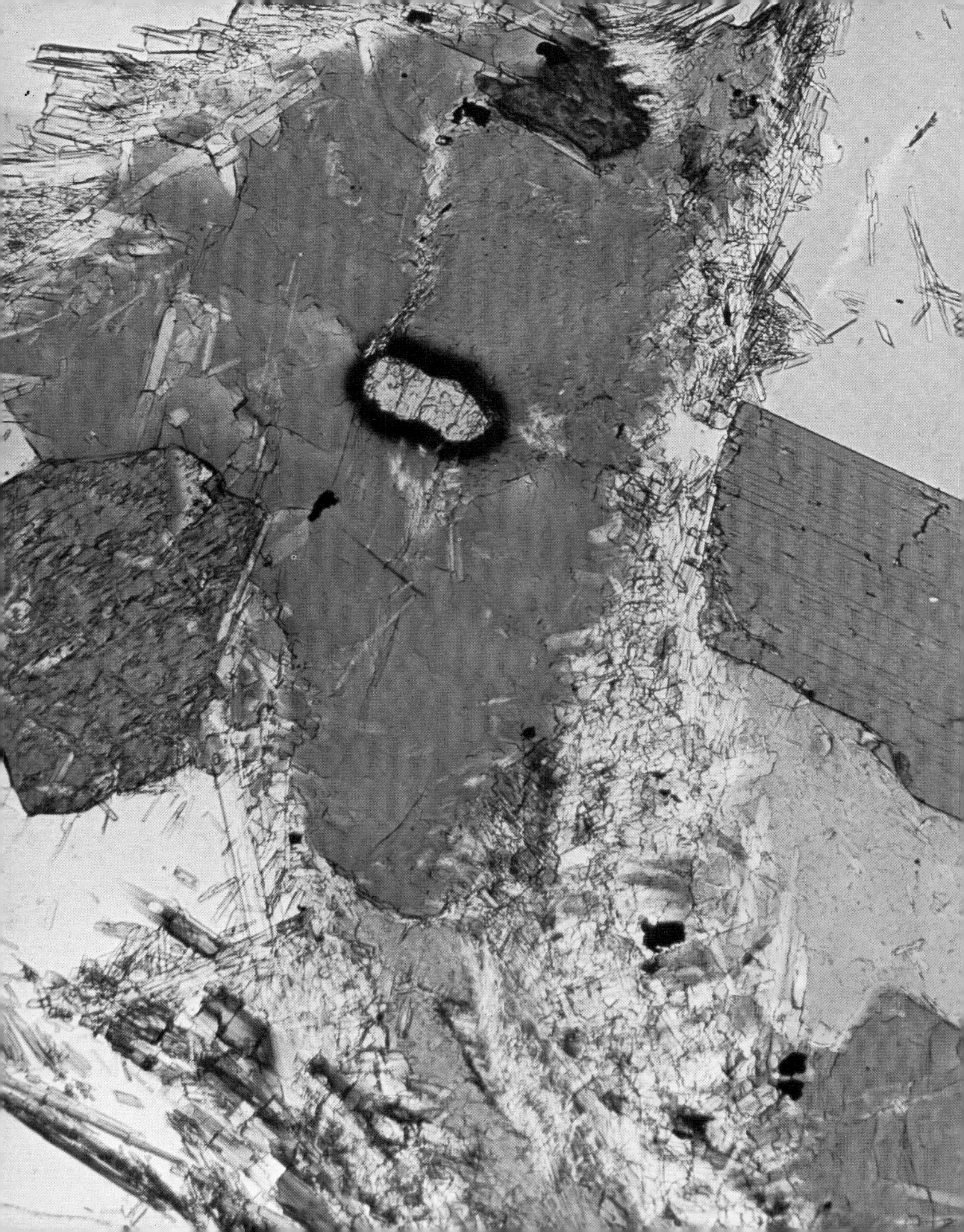

Identifying Minerals

Left: A thin section of mica, seen through crossed polarizers. In the centre is a zircon crystal, appearing bluish green

In the Mineral Kingdom chapter of this book details are given of the characteristic appearances presented by minerals, as well as other properties by which they can be identified. The purpose of this chapter is to expand on the range of mechanical and optical tests that can be brought to bear on minerals by the amateur in the attempt to identify them.

Hardness

To the mineralogist *hardness* means the ability of a mineral to resist abrasion by other materials. The celebrated scale of hardness devised by Friedrich Mohs is still in use today. It consisted originally of a series of minerals ranged in order of hardness, and arbitrarily assigned numbers from 1 (talc) to 10 (diamond). Intermediate values such as 2.5, 5.5 and so on are now employed as well. The only meaning of the Mohs values is that a mineral will scratch any other mineral placed lower on the scale.

A scratch test should, obviously, be carried out on an inconspicuous part of a specimen. The best way is to use a special 'hardness pencil', a holder in the end of which is mounted a mineral of known hardness. The result of the test can be observed with the aid of a microscope. Soft specimens will be scratched by a fingernail, which has a hardness of about 2.5. Slightly harder ones can be scratched with a penknife blade, whose hardness is 5 to 5.5.

Some crystalline minerals display directional hardness, and this should be looked for. Kyanite, for instance, has a hardness of 4 when scratched along the length of the crystal, but of 7 when scratched across it.

Some mineral classes are generally harder than others. Oxides and silicates, for instance, tend to be the hardest apart from diamond. Sulphides, halides, borates, sulphates and phosphates tend to be soft; so, too, do native metals.

Other scales attempt to give a quantitative value to mineral hardness. The Knoop scale, for example, is based on the size of indentation produced in a material by a controlled load. It assigns vastly different values to diamond and corundum, which are respectively at 10 and 9 on the Mohs scale. The Knoop value for diamond is 8,000–8,500 and for corundum is 2,000. Quartz (Mohs value 7) is placed at 660–900 on the Knoop scale.

Anyone dealing with gem materials will know that comparatively few are able to resist abrasion by airborne dust. The stones that are most suitable for jewellery are therefore the harder ones, those with Mohs hardness greater than 7. These include corundum (ruby and sapphire), topaz, emerald and tourmaline.

Tenacity

Closely linked with the hardness of a mineral is its *tenacity* or toughness. Jadeite has a Mohs hardness of 6.5, but the arrangement of its minute interlocking crystals makes it very tough. It is in consequence one of the most difficult of minerals to fashion. Tenacity shows itself in various ways, which we now consider.

A mineral is classed as *sectile* when it can be pared with the blade of a knife without powdering; gypsum is an example. If the specimen does powder easily, it is *brittle*, like calcite or apatite. It is *malleable* if sections of it can be hammered without powdering; such minerals include most native metals. If thin sections of the mineral bend without breaking and remain bent, like selenite or chlorite, the mineral is said to be *flexible*; if the piece springs back into shape, it is described as *elastic*.

Streak

When a specimen is drawn along a piece of unglazed porcelain a line of colour may result. This is called the mineral's *streak*. This is one of the easiest tests to perform in the field. Many minerals that appear dark in the mass may give a much lighter streak. Hematite, for example, may appear black when massive, but yields a red streak. If the mineral is powdered, its powder will look the same colour as the streak.

The streak test is of no great value for identifying silicates and carbonates, many of which give a white streak. Many other minerals, however, especially the iron oxides, have very characteristic streaks.

Specific gravity

An important property of a mineral is its *specific gravity* (written SG; some authors use the symbol G). It is defined as the ratio of the weight of a given volume of the mineral to the weight of an equal volume of water. The copper ore chalcopyrite, for example, has a specific gravity of 4.1 to 4.3 (varying from specimen to specimen). This means that a piece of it weighs between 4.1 and 4.3 times as much as an equal volume of water. Being a ratio, SG is a pure number without associated units. To give some idea of the arithmetical values of mineral SGs, that of diamond is 3.52 and that of gold is 19.3.

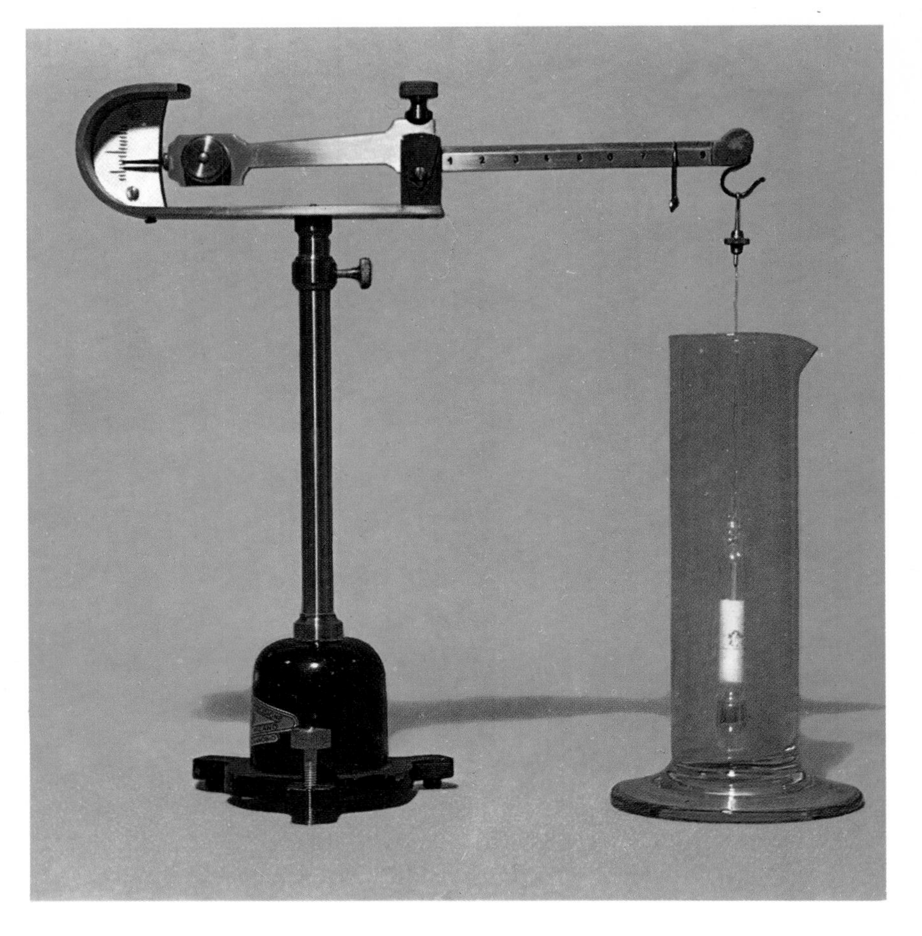

Above: A Westphal balance, used to determine the specific gravity of mineral specimens

An experienced collector can get some idea of the SG of a mineral in the field by 'hefting' it—just lifting it in his hand so that he can judge its weight compared with other mineral specimens of similar size.

A convenient method of determining the SG of a reasonably large specimen is to weigh it, first in the normal way and then suspended in a liquid. We shall discuss this first on the assumption that the liquid is water.

The specimen's weight in water, W, will be less than its weight in air, A. The difference between these, $A-W$, is called the water's upthrust on the specimen. According to Archimedes' Principle, this upthrust is simply equal to the weight of the water that the specimen displaces when immersed. Hence:

$$SG = \text{wt of specimen}/\text{wt of equal volume of water} = A/(A-W)$$

If the weight L is recorded when some liquid other than water is used, then the figure obtained by a corresponding calculation must be multiplied by the liquid's own specific gravity, G:

$$SG = (G \times A)/(A-L)$$

Archimedes is reputed to have hit on this principle when his bath overflowed as he climbed into it. He then used it to show that a crown

Right: A specific gravity balance that can be made by the amateur from the simplest materials

A home-made SG balance

The balance has a beam about 30 cm long, made from a metric rule or a piece of plain wood to which a strip of graph paper is glued. The pivot is a steel pin or needle set about one-third of the way along the beam. The pivot should be as nearly horizontal as possible. A cut-away plastic tube or phial provides the pivot's support. The pivot should rest on a level surface, *not* in a groove. A slotted piece of wood limits the beam's swing, and a reference mark indicates the balance point. Pieces of bent stiff wire hold the specimen and the counterweights, such as machine nuts. To make an SG determination, choose a counterweight such that the specimen is balanced about halfway along the scale. Note the specimen's distance from the pivot, L_A. Then immerse the specimen in cooled boiled water with a tiny drop of detergent, and brush off all the bubbles. Balance the immersed specimen in a new position and take the reading, L_W. These two distance readings are *inversely* proportional to the specimen's weight in air and water respectively. The specific gravity is given by the formula (weight in air)/(weight in air—weight in water), which after a short calculation becomes $L_W/(L_W - L_A)$

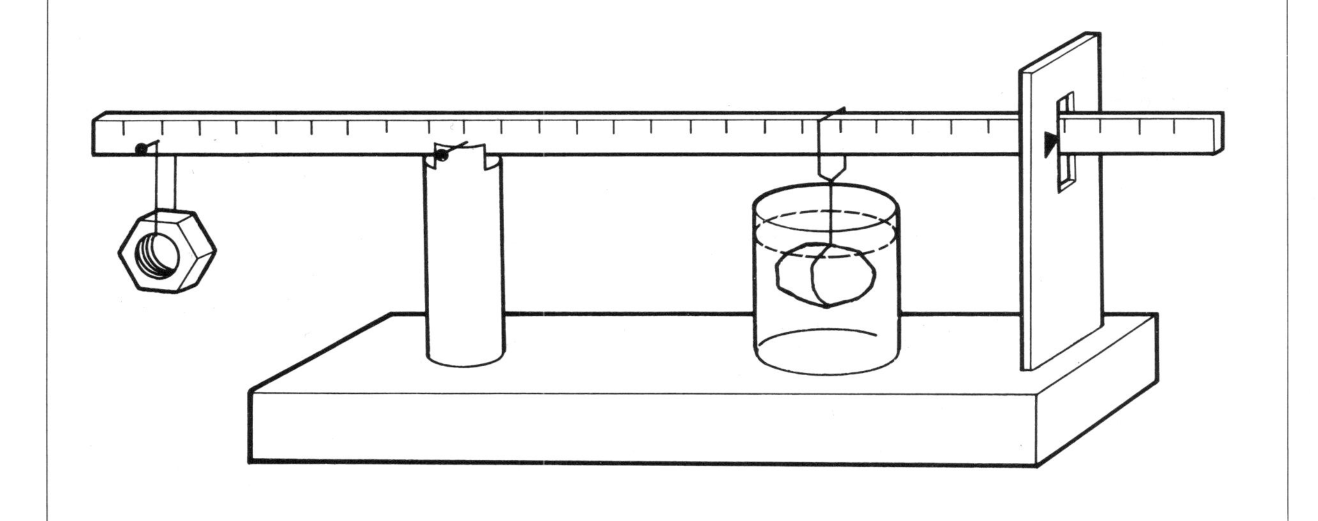

supposedly made of pure gold was in fact adulterated with silver, by measuring its SG and finding it was lower than that of gold. Those who do not have golden crowns to experiment on can nonetheless use the principle with the aid of the simple home-made balance described here, which should allow determinations of SG to at least one decimal place.

Another way of establishing the SG of a mineral is to observe its behaviour in 'heavy' liquids whose SG is known. Such liquids can easily be bought; they tend to be expensive and should be treated with caution as many of them are toxic. If care is taken not to contaminate them they should last for a long time.

Testing is very simple; the specimen is placed in a liquid, and if it floats it has a lower SG, whereas if it sinks it has a higher one. The rate at which it sinks or floats upwards gives some idea of how close its SG is to that of the liquid. The result of such is a test is to 'bracket' the mineral between the SGs of the lightest liquid in which it will float and the heaviest liquid in which it will sink.

The test is commonly used as a quick way of testing the SGs of cut gemstones. Suitable liquids in frequent use in gem-testing laboratories include Clerici solution with a specific gravity of 4.15, made up of equal parts of thallium formate and thallium malonate. This liquid is highly poisonous and should only be used under laboratory conditions. If it is necessary to dilute it, distilled water must be used. Methylene iodide (diodomethane), CH_2I_2, has an SG of 3.3; it can be diluted to lower SGs with liquids such as acetone. Bromoform, $CHBr_3$, is similarly useful; its SG is 2.8, though it, too, can be diluted with acetone.

OPTICAL EXAMINATION

Most collectors will possess some kind of magnifying instrument, ranging from the 10× lens (the standard magnification used in the gem trade) to a complicated microscope. When using a lens the best technique is to hold it close to the eye and to bring the specimen up closer until it is in focus. Most modern good-quality lenses are *aplanatic* (free of edge distortion) and *achromatic* (free of colour fringes).

An important requirement for a microscope employing ordinary (unpolarized) light is that it should have a large working distance—that is, the space between the main or *objective* lens and the specimen.

In advanced work, polarized light is used to examine thin sections of specimens (see The Crystalline State in this volume). Here we shall confine ourselves to what can be seen with ordinary light; the use of polarized light is discussed later in this chapter.

Inclusions

Either a 10 × lens or good binocular microscope will be sufficient to study inclusions and surface features in gemstones. Among the *protogenetic* inclusions (formed before the surrounding crystal) often found in gem materials (the only ones transparent enough to show them clearly) are:

actinolite in emerald and in some of the garnets;
apatite in garnet, spinel and corundum;
diamond in diamond;
diopside in corundum;
epidote in quartz;
hematite in topaz.

Inclusions that are formed at the same time as the crystal containing them are called *syngenetic*. Among these we may encounter:

calcite in ruby, especially from Mogok, Burma;
chrome diopside in diamond;
chromite in peridot;
feldspars in corundum, especially from Thailand;
pyrite in emerald and fluorite.

Inclusions that appear after the formation of a crystal are called *epigenetic*. They usually consist of bubbles of gas or liquid. During earth movements crystals break and the cracks become filled with liquids, which sometimes 'heal' the fracture. This healing can be seen as a plane of minute bubbles.

A different kind of epigenetic inclusion, however, consists of solid crystals. If they have definite orientations they can give rise to chatoyancy or asterism.

Such phenomena as colour banding and zoning can be classed with inclusions for their value in identification. The angular banding of colour in natural corundum, for example, helps the gemmologist to distinguish it from its synthetic counterpart, in which the banding is curved.

Refractive index

The degree to which a transparent substance bends the path of light passing through it is measured by its *refractive index* (RI). There are various ways in which this can be measured in order to gain clues to a mineral's identity.

A very simple method can be used with flat-sided specimens. A microscope is focused on a spot marked on a piece of paper. The specimen is then placed over the spot and the microscope is moved upwards until the spot comes back into focus. Finally the microscope is raised until the upper surface of the specimen is in focus. All three of the instrument's positions are recorded. From them it is a simple matter to work out the spot's real depth below the upper surface (that is, the thickness of the specimen) and its apparent depth viewed through the material. The specimen's refractive index is then given by the ratio of the real to the apparent depth.

A very simple type of refractometer can be used with flat-sided specimens. A face of the material is put in contact with a glass prism. (A transparent liquid of high refractive index is used to coat both the faces, in order to provide good optical contact.) Light is passed into the prism and falls on the interface. Light striking

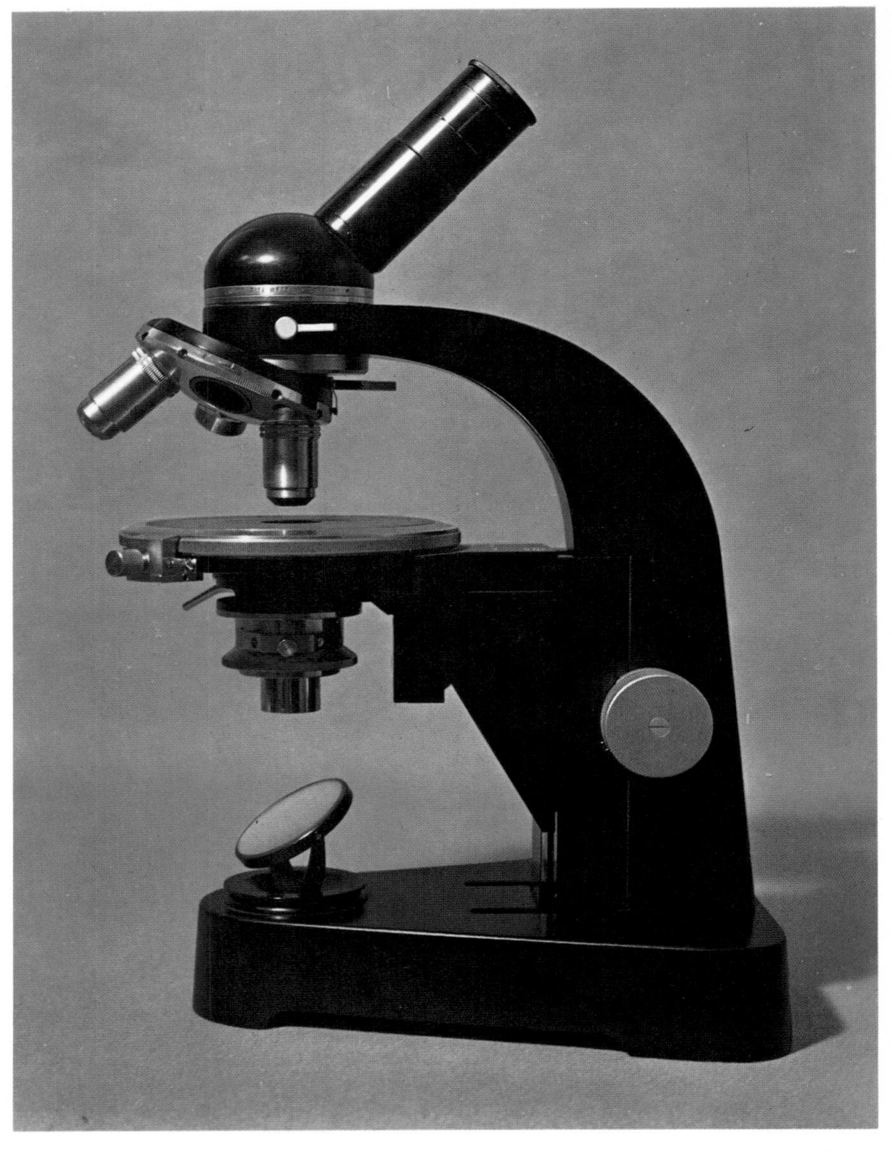

Above: A polarizing microscope. Note the rotatable stage and the three objective lenses, allowing three different levels of magnification

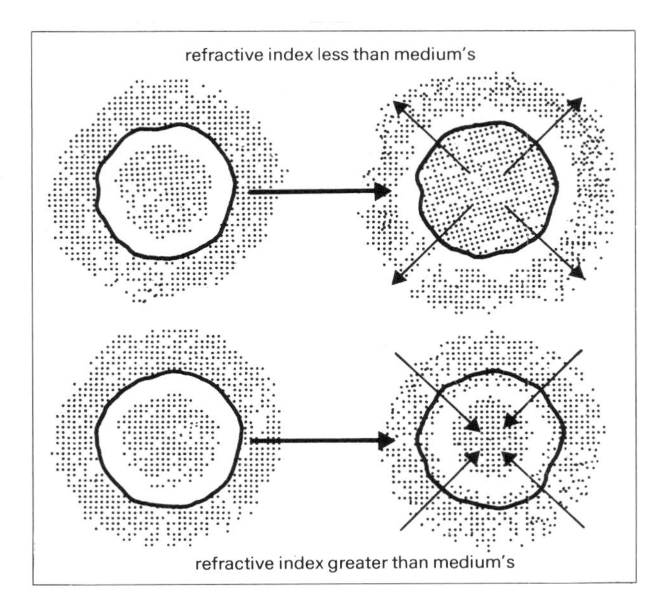

Right: The Becke test. When the microscope is racked up from the focused position, the bright line fringing the mineral grain moves into the medium of higher refractive index

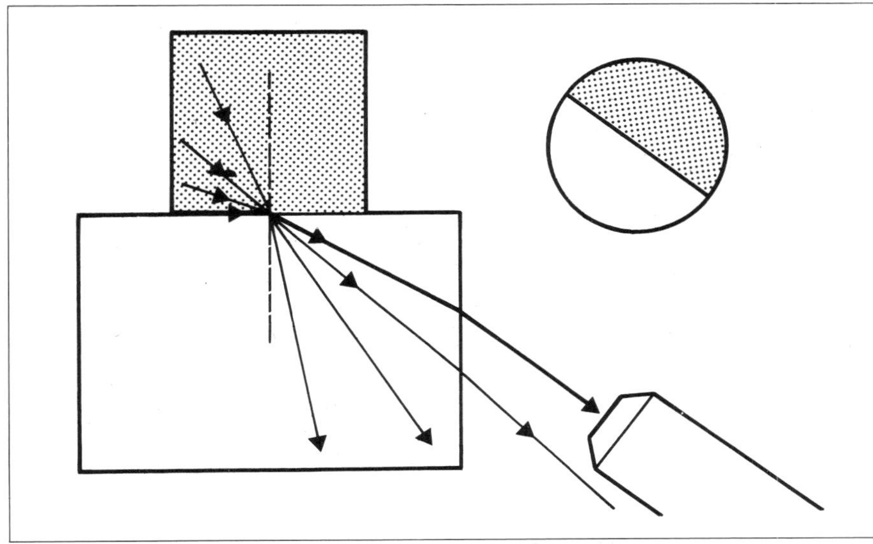

Above: The critical angle refractometer. The specimen rests on a glass block. It is illuminated from above, and the observer sees a light and a dark field. The boundary corresponds to rays striking the interface at a grazing angle; from its position the RI can be calculated

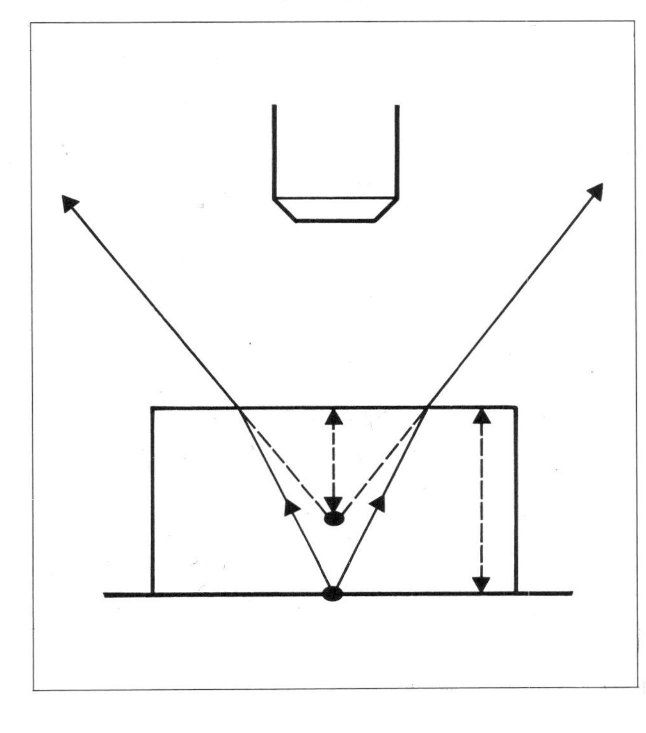

Right: Real and apparent depth. A dot beneath a transparent specimen appears raised by refraction. The ratio of real and apparent depths equals RI

the interface at relatively small angles of incidence (that is, nearly at right angles) passes through. At large angles of incidence (that is, nearly grazing the interface) the light is totally reflected. This reflected light is observed with a small viewing instrument. The viewer is moved until, at a certain angle of observation, a boundary between a light and a dark field is seen. Beyond this position reflected light is not seen, because at the corresponding angles of incidence light passes through the interface. The angle of incidence that the boundary represents is called the *critical angle* and from it the refractive index of the specimen can be calculated.

The methods described are normally used only for cut gemstones, since only these are likely to have sufficiently flat faces. Mineral specimens are often in the form of small grains, and the refractive indices of these can be found with the aid of transparent liquids whose own refractive indices are known.

If the RIs of the liquid and the specimen are similar the edge of the specimen will appear much less sharp and may even seem to vanish, if the specimen is colourless. This, however, clearly permits only a very rough estimate of RI.

If a microscope is focused on a grain immersed in liquid, a bright line will be seen fringing it. If the microscope is raised slightly, the Becke line (as it is called) will move either outwards or inwards. The line moves *into* the medium of *higher* refractive index. If the test shows that the mineral has higher RI, the liquid can be mixed with another one to raise its RI. This is continued until the Becke line no longer appears. The RI of the liquid mixture is measured, which is a relatively simple operation, and the RI of the mineral is then known.

Pleochroism

The property of displaying different colours when viewed in different directions is called *pleochroism*. It is visible in ordinary light, but can be much enhanced with the aid of an instrument called a *dichroscope*. In its simplest form this is a metal tube with an eyepiece at one end containing a calcite rhomb. At one end of the calcite there is an aperture against which a crystal is placed to be viewed. At the other end is an eyepiece through which the aperture can be viewed. Light entering the calcite from the aperture is split into two refracted beams, of opposite polarizations. Looking through the eyepiece the observer sees two images of the aperture, and if the crystal is pleochroic, these will in general be of different colours. Some pleochroic minerals will need to be observed from a range of directions if all their colours are to be seen. A dichroscope can also be constructed from a cardboard tube and a piece of Polaroid cut into two pieces, with their polarizing directions at right angles to each other.

The dichroscope is somewhat limited in use, but good results are obtained when ruby or sapphire are observed. Ruby shows crimson and orange, sapphire shows dark and light blue. Synthetically produced rubies can be cut so that their pleochroism is seen when they are viewed through the table facet. Natural ruby is usually cut in a different orientation, which does not display the effect.

The polariscope

Polarized light can be used to distinguish between singly refracting material and the various doubly refracting materials. The *polariscope* is used for this purpose, either alone or built into a microscope (called a petrological microscope because of the importance of this technique in examining thin rock sections).

The polariscope consists of two pieces of Polaroid mounted one above the other with a few inches gap into which a specimen can be inserted. The device is illuminated from below. The polarizers are permanently in the crossed position. This means that no light can get past the second filter when there is no specimen in the polariscope. The same is true when singly refracting materials are examined, for these do not rotate the direction of polarization of the light. Doubly refracting materials, however, do in general rotate the polarization direction. When such a specimen is placed in the polariscope and rotated, it will allow light to pass at four positions in each rotation. At the intermediate positions it will block out light to some extent.

Many amorphous and some cubic-system materials such as synthetic spinel and some garnets display a striped effect known as anomalous double refraction which can be quite characteristic.

Luminescence

Both the mineralogist and the gemmologist find *luminescence*—the emission of light by a specimen when it is illuminated by light of a different wavelength—extremely useful in testing.

When the stimulated emission is short-lived—that is, vanishing virtually instantly when the stimulating radiation is turned off—the effect is called *fluorescence*. When it lingers for some time afterwards, it is called *phosphorescence*.

C G Stokes first developed the theory of 'crossed filters'. He illuminated chromium-bearing stones with blue light, obtained by passing white light through the first colour filter—which could be a copper sulphate solution, for example. He then viewed the illuminated stones through a red filter and found that they glowed a bright red. The red light that they emitted was not present in the illuminating light, and so originated in the specimens themselves (specifically, in their chromium atoms).

Chromium-bearing stones that will glow when submitted to this test include ruby, red spinel, and some alexandrite and emerald. But other materials will also respond to it. Pearls that are naturally black will glow faintly red, whereas those that have been stained black with silver nitrate will not. If a yellow filter is used in place of a red one, some stones will be seen to fluoresce yellow, especially yellow sapphire from Sri Lanka.

A useful device known as the Chelsea Colour Filter allows both red and green light to pass. It was originally designed to test emerald and identify synthetic products. When it was first developed, synthetic emeralds would glow very brightly when seen through the filter because they had a high chromium content. More recent productions to which iron has been added do not glow quite so brightly. Glasses coloured with cobalt, which appear dark blue to the naked eye will appear red through the filter. Rubies, both natural and synthetic, glow a bright red and cannot be distinguished by this test. They can, however, be distinguished from other red stones such as garnets and tourmalines, which glow more faintly.

Ultraviolet (UV) light is the radiation most commonly used to stimulate luminescence in minerals and gemstones. The UV spectrum is divided into short- and long-wavelength regions. There is a variety of UV lamps on the market, some conveniently portable. Provided care is

Right: Interference figure produced by a thin mineral section placed between crossed polarizers. Carefully interpreted, these complex patterns provide evidence about the mineral's crystal system

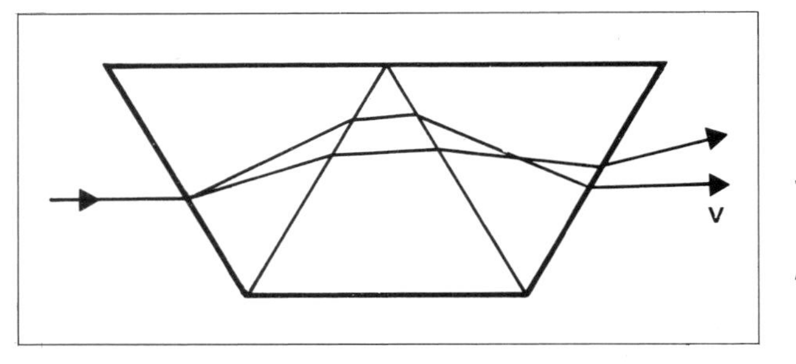

Left: Arrangement of prisms in a direct-vision spectroscope. Red light is deviated least at each refraction, while violet light is deviated most

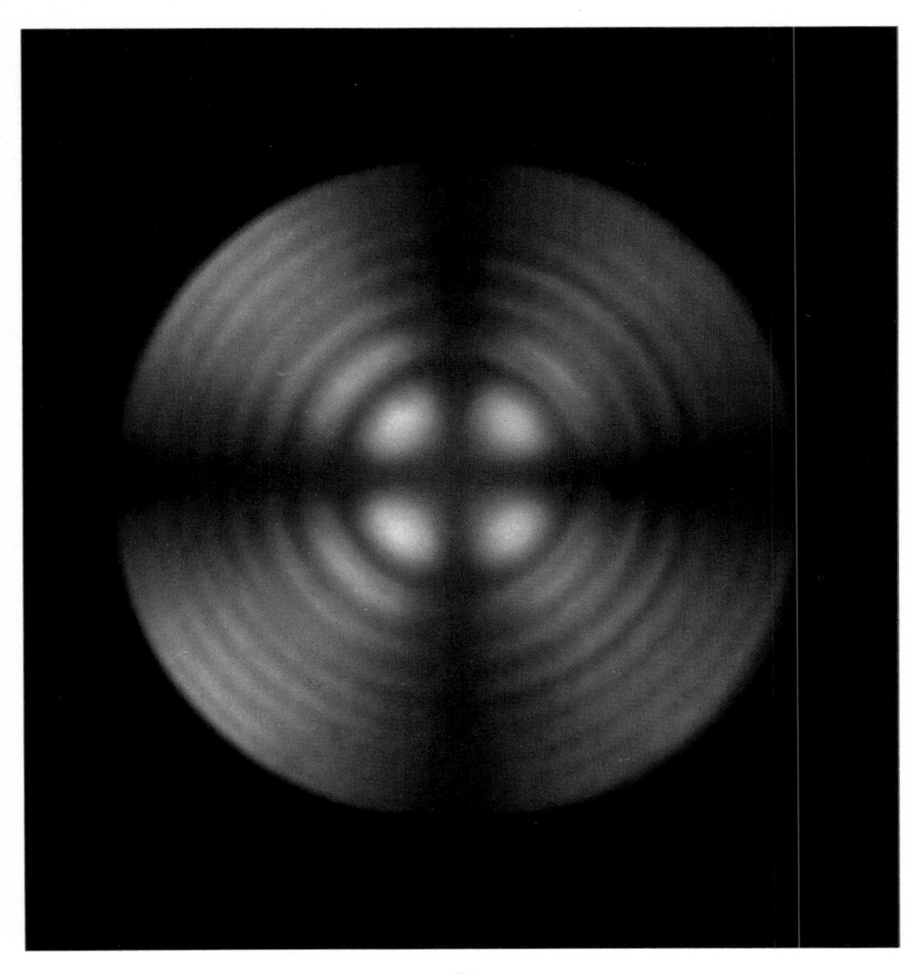

taken to shield the eyes all types are safe to use.

A section of the Identification Tables at the end of this book lists minerals according to the colour of the light they emit under UV illumination.

UV lamps are very useful in collecting; in dark areas such as old mine workings, fluorescent minerals show up clearly.

The spectroscope

The characteristic *spectra* of minerals—the pattern of wavelengths of light that they reflect or transmit—can be identified with the aid of the hand spectroscope, a metal tube containing a train of prisms, with an adjustable slit at one end and an eyepiece at the other. A ribbon-like spectrum of continuous rainbow colours is seen when white light is examined through the instrument. When a gemstone is examined, dark lines or bands are seen crossing this spectrum. These represent 'missing' wavelengths of light absorbed by the material.

The instrument provides a convenient means of identifying a number of stones quickly. Specimens should not be too small since this makes observation more difficult. Observations can be made with transmitted light, when the spectroscope can be mounted on the body-tube of a microscope; or by reflected light, when there need be no additional equipment, though there are various kinds of stand on the market. No difference of technique is needed for testing rough and cut stones, and this is a great advantage.

The pattern of lines and bands shown by some elements is sufficiently distinctive to make recognition easy. Chromium lines are relatively sharp, while the bands and lines due to iron are more diffuse. This accounts for the brightness of colours due to chromium and the more subdued quality of colours caused by iron. Chromium can also be recognized by the presence of emission (bright) lines, which are not normally shown by gem materials coloured by other elements.

A number of synthetic gemstones are coloured by the so-called rare-earth elements and these often show a very complex absorption spectrum. Listed below are the characteristic line patterns of the transition elements:

chromium: sharp lines; doublet (two lines close together), either absorption or emission, at the deep red end of the spectrum; broad band in part of the yellow or green; usually two or three lines in the blue and general absorption in the violet; stones include ruby (lines in the blue), red spinel (no lines in the blue), alexandrite, emerald, green jadeite;

iron: bands rather than lines, in the green and blue; stones include blue, green and yellow sapphire, yellow-green chrysoberyl, peridot, blue spinel, aquamarine and almandine garnet;

manganese: colours are usually pink or orange and there are bands in the blue and violet; stones include spessartine garnet and rhodonite;

copper: two narrow bands in the violet—these are quite hard to see, but provide identification of turquoise;

vanadium: no very characteristic absorption, though the synthetic corundum said to imitate alexandrite shows a sharp line in the blue, which is diagnostic for this material.

cobalt: bands in the orange, yellow and green with slight variations of position in different minerals; best seen in blue cobalt glass and synthetic blue spinel.

Other transition elements show no characteristic absorption spectrum. Uranium gives a spectacular absorption spectrum in some zircons, especially those from Burma, which are greenish-brown in colour.

The Fashioning of Stones

Left: A selection of gemstones cut in various styles. They include a fine Burmese star ruby cabochon at the centre, and a star quartz (centre left) Tourmaline is represented in several colours: pink (upper left), red (upper right), and dark red (lower left). The ring is of green jadeite, and there is a cabochon of the same material at upper left. The large green tablet is the jade-albite mixture maw-sit-sit. Just above the central star ruby is a faceted zincite (orange), a rarely cut material. The fine blue stone at upper left is a Tanzanian sapphire

What is a gemstone? What are the qualities that have led men to value some minerals above others, and to expend so much skill and labour on fashioning them into objects of ornament?

It is evident that throughout history man has prized certain substances for their beauty and their durability, and set a high cash value on them when they are rare. Those that are hard enough to be cut and polished, so that their beauty is enhanced, are called gemstones—as opposed to softer materials, such as gold and silver, that must be worked in other ways.

High-quality gemstone is certainly rare enough. Although a gem mineral may exist in large ore bodies, flawless gem-quality specimens are few. Spodumene, for example, one of the major sources of the metal lithium, occurs in masses sometimes weighing many tons; but only infrequently as gem-quality kunzite or, very exceptionally, hiddenite.

A gemstone's hardness permits it to retain its smooth, freshly polished appearance for a very long time. It is also often the reason for any non-ornamental value that the gem may have. Many engineering processes would be much slower and more expensive if diamond, the hardest natural substance known, could not be employed in drilling and polishing. Diamond and sapphire are used in record-player styli because of their hardness. Synthetic ruby, on the other hand, is used in producing laser beams, not because of its hardness but for the regularity of its atomic structure.

On the whole, however, the value of gemstones to man lies not in their utility but in their enduring beauty. The task of the lapidary is to take a stone that may be of very unpromising appearance, and transform it into a jewel.

Why cut gemstones? To answer this question consideration must be given to the way in which cutting and polishing will improve a material's appearance, and then to some of the ways in which a stone can be cut. Many ornamental substances look unattractive when first mined or collected; agate nodules for instance, often resemble potatoes, and at first glance it would be difficult to imagine their internal beauty. Much gem rough occurs as rolled pebbles, shaped by abrasion during the thousands of years that have elapsed since the crystals were eroded from their parent rock. The majority of these pebbles are found in long-since dried-up river beds far below the places from which they originated. These are called *alluvial* deposits; a large proportion of the world's gemstones occur as alluvial material. The rubies and sapphires that occur in alluvial deposits are hard enough to have remained relatively unscathed in the course of erosion, but many softer gemstones, such as peridot and moonstone, are found as rounded pebbles with frosted surfaces. Some diamond crystals are found as clear octahedra and from Roman times, if not earlier, have been set in jewellery without being shaped. But even diamond often occurs with a frosted surface, though this is not usually due to abrasion. Cutters have often been surprised by the perfection of the material found under the skin of even the darkest, most unattractive crystals.

So, by inference, at least part of the question has been answered: cutting an often unattractive gem rough will considerably improve its appearance. Usually this improvement is so striking that a person seeing the rough and then the cut stone would probably not believe they were the same thing. To produce the best possible results cutting the stone should be carried out with due consideration to the physical and optical properties of the material being worked. If the material is of faceting quality, the relevant properties encountered are likely to be: hardness; dichroism or pleochroism, the property of appearing different colours when viewed in different directions; cleavage; and dispersion, the breaking up of white light into its constituent colours. If the material is of the type that will produce a star or cat's eye effect, the correct orientation is essential if the minute inclusions causing these phenomena are to give a properly positioned effect. If the material is suitable only to be made into a cabochon—a rounded stone—then the rough must be cut in such a way as to make use of any available zoning or surface markings, such as the light- and dark-green banding of malachite, for instance.

A faceted gem is usually cut to produce as much liveliness and brilliance as possible; this is achieved primarily by the correct positioning of the facets around the stone, at the correct angles to produce total internal reflection. This ensures that the light that enters the stone through the top half, or *crown*, is reflected around inside the stone, picking up more body colour than if it had been reflected straight out again. The gem's 'fire' is due to the splitting up of the light into its component colours when it enters and leaves the stone. Again, the skilful angling of the facets can enhance this effect.

We now turn to the practical details of polishing and cutting stones. Great care must be taken over the selection of the material before cutting is commenced, in order to detect cracks, flaws and, especially, incipient cleavages. Failure to carry out this inspection can result in great disappointment for the cutter, for the stone can chip or fall to pieces while being worked. The easiest way to study the material for such features is by immersion in a suitable liquid, provided the stone does not have a frosted surface. If the refractive index of the liquid is close to that of the stone, the worker will be able to see into the stone and note the position of any such flaws. It is quite usual when dealing with frosted crystals, particularly diamonds, to grind and polish a 'window' on the surface to facilitate this inspection. Some gem rough, such as opal, is recovered with the use of explosives, causing considerable damage by cracking pieces of rough that would otherwise be flawless.

Above: A piece of facet-grade rose quartz

Left: The same, immersed in an optically dense liquid, allowing flaws and inclusions to be seen more easily

Right: Some examples of easily obtainable material for tumble-polishing, shown with their polished counterparts

Apart from the natural hazards to be encountered during the selection of gem rough, there are several ways that purchased materials, like lapis lazuli, turquoise and jade may be 'improved'. Lapis may be oiled; soft, crumbly turquoise can be impregnated with a hard-setting resin to make it workable; jade may have its colour intensified or altered by staining. Fortunately these treated substances can usually be detected if a little time is given to examining them before the purchase is made.

THE FASHIONING PROCESSES

Having recognized that there is a need to give a polished surface to our rough rock, whether it be expensive facet-grade material, or just attractive beach pebbles from a recent holiday, it must now be decided which of several different methods will be used to produce this polish. There are three basic methods available to the stone-fashioner, namely: *tumbling*; *cutting a cabochon*; and *faceting*.

Tumbling is the simplest of these methods. A tumbler can be made at home from a small electric motor, a plastic food container, and an engineer's V-pulley, or it can be purchased from suppliers of gemcraft equipment.

The rough stones are placed in the tumbler barrel with silicon carbide grit and water. The barrel is rotated by the motor at about forty revolutions per minute until the stones are smooth and free from surface imperfections—a process taking up to two weeks. At this stage the grit is changed for one of finer grade and the tumbling process continued for about a week. Subsequent grit changes are necessary; the process ends with a grit so fine that the stone's surfaces become highly polished. The total tumbling time is usually from five to eight weeks, depending on the hardness of the materials that are being polished.

A stone that has been sawn or ground into a predetermined shape may be polished in the tumbler barrel. Apart from the basic tumbler, grits and polish, other equipment is needed, and can prove to be quite expensive. Firstly, a diamond-bladed saw will be required to cut the rough rock into manageable pieces. Secondly, a grinding machine with a soft-bonded lapidary wheel is needed to preform particular shapes—hearts, crosses, and so on. For simple shapes, however, the saw may be used alone to 'trim' the slices before tumble finishing. This method is much quicker than tumbling alone, because the major surface imperfections are removed before tumbling begins and the very coarse grit may be omitted. The tumbling process proceeds as described above.

A grinding machine is needed for cutting cabochons; it will need both coarse and fine abrasive wheels. Fine 'wet and dry' emery paper and a hard-felt wheel for polishing are also required; a diamond saw is needed for cutting slices of rock. Suitably sized pieces of the slice may be shaped freehand or, preferably, the shape may be marked out on the surface of the slice using an aluminium 'pencil'. Accurately shaped cabochons can be produced if a metal or plastic template with an appropriate cutout shape is used as a guide.

The roughly shaped piece of rock is usually fixed with wax on to a *dop stick*, a piece of quarter-inch wooden dowel about six inches long. Dopping gives great control over a stone, allowing it to be moved freely in all directions during the grinding and polishing stages without risk to the operator's fingers. The stone is first given the basic shape with the coarse wheel; then the fine wheel is introduced to give a reasonably smooth surface; next fine emery paper is used, which gives the stone a very smooth, dully polished appearance. The final polishing of the cabochon is carried out on the hard-felt wheel using cerium oxide or similar powder as a polish.

With chatoyant or star-stone material such as tiger's eye or moonstone, and star-rose quartz or star ruby the stone must be correctly orientated; otherwise, when the star or cat's eye appears after polishing, it will prove to be misaligned.

The most usual type of cabochon material, such as agate or malachite, is opaque or semi-transparent. It is, however, feasible to cut *en cabochon* gem rough that is too badly flawed for faceting; some amethyst, citrine and garnet falls into this category. Almandine garnet is quite often cut as a hollow cabochon—one whose base has been hollowed out—when its colour is so deep that it would otherwise look dark and unattractive.

The more advanced worker may wish to try out his skill by cutting composite stones, such as opal doublets or triplets. In a doublet the layer of precious material is backed by 'potch' (low-quality) opal or obsidian. An opal triplet is similarly backed but is also topped with rock crystal. These layers may be quite conveniently cemented with one of the normal commercial epoxy resins.

Faceting is a more complex process than cabochon-cutting and requires considerably more patience and skill. Faceted stones are usually transparent with highly polished and regularly positioned flat faces, the facets. The means by which these faces can be cut will be described briefly now but in more detail later in the text.

The equipment needed may either be developed from that already used for cabochon cutting, with the addition of some specialized accessories, or may be purchased as a completely independent unit.

In addition to the equipment already described, a *lap* will be required. This is a flat metal plate that is rotated by a motor in order to produce a flat surface on a stone. It may be of aluminium or cast iron if the grinding is to be done with carborundum grit, or of diamond-impregnated copper. Carborundum grits tend to fly off the lap and must be constantly renewed, so the diamond-impregnated lap is the more convenient and cleaner type.

An attachment will be required to hold the dop stick, which is usually of brass, rigidly at fixed angles. One such device is a jamb peg, a piece of wood or metal pierced with rows of holes, which is fixed in a vertical position

Left: Amazonite, the green microcline feldspar often cut en cabochon or tumbled. This specimen shows evidence of its perfect cleavage

Right: A commercially produced tumbler, just one of many types available. Here the three barrels make it possible to have different grinding stages tumbling at the same time. Alternatively each barrel may be used for just one type of material

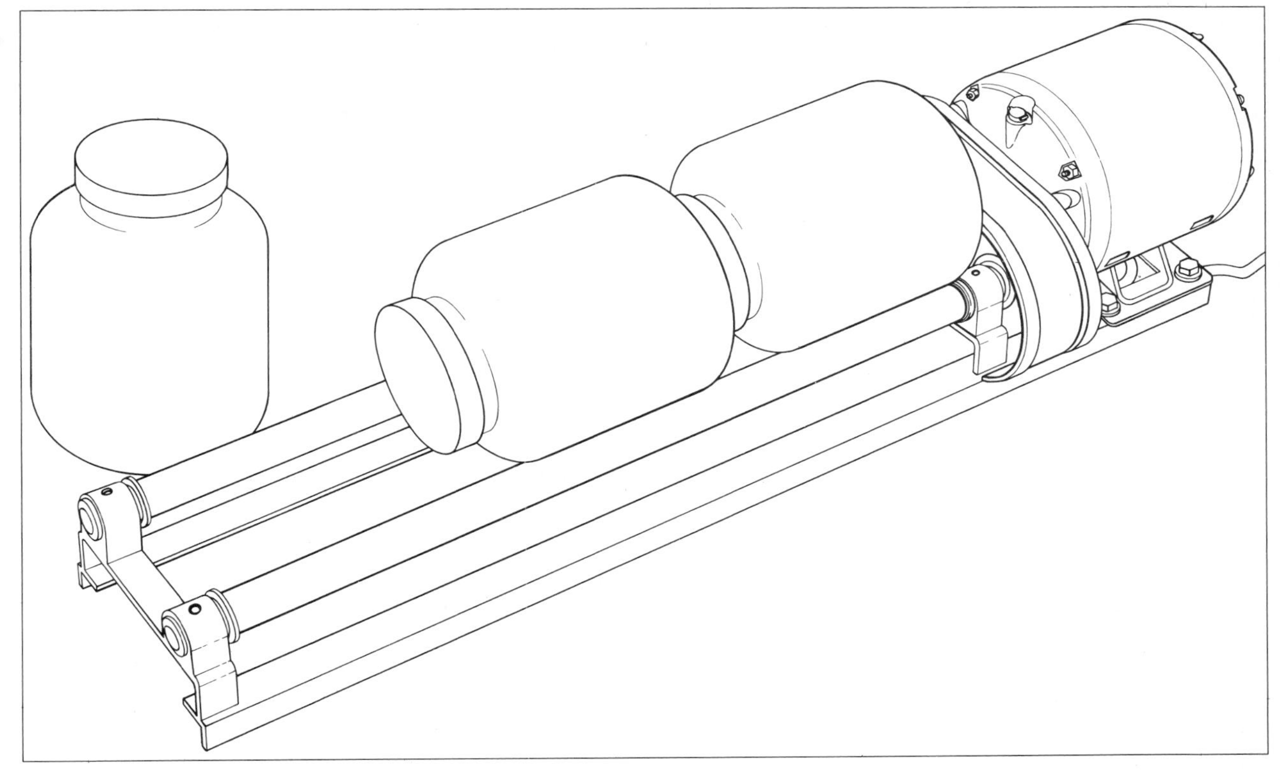

Left: Gem-quality pink tourmaline from Pala, California

alongside the lap. One end of the dop stick is pushed into a hole and the other end, carrying the stone, rests against the lap at an angle. This angle varies according to which hole the dop stick is inserted into. The same task may be performed by a *faceting head*, which carries a rotatable dop arm with a protractor attachment for facet-angle settings.

Finally, a *transfer jig* is needed, which permits the stone to be transferred from one dop stick to another, being reversed in the process, without becoming misaligned.

Having selected the rough, taking a careful note of the position of any flaws, the lapidary grinds it into the approximate shape desired for the finished stone. The part of the stone that is to become the table facet—the central top facet—is ground sufficiently flat to allow contact with the flat end of the dop stick. The preform is then attached to the dop stick, which is set at the appropriate angle and the girdle—the 'waist' of the final gem—is ground until it is perfectly symmetrical and of the correct depth. The dop stick is then set to the correct angle for the pavilion, or lower, facets. A series of flat faces is now rough-ground at this new angle, meeting at a point in the centre of the stone. When all the pavilion facets have been established at the correct angles, they are fine-ground and polished in the same sequence. Then the dop stick is removed from the faceting unit and placed in the transfer jig, and a second dop stick is attached to the newly faceted pavilion. Rough grinding, fine grinding and polishing of the crown, or upper part of the stone, are carried out in a similar manner to that just described. When finished the stone is washed in a suitable cleansing spirit.

Detailed faceting instructions in step-by-step stages for the standard brilliant cut and the trap cut will be found in the main section on faceting.

TUMBLING

Virtually any hard substance, if handled correctly, can be tumbled, but there are some materials that do not merit the time that would have to be spent on them.

All rocks and minerals fall into one of three categories. *Igneous* rocks are formed, as the name implies, by the action of heat, and are of volcanic origin. Some volcanic magma, when cooled very quickly, produces a glass-like substance known as obsidian, which is reasonably hard and quite suitable for all forms of cutting and polishing. Magma that does not reach the earth's surface but is cooled very slowly underground, can form large *pegmatite* crystals. Most of these are suitable for all types of polishing. Among these, the quartz family and the feldspars, together with beryl, topaz and tourmaline form a large proportion of the world's commercial gemstones.

Rocks of the second category are of *sedimentary* origin. They are composed of fragments of rock material that have accumulated over the course of many thousands of years as the result of erosion by wind, ice or water and subsequent deposition in rivers, lakes and seas. Such fragments eventually build up into layers many feet thick, and the weight of overlying material compresses the lower parts into very hard masses. Alternatively, the particles may be cemented together by such substances as calcium carbonate. The majority of sedimentary rocks are not suitable for polishing. However, water has the ability to dissolve very small amounts of silica, and this very weak solution is deposited in cavities in chalk and limestone. Upon hardening it becomes flint and together with its companion chert is very good tumbling material.

Lastly, because both the igneous and sedimentary rocks are sometimes subjected to further volcanic activity or geological upheaval, and therefore tremendous heat and pressure, alteration can take place. Rocks changed in such a way are called *metamorphic*. Quartzite and marble are examples of this type of rock, and they are capable of receiving a substantial degree of polish.

This brief incursion into geology gives some idea of the types of rock from which good tumbling material may be obtained. Those substances that may reasonably be expected to be found in most rock and mineral shops are agate, beryl, chalcedony, all varieties of crystalline quartz, the jades, sodalite, lapis lazuli, marble, serpentine, turquoise, tourmaline, peridot, garnets, and so on.

To achieve the best results from what may be the beginner's first batch of tumbled stones, there are some precautions to be taken before tumbling starts. The stones selected for the barrel must be as free from flaws as possible; otherwise, in the later stages of tumbling, fragments may break away, producing sharp edges that could quickly spoil the rest of the stones. In addition, stones being tumbled together should be of roughly the same hardness. It would be sensible to observe this precaution, at least in the first batch; details will be given later on how to tumble together stones with considerably different hardnesses.

The tumbling technique

If the reader does not possess a tumbler, but wishes to make or buy one, he should turn to the end of this chapter where the relevant details will be found.

In addition to the tumbling machine, silicon carbide grits will be required in the following sizes:

60 or 80 grit for coarse grinding;
220 grit for intermediate grinding;
400 grit for fine grinding.

It is desirable, too, to have 800 or 1,000 grit for prepolishing, and cerium oxide for polishing. There are one or two optional items to which reference will be made where necessary.

Filling the barrel. Rough stones should not be above a certain maximum size, determined by the size of the tumbler barrel. Minimum size is not so important, except that stones that are half an inch or less across could be totally ground away by the end of the process. For barrels with a water capacity of 3 pints or less it is recommended that the stones should not exceed 1 inch in the largest direction. For barrels from 3 pints to 1 gallon in capacity, the maximum size should not be much larger than 1½ inches; for barrels up to 2 gallons the size should not greatly exceed 2 inches. The sizes quoted should produce well-polished stones without undue trouble in a

Above: A collection of some of the most popular stones for gem-cutting.

1, 2. Turquoise
3, 4, 5. Amethyst
6. Rhodonite
7. Rhodochrosite
8. Apache tear (obsidian)
9. Blue goldstone (man-made)
10. Jasper
11. Agate
12. Banded agate
13. Red tiger's eye
14. Aventurine quartz
15. Citrine
16. Turritella agate
17. Jasper
18. Crazy-lace agate
19. Fossilized wood
20. Blue tiger's eye
21. Yellow tiger's eye
22. Moonstone

23. Obsidian
24. Flame agate
25. Lapis lazuli
26. Carnelian
27. Eye agate
28. Opalized wood
29. Rose quartz
30. Moss opal
31. Moonstone
32. Chalcedony with included pyrite
33. Hematite
34. Obsidian
35. Amazonite
36. Chalcedony with included pyrite
37. Flame agate
38. Mother of pearl
39. Abalone
40. Rose quartz
41, 42. Moss opal
43. Banded agate
44. Blue lace agate
45. Carnelian
46. Banded agate

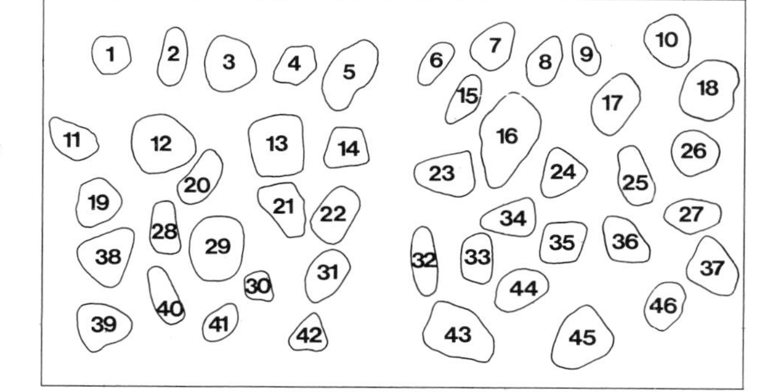

reasonable period of time. The description that follows applies to a tumbler with a 2-pint capacity, 4 inches in diameter by 6 inches long.

Place approximately $2\frac{1}{2}$ pounds of rough stone, comprising varieties of crystalline quartz, agate and carnelian into the barrel. The stones should occupy approximately three-quarters of the barrel but not more. Cover them with water and place two rounded tablespoons of 80-grit silicon carbide on top. Replace the lid of the barrel tightly to ensure that there is no water leakage.

Stage 1 : coarse grinding. Place the barrel on the tumbler, and switch on. The noise from the tumbler may be disturbing, and it should therefore be kept in the garden shed or garage.

After forty-eight hours, remove the barrel and look inside it. Do not be disappointed if very little appears to have happened. Upon closer inspection, you will see that some of the sharp edges of the stones have become slightly rounded. You may also safely assume that most of the grit has been ground too fine to be of much further use; put a further two tablespoons of grit in the barrel, closing the lid tightly. Tumbling may now continue for a further three or four days.

The next inspection should be a little more thorough. Put the barrel contents carefully into a plastic washing-up bowl or similar container and wash the stones. You may notice that there appear to be fewer and smaller stones than you began with and that the grit and the small particles of abraded stone now comprise a thick grey sludge. This sludge may be thrown away, but not down the sink or any other household drain, as it will very quickly cause an obstruction. You may safely dispose of it by burying it in the garden.

Coarse grinding should continue for at least a further week, but certainly until the stones are perfectly round and smooth, free from surface pitting or irregularities. Before going on to a finer grade of grit, you should examine each stone for possible fractures, and discard any doubtful material.

Stage 2 : medium grinding. When these criteria have been met both the stones and the barrel must be cleaned thoroughly to remove all particles of coarse grit, especially those trapped in and around the lid. The stones are once again placed in the barrel and just covered with water. The barrel is charged with about two rounded tablespoons of 220-grade grit, sealed and allowed to tumble for a week. The stones should now appear to be perfectly smooth, and when dried should have a plain matt surface of fine texture. If they do not, they should be further tumbled with 220 grit. When the stones are smooth, the sludge must be disposed of and the barrel washed out.

Stage 3 : fine grinding. When you once again place the stones in the barrel, you will notice that their bulk has been reduced by fifteen to twenty per cent. This loss may be made up at this stage with small round pebbles purchased specifically for this purpose or collected from a beach or river bed. It is important that these stones are very smooth and free from surface blemishes. Remember, this filler material can be used over and over again.

Fill the barrel with stones to approximately the original level, and add water together with two tablespoons of 400 grit. Tumbling should continue for approximately a week to ten days, with one intermediate change of grit. The stones' appearance at the end of this stage should be smooth and rounded, and they should have a very fine matt surface when dry. Wash out as before.

Stage 4: prepolishing. This time use the prepolishing grit. Now the greatest care must be taken to ensure that no grit is carried over from any of the previous stages and that all the stones are free from fractures and flaws, which could carry old grit. Prepolishing should continue without a change of grit for approximately ten days. A small quantity of a cellulose wallpaper paste may be added to the water to thicken up the mixture, so that the stones are not tossed around inside the barrel as violently as they were previously. At the end of this stage the stones should have a dully gleaming surface. Wash the barrel out and then, just to be on the safe side, wash everything again in hot soapy water.

Stage 5: polishing. Reload the barrel, add water and approximately one heaped tablespoon of cerium oxide. It is recommended that some of the stones previously added to make up the bulk are removed and replaced with pieces of broken-up expanded polystyrene tiles or similar material; even half-inch sections of old leather dog-lead have been used successfully. This small quantity of softer material serves two purposes: it helps to stop stone damage, and it assists in the actual polishing by trapping the minute particles of polishing compound in its pores, so acting like the hard-felt wheel used in polishing cabochons. The stones should be allowed to tumble for about a week, after which time it will be fairly obvious whether they are properly polished or need a few more days.

Stage 6: washing. If the stones are ready, wash them and clean the barrel; return the stones to the barrel, add water and about 10 ml of liquid soap and set the tumbler once again in motion for about twentyfour hours. This final washing process is carried out to free the last traces of polishing compound, which can be very persistent, from tiny remaining surface imperfections.

Tumbling hard and soft material

When a wide variety of stones is tumbled to make cheap gift jewellery the batch is likely to include a few pieces of material considerably softer than the bulk. For example, in a batch of stones consisting mainly of agates, crystalline quartz and jade, it may be desired to tumble some jet, Blue John and mother-of-pearl or abalone. This may be carried out quite easily if the following procedure is observed:

1. The barrel should be charged as normal using a proportion of bulk make-up stones equal in volume to the softer material to be tumbled;

Below: Unpolished preforms and their tumbled counterparts (bottom). The top group have been roughly shaped and are ready for tumbling while the bottom group have just been taken from the final polishing stage

this softer material should not yet be placed in the barrel. Start the rough grind as normal.

2. When the rough-grinding stage is well advanced, interrupt the tumbling and substitute the softer stones for the bulk material. Continue tumbling as normal to the end of stage 1. Inspect the new material during the medium grind of stage 2 to ensure that abrasion of the softer material is not carried too far.

3. When the new material reaches the desired shape and smoothness it should be removed from the barrel. You will find that some material will be ready before the rest; therefore, to avoid excessive reduction in size of those pieces, they should be removed as soon as possible after becoming smooth, even though the remainder of the soft material has not yet reached this stage. As pieces are removed, bulk make-up material has to be added again. But the bulk make-up stones originally removed may be too rough to be included at this stage, and therefore either polished bulking stones from a previous batch, or small pieces of wood, plastic or similar material should be used.

4. When most of the stones have reached the prepolished stage, the softer material should once again be added, and the bulk make-up removed, so that the barrel level stays the same. Tumbling should now continue exactly as described previously, except that it may be necessary to prolong the prepolish stage for an extra day or two to complete the finish on the softer material. The polishing stage is as before. Thorough washing is essential, especially for abalone, which has numerous small holes that provide excellent traps for the grits and polish.

Using preformed shapes

The operator must possess a diamond trim saw or a lapidary grinding machine or both if he wishes to make a contribution to the shape of the finished tumbled stone. This method not only produces stones of the desired shape for jewellery, but significantly reduces the tumbling time.

Material of tumble or cabochon quality may be sliced with the trim saw and then cut into a variety of different shapes, such as stars, crosses, rectangles and so on. When choosing the rough material it is important to remember that a typical trim-saw blade, six inches in diameter, can cut to a depth of not more than about two inches, because of the bulky shaft supports at its centre. The rough should then be less than about two inches in diameter, or it will need to be turned, and cut again.

When the desired shapes have been obtained, whether with the saw alone or with the grinding machine as well, they are polished in a tumbler barrel by themselves or as part of a normal tumble batch. If they are tumbled by themselves remember that for best results the barrel should be about three-quarters filled. If they are to be part of a normal tumble batch, in which the remainder of the stones are in a completely rough state, it is a good idea not to introduce the shaped stones until the major rough grinding has occurred; not only does this spare them excessive grinding, but it also allows them, rather than rounded pebbles, to be used as the bulk make-up material.

If preforming is by the use of the saw alone, a raised lip will be produced on one half of the sawn stone as the blade breaks through. This problem may be lessened if the pressure of the stone against the blade is reduced to a minimum as the cut nears completion. Another problem frequently encountered when using the diamond saw is the deep circular cutting marks produced on the cut surfaces. These may be due to excessive pressure of the stone against the blade, which tends to make the blade run out of true; alternatively, the blade itself may not be running exactly true on its shaft. When a grinding machine is available it is usually a matter of a few moments' work to remove the break-through lip, but it may take quite a lot longer to remove the cutting marks from large areas of flat surface.

Large quantities of cabochons may be produced simply by shaping them roughly and then tumble-finishing them; this is a method often used commercially. Accurately calibrated stones are difficult to produce by tumbling.

Providing that excessively large break-through lips, surface irregularities and saw grooves have been removed, tumbling may commence with the 220 grit and should continue through all subsequent stages as for a normal batch of stones. The advantages of this method are mainly in saving time; since there are no very rough areas to coarse-grind, tumbling should be completed ten to fourteen days sooner than with an ordinary tumble batch. There is also the added saving of a few ounces of coarse grit, less wear and tear on the tumbler and a slight reduction in the electricity bill.

Care should be taken to ensure that the sawn material has no obvious faults or fractures, and that the shapes are not cut excessively thin, which would result in their breaking-up during tumbling. However, some materials such as jadeite and nephrite are very tough; delicately shaped pieces of these materials are far less likely to suffer damage than, say, obsidian, which is rather brittle, and amazonite, which has at least one direction of easy cleavage.

Lastly, remember that, at the end of any stage of the tumbling process, if the stones are not ready for the next stage, even prolonged further tumbling will be unsuccessful. It is therefore far better near the end of each stage to be over-cautious and tumble for too long than to be impatient and risk spoiling the stones. A useful tip that will help to determine the point at which a tumbled stone is ready for the polishing stage is to smear a small quantity of cerium oxide on to a piece of hard damp felt and rub the stone vigorously for a minute or so. A good polish should be obtained; if, after this check, the surface appears to be still slightly pitted the prepolish stage should be repeated or the whole batch returned to the fine-grinding stage.

Points to watch

Now that the various stages of the tumbling process have been described, certain aspects may require clarification—sizes of carborundum grit, tumbling speeds and barrel shapes.

To obtain good results from the tumbler, it is important that the barrel rotates at a speed that will allow the stones to grind gently. If it turns too fast, the stones will cascade violently, or even be pressed to the wall of the barrel by centrifugal force without any grinding action taking place. If the barrel turns too slowly, the stones will remain in the lower part of the barrel, sliding over each other and becoming flattened. In calculating the barrel's speed to obtain the best tumbling action, several factors should be taken into consideration: barrel diameter, drive-motor speed and the diameter of the bearing shafts. If the unit is home-made the speed should be calculated before final assembly of the components. Once the speed of the electric motor is known it is just a matter of adjusting ratios to produce a barrel speed of approximately 45–50 rpm when using a 4-inch barrel. Further information on tumbling speeds will be found in the section on making a tumbler. It can be expected that a manufactured tumbler's speed is appropriate to the size of the barrel provided, and for this reason different-sized barrels should not be used without re-checking the tumbling speed. A quick method of carrying out this check is to stick a small piece of coloured adhesive tape to the outside of the barrel and count the number of rotations in one minute.

There are two basic types of barrel cross-section, round and hexagonal. Both types have their devotees but there is little difference in the final result. It is said that the hexagonal barrel's flat internal surfaces help to carry the stones to a higher point before they fall gently away. This shape of barrel requires some form of circular rubber 'tyre' round each end to allow smooth rotation on the drive shafts.

Plastic or metal household containers have been quite easily adapted to make cheap and effective barrels. Before large quantities of manufactured tumblers came onto the market, tumbler barrels were made from empty paint tins, glass jars and round food containers. It is necessary to line metal and glass barrels with rubber sheeting or some other protective material; quite often sections cut from a car-tyre inner tube have been used for this purpose. Some quite satisfactory stones have even been produced using an old car tyre as a barrel.

When a metal barrel is used, the gas pressure inside it gradually increases. This is probably due to a reaction between the contents and the barrel. This pressure should be released twenty-four hours after beginning each stage and at least once more in each stage. Failure to observe this precaution could easily result in the rupture of the barrel, which would not only make a considerable mess, but also allow some of the loose grits to find their way into the electric motor, causing irreparable damage.

Some barrels are made of rigid plastic tube with moulded semi-rigid plastic end-caps. These caps are completely airtight and hence difficult to put on (since an extra quantity of air has to be compressed into the barrel) or to take off (since a partial vacuum has to be overcome as the cap is slid off). An easy way of overcoming this problem is to drill a small hole through each

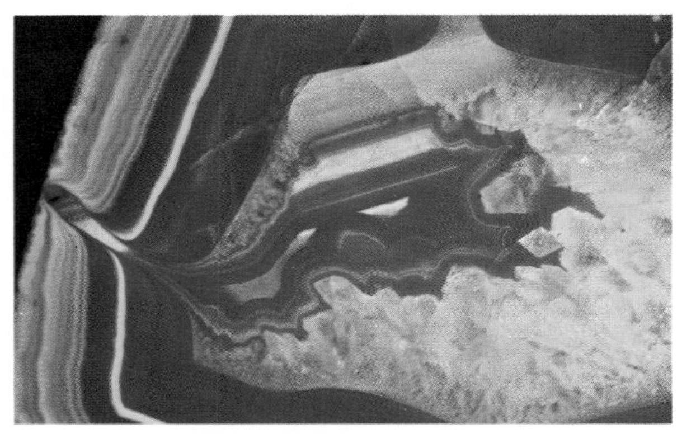

Below left: An agate slice, its banded structure enhanced by polishing. The red areas are carnelian; crystals of translucent chalcedony can also be seen

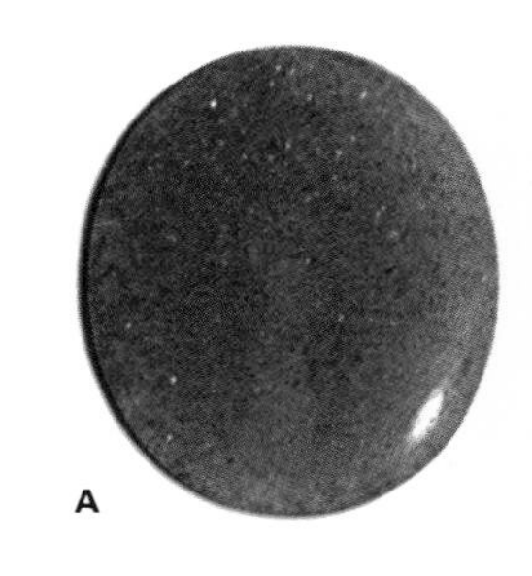

A

B

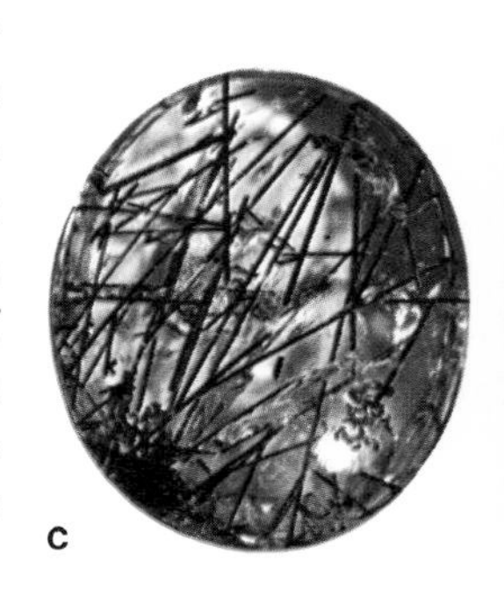

C

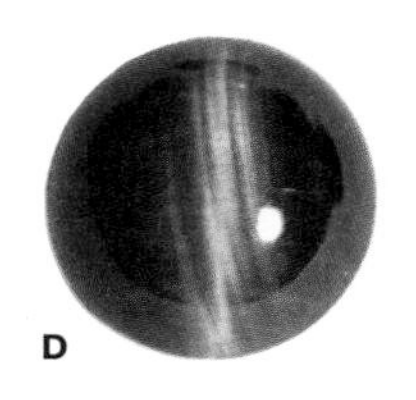

D

Below: Cabochons of four varieties of quartz. From the top: chrysoprase; agate; clear quartz with tourmaline inclusions; and tiger's eye

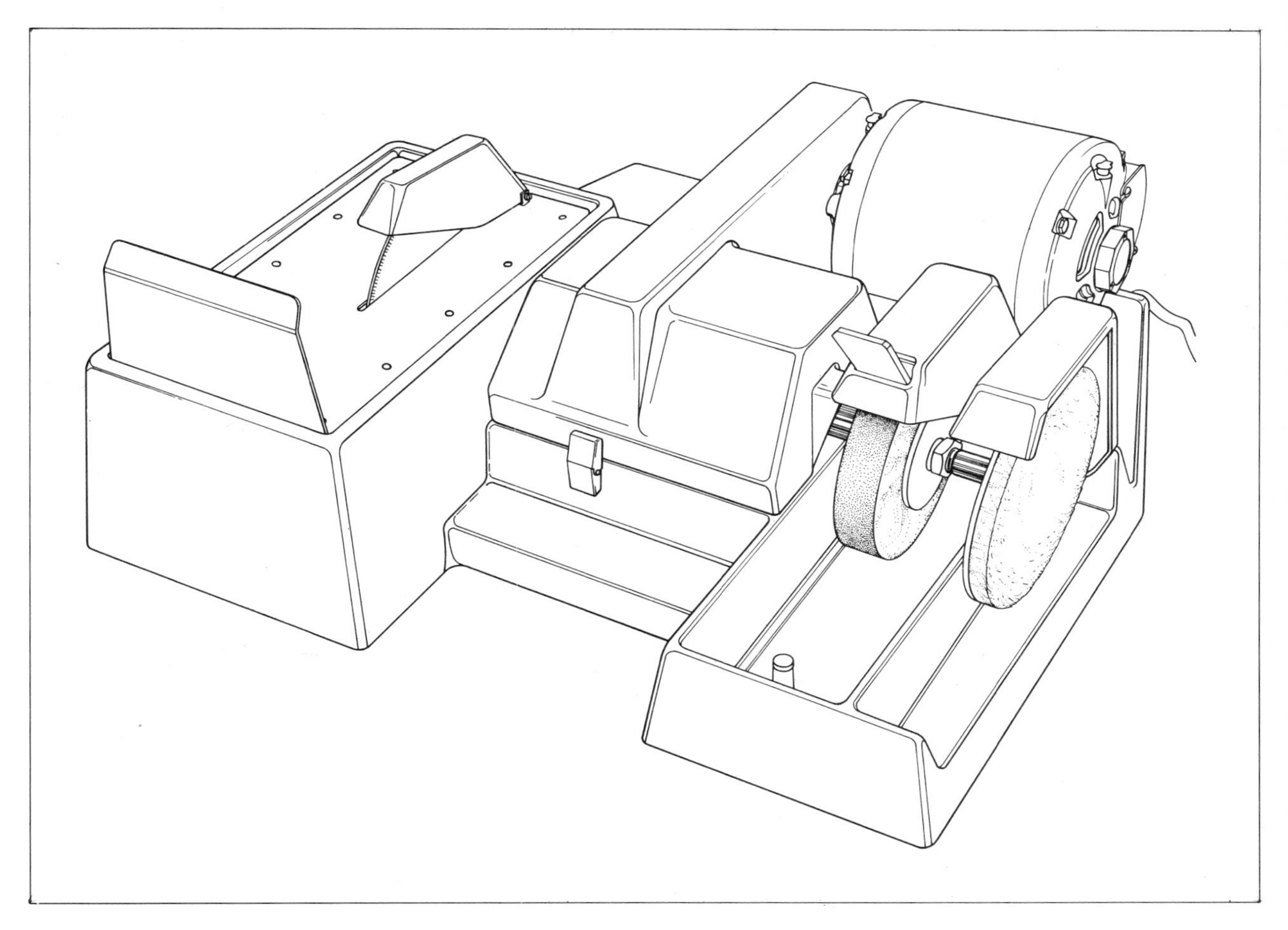

Right: A commercially made multi-purpose grinding and polishing unit. It combines facilities for sawing, grinding and polishing on a common axis. A range of turning speeds is available, provided by interchangeable pulleys

cap. During tumbling the hole is sealed with a small piece of adhesive waterproof tape. This is removed while the cap is put on and taken off. If the tape is placed off-centre, it provides a convenient marker for counting the barrel's revolutions.

Why is it necessary to replace the grit during each tumbling stage? In principle tumbling could be completely successful if carried out with a single large load of coarse grit and continued long enough. But it would take even longer than the procedure recommended here. As tumbling proceeds grit particles are worn smoother and smaller, so that their effectiveness is reduced. Renewal of grit speeds the abrading process. It is not necessary to renew the cerium oxide during the polishing process because the reduction of particle size is in this case beneficial.

CUTTING A CABOCHON

Cabochons may be cut from almost any hard material, from a very close-grained wood such as lignum vitae or ebony, through glass and plastics to virtually all gem materials. There are many methods to choose from when cutting a cabochon, and the more important of them will be discussed in this section. Some require very little equipment, which can often be found in craft shops as hand-polishing kits. Usually these contain an assortment of different grades of carborundum grits, packed in small glass vials, pieces of coarse and fine abrasive paper and a polishing compound, often cerium oxide. Usually the instructions in these kits recommend that the pieces of rough stone be worked until smooth with the abrasive paper, and then with a piece of hard felt or similar material with the fine powders provided. A variation of this method is one that is used for polishing flat surfaces such as those of agate slices, or the rims of geodes. The slice is ground with a range of grits and polished on a sheet of plate glass until the desired finish is obtained.

Almost all other ways of getting the same result employ some form of mechanical process. This may involve a straightforward horizontal or vertical grinding machine with attachments to permit sanding and polishing, which is the method that will be described in greatest detail in the text; or specialized cabochon-cutting accessories may be attached to lapidary machines. One important type of attachment consists of specially prepared diamond-impregnated copper cups, which replace or are combined with the normal lap unit. Cabochons may also be cut with a specially prepared vertical grinding wheel having a grooved cutting surface of the same curvature as the cabochon it is forming. An alternative to this latter type is a horizontal aluminium or cast-iron lap with specially prepared deep grooves of round cross-section. The stone is held in the groove and ground with loose carborundum grits to produce the desired shape.

Whichever of the above methods is used it is important to remember that certain types of material need handling in special ways. Cutting a cabochon invariably generates a considerable amount of heat, especially during polishing; for this reason it is a good idea to obtain as much information as possible about the physical characteristics of the stone before making a start. One problem stone is opal, which can crack or lose colour from overheating. For this reason some people prefer to cut them by hand rather than by one of the mechanical methods, even though these are quicker.

The following list of equipment may be considered as the minimum amount necessary to produce a good-quality stone suitable for mounting:

a horizontal or vertical combination lapidary grinding machine, complete with coarse (80-grit) and fine (220-grit) soft-bond lapidary wheels;
a wooden-backed rubber pad for use with wet-and dry paper or emery cloth;
a hard-felt wheel;
dop sticks;
dop wax;
cerium oxide;
an adequate water supply for lubrication.

Some types of multi-purpose machine are fitted with a water reservoir; where this is lacking it will be found that an adequate supply of water may be obtained by using a dish-washing detergent container or something similar. Whichever is used it is of prime importance that the water is scrupulously clean and free of grit contamination, as even one small particle of silicon carbide can wreak havoc during the polishing stage. Tap water will usually be quite adequate.

Selecting rough material

When cutting a cabochon for the first time, it is wise to choose a material that is reasonably hard, easily obtained and inexpensive. Such a stone would probably be one of the massive varieties of quartz, such as agate or carnelian, bloodstone or jasper. For the more advanced worker there are certain gem minerals whose inclusions or physical make-up produce special optical effects when the stone is cut in certain ways. The first of these effects to be discussed is *asterism*. As the name implies, material of this type will show a starlike pattern of reflected light when properly orientated. In the majority of cases the effect is caused by the reflection of light from large quantities of inclusions, usually taking the form of needles of rutile (titanium dioxide). Sometimes the inclusions take the form of minute hollow canals left behind when rutile needles have been dissolved away by geological action. They are orientated in certain specific crystallographic directions; in the case of corundum, beryl and quartz these are the three directions parallel to the lateral axes. Some spinels and garnets also show stars, but since these are both members of

Left: A good example of the labradorite variety of feldspar (top), displaying a fine play of colour. Beneath it are two cat's-eye stones, cut en cabochon to give the band of light seen from end to end. This effect is produced when light is reflected from tiny inclusions within the stone

the cubic crystal system, the needles are parallel to the faces of the dodecahedron. (It must be remembered that gem material frequently does not outwardly display the geometric forms characteristic of its crystal system.) In the case of garnets, the star may be seen either by reflected light (*epiasterism*) or through the stone by transmitted light (*diasterism*).

In order that the rough material may be orientated to produce a properly positioned star, the inclusions that cause it must lie parallel to the base of the cabochon. Fortunately, stones in the hexagonal system always have these inclusions at right angles to the 'vertical' or c-axis. It is then simply a matter of aligning the section of crystal correctly on the dop stick before cutting commences. If the material is massive or in irregularly shaped pieces, orientation is a little more difficult, but usually a definite sheen may be seen when the observer looks along the c-axis. When the cabochon has been dopped and preformed the exact position of the star may be determined by moistening the surface of the stone and holding it under a fairly strong, single-point light source.

Inclusions are also the cause of *chatoyancy*, but unlike those in star stones they lie in a single direction, usually parallel to the vertical axis. Stones cut from rough of this type show 'cat's-eye' effects, hence the name chatoyancy. The most important cat's-eye stone is cymophane, a variety of chrysoberyl; its inclusions are usually short hollow canals. Chatoyancy is also best seen under a point light-source and will appear as a band of light crossing the cabochon. An oval cabochon should be cut in such a way that the band of the cat's-eye lies parallel to its length. Other stones displaying chatoyancy are tourmaline, apatite, diopside, some varieties of quartz, including tiger's eye, and scapolite.

The remaining important optical effects are to be seen in certain members of the feldspar group; in some instances the effect is named after the stone in which it is most often seen. *Labradorescence* is probably the most important, and is seen mainly in the labradorite variety of plagioclase feldspar; it is a play of colour, usually greens and blues, and occasionally purple, gold and red against a background colour of dark grey. The effect is due to the interference of light from the thin layers in which labradorite occurs due to polysynthetic twinning, and from platelets of magnetite as inclusions. Some feldspar, especially oligoclase, has minute inclusions of hematite or goethite arranged parallel to certain crystallographic directions, producing flashes of red and orange; this effect is usually termed *aventurescence*. However, a variety of massive quartz with inclusions of small plates of fuchsite mica is called aventurine quartz though it is of a green body-colour. The silvery or bluish sheen seen in the moonstone variety of orthoclase feldspar is often described as *adularescence*. In this case the sheen is due to alternating layers of orthoclase and albite feldspars, which reflect back light rays that interfere with each other.

In all cases of optical effects seen in feldspars,

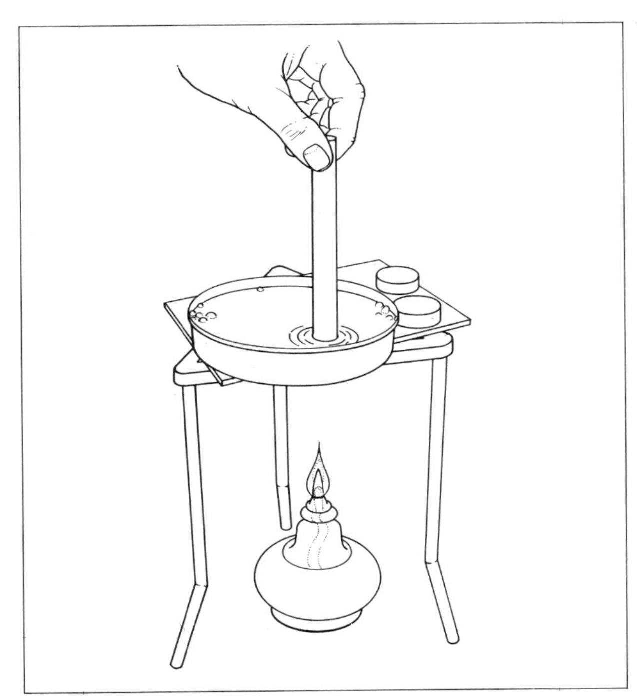

Right: Tipping a dop stick with wax

Far right: Dop sticks, tipped with wax and ready for a stone. Differently sized dop sticks are used for different sizes of rough material

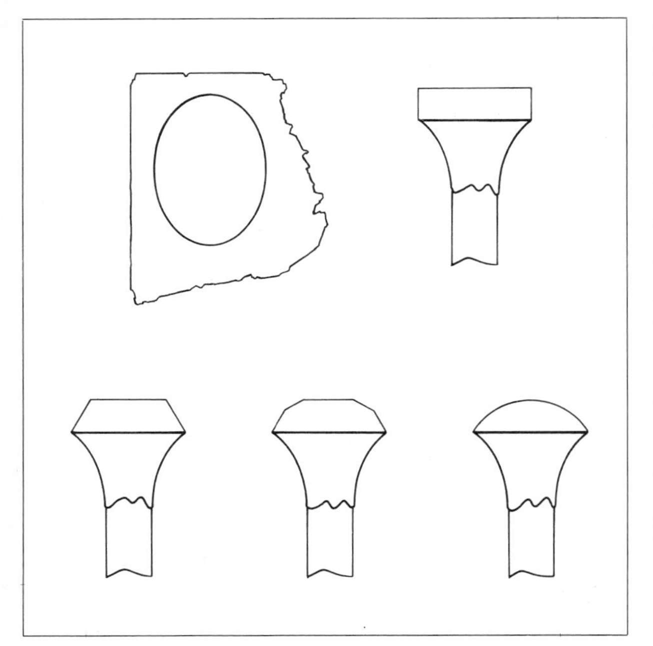

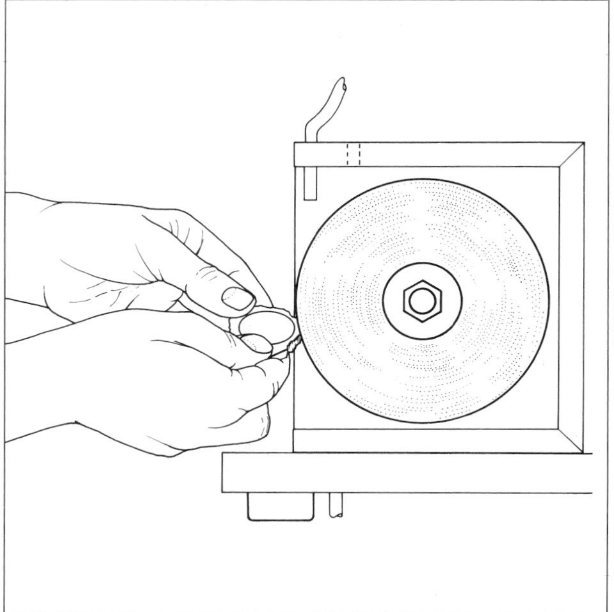

Far left: Five stages in the production of a cabochon. First, the outline of the cabochon is drawn on the slice of rock with an aluminium pencil. The rough is trimmed down to the outlined shape and then dopped. The sketches show how subsequent grinding produces the final shape

Left: The ideal position for the hands and the piece of rough while the outline is being ground to shape. The stone is more manageable at this stage without the dop stick

it may be safely assumed that the inclusions, twinning or layering are parallel to certain crystal directions which are, fortunately, also directions of easy cleavage. Correct orientation prior to cutting may be made by viewing the stone under a strong single-point light source, with the eye as nearly as possible over the top of the stone.

Many other materials suitable for cabochons have an appearance that makes it desirable to orientate them in special ways. In the majority of cases it is zoning or banding, as in rhodochrosite or malachite. In these cases details will be found at the end of this chapter. In the Mineral Kingdom section information on orientation and cutting is given. In the following description of cutting and polishing a cabochon it will be assumed that the stone is a piece of bloodstone.

Trimming and dopping

The first step is to reduce what may be quite a large piece of rough to a more convenient size. If a diamond-bladed trim saw is available the rough may be sliced and trimmed until the piece remaining is slightly larger than that needed for the finished stone. Any desirable markings should be correctly positioned, usually centrally. If there is no trim saw available the rough will have to be preformed by grinding with a coarse lapidary wheel until the approximate required shape is obtained. Usually this is done by holding the stone against the grinding wheel with the fingers, but if it is very small the stone may be temporarily dopped.

Cabochon profiles may vary considerably, and it is important to decide just what is required before beginning cutting. Opal often occurs in thin seams, in which case a high-domed stone cannot be cut from it. A garnet of deep colour, should be cut quite shallow or as a hollow cabochon. Many materials, on the other hand, especially those with certain optical effects, require a fairly steep-sided stone—for example, tiger's eye or moonstone.

A stone can be cut to a given size by marking the outline on to the slice of rock. Templates are available for the convenience of the cutter, with accurate outlines of virtually any desired shape, ranging from ovals through squares and rectangles to hearts, crosses and so on.

When the slice of rock from which the cabochon will eventually be formed has been cut, the desired outline in the template is positioned over that part that is to be used for the stone. The outline may now be drawn on to the slice with an aluminium pencil. If ink is used, there is always the possibility of blotches or runs, whereas the aluminium marker's line is fine and clear, requires no drying time and will not normally rub off. It is now just a matter of making sure, when grinding, that the stone's shape conforms to this outline. The aluminium pencil should be kept quite sharp, since a thick line may confuse the cutter and could result in a spoilt stone.

This is perhaps a good point at which to describe dopping in some detail. Whether home-made or purchased, dop wax consists basically of a mixture of shellac and sealing wax, although

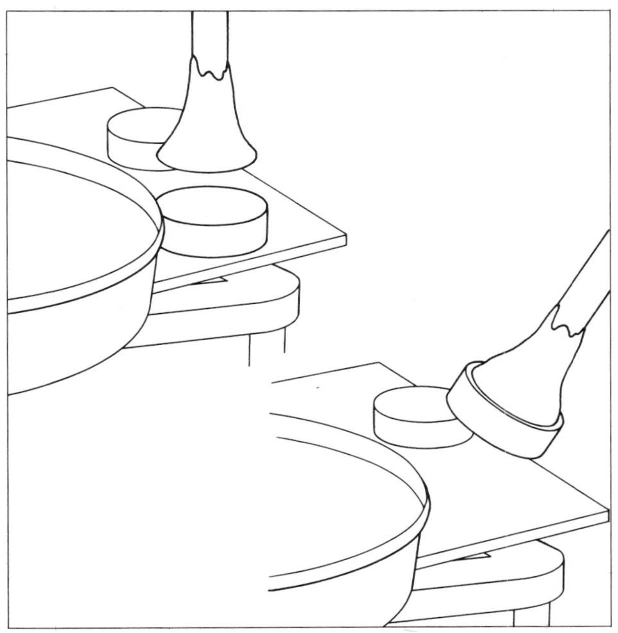

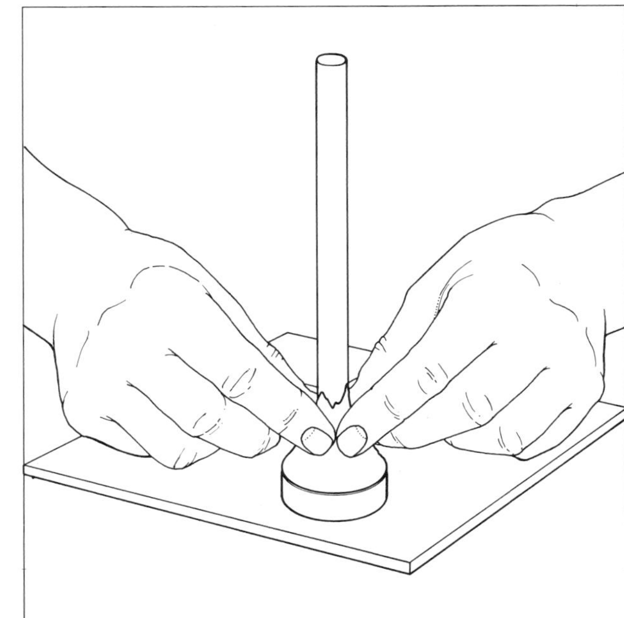

Right: Fixing the stone to the dop stick. When the stone is hot enough to melt the wax, the waxed dop is lowered onto the stone. The combined unit is then lifted off to cool

Far right: Making sure that the dop stick and the stone are correctly aligned before the wax sets hard

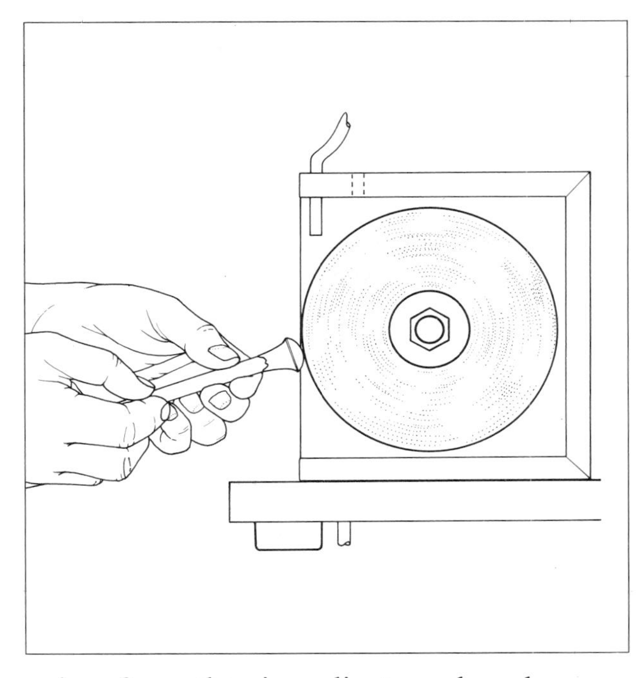

Right: The correct way to grind the stone. Notice the position of the hands allowing axial rotation even though the stone is held quite firmly

quite often other ingredients such as beeswax and clay are added. Certain stones are susceptible to damage by heating, and for these a lower-temperature dop wax should be used; it is suggested that pure shellac be used rather than risk damaging what may be quite a rare stone. Apatite, fluorite, opal and spodumene are typical examples of heat-sensitive stones. Details of heat-sensitivity are given in the relevant entries in The Mineral Kingdom.

When choosing a dop stick, which for this purpose may be just a length of wooden dowel, care should be taken to select a rod thick enough to give good support to the stone being worked. To dop the stone correctly the following procedure is recommended. Firstly, the dop wax should be placed in a small metal dish, which is placed over a source of heat. A small spirit lamp is ideal for this purpose, the dish being supported on a laboratory stand. Having melted the wax, take the dop stick and dip one end into it, to a depth of approximately three-quarters of an inch. Ensure that there is sufficient wax on the stick to cover the base area of the stone. It is a good plan at this stage to twirl the stick through the spirit-lamp flame in order to make the wax round and smooth. It should then be stood upright on the wax, on a cool flat surface such as a small piece of plate glass. It is a good idea to prepare several dop sticks of different sizes in this manner to save time at a later stage.

The stone, which has already had a good deal of shaping and should have a flat surface on the least desirable portion, is now placed on a piece of sheet metal with its flat face uppermost. This is in turn placed over the spirit lamp alongside the dish of dop wax. The stone thus becomes gently heated, which is far better than holding the stone directly into the flame with a pair of pliers, at the risk of damaging the stone by sudden local heating.

As soon as the stone is hot enough, touching its surface with the wax on the dop stick will cause the wax to melt and stick to the stone. At this stage the stone may be picked up gently, either with a pair of pliers or with the dop stick. Taking care not to burn the fingers, manoeuvre the stone into a central position on the dop stick. The wax must form a support, covering at least two-thirds of the cabochon. When it is certain that the stone has been placed symmetrically the wax and stone may be allowed to cool. A good way to ensure that the wax has set properly is to immerse it for a few seconds in cold, but not icy, water. Forming and polishing the cabochon may now start, with the 80-grit grinding wheel on the machine.

Grinding the stone

It is important when cutting cabochons on the grinding wheels to maintain a constant and adequate supply of lubricant, usually water. The surface of an inadequately lubricated wheel will soon become clogged with debris, hindering the cutting process and possibly causing uneven wear on the wheel. It is also advisable to maintain strict cleanliness when handling the hard-felt wheel used for polishing, as the slightest amount of abrasive grit could easily ruin the final polish of an otherwise perfectly cut stone.

If the cutter normally produces only one or two sizes to suit specific items of home-made jewellery, he can make a template from a thin sheet of flexible plastic, such as the flat side of an old food-storage container.

Stage 1: coarse grinding. To obtain the best results from grinding wheels it is recommended that they run with a circumferential speed of approximately 4,000 feet per minute. This means that if a horizontal grinding wheel six inches in diameter is being used, and most of the grinding is carried out near its edge, it should rotate at approximately 2,500 rpm. The stone should be brought into contact with the grinding wheel in such a way that the dop stick is almost horizontal. It is recommended that, when bringing any stone into contact with a revolving grinding wheel, the stone be moved continually over the surface of the wheel, preferably in gentle arcs towards the edge; this continual movement ensures that the surface of the wheel remains relatively flat. If this is not done the grinding wheel will very soon have deep grooves or scores in it, which not only shorten its life but make it very difficult to grind flat surfaces.

While in contact with the wheel, the dop stick should be continually rotated about its own axis. This ensures that grinding produces a smooth, even outline around the stone. If care is taken when rotating the dop stick, the hand will rise and fall automatically with the contour of the stone.

When this initial grinding has been carried out satisfactorily, the angle of the dop stick must be changed from horizontal to approximately 30°. Grinding should continue exactly as above until the stone has a bevelled edge to approximately three-quarters of the depth of the stone. It is important that the stone be rotated evenly at a fairly constant angle; otherwise the finished stone will have an irregular setting edge. When this edge is completed the angle must again be altered, this time to approximately 65°–70°, in order to give the stone a roughly domed appearance, though there may still be an area in the centre that has not yet been ground.

Once these major bevels have been produced, it is a good plan to grind quickly over the edges of the bevels to produce a more rounded outline. Before going on to the next stage, hold the stone up to the light and make sure that the dome is central. Correction may be more easily carried out now than later.

Stage 2: fine grinding. Remove the coarse-grinding wheel from the machine and substitute the fine wheel. After making sure that the lubricant is flowing freely, bring the stone into contact with the wheel so that grinding commences with the area of the stone just above the setting edge. Pressure on the wheel should be fairly light, just sufficient to grind away irregularities rather than create new flats. The dop stick must be rotated about its own axis as before, but this time, instead of grinding one complete path around the stone, after every two or three complete revolutions the dop stick angle should be increased by about 5° continuously from almost horizontal to almost vertical. When the very top of the stone is being ground, be careful not to keep the stone on the wheel for too long. This could result in a flattened top, which is most undesirable if the stone is to yield a special optical effect.

Fine grinding should continue until such time as a perfectly regular dome of the required proportions has been obtained; this is important, for the subsequent sanding and polishing operations remove insufficient material to alter the shape of the stone.

Stage 3: coarse sanding. The fine-grinding wheel should be replaced by a rubber supporting disc. If this accessory is not obtainable from a local supplier it may be made quite easily. Cut a six-inch disc of marine-ply wooden sheet and drill a hole in the centre the same size as the machine's main shaft. On to this should be cemented a quarter-inch-thick sheet of foam rubber. Rather than try to cut the foam rubber into a disc before cementing to the plywood it is

Above: A selection of hand-cut cabochons and shapes. The more unusual shapes have been cut entirely by hand without a dop stick, unlike the round and oval stones which were dopped to give a more regular outline

a good idea to cement it first, and cut it when set. The sanding discs, which may be purchased or home-made, should be of 320-grit wet-and-dry paper to begin with, followed by 400- to 600-grit. They should be fixed to the backing disc, being held in place by a large washer and a locking nut. Alternatively they may be affixed to the supporting disc using a photographic latex adhesive that will allow easy removal.

Sanding speed may be the same as for grinding, but sometimes it can be beneficial to use a lower one. One way is to use direct drive from the motor, which in most cases runs at approximately 1,400 rpm. This is achieved simply by having pulley wheels of the same size on both the motor and the lapidary unit. The advantages of slower speeds are apparent when relatively soft or small stones are being sanded; fast speeds can remove material too quickly in both cases.

Sanding may be carried out with virtually the same movements of the dop stick and stone as in grinding. This operation should continue until the stone is absolutely smooth and perfectly regular in shape. Remember that fresh sanding discs are very abrasive; it is advisable to hold a piece of agate against a new disc for a few seconds to reduce the harshness of the surface. Pay special attention to the top of the cabochon, as this area is the part most often neglected. It is a good idea to inspect the stone frequently, preferably with a watch-maker's eye glass, to make sure that there are no areas left unsanded.

Stage 4 : fine sanding. This operation is similar to coarse sanding, except that the abrasive paper is much finer; 1,000-grit, if obtainable, is the most suitable. In addition to the rotary movements of the stone, it is a good plan to make rocking movements of the stone backwards and forwards without rotation, so that the abrasive paper makes contact with the stone in swathes from side to side of the stone. Fine sanding should continue until all the surface that is to be visible is entirely free from surface imperfections and has a dully gleaming lustre.

Stage 5 : prepolishing. According to the quality of the finish obtained during fine sanding, this stage may be omitted at the cutter's discretion; but it can only help to produce a good finish. For this operation a hard-felt wheel and a 1,200 to 1,500 grit will be required. The rubber supporting disc and abrasive paper should be replaced with the hard-felt wheel. When the motor is started the water should be allowed to flow onto the wheel, continuing to run until the surface of the wheel is moist all over. The water is turned off, and then the motor, in that order. When the wheel is stationary the prepolish powder should be sprinkled evenly over the surface of the wheel and rubbed in gently with the fingers.

Some cabochon material, especially when it includes areas of slightly different composition, or is veined with stone of different hardness, may be subject to undercutting unless considerable care is taken in choosing the final abrasive and polish. Good examples of this are some types of jade, which were at one time distinguished by an 'orange-peel' surface caused by the differing orientations of the minute interlocking crystals. Now with the advent of ultra-fine diamond compounds for polishing, a surface may be produced that is virtually free from this effect.

An alternative method of applying this compound is to put a few grams of the prepolish into a squeezy bottle, into which is then poured sufficient water to make a very thin creamy mixture. This may then be applied to the wheel when it is in motion, without the wheel requiring to be stopped. Remember that sufficient lubrication will still be required, and as this may not necessarily be provided by the prepolish compound, a continuous gentle flow should be provided from the normal water source. The stone should be both rotated and rocked back and forth until the stone has a good all-over polish.

Stage 6 : polishing. This stage will also require a hard-felt wheel, but under no circumstances should the same wheel be used for both prepolishing and polishing. Polishing may be carried out in exactly the same way as prepolishing except that the polishing medium is different, in this case being cerium oxide, tin oxide, Linde A, or some other suitable agent. Polishing should be conducted with similar motions to those of prepolishing and should be continued until the stone acquires a very high degree of polish.

Stage 7 : grinding the back. The stone should be removed from the dop stick by gently heating it and prising it free with a knife blade, or by placing the stone and dop stick in the freezing compartment of a refrigerator very briefly. This should cause sufficient contraction to allow the stone to be broken loose quite easily. (The stone must not be left in the freezer for too long; it could be made brittle.) The stone must now be reversed on the dop stick in a manner similar to that of the original dopping.

If the stone was given a reasonably flat back when it was being roughed out, it should now be given a small bevel on the underside. If it did not have one, it should be held against the fine wheel until a smooth flat base is obtained. The shallow bevel may then be ground onto the base. The stages described above should be worked through to obtain a polish of whatever quality is required. Quite often the back of a cabochon is not seen, according to the style of setting. The cutter must decide for himself the finish that he requires.

Composite stones

The advanced worker may wish to cut doublets, consisting of a thin layer of precious material with a backing of poorer quality, such as obsidian or potch opal as in an opal doublet; or triplets, which are topped with a layer of colourless transparent material such as rock crystal. Either of these types may be made quite simply by cutting thin slices of stone, making sure that they are absolutely flat, and then cementing them together with a suitable adhesive. Two-part epoxy resin is usually the most suitable for this purpose. It is important that the mixing instructions for the resin are carried out correctly. It may be an advantage to heat the

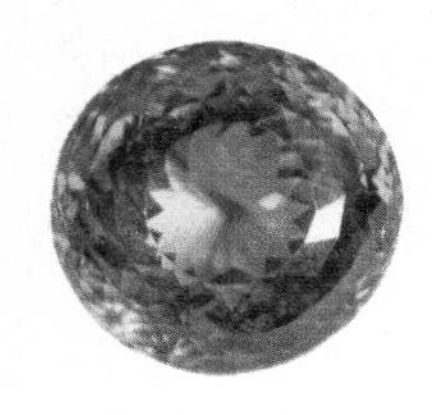

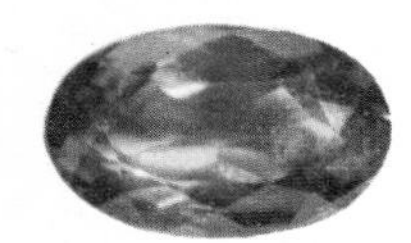

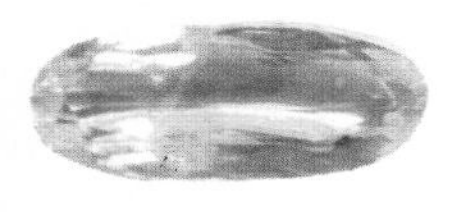

Above: Three examples of the gem-faceter's skill. The top stone is a superb piece of workmanship, having 289 facets, all of which are perfectly positioned. Below it are two of the rarer gem varieties: apatite (centre) from Lake Baikal, and hambergite (bottom) from the Malagasy Republic

Left: A fine gem-quality emerald crystal on quartz and dolomite

sandwiched pieces gently to approximately 65°C, so that the cement cures, not only more quickly, but as a transparent layer; some epoxy resins tend to become cloudy if cured at room temperature. When the cement has set, the cabochon may be cut in the normal way, but care should be taken to avoid undercutting of the edges of the layers owing to their different hardnesses.

A quite effective composite cabochon can be made by using a slice of abalone instead of opal; even butterfly wings have been used on rare occasions.

CUTTING A FACETED STONE

In the introduction to this chapter we saw that faceted stones are generally cut from transparent material; some opaque substances, however, such as hematite, are also faceted, though usually only on top. We also discussed those physical and optical properties most likely to influence the stone's final appearance. The object of faceting a stone is to allow light entering through the crown to be internally reflected and, in travelling around inside the stone, to pick up as much body colour as the depth of stone will allow. In addition, rays of light entering and leaving the stone are broken up into their component colours in much the same way that sunlight passing through rain drops is broken into the colours of the rainbow. The reason for this is simply the difference in the refractive index for the various component wavelengths of white light. This property is used to greatest advantage with colourless stones, especially diamonds and diamond simulants, for the desirability of most of these stones depends heavily on their characteristic fire. Some stones, such as demantoid garnet and some zircons, have this property of 'dispersing' light to a high degree, but are coloured, so that fire is not so noticeable, although it does add to the stone's general liveliness.

Another important property is *dichroism*, found in doubly refractive stones. Two rays are produced on refraction and, owing to selective absorption of different wavelengths, rays originally white become tinged different colours. In some instances dichroism may lead to only a slight difference in shade of the two rays; for example aquamarine may look deep blue and pale blue when viewed in different directions. Dichroism is more prominent in some stones and can often be used as an aid to identification. Ruby, for example, can easily be confused with spinel, another red stone. But they can be distinguished by the dichroism of ruby, which looks orange when viewed in some directions and crimson when viewed in others. Spinel is the same red shade from all angles.

The term dichroism should only be used to describe stones that can show two colours. Stones that can show three colours are known as *pleochroic*. A classic example of pleochroism is tanzanite, the newly discovered variety of zoisite. This stone belongs to the orthorhombic system; the colours seen are reddish-violet, deep blue and yellowish-green. Usually this material is heat-treated to alter the yellowish-green component to deep blue, and is then marketed as a sapphire simulant.

Fortunately, providing the rough has been identified, it is an easy matter to determine whether it is dichroic or pleochroic. Only crystals of the tetragonal and hexagonal systems are dichroic. Stones of the cubic system are singly refractive. The three other crystal systems, the orthorhombic, monoclinic and triclinic, yield pleochroic stones.

The Mineral Kingdom provides information about dichroism or pleochroism where applicable. This property should be borne in mind when starting to facet any multiply refractive material; the cutter must decide which of the colours he desires for the material he is cutting, and this will require careful orientation of the rough before preforming.

A large number of gem crystals have certain directions along which the atomic bonding is weak, and along which the material parts easily. This property is known as *cleavage* and is responsible for many of the cutter's difficulties. Even though it is extremely hard, a diamond can easily be made to part along the octahedral plane. Other stones with this property are topaz, where the cleavage is parallel to the basal plane, and spodumene, which has two directions of cleavage parallel to the c-axis. Further common materials suitable for faceting that have prominent cleavages are calcite and fluorspar. It is often difficult to obtain a good polish if the table facet is orientated parallel to a cleavage direction. In order to overcome this it is necessary to facet the stone so that the table is tilted a few degrees off the cleavage direction.

Now that we have reviewed several of the features likely to be encountered with facetable material, we should give consideration to choosing a suitable stone for faceting. Probably the most significant aspect is availability. It is

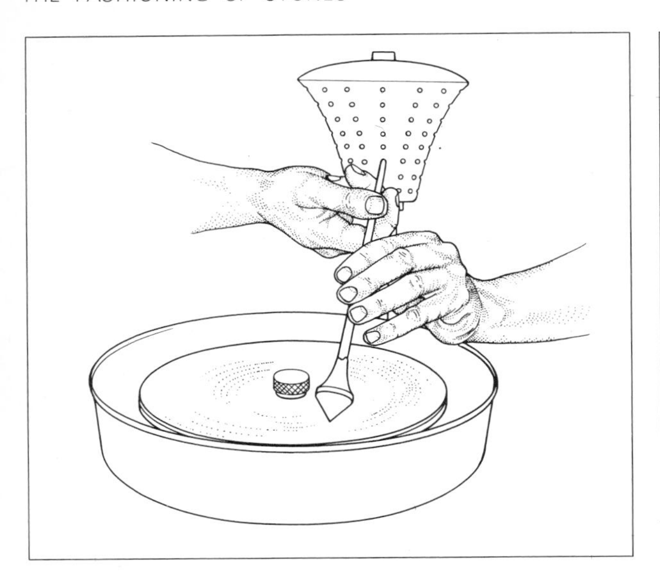

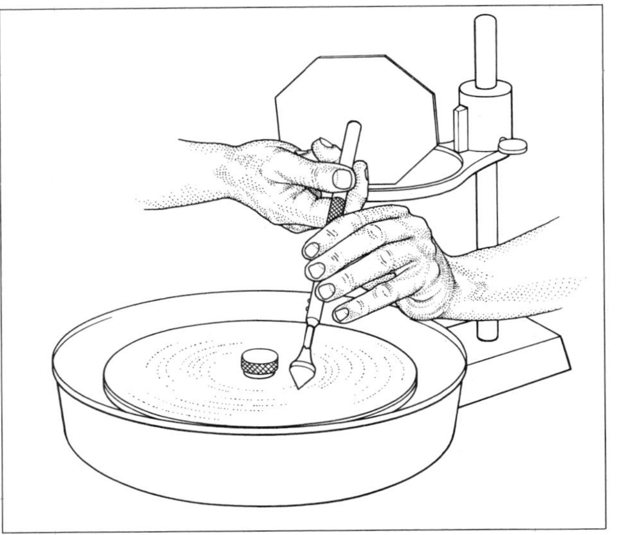

Far left: The jamb peg in use. The faceting angle may be varied simply by inserting the tip of the peg into a new hole

Left: A template in use. The outline shape of the stone is controlled by the octagonal template, while the faceting angle is altered by raising or lowering the column-mounted table

suggested that the beginner practise with cheap and easily obtained material until he is familiar with his machinery and the faceting process; even pieces of glass will suffice until the basic techniques have been mastered.

There are certain varieties of stone that are readily available in all places; typical of these are the transparent quartzes such as rock crystal, amethyst, citrine and smoky quartz. Beryl, topaz, tourmaline, garnet and peridot are also often readily obtainable. Certain stones are sufficiently rare to be collectors' items only, such as andalusite, apatite and brazilianite. In the account of faceting that follows, we shall assume that rock crystal is being used.

Selecting rough material

Before beginning faceting, look carefully at the available rough material with a view to choosing a piece that not only will produce as good a colour as possible, but may also be preformed into the desired shape with minimum wastage. If the rough is broad but thin, even if it has good colour, only small stones can be cut from it. Kunzite is a good example of this; its crystals are wide and long but thin. As kunzite shows the best colour along the vertical axis, it is usually only suitable for oblong step-cut stones. On the other hand the barrel-shaped crystals of sapphire are ideal for round brilliant-cut stones.

Having decided what shape of stone may be expected from any given type of rough, the next step is to choose a piece satisfying the preceding criteria and that will produce the largest possible cut stone. If a piece of rough is badly marked or flawed but includes a good cuttable portion, do not forget that the price, if the stone is purchased by weight, will include the flawed part which will almost certainly be discarded. Remember also that some gem crystals have very pronounced cleavage. Fortunately, however, it is often possible to recognize a cleavage plane in rough material; there may be typical incipient cleavage signs, such as rainbow-coloured Newton's rings, or there may already be a cleavage face, recognizable by its pearly lustre.

The standard brilliant cut

Preforming and dopping. The rough stone is ground on a coarse silicon carbide wheel until its shape approximates that desired for the finished stone. Since the stone is to become a round brilliant it must be preformed so that it

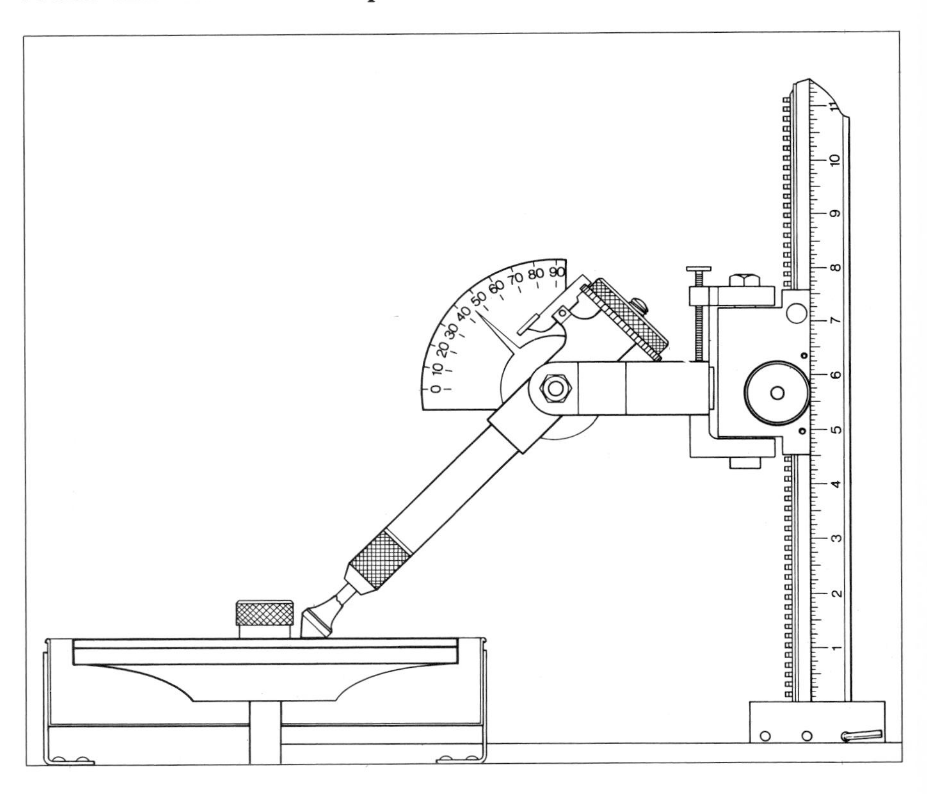

is round, viewed from above. The part that is to become the crown has a bevel at 45° to the central axis of the preform, and a flattened top that has sufficient width for a table facet of half the stone's diameter. The part that is to become the pavilion is ground into a conical shape, whose sides are at about 45° to the central axis.

When the preform has been satisfactorily shaped it should be examined closely with a lens to make quite certain that all the surface imperfections have been removed and that the stone has not been chipped during grinding. The stone is now ready for dopping.

As in the cutting of a cabochon, the stone preform must be fixed to the dop, using dop wax. The dop for the crown of the stone has a flat but scored surface; the dop used for the pavilion has either a conical recession for round stones or a v-section slot for step-cut stones. These surfaces are scored too, to help key the wax to the metal.

The warmed end of the dop stick is inserted into the molten wax. The preform is gently heated. Again care must be taken to ensure that neither the wax nor the stone overheats, for this would tend to make the wax very brittle, with some loss of adhesive power, and the stone could develop flaws. The crown of the stone is firmly and centrally fixed to the dop stick and the unit is allowed to cool gently. This should not be done by immersing the stone in water or resting it on a cold glass or metal surface, for a rapid change in temperature could crack it.

Left: A stylized index-type faceting head, cutting a pavilion facet. Both the facilities for changing the index number and for altering the cutting angle by raising or lowering the head can be seen

The pavilion facets will be cut first. This has the advantage for the beginner that if, owing to poor dopping or simple accident, the stone should break away from the dop stick and become damaged, less re-cutting will be needed than if the crown were cut first and damaged. Cutting the pavilion first also ensures that its centre, the culet, can be well formed.

Stage 1 : cutting the girdle. The faceting machine is fitted with the copper lap impregnated with 360-mesh diamond powder. The angle of the dop arm is adjusted to 90°, so that its long axis is parallel to the plane of the lap. The faceting head is lowered until the stone just makes contact with the lap. Before starting the motor, lift the dop arm out of the cutting position to avoid any sudden snatch as the lap starts to rotate. Lubrication is by means of a water-soaked cloth or wad of cotton wool, which should be held gently against the lap whenever the stone is being worked. With the motor running the dop arm is lowered to bring the stone into contact with the revolving lap. Grinding the girdle is now a straightforward matter of rotating the dop and stone against the diamond lap, lowering the faceting unit as necessary, until the outline of the preform is perfectly round. This operation should continue until the girdle area is of such a width all round the stone that a sufficient depth for the setting edge will remain after the facets have been ground. It is recommended that the dop unit be passed in a gentle arc across the lap while being rotated, to prevent excessive wear in any one area of the lap; a similar movement was used when grinding a cabochon.

Stage 2 : cutting the pavilion facets. Having first checked that the grinding of the girdle has removed any small flaws, change the angle of the dop arm to that recommended for the pavilion facets, in this case approximately 43°. Ensure that the correct index wheel – in this case a 64-toothed wheel – is fixed to the dop-arm assembly and that the trigger is locked at facet index number 64 (which is also number zero on a 64-toothed wheel). The dop arm is lowered so that the stone comes into contact with the coarse-diamond lap, and, with the same arcing movements, lowering the faceting unit as necessary, the stone should be moved across the lap until a flat has been ground from the girdle to the centre-line of the stone. When this flat has been ground properly the dop arm should be lifted, and the index wheel reset at number 32. This will allow a similar flat to be ground opposite the first. This new flat should be ground away until it meets the other centrally at the base of the pavilion.

Once again the index wheel should be altered, first to 16 and then to 48, to allow flats to be cut at 90° to the first two. If this has been done correctly the stone will now have a circular girdle and four flat faces meeting at the culet.

The next step is to cut another four faces, each centred on the line between a pair of the first four facets, meeting once again at the culet. Stop the motor and lift the dop arm out of position. Change the lap for the one impregnated with diamond compound of approximately 1200 mesh. Make sure that the underside is perfectly clean and free from small particles of grit, which could upset the true running of the lap and spoil the facets. Set the index numbers to 8, 24, 40 and 56 successively for these faces. Great care should be taken to ensure that these new facets meet exactly at the culet without spoiling the point, as could quite easily be done if the cutter were too hasty. At the end of this operation the pavilion should have eight equally sized and spaced facets, the *pavilion mains*.

Keeping the dop arm at the same angle, go over each of the facets cut so far, in order to remove all the small surface irregularities and scratch marks produced by the coarse lap. Once again ensure that cutting does not continue unnecessarily, producing irregular facets.

With the fine lap still on the machine, the time has come to cut the remainder of the pavilion facets. These are known as *break* facets or *lower girdle* facets. In order to cut these it is necessary to alter the angle of the dop arm to a few degrees more than was previously used. Reset the dop arm angle to 45° and adjust the height of the faceting head until the stone just makes contact with the lap. Before starting the motor again, the index wheel should be set at either number 62 or number 2, so that facets that are placed symmetrically with respect to the main facets may be cut. Similar facets should be cut, two to each main facet, right the way round the stone using the index numbers listed in the chart.

When each of these lower girdle facets has been satisfactorily positioned they must all be polished. To do this, remove the diamond lap and replace it with a lap suitable for the polishing medium – a copper or tin lap with ultra-fine diamond powder, or a plastic lap with cerium oxide, tin oxide or Linde A. For the rock crystal that is being cut here a plastic lap with cerium oxide will be used. It will be an advantage to have the cerium oxide mixed with water to a thin creamy consistency in a small dish. A piece of cotton wool or a fine brush may be dipped into this so that the polishing medium may be transferred to the lap. Alternatively, a squeezy bottle may be used. The best vehicle, however, is the cotton wool wad.

Polish each of the twenty-four facets on the plastic lap, using the same index numbers and cutting angles as in fine grinding. On most good-quality facet heads there is a device that allows slight adjustments to be made to the position of the stone in relation to the lap, so that very accurate positioning may be made. The facets should be inspected frequently with a lens; they must be completely free from fine scratches and have a high polish all over.

If it is desired to cut a culet the stone and dop should be placed into a 45° dop-arm adaptor, which allows a facet to be cut perpendicular to the axis of the stone. The cerium oxide and plastic lap may be used if a very small amount of stone is to be removed, just sufficient to help prevent point damage; if a definitely visible culet is required, the fine-diamond lap should be used before polishing commences.

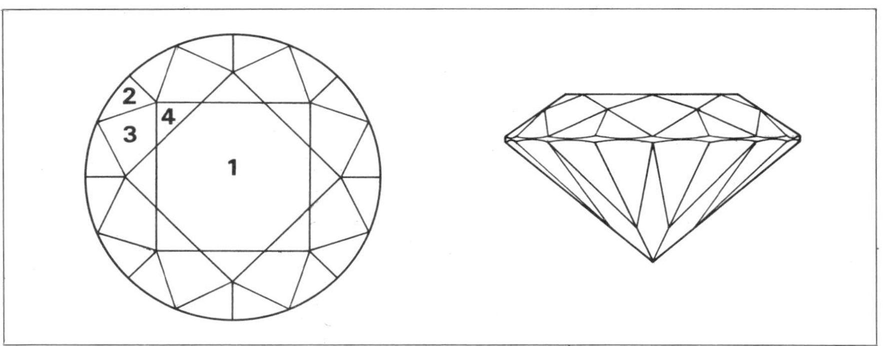

Top: Versions of the brilliant cut

Stage 3: reversing the stone. In order that the crown facets may be cut and polished it will be necessary to turn the stone round. If the stone were a cabochon, it would simply be a matter of warming the wax, removing the stone and re-dopping it the other way up. This cannot be done with a faceted stone, since it is necessary to align the crown facets accurately with the pavilion facets. The majority of metal dop sticks have a special positioning device, either a locating groove or a fixed pin. To carry out the reversal it is necessary to resort to a transfer jig. This is a piece of aluminium or steel with an accurately cut groove along its length and a screw-locking device at each end; two dop sticks lying in the grooves and facing each other may be accurately aligned and then held rigidly. In order that the stone may be satisfactorily reversed, proceed as follows.

1. Clamp the dop stick holding the stone firmly in the transfer jig. Clamp the second dop stick, which has a conical recess, already coated with

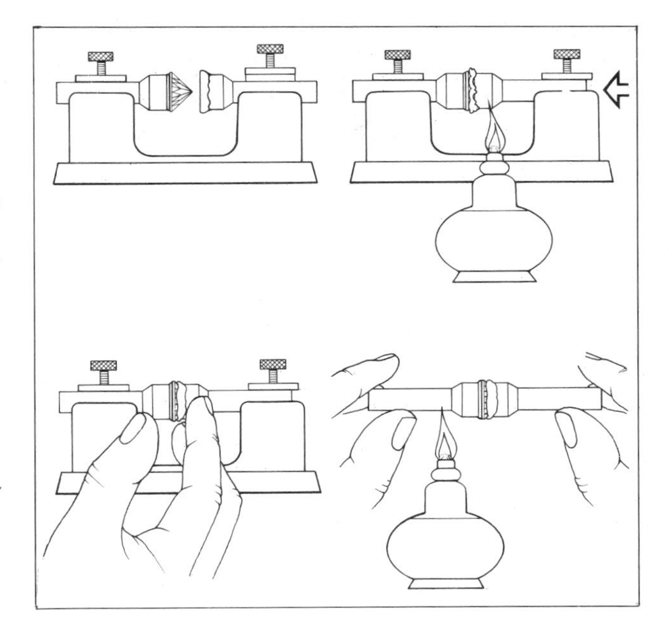

Right: The transfer jig. The dop stick carrying the stone is clamped in the jig opposite a stick shaped to take the pavilion, and already dipped in dop wax. The second dop stick is heated, pressed against the stone and clamped. The wax is shaped around the pavilion and allowed to set. Then the first dop stick is removed by heating

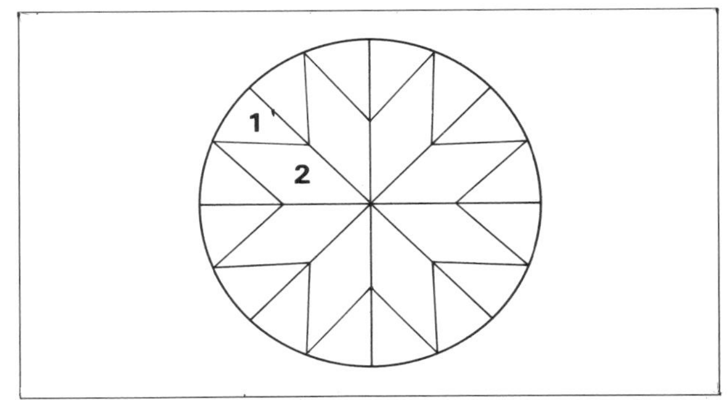

Left: Crown and side view of a brilliant-cut stone. 1: The table; 2: upper girdle facet; 3: crown main facet; 4: star facet

Right: The pavilion. 1: Lower girdle facet; 2: pavilion main facet

hot wax, loosely in the other side of the jig, so that the pavilion fits snugly into the waxed cone.

2. Clamp the second dop stick firmly, and then gently heat its metal body until the wax once again becomes soft. While the wax is still soft it may be moulded carefully into shape around the stone; make sure that it does not cover the girdle. Set the transfer jig aside and allow it to cool completely.

3. When the wax and the second dop are both cold, the two dops, with the stone between them, may be removed from the transfer jig. The end of the first dop should be heated in a flame until its wax softens, whereupon the stone and the old dop can be induced to part company. It is important that the original dop be heated quickly so that the wax melts before the heat is transferred through the stone to the new wax, as this could allow the stone to become misaligned.

4. The stone should now be freed of all the unwanted wax covering the crown area. This may be carried out most easily by wiping the stone gently with a piece of cloth soaked with methylated spirit, taking care not to get it too close to the spirit lamp flame.

Stage 4: cutting the crown facets. Place the new dop stick into the dop arm and set it to the crown angle, which, as already stated, should be between 40° and 50° for quartz. In this case, matters are made easier by the fact that an angle of 45° has already been set on the dop arm from the last facets polished on the pavilion (unless a culet facet was polished). Eight *crown main* facets, corresponding to the eight pavilion main facets, will now be cut.

Assemble the coarse-diamond lap on the machine and set the index wheel to 64 so that the new facet will be exactly opposite its pavilion counterpart. Start the motor and, lubricating in the usual way, cut the first crown facet. This should be ground until the girdle, at the end of the facet, is of the thickness required for the setting edge, and the other end of the facet is rather less than a quarter of the stone's diameter from the centre. Now cut the other corresponding facets using the index numbers 32, 16 and 48. When these facets have been properly cut the centre of the crown should form a square, although it may have rounded corners.

Next cut the intermediate facets in a similar manner, using index numbers 8, 24, 40 and 56. Unlike their counterparts on the pavilion, these should be ground with the coarse lap, since there is more material to remove than previously. When completed these eight facets should all be of the same shape and size, and should all extend to the same distance from the girdle. If this is not so, those that are faulty should be reworked. When all are satisfactory, they should be smoothed with the fine diamond lap. These facets may be polished now or after the remaining facets, other than the table, have been ground. It may be an advantage to polish all these facets together later, for they will be considerably reduced in size, although it will mean judicious use of the micro-adjuster.

Stage 5: cutting and polishing the table. Remove the fine lap and replace it with the coarse. Remove the dop stick and fit it in the 45° adaptor, already used to cut the culet if there was one. Instal the combination in the dop arm and set the dop arm angle at exactly 45°.

With the dop arm slightly too high, so that the stone does not quite touch the lap, start the motor. Grind the table by slowly lowering the faceting head, while using the normal arcing movements described previously. The facet unit should continue to be lowered slowly until the table facet has been ground to approximately half of the width of the stone. If the crown main facets have been positioned correctly and if the grinding of the table facet has been carried out exactly perpendicular to the stone's axis, it is a simple matter of comparing the diameter of the stone and the distance across the table facet.

After changing the coarse for the fine lap, the table facet should be ground just a little more to remove scratches. Then it may be polished on the plastic lap with cerium oxide. Make any necessary alterations with the micro-adjuster on the dop arm so that the table is polished evenly all over.

Stage 6: cutting and polishing the remaining crown facets. Having finished the table it is now necessary to grind and polish the remaining facets, the eight *star* facets, around the table, the eight *crown main* facets, also called *kites* from their shape, and the sixteen *upper girdle* facets, also called *skill* facets, because they are said to require a great deal of skill.

The 45° adaptor should be removed, the dop replaced in the dop arm and the fine lap placed on the machine. Alter the dop arm angle to approximately 25° or 26°, and ensure that one of the eight main facets is correctly aligned with the lap. Set the index wheel so that the star facet will be positioned centrally on the edge of two main facets and just cutting into the octagonal table facet. The index numbers for the star facets are 4, 12, 20, 28, 36, 44, 52 and 60. When these facets have been properly cut the table will be surrounded by eight triangular facets of uniform size and regular position.

The sixteen upper girdle facets are ground next. They should be cut with a dop arm angle of approximately 48°, but the cutter will find that the exact angle depends mainly on the relationship between the size of the table and the diameter of the stone; all that is really necessary is that they should split each of the crown main facets, which at this stage will be somewhat bell-shaped. They will become kite-shaped. The index numbers for these sixteen facets should be 62, 2, 6, 10 and so on. For a complete list of index numbers for each type of facet refer to the cutting chart. All these facets may now be polished in the normal way using the plastic lap and cerium oxide.

If it is desired, the girdle may be polished by resetting the dop arm to 90° and lowering the facet head so that the girdle comes into contact first with the fine-diamond lap, and then with the plastic lap; the index trigger should be released,

Standard brilliant cut

	cutting order	index number	polishing order	angle
pavilion main	1	64	1	43°
	2	32	2	..
	3	16	3	..
	4	48	4	..
	5	8	5	..
	6	24	6	..
	7	40	7	..
	8	56	8	..
lower girdle	9	2	9	45°
	10	62	10	..
	11	6	11	..
	12	10	12	..
	13	14	13	..
	14	18	14	..
	15	22	15	..
	16	26	16	..
	17	30	17	..
	18	34	18	..
	19	38	19	..
	20	42	20	..
lower girdle	21	46	21	..
	22	50	22	..
	23	54	23	..
	24	58	24	..
crown main	25	64	34	45°
	26	32	35	..
	27	16	36	..
	28	48	37	..
	29	8	38	..
	30	24	39	..
	31	40	40	..
	32	56	41	..
table	33	—	25	0°
star	34	4	26	25°
	35	12	27	..
	36	20	28	..
	37	28	29	..
	38	36	30	..
star	39	44	31	..
	40	52	32	..
	41	60	33	..
upper girdle	42	2	42	48°
	43	62	43	..
	44	6	44	..
	45	10	45	..
	46	14	46	..
	47	18	47	..
	48	22	48	..
	49	26	49	..
	50	30	50	..
	51	34	51	..
	52	38	52	..
	53	42	53	..
	54	46	54	..
	55	50	55	..
	56	54	56	..
	57	58	57	..

allowing the stone and dop to rotate freely. Alternatively the girdle may be surrounded by a series of oblong flat facets corresponding in position to each of the upper and lower girdle facets. It may be a distinct advantage when cutting certain stones, zircon for example, to cut an extra row of pavilion facets, extending from the culet halfway up each of the pavilion mains. If it is decided to cut these extra facets, the index numbers will be the same as for the upper girdle facets; but different angles will be required, best decided by trial and error.

Stage 7: cleaning the stone. When all the cutting and polishing has been completed satisfactorily the dop may be removed from the facet head and warmed gently; then the stone may be removed from the wax. In order to avoid damaging the stone by scraping off the warm wax with a knife blade, most may be removed with a thumb nail and then, if the stone is rubbed vigorously with a cloth moistened with methylated spirit, the final traces of wax will be removed. The stone should now be washed carefully in warm soapy water and dried with a soft cloth to remove the last traces of grease and dirt in order to reveal the full beauty of the newly faceted stone. If all the steps have been carried out properly and carefully the cutter can justly feel proud of his achievement.

The step cut

When the principles of faceting have been mastered, the cutter will no doubt wish to experiment and extend his repertoire. There are many ways in which this can be done; there are at least a dozen different cuts based upon the round stone, besides the brilliant; and they may be modified so that oval stones may be similarly fashioned. One alternative is the *step*

Step cut

	cutting order	index number	polishing order	angle
girdle	1	64	1	90°
	2	32	2	..
	3	16	3	..
	4	48	4	..
pavilion main 1	5	64	17	67°
	6	32	18	..
	7	16	19	..
	8	48	20	..
pavilion main 2	9	64	13	59°
	10	32	14	..
	11	16	15	..
	12	48	16	..
pavilion main 3	13	64	9	51°
	14	32	10	..
	15	16	11	..
	16	48	12	..
pavilion main 4	17	64	5	43°
	18	32	6	..
	19	16	7	..
	20	48	8	..
crown main 1	21	64	30	45
	22	32	31	..
	23	16	32	..
	24	48	33	..
crown main 2	25	64	26	39
	26	32	27	..
	27	16	28	..
	28	48	29	..
table	29	—	21	0
crown main 3	30	64	22	33
	31	32	23	..
	32	16	24	..
	33	48	25	..

Below: Crown and pavilion plans and profile of a trap-cut gemstone

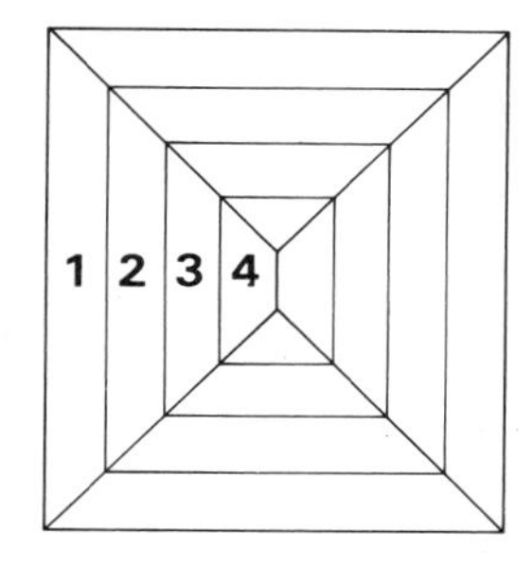

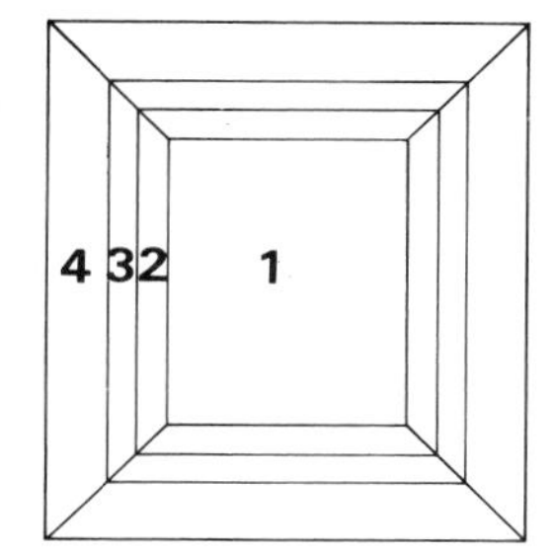

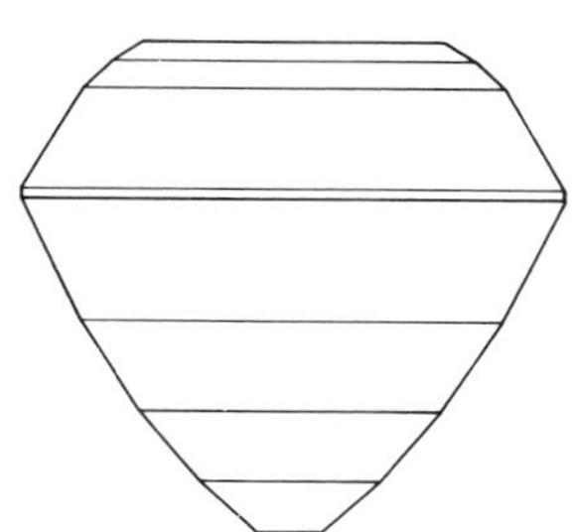

cut, so named from the shape of its facets, which are simple parallel rows of rectangular facets on both crown and pavilion. It is also called the *trap* cut because it has the ability to trap the light entering the stone. Although this style may be cut with widely varying proportions it should be borne in mind that the best effects are obtained when the bottom row of facets on the pavilion and on the crown are at the recommended cutting angles for that particular material. There may be widely varying numbers of parallel rows of facets, cut in a wide variety of angles, from, say, two rows on the crown and two on the pavilion, to as many rows of facets as the cutter can get on the material. The angles between the facets generally change in equal increments, which may be anything from 2° to 10°, though steps of 5° and 10° are normal.

Preforming and dopping. Normally any gem material may be cut as a trap-cut stone but there are several gems whose crystal outline is especially suitable. For example, the rounded-triangular cross-section of a tourmaline crystal allows stones to be cut in this form with minimum wastage. The outline of the preform should be like a long box, with a square section. From this basic shape a preform should be fashioned whose crown is approximately one-third of the thickness of the stone in depth, leaving approximately two-thirds of the thickness for the pavilion, as shown in the diagram.

There are several modifications to this basic cut. For example, the corners can be cut off to give an octagonal outline; this is the *emerald* cut. Certain other outline shapes are commonly used, including the baton, kite, lozenge, triangle and trapeze.

Apart from the difference in shape, the preforming is carried out in almost exactly the same way as for the round brilliant. Except that a metal dop with a square section and flat top is used, the dopping of this type of preform is as previously described.

Stage 1: grinding the girdle. With the coarse diamond lap on the machine and the dop arm angle set at 90°, adjust the height of the faceting unit until one flat side of the preform just touches the lap. Lift the dop arm out of position, start the motor and, ensuring that an adequate supply of lubricant is available, lower the dop arm until the stone again touches the lap, allowing cutting to commence. The indexing wheel should preferably be set to 0, or 64, with the trigger locked so that the stone remains square on the lap. Lower the facet unit until sufficient material has been removed from the preform on this side of the stone. The next operation is to reset the index wheel to number 32 so that a similar flat face may be cut on the opposite side, followed by further changes to index numbers 16 and 48 in order to complete the girdle. The facet unit's height must be changed as necessary to keep the facets parallel to the stone's axis.

Stage 2: cutting the pavilion facets. Ascertain the correct pavilion angle for the material being cut. In what follows we shall assume that it is rock crystal, for which the angle is 43°. Remember that this angle of 43° refers specifically to the last or *point* row of facets and therefore all other rows of facets on the pavilion must be at greater angles, at equal intervals between the pavilion angle and 90°. For example, if it is decided to cut four rows of facets the angle of the first row, underneath the girdle, will be 43° plus the incremental angle multiplied by three. If it is decided, in this instance, to have angles of 8° between the rows then the first row cut will be at an angle of 43° plus 3 × 8° – namely 67°. This angle should now be set on the dop arm.

Still using the coarse lap, four equal-sized faces should be ground around the stone. These are followed by a similar row of faces cut on the same lap, with the dop arm angle now set to 43° plus 2 × 8°, making the angle 59°; then the angle should be altered to 51°, and for the point row, to 43°. Obviously the index wheel numbers will be the same as when cutting the girdle. These rows of facets should now be reground using the fine lap to remove surface imperfections; if necessary, the micro-adjuster should be used to obtain perfectly parallel facets. Polishing may be carried out on a plastic lap using cerium oxide.

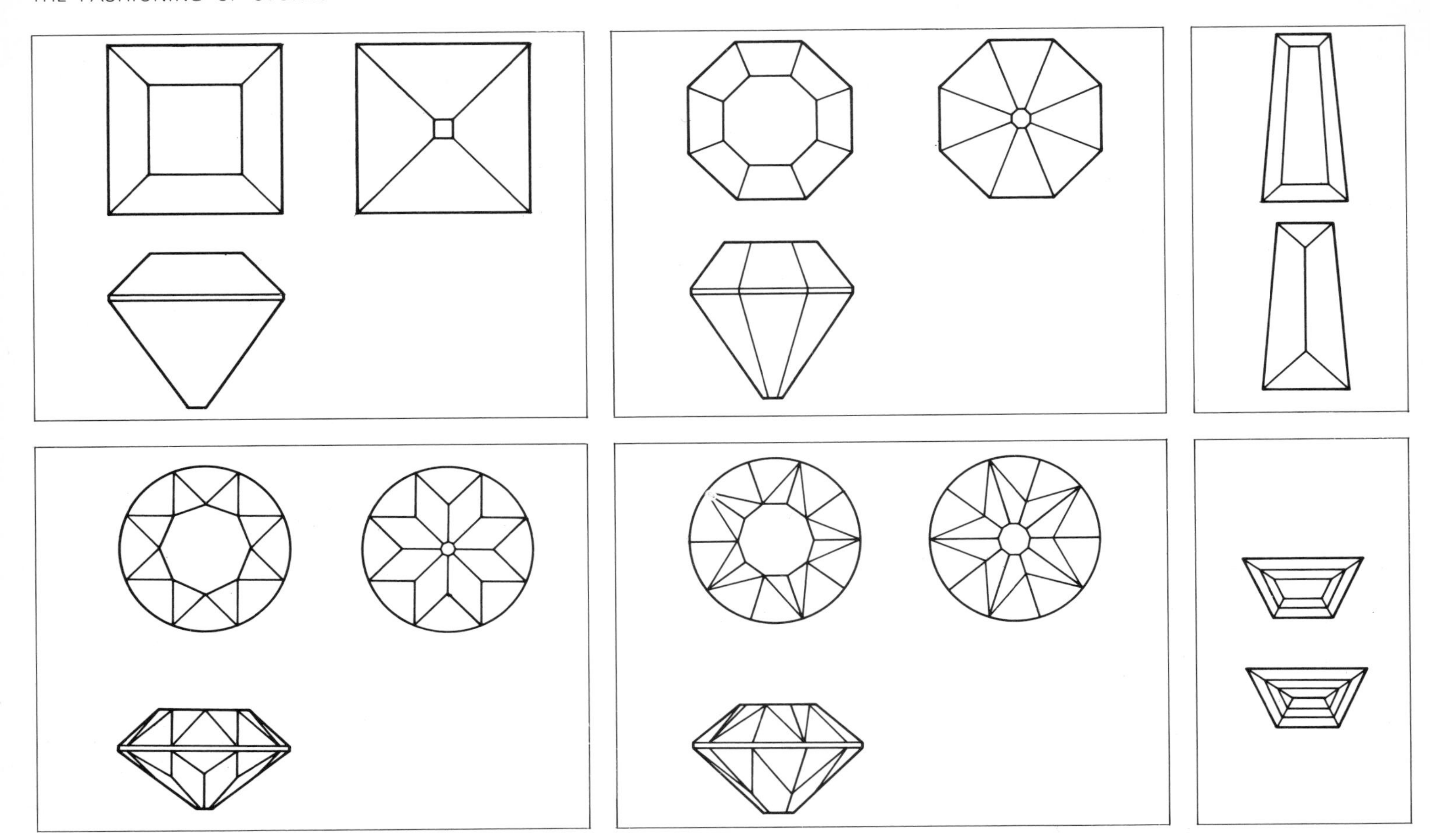

Stage 3 : reversing the stone. Once again, in order to cut the crown facets, the stone must be reversed, using the transfer jig in the manner described previously. A second dop with a square section and V-groove is used instead of the dop with a conical recess. All other aspects of reversing the stone are as already described.

Stage 4 : cutting the crown facets. First decide how many rows of facets there will be, and then reset the dop arm angle to the recommended crown angle for the first row. The crown angle has the incremental angle deducted from it for the second row, the incremental angle deducted again for the third row, and so on for each row to be cut. Our stone will have three rows of crown facets; the first row will be at an angle of 45°; rows two and three are at angles of 39° and 33° respectively. Apart from the table facet, which will be dealt with next, all crown facets are cut and polished in exactly the same way as their pavilion counterparts.

Stage 5 : cutting and polishing the table. For this facet the 45° dop adaptor should be used, and the dop arm angle should be set at 45°. The facet unit should be lowered while grinding until the table facet is approximately half of the girdle width. Unless there is a large quantity of material to be removed to obtain this table width, the fine lap only should be used, except for polishing. When you have successfully cut and polished the complete stone, it may be removed from the dop and cleaned as usual.

The development of faceting

With the exception of diamond, little was done to improve fashioning beyond the cabochon or the polishing of natural crystal faces until about the 17th century. Simple fashioning of diamonds, however, was in evidence in the mid-14th century. Both Indian cutters and those in some parts of Europe were polishing away a point of a perfect or near-perfect octahedral crystal until a table was formed. This became known as the table or point cut. Until this time it was standard practice to grind away flaws and inclusions on the rough diamond, thereby covering it in a multitude of irregularly shaped flat facets. A typical example is the famous Orlov Diamond, cut originally by Indian lapidaries.

Eventually the table cut was modified so that further corners were ground and polished. This in turn led to considerable experiment to improve the appearance of the cut stones, until the brilliant cut was developed. Even this has

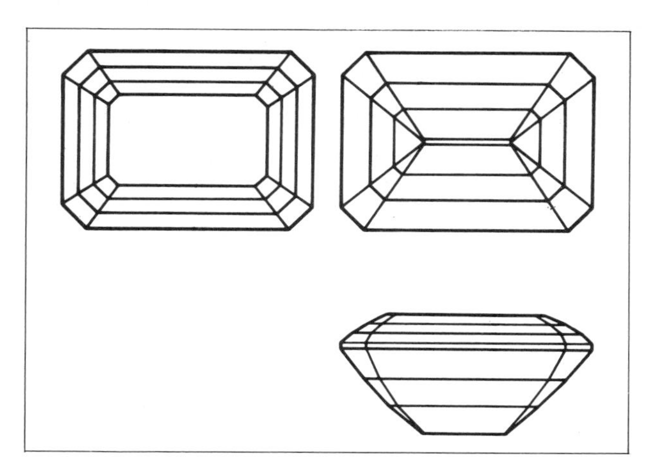

Left: A selection of fancy cuts based on the trap cut. They are the square cut (far left), the eight cut (centre left) and the tapered baguette

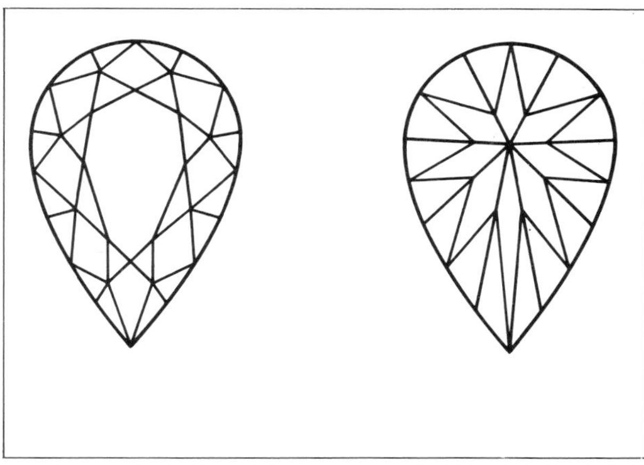

Right: The emerald cut (near right), which is a variation on the standard trap cut, and the pendeloque or pear-shaped brilliant

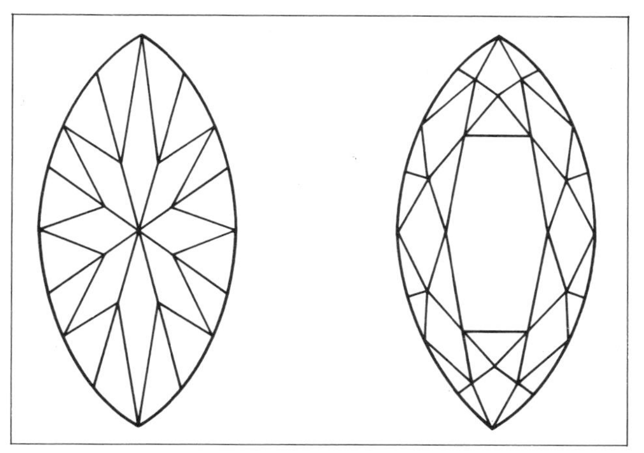

Left: Further fancy cuts. The Swiss cut (far left) and the split brilliant cut (centre left) are both based on the brilliant cut. The trapeze cut is based on the trap cut

Right: The marquise or navette, based on the brilliant cut

since been improved, so that the latest style, containing 57 facets (58 if one includes the tiny culet), is considered to be the shape that provides the maximum amount of brilliance and fire.

When the cutting and grinding of diamonds became possible after the introduction of diamond powder as an abrasive, the coloured gemstones, such as emeralds and rubies, became less important, although very fine specimens have always been held in high esteem.

Probably the most important step forward was to fashion stones other than diamonds in regular designs that included the correct angles for the production of total internal reflection. These allow even those stones with small amounts of dispersion, and with only pastel shades of colour, to become as popular as their richly hued companions the emerald, the ruby and the sapphire.

There now exist many important gem cutting centres throughout the world, cutting commercial gems in about six different styles: the brilliant, the modified brilliant for zircons, the trap and emerald cuts, the scissors cut, the mixed cut (with brilliant-cut crown and step-cut pavilion), and the cabochon. It is significant, however, that even now dedicated lapidaries are experimenting with special shapes and effects. Some of these cuts may have two hundred or more facets.

Considerable research has also been going on in the diamond-cutting industry to produce different cuts that enable more of the awkward shapes of rough to be cut economically. Probably the best of these is the profile or princess cut, which consists of a series of regular V-shaped grooves on the back of a thin diamond with a wide table. It gives a much greater effect for the stone's weight than could be obtained by conventional cutting. Another advantage is that much larger stones can be cut from flat crystals.

Surface effects of polishing

When certain types of gemstones are polished a layer at the surface about one micrometre thick melts and flows across the surface of the stone as a result of the heat generated. This material solidifies to form a *Beilby layer*. In some instances, especially in the quartz family, this layer recrystallizes on cooling, but in some stones it does not. Usually when a gem variety is fused and allowed to cool quickly, so that it forms a glass, the physical and optical constants are lowered. For example, beryl glass has a refractive index of about 1.51, whereas crystalline beryl has a mean refractive index of 1.58. It may be thought that a stone that produces an amorphous Beilby layer would actually have the refractive index typical of the layer; but this layer is in fact far too thin to affect any of the normal optical characteristics.

In some stones, especially diamonds, the quality of the polish depends solely on the fineness of the abrasive used for polishing. It is thought that in these cases the temperature required to form a layer is greater than the temperature obtained during polishing.

Diamond cutting

Generally speaking, diamonds are not cut and polished by anyone other than a professional diamond cutter, but this does not mean that, using the equipment available to him, the amateur or professional lapidary cannot make the attempt. Difficulties do arise, however, for there are certain operations normally only carried out in a diamond-cutter's workshop that may be necessary in order to remove relatively large quantities of unwanted material. Normally these operations need not be undertaken by the lapidary, for most gemstones are sufficiently soft to enable the faulty or excess material to be ground away. More for interest to the amateur and the lapidary than as a step-by-step instruction series, there follows a brief outline of the various processes involved in the production of a brilliant-cut diamond from a typical octahedral crystal.

Before cutting can begin there are several operations that must be undertaken. Firstly, the crystal will normally have to be reduced in size, either by cutting it in half to produce two pyramid-shaped pieces, or by cutting it to produce one large stone and one small one.

Sometimes it is necessary to remove inclusions

Left: A selection of rough diamond crystals, weighed and ready for cutting

Right: A closer look at raw uncut diamonds, showing a great variety of crystal shapes and colours. Some of these stones are fit only for industrial uses by virtue of their poor colour or inclusions, while others will be cut to produce flawless gems

Left: Four splendid diamonds. The Eugénie Blue (top left), a blue stone of 31 carats, probably belonged at one time to Eugénie, wife of Napoleon III and Empress of France. The Oppenheimer (top right), named after the late Sir Ernest Oppenheimer, was found in 1964. It is an uncut yellow diamond of 253 carats, probably the best specimen of its type ever discovered. The Regent (bottom left), is a cushion-shaped white stone of 140 carats. The Earth Star (bottom right), weighing 111 carats, is probably the largest brown diamond ever found

in the crystal, which is done in one of two different ways. *Cleaving* is carried out if the imperfections are in a position where they may be removed without undue loss of material. The cleavage plane in a diamond is always parallel to the octahedral face. If the inclusions to be removed are near the surface a layer containing these inclusions is cleaved away. To do this, the stone is cemented to the end of a special stick with a particular type of cement based upon the lapidary's dop wax formula of rosin and shellac. Next a small notch or *kerf* is ground in the stone by hand, using another diamond, which will support the special cleaving blade. After considerable inspection to ascertain that this kerf is in exactly the right position, the blade, which is generally of thick steel with an edge that will act as a wedge rather than a knife, is dealt a blow with an iron bar or mallet. If the positioning of the kerf is correct the stone will fall neatly into two pieces. If it is not correct the stone can easily shatter.

The inclusions, if near the centre of the octahedron, may alternatively be removed by sawing a slice from the stone. A diamond is normally only sawn in one of two possible directions, parallel to cubic or dodecahedral faces; in other crystallographic directions the stone is too hard.

Sometimes inclusions may be removed during *bruting* rather than by either of the previously described methods. Bruting is the method used to produce the diamond preform and is totally different from that used to produce other gemstone preforms. The diamond is fixed on a special dop, which can be fitted into a small lathe known as a bruting machine. The diamond is revolved whilst another diamond is held against it, in much the same way that a piece of metal is turned on an ordinary lathe while another piece of metal is held against it. In order to save time another stone that has to be bruted is chosen as the stationary stone.

Once the preform, or bruted stone, has been re-dopped using either a mechanical dop or the more traditional solder dop, it is ready for the grinding and polishing of the facets. The lap used is generally a disc of cast iron approximately twelve inches in diameter; its axle either passes through it and rests in bearing cones above and below it, or merely rests, in a ball or roller bearing below. The lap rotates at approximately 2,500 rpm. The surface is scored and charged with diamond paste, which is worked into the surface of the lap by pressure from the stones as they are being ground and polished. Often three or four diamonds are ground and polished on the lap at the same time. Usually the diamond-cutter relies upon experience and expertise alone to produce facets of the correct size and at the correct angles.

Generally the table facet is ground first, followed immediately by one of the crown main facets. As soon as the cutter is satisfied, three further crown main facets, which are at 90° to each other, follow. The corresponding four pavilion mains, and then the remaining four crown facets are cut. These are followed by the remaining four pavilion facets and the culet.

The stone is then returned to the bruting machine where the girdle of the stone is added. Then the facets are polished. All the polishing and grinding, apart from bruting, has been carried out by an operator called a *cross-cutter*, but now the stone is passed to the *brillianteerer* who cuts and polishes the remaining crown facets – eight star facets around the table, and sixteen upper girdle facets. The stone is then reversed and the corresponding pavilion facets are ground and polished. The finished stones are now bathed in acid to remove all traces of grease and dirt.

It will be remembered that in cutting the rock

crystal brilliant it was not important that the cutting angles were exactly accurate. In the best-quality diamonds of the same shape, the facet angles must be very accurate; the pavilion angle must be close to 40° and the crown angle should be close to 34°.

NOTES ON GEM MATERIALS

In the following pages there are general notes on the main gem minerals of interest to the lapidary. More specific details are to be found in the entries for these and other minerals in the Mineral Kingdom section. Some synthetic materials are treated separately at the end of the section.

Abalone. A warm-water shell of bright colours, often producing pearls with the same appearance. Owing to its softness it may be worked either by hand, using steel tools, or shaped by grinding and sanding. Avoid breathing the dust which contains keratin (a protein) as this may be harmful. Abalone may be tumble polished.

Actinolite. An amphibole mineral which is one end-member of the actinolite – tremolite isomorphous series, two varieties of which are asbestos and nephrite jade. Actinolite may be faceted when the rough is clean, but great care must be taken, for not only is the material very brittle but it possesses several directions of perfect cleavage.

Agate. A cryptocrystalline quartz belonging to the chalcedony family. It generally shows a series of concentric or straight bands of various colours but may also show dendritic patterns. It is very tough and takes a good polish using normal techniques. The many different varieties of agate make it ideal material for tumbling.

Amber. A fossilized hydrocarbon resin some thirty million years old. It may be hand-worked using steel tools or fashioned using normal lapidary equipment. Care should be taken when dopping, for amber softens at about 150°C, and plenty of water must be used when polishing, for the same reason.

Amblygonite. A lithium mineral, which in gem-quality rough may be difficult to obtain. It is fairly soft and sufficiently heat-sensitive to warrant a low-temperature dop wax.

Andalusite. A very attractive faceted stone when correctly orientated, due to the very strong pleochroism; the stone may appear green with reddish-brown highlights at the ends. No particular problems involved in cutting or polishing. A variety of andalusite is chiastolite which shows a change of pattern with each successive slice across the crystal.

Anglesite. A very soft, very fragile and highly heat-sensitive substance. It may produce very attractive cut stones when rough is available.

Anyolite. A massive ornamental material composed of ruby crystals embedded in green zoisite. Unfortunately the rubies are usually opaque, so the material is generally used for carvings or decorative slabs; however, small ruby crystals with a green surround make attractive pendant pieces and so on. Not usually suitable for tumbling, due to the extreme hardness of the ruby.

Apatite. Although very brittle and heat-sensitive, very attractive bright stones may be produced. The rough, which is usually yellow or yellowish-green, also occurs violet, olive green and blue. The large crystals are generally flawed.

Apophyllite. This rather fragile gemstone is occasionally faceted, but cabochons are more usual, the crystals often displaying a pearly lustre. This is mainly a collector's gem and rough may be difficult to obtain.

Aragonite. Often used to produce cabochons showing brown and cream stripes. This material may also occur sufficiently transparent to produce faceted stones although the crystals are usually sold in groups for mineral collectors.

Axinite. This strongly pleochroic stone is seldom found as flawless rough although a new locality in California is said to produce large flawless faceting material.

Azurite. Alternatively named chessylite after its French source of Chessy, this generally massive material, often found in association with malachite, produces attractive cabochons. When found in transparent facet-grade crystals the stones should not exceed about an eighth of an inch in thickness because of the very deep colour. A small quantity of dilute hydrochloric acid added to the polishing medium and mixed with water to a thin creamy consistency may help to polish massive azurite.

Baryte. This material is of worldwide occurrence, but the best gem quality comes from Colorado, Dakota and northwest England. An interesting massive brown variety is found in Derbyshire. Baryte is very fragile and heat-sensitive and has several directions of perfect cleavage.

Bayldonite. This rare, massive, green material produces interesting bright cabochons although it is doubtful whether it will be found for sale at mineral dealers. Occurs in Tsumeb, South-

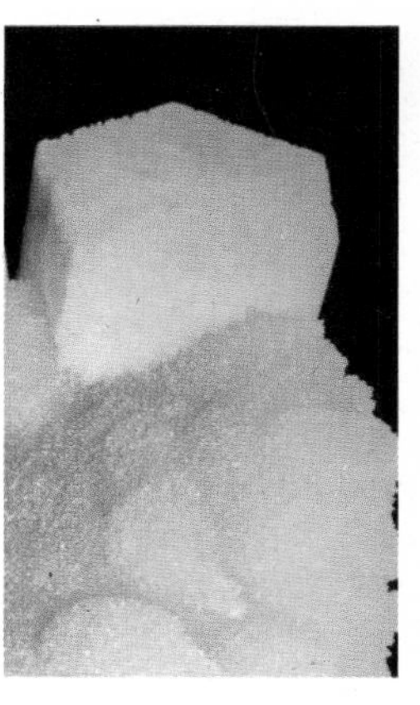

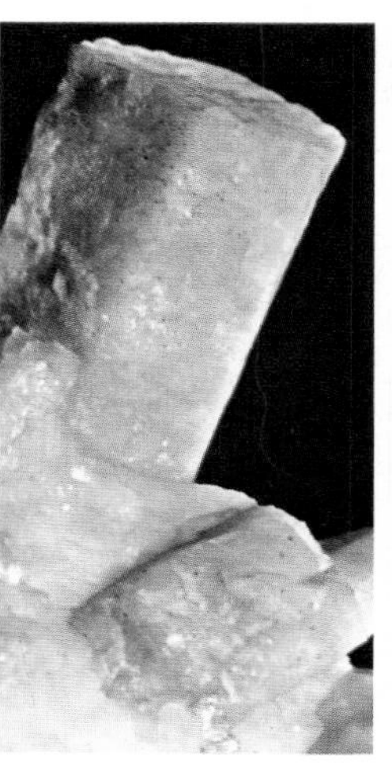

Top: A crystal of aragonite showing typical pseudo-hexagonal twinning

Above: Gem-quality aquamarine, the blue variety of beryl, displaying typical prismatic habit

Above right: Chalcedony showing an interesting mamillary formation

Right: A tabular prism of baryte, sometimes known as heavy spar

West Africa and Cornwall.

Benitoite. This very rare gem material is found only in San Benito County, California. Facet grade material for the amateur is almost unobtainable and will almost certainly be badly flawed.

Beryl. Apart from emerald, which is generally badly flawed since the best material is cut at the mines, the several varieties of beryl of faceting quality may be obtained in large clean pieces, especially aquamarine. No difficulties should be experienced in cutting and polishing although with emerald it may be necessary to seal the surface flaws to prevent ingress of polishing compound. Alternatively the compound may be dissolved after polishing by immersion in nitric acid.

Beryllonite. This is a rare collector's item found only in Maine. Cut and polished with difficulty because of cleavage, it is also brittle and heat-sensitive. The gems may be colourless to pale yellow.

Boracite. This pale green material is mainly found in Germany. Little has been done to promote it although its physical and optical characteristics suggest that it would provide a perfectly reasonable gem. Boracite is not usually offered for sale to the public as gem rough.

Bowenite. This is a hard variety of serpentine sometimes resembling jade. Although relatively common the better varieties are not often seen for sale. Usual colours include green and mottled yellowish-green. Bowenite is sometimes stained to improve its resemblance to jade.

Brazilianite. This is a comparatively new gem material, discovered in Brazil in 1944. Most rough material offered for sale is flawed, and when cutting and polishing care should be taken to avoid the perfect cleavage.

Calcite. The stalagmitic massive material known as 'onyx marble' and the normal veined or pure-white marble are both varieties of calcite and are sufficiently soft to allow hand-working with steel tools. Clear crystalline calcite presents several problems to the faceter, however, as the perfect cleavages and heat-sensitivity create difficulties. These, fortunately, are not insurmountable. Faceted stones are usually step-cut. Although soft the massive varieties may be tumbled. Difficulty may be experienced however in obtaining a good final polish.

Cancrinite. This orange or yellow mineral only very rarely yields transparent faceting-quality material, and even then it is generally flawed and only usable in small pieces. Differing hardnesses in different areas may cause undercutting unless care is taken.

Cassiterite. Faceting-quality material is generally very tough. Colourless material is the most desirable but is very rare; other colours are black, brown, red and yellow.

Celestine. This heat-sensitive fragile material occurs generally in small blue or colourless crystals although red and orange are sometimes found. Its cleavage and softness make it a difficult stone to cut but pleasing gems may be produced.

Cerussite. A collector's item only, for this mineral is very soft and highly sensitive to heat. May occasionally produce cat's-eye stones.

Chalcedony. A cryptocrystalline variety of quartz that is generally very tough. Although mainly without a distinctive pattern chalcedony occurs in many colours; carnelian is an orange or reddish variety and bloodstone is deep green with red flecks. Pale material is often stained but the colours generally appear unnatural. All varieties of chalcedony are ideal material for tumbling.

Chondrodite. A difficult stone to cut and polish; deeply coloured rough should be kept as shallow as possible. Rough material is exceedingly rare.

Chrysoberyl. A large percentage of rough chrysoberyl is flawed and often occurs as waterworn pebbles. The alexandrite variety is very rare and cymophane rough is virtually unobtainable. Care must be taken during polishing to ensure that the stone is kept cool for the heat produced could soften the dop wax.

Chrysocolla. This bluish-green or turquoise cryptocrystalline material is capable of producing very attractive cabochons. The mineral itself may be found either impregnating quartz and opal or by itself as a hydrous copper silicate with various impurities such as silica and iron oxide. Depending upon whether it occurs free or with quartz the hardness may vary from 2 to 7 on Mohs' scale.

Clinozoisite. A fairly fragile material with easy cleavages. Clinozoisite is generally a lighter shade of green than epidote, to which it is closely related. This is mainly a collector's item and rough may be difficult to obtain.

Colemanite. A fairly rare collector's gem that is not only quite soft but also brittle; however, few difficulties should be experienced during cutting and polishing. Material of faceting quality is scarce.

Coral. This organic gem material, consisting in the main of skeletons of the coral polyp, occurs in colours ranging from red through shades of pink to white and black (though this latter type consists mainly of conchiolin, a dark brown organic material). Cutting and polishing is generally carried out by hand, using steel tools and abrasive papers.

Corundum. Varieties of all colours other than red (ruby) are called sapphire for example, green and yellow sapphire occur. When needle-like inclusions are present parallel to the lateral axes, cabochons may be cut showing a star effect. Corundum is cut with difficulty owing to its hardness, but produces highly lustrous gems. The hardness may vary a little with the direction of the crystal.

Crocoite. Small very soft fragile crystals may be occasionally faceted for collectors, although the rough seldom occurs in pieces clean enough for cutting. This material is also quite sensitive to heat and should therefore be treated with great care.

Cuprite. A very deep red, soft and comparatively rare gemstone having a very high refractive index and almost metallic lustre. The faceted stones are usually shallow to reduce the depth of colour, and are mainly cut from rough found in New Mexico and South-West Africa.

Danburite. The rough material is usually colourless although good wine-yellow and pale pink material is known. The brilliant cut is generally used and produces clean attractive stones.

Datolite. A gem whose chemical composition is quite close to that of danburite. The material is generally only cut for collectors, although massive material is sometimes cut *en cabochon* from cream-coloured or brown rough.

Diamond. This is the hardest natural substance known to man, although there are synthetic products said to be at least as hard if not slightly harder. This gem is not normally cut by anyone outside the diamond trade. Good-quality rough material is not only very difficult to obtain but is also beyond the pocket of all but a few collectors.

Diopside. This gem mineral, occurring in colours from pale green to a bright chrome green, and from colourless to bluish-violet, can make very attractive cut stones, and occasionally occurs sufficiently fibrous to allow cat's eyes to be fashioned.

Dioptase. A gem rough that generally looks better as a crystal group than as the very small green stones that usually are all that can be cut from it. A crystal group may be very expensive, whereas single crystals are sometimes offered for sale for faceting at quite reasonable prices.

Dumortierite. This mineral occurs only rarely as faceting-quality rough. Usually it is found impregnating quartz and produces attractive blue cabochons. This type of material generally has the physical properties of the quartz and should be worked accordingly. It is often tumble-polished.

Eilat stone. This material, composed of varying amounts of copper minerals, including turquoise and chrysocolla, produces attractive cabochons of veined blues and greens. It is said that this mineral originates from King Solomon's copper mines near Eilat on the Red Sea. It is made into cabochons and is quite often tumble-polished.

Enstatite. When of gem quality, this pyroxene mineral usually occurs as pebbles, in shades of green and brown. One variety with a high iron content may produce a star effect on a bronze background, and is called bronzite. An attractive green gem may be cut from the rough that occurs as an accessory mineral with diamond in the South African 'blue ground' kimberlite.

Epidote. Pistacite is an alternative name for this gem mineral, owing to its pistachio-green colour. The gem quality rough has strong pleochroism and should therefore be orientated correctly before cutting. A large percentage of rough is so dark that stones should be kept shallow to improve the colour.

Euclase. Many workers consider that this material is difficult to cut due to the perfect cleavage, although no real difficulty should be experienced if care is taken. The rough material is quite scarce, but when available produces interesting pale blue and green gems somewhat similar in appearance to aquamarine.

Feldspar. This important rock-forming group of minerals produces material suitable for faceting, as in the yellow orthoclase or straw-coloured labradorite, or for cutting cabochons or tumble-polishing. Although fairly soft and with perfect cleavage, little trouble should be experienced when handling this material.

Fluorite. This heat-sensitive mineral, which, in the crystalline variety, has four directions of perfect cleavage, produces attractive faceted gems in many colours. The massive variety from Derbyshire known as Blue John is composed of blue or purple layers in a background of yellow or colourless material. This latter type is often carved or turned on a lathe to produce bowls or vases, and much of it has to be resin-bonded to stop it breaking up during fashioning. Blue John may also be tumble-polished.

Garnet. Generally thought to be of a red or brown colour, this isomorphous series of gem-

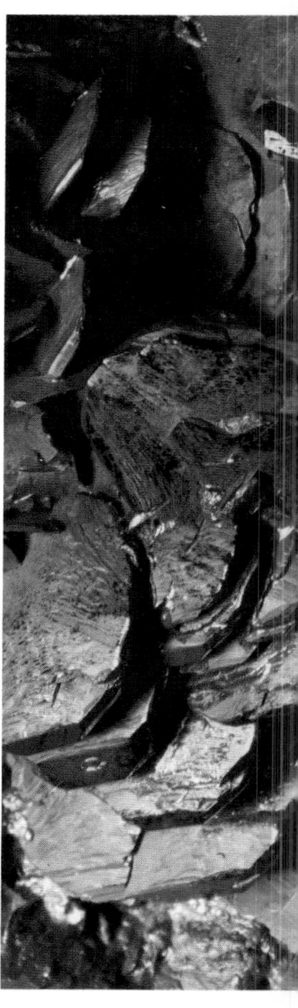

Left: The typical garnet habit is displayed by this specimen of grossular garnet

stones does in fact produce material in various other hues from black through bright greens and yellows to brown and violet, and recently even a colour change variety from Tanzania. No real problems are likely to arise during fashioning although very dark almandine is often cut as a hollow cabochon to reduce the colour density. The more common varieties are also tumble-polished.

Glass. There are very many types of glass, the majority of which are man-made and are usually used to simulate natural gems, such as goldstone, which has included copper crystals so that it imitates reddish-brown sunstone. Some natural glasses may be of extraterrestrial origin, such as the australites and billitonites. Owing to the usually low refractive indices of glass the faceting angles should be between 40° and 50° for the crown angles, and about 43° for the pavilion. The majority of glasses are not sensitive to heat although they should be handled with care owing to their brittleness. Glasses may also be tumble-polished.

Above: Octahedral cuprite. This red copper mineral may be cut into attractive gems when the crystals are of sufficient size

Left: Hematite in lenticular form, one of the interesting habits of this important iron ore

Gypsum. This very soft mineral occurs not only as alabaster, a granular pale translucent material very suitable for the manufacture of clock cases, paperweights and ash trays, etc, but also as satin spar, a fibrous massive material that is extensively used for animal carvings, beads and other ornamental objects. Both materials are soft enough to be cut and polished with hand tools and abrasive paper.

Hambergite. A rare collector's item that is somewhat difficult to work due to differences in hardness with direction, the cleavage and the fact that it is also quite brittle. Rough material is apparently only available now from existing collections as the major source, the Malagasy Republic, is now exhausted.

Hematite. The dust created in cutting and polishing this commercially important iron ore is blood red and it is from this effect that the name is derived. It is used mainly for intaglios or cabochons, some of which are semi-faceted; it is also extensively tumble-polished. When hematite is powdered and purified it becomes the jeweller's rouge often used as a polishing agent.

Howlite. This mineral is often stained blue to imitate turquoise, or it may be used in its natural white or delicately black-veined state. It is well suited to carving, since it occurs in masses weighing 2 kilos or more.

Hypersthene. This is the iron-rich end member of the enstatite–bronzite–hypersthene series. Although normally too dark and opaque to produce faceted stones it may be cut to produce attractive cabochons. The material should be treated with some care as differences in hardness and brittleness make it difficult to saw, grind and polish.

Idocrase. This mineral occurs in several varieties: vesuvianite, an attractive transparent greenish or brownish stone, sometimes faceted for collectors; more importantly, californite, a massive jade-like material extensively used for carvings, and as a substitute for jade; and cyprine, a comparatively rare massive blue variety. No particular difficulties should be experienced when working with any of these materials.

Iolite. Also called cordierite after the French geologist P L A Cordier, or dichroite, owing to the very strong pleochroism. The main colours it shows are yellow, blue and violet. No particular difficulties should be encountered, but the cutter should remember to orientate the rough so that the blue (or violet) colour is on top. The rough material is readily obtainable but nearly always flawed.

Ivory. This material is used extensively for carvings, beads and buttons. It may be of the dentine type, mainly from the tusks or teeth of the elephant, hippopotamus or walrus; vegetable ivories come from the white nuts of certain palm trees, especially the ivory palm from Peru and the doom palm from Africa. Fashioning may be carried out with steel carving tools and abrasive papers, with Tripoli abrasive mixed with oil as a final polish.

Jadeite. This is an exceptionally tough member of the pyroxene group, extensively used for carvings and occurring in several attractive colours. Although not difficult to work, care should be taken with the polishing to prevent undercutting of the minute interlocking crystals which creates a typical 'orange-peel' effect.

Jeremejevite. Newly discovered, this gem is still a rare curiosity and will almost certainly be unobtainable for some time. In case a lucky collector should obtain a small quantity the facts as known so far are: it occurs in South-West Africa and Russia as small blue crystals; the refractive indices 1.639, 1.648 suggest cutting angles of 40° for both the crown and the pavilion; and dichroism is strong, the hues exhibited being blue and colourless. No other data relevant to cutting and polishing are available.

Jet. This soft fossil wood is related to coal. At one time it was used extensively for mourning jewellery. It may be worked by hand using steel tools and abrasive papers and easily takes a very high polish. The best-quality material is found at Whitby in England. Jet is sometimes tumble-polished.

Kornerupine. Apart from possible cleavage, no problems should be experienced when cutting and polishing this fairly rare collector's gem. Material for both star- and cat's-eye stones have been found, but for normal faceted stones the table should be placed parallel to the length of the crystal.

Kyanite. One of the serious problems in working with this attractive, though usually badly flawed, material is the very great difference in hardness that exists in different crystal directions, from 4.5–7.5 on Mohs' scale. When sawing or grinding very great care should also be taken to avoid opening up incipient cleavages.

Lapis lazuli. Also known as lazurite, this material is in fact a mixture of several minerals; the most important of these are lazurite and calcite, usually combined with sodalite, haüyne and noselite, and often containing bright crystals of pyrite. The material is fairly soft but should be worked carefully to avoid undercutting when a large proportion of the harder pyrite is present. Polishing may also be accomplished in the tumbler.

Lazulite. Apart from the very small crystals from Brazil, the majority of this material is only suitable for cabochons. This is generally accepted to be a collector's item due to its rarity, although no problems in cutting and polishing should be encountered if and when rough is available.

Lepidolite. This lithium mica, usually in shades of pink or rose red, is only rarely found in faceting quality. The material is generally used for ornaments and carvings. Although care must be taken, it may be worked and polished using steel hand-tools, abrasive papers and cerium oxide.

Magnesite. Great care should be used when cutting and polishing this material, for not only is it heat-sensitive, requiring a low-temperature dop wax, it also has several directions of cleavage. Magnesite is of world-wide occurrence, but faceting-grade rough comes only from Brazil.

Malachite. Due to its beautiful markings in pale and dark green, this major ore of copper has long been used as an ornamental stone. Care in polishing, to prevent undercutting between the light and dark bands, will produce very attractive cabochons, ash trays, beads, and so on. It may also be tumble-polished, although for best results the complete tumble batch should consist entirely of malachite. Hand-polishing may be assisted by adding a small quantity of vinegar to a thin cream of polishing compound.

Marcasite. This mineral, composed mainly of iron and sulphur, is similar to pyrite in its lapidary qualities in that the fairly high heat sensitivity and brittleness call for careful dopping and cutting. The 'marcasites' set in cheap jewellery are often cut from pyrite or occasionally even faceted steel or other metals.

Maw-sit-sit. A material composed of a mixture of jade and albite is given this name by the Burmese, from the place in which it occurs. Normally cut as cabochons or beads it should be treated in a similar manner to jadeite; the risk of undercutting may be more pronounced. The material comes from Upper Burma and is not usually obtainable through retail outlets.

Meerschaum. Alternatively called sepiolite, this very soft creamy-coloured mineral has often been used for pipe bowls and ornate carvings. It may be hand-worked with steel tools, abrasive paper and a suitable polishing compound. Meerschaum is obtained from Asia Minor and was at one time used in Morocco as a substitute for soap.

Microlite. This dense, comparatively soft, rare mineral may occasionally be obtained in faceting quality, although it normally only occurs in small, poor grade crystals. Microlite is a collector's gem and rough is almost unobtainable.

Natrolite. Heat sensitivity and perfect cleavage make this gem potentially hazardous for the cutter, but providing care is taken, cutting and polishing should be virtually trouble free. Faceting-grade natrolite is scarce, but material may be obtained more readily for cabochons and so on.

Nephrite. This is one of the jade minerals, and is an exceedingly tough member of the amphibole series, related to actinolite and tremolite. Composed of minute interlocking crystals, this structure may cause undercutting during polishing unless care is taken to sand it thoroughly. A large quantity of nephrite occurs as boulders and pebbles with a brownish skin, probably due to weathering, and this is often made use of in carvings incorporating a two-colour design. The best material comes from Siberia, New Zealand and Alaska, but lower-quality material is generally available from many parts of the world, and may be tumble-polished.

Obsidian. A volcanic glass cut and polished for jewellery and often tumbled. This material occurs in a variety of colours from transparent

Above: The rare ore of silver, proustite. This may darken when exposed to daylight, but is capable of being fashioned into attractive stones

Right: Prehnite, a mineral which can sometimes produce a reasonably good jade simulant. The material seen here is lenticular

smoky-brown faceting grade through brown and red (with elongated inclusions that produce a sheen effect), to black (with or without inclusions resembling snowflakes; in fact this latter type is often called 'snowflake obsidian'). Somewhat heat-sensitive and brittle, care must be exercised when fashioning. Faceting angles are the same as for glass: crown 40°–50°, pavilion 43°.

Opal. An amorphous silica mineral; it may show a play of colour, when it is called precious opal, or it may be transparent with body colour, as in fire opal, or it may be translucent with dendritic inclusions, when it is called moss opal. Somewhat heat-sensitive, this material should either be dopped very carefully or be held by hand during fashioning of the cabochon forms. Moss opal and the poor grades of precious opal are sometimes tumble-polished.

Pearl, mother-of-pearl, shell. Pearls may be obtained from both freshwater and saltwater molluscs, although those of commercial importance are restricted to certain types of oyster, clam and mussel. Apart from drilling or partial drilling for necklaces or setting, very little work of lapidary significance is carried out on pearls. Certain pearl oysters having a desirable iridescent internal lustre are made use of in lapidary under the general heading of mother-of-pearl, which is often utilized for inlay work, buttons, carvings, etc. The shells of other sea creatures are also employed and find an important use in the production of cameos, the two most important of which are the helmet shell (*Cassis madagascariensis*) and the queen conch (*Strombus gigas*). Another interesting natural saltwater material sometimes used in jewellery is the operculum, or trapdoor, of the winkle-like mollusc *Turbo petholatus*; these are sometimes also called Chinese cat's eyes because of their appearance, a cream-coloured dome with a top of green or brown.

Pectolite. A greyish-white fibrous massive mineral sometimes cut *en cabochon* for collectors and prized for its sheen. Fairly soft and somewhat heat-sensitive, this material should be handled with care. When available good specimens come from the USA, Scotland and Italy.

Periclase. This magnesium oxide mineral, which was at one time quite scarce in facet grades, is now synthesized and may be available from time to time. Once cut and polished, the surface may become dull and lifeless. One way to overcome this is to place the cut stones in a suitable transparent plastic container to prevent deterioration.

Peridot. The centre member of the fayalite–forsterite series, this pleasing oily-green gem may be cut and polished relatively easily, although care should be taken during the final grinding of the facets to prevent pitting. Gem-quality rough is readily available in small pieces, and although larger specimens are often obtainable they are usually quite expensive. Smaller pieces are often tumble-polished and drilled for necklaces.

Petalite. No great difficulty should be experienced when working with this generally massive material. Small facet-grade pieces are sometimes available and the pink cabochon-quality material makes interesting weak cat's eyes. The faceted stones are mainly collector's items. The massive pink variety may be tumble-polished.

Phenakite. This material when faceted produces somewhat glassy cut stones. Gem rough is not readily available and even then will only produce small stones as the material is generally flawed. A collector's item.

Pollucite. A rare collector's gem, this mineral generally produces colourless faceted stones although pastel shades of yellow and pink are known.

Prehnite. Generally of a pale oily green, and mainly translucent, this material occurs in nodular masses. When cut and polished it may resemble jade although most material is too pale to be a convincing substitute. Cuts and polishes easily except for possible parting along the direction of the fibres. Facet-grade material is rare but cabochon-quality material is quite common. Although soft, prehnite may be tumble-polished.

Proustite. A rare, soft, silver-ore mineral which is occasionally cut into small, attractive faceted stones. Very difficult to cut and polish, this deep-red mineral will also become considerably darker when exposed to daylight and for this reason should be kept in a suitable dark container. This is definitely a collector's gem only, although of widespread occurrence.

Psilomelane. This shiny black manganese mineral may easily be confused with hematite, and is often used for similar purposes, such as making intaglios, semi-faceted stones and cabochons. Psilomelane is also quite suitable for tumble-polishing.

Pyrite. This brass-yellow mineral is the material commonly used for commercial 'marcasites' although it may, when of suitable quality, be used in its own right as jewellery. The cube-shaped crystals are often, for example, sawn in half and used for making cuff-links. When sawing, grinding and polishing, care should be taken as this material is heat-sensitive and brittle.

Quartz. This very common mineral, occurring as it does in several different forms, may usually be cut and polished quite successfully without resorting to any special techniques; the tiger's-eye and hawk's-eye varieties may undercut during the final stages of sanding and polishing. A small quantity of synthetic quartz is now available through retail outlets for the faceter, and some very attractive stones may be produced in unusual colours, blues and greens especially. This synthetic material is reputed to be quite brittle and care should be taken when fine-grinding near the culet to avoid damaging the stone. Because most varieties are of similar hardness, quartz is eminently suited to tumble-polishing, although the cryptocrystalline varieties, which are tougher, will normally polish before the crystalline types.

Rhodochrosite. This rose-red manganese carbonate generally occurs in nodules, and is normally suited to the fashioning of cabochons and beads. Crystals, which are generally small and usually flawed, sometimes produce an acceptable faceted gem but good facet rough is very rare. Although quite soft this material is often tumble-polished.

Rhodonite. Somewhat similar in colour to rhodochrosite this manganese silicate mineral is of widespread occurrence in its massive form, and as such is normally used for cabochons, carvings, beads, etc. Very rarely occurs in pieces clean enough to facet, and when suitable material is obtained it must be treated with the greatest care. The massive variety is quite suitable for tumble polishing.

Rutile. Although this material occurs in faceting quality in reds and browns, the majority of faceted stones are produced from synthetic material made in a modified Verneuil furnace. The 'boules' so produced are black and opaque and must therefore be subjected to further treatment in an oxygen-rich atmosphere. According to the time taken for this process the final colour may be yellowish-white, yellow, orange, red or blue. Pitting is likely to occur during faceting, but may be overcome by changing the direction of fine grinding.

Scapolite. This mineral is a mixture of marialite and meionite and occurs in Burma, Brazil and Madagascar. Cat's eyes and ordinary cabochons are usually produced from the rough material, although facet quality is sometimes available. Cutting and polishing should be trouble-free providing care is taken to avoid the cleavage directions.

Scheelite. Gem-quality natural material comes mainly from Mexico, California and Arizona, but will seldom be seen offered for sale as faceting rough. Material of greater interest to the lapidary is synthetic scheelite, which may be doped with various rare-earth elements to produce yellow, brown and lavender. No problems should be experienced during cutting and polishing.

Serpentine. Under this general heading may be found several minerals, all closely related, that provide the lapidary with interesting material. The mottled varieties from the Lizard in Cornwall are usually reddish-brown, green or grey with veins or patches of the other colours. The much less common williamsite is a bluish-green chromium-rich variety often obtained from Maryland and California. Other varieties are verd-antique, a green serpentine generally mottled or veined with paler greens, yellows and white, and bowenite (*q.v.*), which is the name of a series of serpentines that are distinguished from the others by the much greater hardness of 5–6 on Mohs' scale as against 2.5–4 for the others. Cutting and polishing may be undertaken by hand using steel tools and paper abrasives, or they may be formed into cabochons in the normal way. The serpentine from the Lizard, Cornwall, is turned on lathes into souvenirs, ashtrays, etc. Although fairly soft, serpentine may be tumble-polished but the whole batch should be of the same material.

Sillimanite. Also known as fibrolite, this violet-blue or greenish material, from which cabochons or tumble-polished stones are cut, very rarely produces rough clean enough to facet. Some material, especially that from the Sri Lanka gravels, produces good cat's-eye stones. In all cases care must be taken during cutting and polishing, due to the cleavage.

Sinhalite. It was not until 1952 that this mineral was recognized as being entirely different from peridot, with which it had always been mistakenly grouped. Although the physical and optical properties of the two gems are similar, sinhalite is generally much more brownish green or golden brown and bears a strong resemblance to chrysoberyl and zircon of a similar colour. No problems should be experienced when cutting and polishing this material. The only significant gem deposit is in Sri Lanka and the mineral's name is derived from the Sanskrit name for the

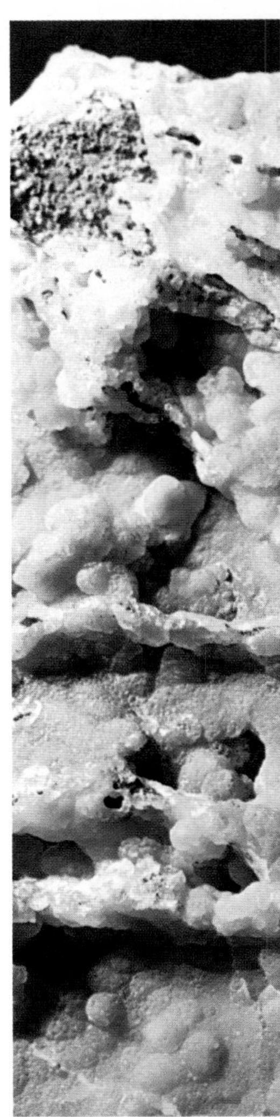

Above: Iron pyrite or fool's gold. Notice the typical striations

Left: Concretionary smithsonite. It is awkward to work, but produces an attractive cabochon

Below left: Sphalerite or blende, an ore of zinc. It is shown here on calcite

Below: Translucent quartz, with dolomite

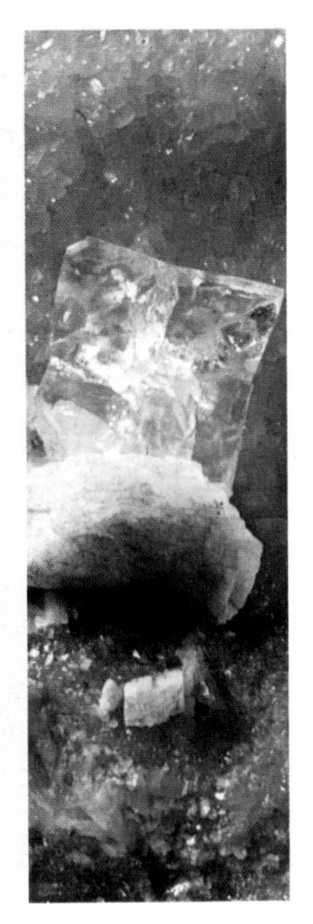

island, Sinhala. Material is not readily available through retail outlets.

Smithsonite. Sometimes marketed under the name Bonamite, this attractive zinc carbonate mineral is of worldwide occurrence. The major sources of lapidary quality are Sardinia, Greece, Spain and the USA, although the best material comes from Tsumeb in South-West Africa; faceting-grade material, which is rare, is found in New South Wales. The only problems likely to occur in polishing are the possible separation of the growth layers in massive material and the possible cleavages of faceting material.

Sodalite. Only the rich blue variety of this mineral has any lapidary significance although it also occurs in greys, greens, yellows and reds. Apart from its important uses as an ornamental mineral for carving, as well as for cabochons and beads, sodalite is one of the major constituents of lapis lazuli. Occasionally material is found that is of faceting quality; the author has handled faceted stones of three to four carats, the stones having a hazy, sleepy appearance. This faceting material is said to be of Canadian origin, as is the massive material which is also found in India, Bolivia and Brazil. Sodalite may also be tumble-polished.

Sphalerite. Also known as zinc blende, blende or black Jack, this important zinc ore of worldwide occurrence is occasionally found in clean facet-grade pieces, generally of a dark brown to yellowish-brown colour, but sometimes in yellow, orange, red and green. Although this material has perfect cleavage, fashioning is usually trouble-free providing care is taken.

Sphene. Titanite is another name for this mineral which produces brilliant faceted stones, which may have an adamantine lustre. Unfortunately most rough, which occurs chiefly in yellows, browns and greens, permits only small clean stones to be faceted, due to the generally bladed shape of the crystals. It is said that the brown material may change colour to an orange hue upon heating to a dull red. No difficulty should be experienced when working with this material.

Spinel. When properly cut, spinels make very attractive stones in a wide variety of colours, including red, orange, pink, blue, purple and green, though the green is generally very dark due to a high iron content. Rare crystals have inclusions that will produce a four-rayed star when properly orientated; these should be cut *en cabochon*. Synthetic spinel is produced in vast quantities in a very wide range of colours, and colourless. Colourless stones do not occur naturally; some natural stones that upon first appearance seem colourless in fact show slight traces of pink. Both natural and synthetic material is easy to fashion and no difficulties should be experienced.

Spodumene. Good coloured rough is capable of producing very attractive faceted stones; mauve, lilac-pink, yellow and yellowish-green are the most common; much rarer are blues and bright greens. Although the rough is difficult to saw and grind due to the strong prismatic cleavage, polishing is trouble-free. Stones should be fashioned with the table facet perpendicular to the length of the crystal for the best colour. Kunzite, the mauve variety, will turn to a bluish green when exposed to x-rays. The colour is stable only providing that the stone is kept away from strong sunlight, in which it will revert to its original colour. The same reversal may be caused by heating to about 200°C. Faceting-grade material is easily obtainable and is generally inexpensive.

Steatite. Also known as soapstone, this very soft mineral is usually used for carvings, and it figures extensively in Chinese art. Occurring in yellows, browns and reds and occasionally in shades of green, it is found all over the world. It may be worked by hand with steel tools, abrasive papers and suitable polishing compounds.

Stichtite. This chromium-rich serpentine-like mineral produces attractive rose-red or lilac cabochons. The rough is fairly rare and is mainly a collector's item, although it may occasionally be possible to obtain rough from Tasmania and Canada. Treat as for serpentine, taking care during grinding and sanding as the material is very soft.

Topaz. Providing care is taken to avoid the one direction of perfect cleavage, there should be no difficulties experienced when cutting and polishing this material. Although most gem-quality rough is colourless, topaz does occur in shades of blue, generally quite pale, brown, sherry-coloured, and pink; these last are heat-treated Brazilian stones, originally of a brownish colour. Occasionally the blue material will produce cat's eye stones when cut *en cabochon*. Topaz often occurs as water-worn pebbles as well as crystals. Facet rough weighing several ounces may be obtained.

Tourmaline. Found in a very wide range of colours, including parti-colours (two or three different colours present in the same crystal), allowing interesting stones to be cut. Care should be taken when cutting and polishing as the material is somewhat heat-sensitive. Apart from this there are no real difficulties. Tourmaline is also tumble-polished.

Tremolite. An end-member of the actinolite–tremolite isomorphous series which, when occurring as a tough interlocking mass of fibrous crystals forms nephrite jade of white or pale shades. As green, fibrous crystals it forms moderately good cat's eyes. It also occurs as a pink transparent faceting-quality material known as hexagonite. Somewhat difficult to fashion because of its perfect cleavage and fragility, it is also said to exhibit different directional hardnesses.

Tugtupite. Discovered in 1960, this cyclamen-red mineral is related to sodalite. Although most of this material is massive, sufficient transparent material has been found to produce faceted stones. The paler areas of the massive material tend to lose their colour in darkness, recovering only upon exposure to daylight. Very little is known of the fashioning characteristics, but because of its closeness to sodalite, fashioning details should be somewhat similar, including faceting angles.

Turquoise. This pale-blue to green massive material has been used as an ornamental stone at least since the origins of Egyptian culture and possibly before. No real difficulties should be experienced when fashioning providing care is taken to dop at low temperatures and to avoid the use of lubricating oils during sawing, as turquoise is somewhat porous and will easily discolour. Rough is available from many sources, but a large proportion may be too powdery to work properly unless impregnated with resin. Staining is also often resorted to in order to improve poor colour.

Unakite. This pleasantly mottled rock composed mainly of pink feldspar, green epidote and quartz is used chiefly in the production of cabochons and is also often tumble-polished. There is a tendency to undercut as both the feldspar and the epidote are softer than the quartz.

Variscite. Somewhat similar in appearance to greenish turquoise, this mineral occurs mainly in Utah and is also often known as utahlite. The compact masses of fibres form nodules, which when sliced and polished provide pleasing display specimens especially when veins of secondary minerals are disseminated throughout. Variscite may also be fashioned into cabochons and beads or tumble-polished.

Verdite. A serpentine-like mineral of deep-green colour often with a light mottling of red or yellow patches. It may be fashioned by hand with steel tools, abrasive papers and a suitable polishing compound.

Wardite. Resembling a bluish-green turquoise, this mineral is found in cavities in variscite. Little is known of its fashioning, but treatment as for turquoise should provide acceptable results when cutting cabochons, etc.

Willemite. Related to phenakite, this zinc silicate mineral is sometimes cut and polished for collectors. Generally of a greenish-yellow colour, although orangeish material is known, this mainly ornamental stone cuts and polishes relatively easily even though quite brittle. Faceted stones are occasionally cut, and facet angles of 40° for both crown and pavilion should give satisfactory results.

Xalostocite. Alternatively named landerite or rosolite, this attractive material is composed of a cream-coloured marble in which large crystals of pink grossular garnet are found. It comes from Xalostoc in Mexico and may, when available, be fashioned into ornaments. Because of the difference in hardness between it and the included garnet, care must be taken to avoid undercutting.

Zincite. Beautiful highly lustrous orangey-red stones may sometimes be fashioned from this rare zinc oxide. Somewhat difficult to cut and polish because of its brittle nature, the facet-grade rough occurs mainly as crystalline fragments, which are so small that cut stones exceeding about one carat are very rare.

Zircon. If one includes the heat-treated stones, this mineral provides cuttable rough in nearly all colours ranging from colourless through yellow to orange, red and brown and to green and blue. Most rough occurs as rolled pebbles, the most prolific areas being Sri Lanka and Indo-China. Heat treatment of the reddish-brown material yields golden, blue and colourless rough. The only real difficulties that may be encountered during fashioning occur during sawing and polishing, for the hardness varies throughout the stone. When cutting zircon to obtain the best effects the modified brilliant cut is used, the so called zircon cut, which has an extra row of facets on the pavilion.

Zoisite. Until 1967 the only known varieties of this mineral were the massive pink thulite, sometimes used for ornaments or cabochons, and the chrome-rich anyolite (*q.v.*), with its ruby crystals, which makes attractive ornaments. Then a new gemstone variety was found in Tanzania, which when faceted produces beauti-

Right: A richly coloured tourmaline, comparatively rare in large sizes

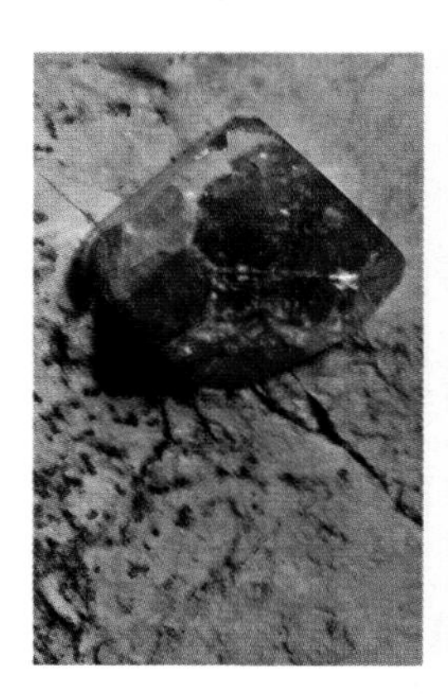

Above: A stumpy pyramidal crystal of zircon. When heat-treated these deeply coloured stones become white enough to act as diamond simulants

Right: Zincite. This rare, highly lustrous gem material is difficult to fashion because of its brittle nature

ful gems. A considerable range of colours became available to the cutter from violet-blue through green to yellow, pink and brown, the most important colour being blue. Heat treatment to about 700°C will apparently alter most colours to blue although some consider that the natural colours are more attractive. Care should be exercised when cutting and polishing as the material is somewhat heat-sensitive and brittle.

Synthetic gemstones

Synthetic gemstones are produced in several different ways, some closely resembling natural processes and others devised specifically for the production of certain 'space-age' materials where very high purity is of paramount importance. Brief descriptions of the commonly used methods follow.

In 1902 the *Verneuil* method for the production of synthetic ruby was introduced. Invented by a Frenchman, Auguste Verneuil, it is still used today, though with more sophisticated machinery and techniques. The process consists basically of allowing a controlled quantity of the ingredients to fall through the intensely hot flame of an oxygen-hydrogen blowpipe. The raw material consists of finely powdered alumina and a colouring agent (chromic oxide for ruby). It is contained in a hopper above the apparatus, which is usually vibrated in order to produce a constant flow of feed powder to the flame. The alumina, which melts at a temperature slightly in excess of 2,000°C, falls onto a ceramic holder or 'candle', where it cools and crystallizes. As the cone of solid material builds up, the ceramic candle is lowered, so that the molten top of the cone is constantly in the flame. Adjustments are made to the rate of powder feed and lowering of the growth zone, until the 'boule' of synthetic corundum is about 20 to 25 mm in diameter and about 50 to 75 mm long. When the desired size has been obtained the flame is shut off and the boule is allowed to cool down. Then it is broken away from the small cone upon which it grew. The breaking away of the boule nearly always causes it to split lengthwise into two pieces of about the same size. When star-stone material is produced, the feed powder contains titanium oxide. The boule then has to be further treated in a furnace at temperatures in excess of 1,100°C, in order that the titanium oxide forms rutile needles. These are orientated in the correct crystallographic directions to provide star stones when cut. As will be seen from the notes below on synthetic materials, other gem substances produced with this method include spinel, strontium titanate, rutile and scheelite.

The *hydrothermal* process produces synthetic gems by crystallizing the material, usually either quartz or emerald, from aqueous solution. Crushed feed material is placed in the bottom of a pressure vessel called an autoclave. Here it is dissolved by water, which often contains sodium carbonate, at a temperature of approximately 400°C and a pressure of about 15 atmospheres. The hot saturated solution rises by convection to a cooler level at approximately 360°C. The cooler liquid cannot hold so much dissolved material, and some of it crystallizes out onto specially prepared 'seed plates' that are suspended from the top of the autoclave. Although synthetic quartz is now available as gem rough, a considerable amount is used in the electronics industry.

The *flux diffusion* or *flux melt* method of producing synthetic gems involves dissolving the basic ingredients in a platinum crucible with a flux, generally lead oxide, potassium fluoride, lead fluoride or boron oxide. The whole melt is raised to a high temperature and then cooled slowly. When the solution falls below the temperature at which it is saturated, crystals grow either on seed plates or on nuclei that appear spontaneously. The *Bridgman-Stockbarger* technique is somewhat similar; it involves producing a melt of pure material in a furnace. Crystals begin to grow as the crucible temperature is lowered.

The *Czochralski* pulling technique also involves the use of pure material as a melt. Crystal growth is started when a seed is gently lowered into contact with the surface of the melt and is then withdrawn slowly. At the same time the seed is rotated at speeds up to 100 rpm. This allows long straight rods of synthetic gemstone to be formed. Although the material so produced may be used as gem rough, its more usual applications are in scientific research.

Beryl. Two main methods are employed to produce synthetic beryl, the hydrothermal process, and the flux-melt process. Both methods produce significant quantities of gem-quality crystals. Groups of these are sometimes available for use as specimens or for setting directly in jewellery, rather than as individual crystals for faceting. It is said that synthetic emeralds are not heat-sensitive and may even be made white-hot without damage. Fashioning techniques are the same as for the natural material.

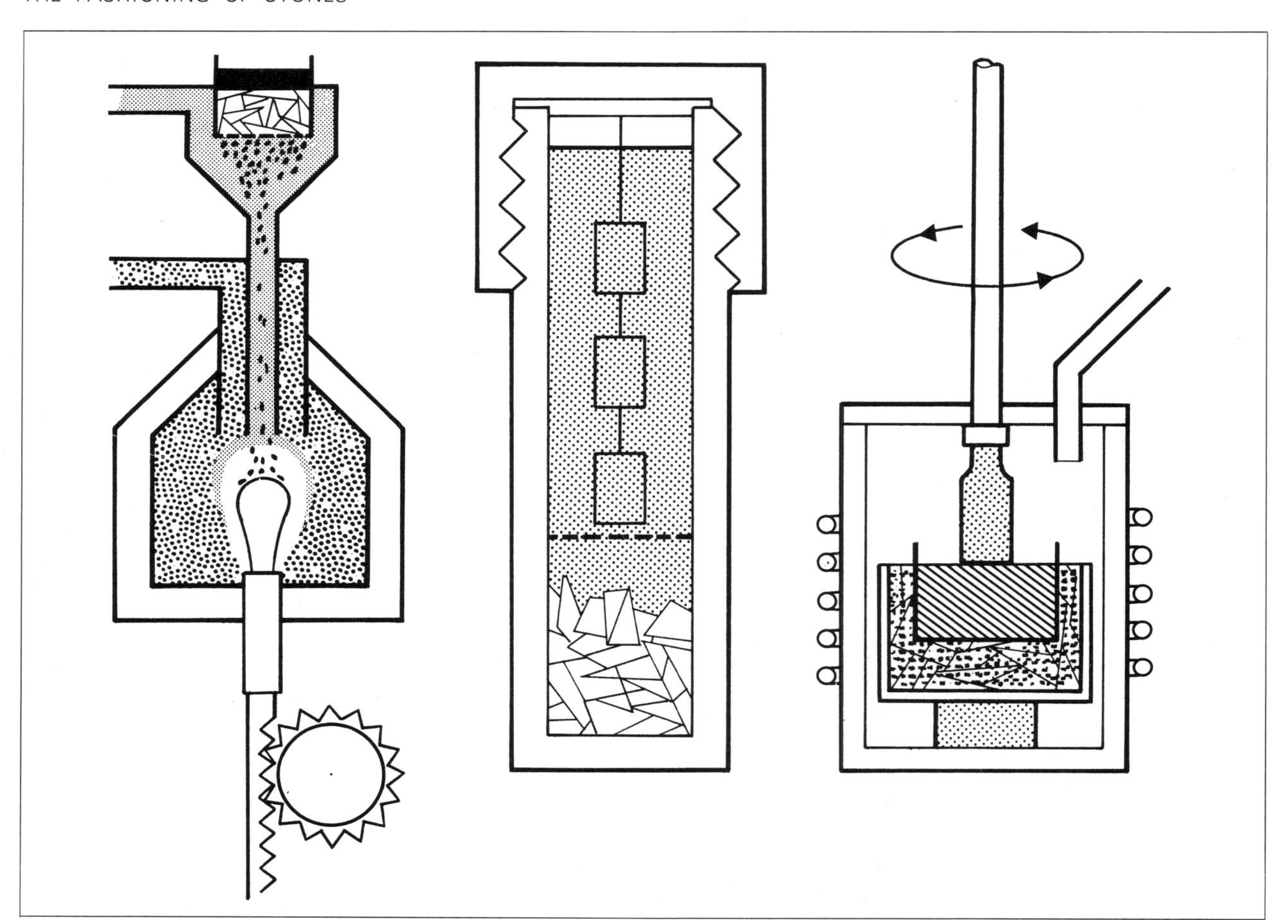

Left: The three most common methods of making synthetic gems. In the Verneuil furnace (far left) a stream of powdered alumina and a colouring agent falls through an oxyhydrogen flame, is melted and drops onto a 'boule' of already crystallized material, to which it adds. In the autoclave (centre) feed material is dissolved in an aqueous solution at a high temperature and pressure. It is deposited on 'seed plates' in the cooler upper parts of the vessel. In the Czochralski process (near left) a seed is slowly withdrawn from a melt while being rotated. New growth occurs at the zone of contact between seed and melt

Chrysoberyl. One of the latest additions to the synthetics repertoire, this is an alexandrite with good colour change – that is, showing a marked difference in colour in natural and artificial light. Little is known about the fashioning of this material and it is unlikely that it will soon become available to the amateur.

Corundum. Ruby and sapphire are produced by the flame-fusion method, utilizing the Verneuil furnace, and stones of a very wide range of colours are available, generally as half-boules. Cutting techniques are similar to those for the natural material although it is somewhat more brittle; it is readily available and inexpensive.

Diamond. Although considerably more expensive than the natural stones, gem-quality diamonds in cuttable sizes are now being manufactured. Amazingly, these synthetic stones can be produced in several different colours. Little is known about the cutting and polishing characteristics, but reports would suggest that difficulties may be encountered in sawing and polishing due to a possible increase in hardness.

Fluorite. Although produced mainly for scientific purposes, this interesting material is sometimes available to cutters. The physical and optical constants are close to those of natural fluorite and it should therefore be treated in a similar manner. A wide range of colours have been made, including reddish, yellow, green and purple. In most cases it is manufactured by growing crystals in a melt of the pure fluorite or by the Czochralski method.

Garnet-type synthetics. Several different substances of faceting quality have been made over the last few years. Although generally known as

Left: Synthetic quartz being lifted from an autoclave

garnets, these stones do not contain silica, which is a major constituent of natural garnets; they are oxides, and resemble garnet only in structure. Produced mainly by the Czochralski pulling technique, several chemically different types exist. The most common is yttrium aluminium oxide, probably better known as YAG, and cut stones of this material have been available for some years under such names as Diamolin or Diamonair. Doping with chromium, cobalt, manganese or certain rare-earth elements produces such colours as green, blue, red or lilac. One of the latest oxides of gem significance is gadolinium gallium garnet, often called GGG. Not only does it have a high refractive index of 2.03, it also has the significantly high dispersion of 0.038, and is therefore a good diamond simulant. The density is very high at 7.05, the hardness is about 6.5. To the cutter, certain facts are important: firstly, most garnet-type synthetic stones crystallize in the cubic system, although a very recent development in the laser industry is a YAG of orthorhombic structure. The hardness of the cubic types is approximately 8.5 and that of the orthorhombic about 8. The density is approximately 4.6 for the cubic yttrium synthetics and 5.35 for the orthorhombic types. The refractive indices for all the cubic types lie between 1.83 and 1.88 whilst the orthorhombic material has indices of 1.93 to 1.95. Bearing this in mind the cutting angles for both types should be: crown angle 40°; pavilion angle 40°.

Lithium niobate. This interesting material, for which nature has no counterpart, is generally produced by the Czochralski pulling technique and is available in many different colours, including red, blue, green, yellow, brown and colourless. It is sometimes marketed under the name Linobate. It is hexagonal in crystal structure, having refractive indices of 2.21 and 2.30, a hardness of approximately 5.5 and a density of 4.64. Apart from the very high double refraction, it has a dispersion of 0.130, almost three times that of diamond. Cutting angles should be between 30° and 40° for the crown, and between 37° and 40° for the pavilion.

Opal. It is only since 1970 that any significant production of synthetic opal has been undertaken. Some of the material is exceptionally beautiful, comparable to the most expensive grades of natural opal. Both black and milk types have been manufactured. The physical and optical properties are similar to those of the natural stones except that the hardness may vary, approximately 4.5 being the minimum.

Quartz. Usually produced by the hydrothermal process, some attractive stones may be cut from this material. The growth is generally made upon the sides of a 'seed' plate of quartz, usually colourless. When gems are cut care must be taken to avoid this colourless patch unless it can be orientated parallel to the table; alternatively the seed may be used to cut an interesting 'striped' stone. The physical and optical constants are the same as for the natural material. The colours available are green, blue, yellow, yellowish-brown and colourless. Some workers have reported that it is more brittle than natural quartz.

Rutile. This material is produced mainly by the Verneuil process and occurs yellowish-white, yellow, orange, red and blue. The refractive indices are 2.62 and 2.90, which is a strong double refraction. The density is 4.25 and the hardness 6.5 approximately. The cutting angles should be between 30° and 40° for the crown and between 37° and 40° for the pavilion. During fine grinding pitting may occur and may be cured by changing the direction of the cut.

Scheelite. Produced either by the Verneuil method or the Czochralski technique, synthetic scheelite is available in several colours, including lavender, yellow, brown and colourless. Scheelite crystallizes in the tetragonal system and has refractive indices of 1.920 and 1.936, a density of 6.1 and a hardness of about 5. Cutting angles should be approximately 40° for both the crown and the pavilion.

Spinel. This material is produced by either the Verneuil method or a flux fusion technique in many attractive colours, some having no natural counterpart. Boules from the Verneuil furnace are the most common. As far as the lapidary is concerned the physical and optical constants are similar to those of natural spinel, although slightly higher refractive indices and densities must be expected in the synthetic material due to a higher alumina content. Cutting angles are as for natural spinel.

Strontium titanate. This material, which has no natural counterpart, is generally produced by the Verneuil process to give very good, water-white diamond simulants, though with a dispersion some four times that of diamond. The refractive index is 2.41, the density is 5.13 and the hardness is 6. Often marketed under the name Fabulite, it must never be cleaned in an ultrasonic bath as the stone will be quickly damaged. The recommended cutting angles are: crown 30°–40°; pavilion 37°–40°.

MAKING A TUMBLER

There is a wide range of commercially made tumblers on the market. Some are quite inexpensive, but nearly all will yield perfectly satisfactory results. There are also more expensive machines using high-frequency vibration rather than a rotational action. These vibro-tumblers are usually excellent and can probably produce as good a result in a week to ten days as conventional tumblers in four to six weeks. However, they are beyond the pocket of most amateurs. It is the intention of this section to describe how a conventional tumbler may be made, as cheaply and easily as possible.

The basic requirements are listed here but alternatives, when they exist, and further articles that may be obtained quite readily from domestic sources, are referred to in the text.

The motor must be of the continuously rated type, so that long runs can be sustained without overheating. Such a motor is generally used to drive the cooling system of a refrigerator, and is usually of $\frac{1}{3}$ to $\frac{1}{4}$ horsepower. Since the same motor could be coupled to other lapidary machines when the tumbler is not required, the extra power output can be useful. Most commercial tumblers with barrels of small capacity – that is, $\frac{1}{2}$ kilo to $1\frac{1}{2}$ kilos – use motors of $\frac{1}{10}$ horsepower or less.

The motor should have slotted fixing holes by which it is screwed or bolted down on its base; the slots allow some alteration to the tension of the belt. The motor should be fixed onto one end of the baseboard with its long axis parallel to a short side of the board. The baseboard itself may be made from plywood, planking, pulpboard or chipboard; the motor should be firmly affixed using either screws or small nuts and bolts. At this stage the motor pulleywheel may be fitted. It should be remembered that a sufficient length of wire must be fitted together with the switch and a suitable plug.

The runners must now be fastened to the baseboard. They should be positioned parallel to the long side and to each other. The distance between them is determined by the length of the barrel to be used. The runners are used to raise the bearings so that the tumbler-drive pulleywheel does not foul the baseboard. The height of the runners is related to the size of the pulley to be used; if, for example, the tumbler pulley is 2″ in radius the centre of its bearing must be at least $2\frac{1}{4}$″ above the baseboard to allow for the thickness of the drive belt. Assuming that the bearings used have a centre-to-base distance of $\frac{3}{4}$″ then runners 2″ high will be sufficient. They should be fixed in such a way that the tumbler drive-shaft pulley is aligned with the motor pulley. It is a good idea to loop the drive belt around the motor pulley and the tumbler-drive pulley, which is held in the hand, and try out various positions for the runner. The chosen position should be marked on the baseboard. The drive belt will have to be loose when fitted and allowance should be made for this when positioning the tumbler-drive shaft.

With the runners fixed to the baseboard the first pair of bearings may now be fitted. The bearings can be of the block type, having phosphor-bronze bushes, or can simply be pieces of nylon or tufnol with holes allowing easy rotation of the tumbler shafts. The first bearing may be fitted in the marked position. Its companion on the other runner may be aligned by fitting one of the steel shafts between them and adjusting the second bearing gently until the shaft runs smoothly. Before finally fixing this second bearing, the shaft should be fitted with its plastic tube in such a way that sufficient shaft extends beyond the first bearing to allow the pulley to be fitted to it; the tubing should be just long enough to cover the inside portion of

Tumbler components

quantity	components
1	baseboard 20″ × 15″ × 1″
1	continuously rated motor $\frac{1}{30}$ hp to $\frac{1}{4}$ hp
2	wooden runners 6″ × 2″ × 1″
2	steel rods 10″ × $\frac{3}{8}$″
2	lengths plastic tubing 8″ long × $\frac{3}{8}$″ internal diameter
1	1″ V-belt pulleywheel
1	3″ V-belt pulleywheel
4	bearing blocks for $\frac{3}{8}$″ shaft
4	bolts, nuts and washers for holding motor
6	screws for holding runners
1	4″ tumbler barrel
8	fixing screws to suit bearing blocks
1	on/off switch
1	V-belt of suitable length

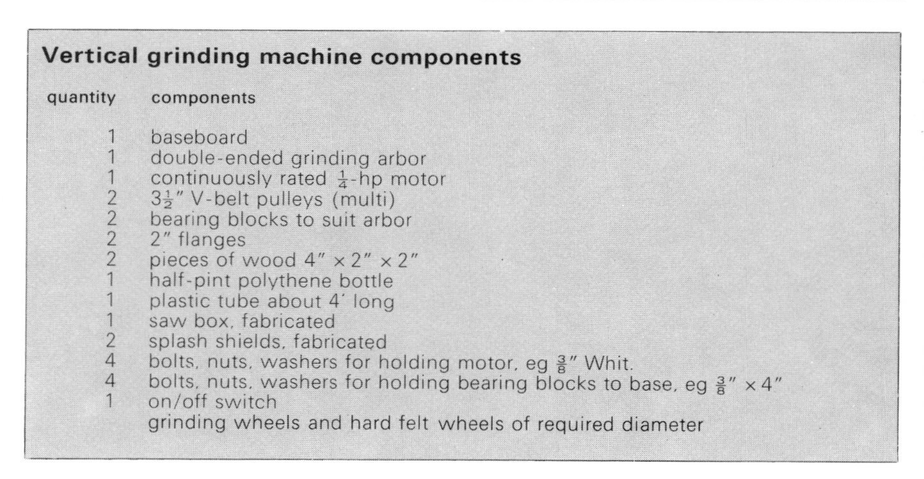

Vertical grinding machine components

quantity	components
1	baseboard
1	double-ended grinding arbor
1	continuously rated $\frac{1}{4}$-hp motor
2	$3\frac{1}{2}$″ V-belt pulleys (multi)
2	bearing blocks to suit arbor
2	2″ flanges
2	pieces of wood 4″ × 2″ × 2″
1	half-pint polythene bottle
1	plastic tube about 4′ long
1	saw box, fabricated
2	splash shields, fabricated
4	bolts, nuts, washers for holding motor, eg $\frac{3}{8}$″ Whit.
4	bolts, nuts, washers for holding bearing blocks to base, eg $\frac{3}{8}$″ × 4″
1	on/off switch
	grinding wheels and hard felt wheels of required diameter

Right: Suggested layout for a home-made tumbler. The barrel here is a paint can. If this type of barrel is used, the tin should be lined with rubber sheet and the lid should fit tightly

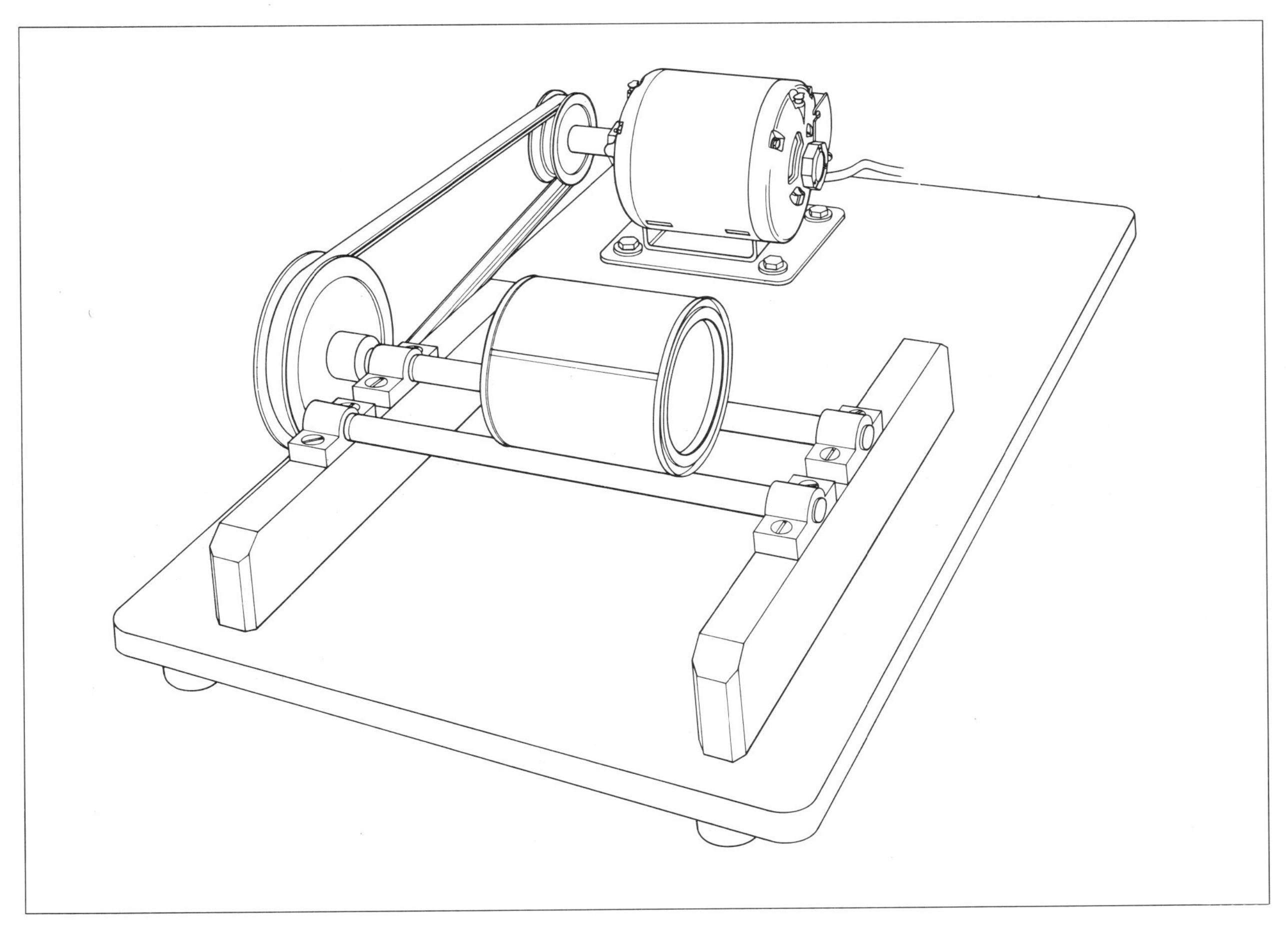

the shaft, leaving a gap between it and the bearing of about one sixteenth of an inch, to prevent possible friction.

The third and fourth bearings should now be fitted, but it is necessary to carry out a series of trials in order to obtain the optimum positions for them, which depends mainly on the diameter of the barrel to be used; the barrel should ride freely above the shafts. It may be an advantage to fix the new bearings loosely to the runners and place a barrel, partially filled for extra weight, onto the shafts. If the tumbler-drive pulley is now turned slowly by hand, the barrel should rotate. If it does not, the shafts should be brought closer together allowing the barrel to ride higher. When this position is obtained, fix the second pair of bearings firmly to the runners; before the last one is finally fixed put the tubing on the second shaft and fit it between the bearings. Then fit the drive belt.

Before switching on the motor to test the tumbler, it is advisable to place a few drops of light oil on the bearings to help them to run freely. Care must be taken not to allow any of this oil to come in contact with the rubber tubing as this will allow the barrel to slip and may even prevent it from turning altogether.

Once the tumbler is running smoothly a loaded barrel should be placed on it and the barrel speed calculated in the manner described previously.

MAKING A GRINDING AND POLISHING MACHINE

Generally speaking the equipment available for cutting, grinding and polishing is expensive. Quite often all that the amateur requires is a small amount of reliable machinery, most of which, for the sake of his pocket, he can make himself. In this section we shall discuss the manufacture of a machine that can be used for grinding and polishing, with the emphasis on keeping the cost at a minimum; with a little ingenuity it may quite easily be adapted to make a trim saw. Basically there are two types of machine available, those that have the working attachments in the vertical plane, and those in which they are horizontal.

The vertical grinding machine

The equipment needed for a vertical machine is listed in the accompanying table, and may be purchased from any good engineering supplier. The size of the pulley wheels is determined by the speed of the motor and the required speed for the grinding wheels. If, for example, the wheels are 6″ in diameter and should have a surface speed of 4,000 fpm, they will have to rotate at approximately 2,500 rpm. If the speed of the motor is 1,425 rpm, the pulleywheel sizes should be in a ratio of almost 1¾ to 1. The pulleywheels on both the motor and the shaft can be of the multi-ratio type, to provide a wide range of turning speeds.

The shaft should be fixed near to the end of the wooden base-board, which should be just wide enough to fit inside the grinding wheels without touching them. The motor should be fitted on the same board, to the rear of the arbor to avoid splashes from the grinding wheels. When the shaft is purchased it may or may not be fitted with flange plates; if not, they must be obtained, for it is these plates, similar to large support washers, which will take most of the load when the grinding-wheel fixing nut is screwed home tightly. Failure to provide adequate flanges may create strain on the wheels, which could shatter. For a six-inch wheel flanges of approximately two inches in diameter should be satisfactory, but it is always best when purchasing the grinding wheels or the arbor to obtain flanges suitable for the shaft or wheel size.

Some form of lubrication for the wheels will be needed and this may easily be provided from a stand placed to the rear of the grinder. On this is fixed a spring clamp to hold an inverted plastic bottle that has a piece of thin tubing sufficiently long to reach from the reservoir to both ends of the shaft. It should be fixed at a point close to the shaft with a piece of stout wire, to allow alterations to its position when the water is supplied for different purposes.

When grinding is under way, with the water flowing, some form of splashguard will be necessary to prevent dirty water and debris not only spoiling clothing and surrounding equipment, but also reaching the motor with possibly disastrous consequences. Splashguards may be obtained from the engineering suppliers, but quite adequate guards may be made from sections of large polythene containers or from thin plywood sheet fixed to the base in such a way that they cover the grinding wheels or other attachments.

Some form of drainage will also be necessary and once again a small piece of tubing may be used, fixed to the base of the splashguard with the other end in a suitable container.

Horizontal grinding machine components

quantity	components
1	baseboard 20″ × 20″ × 1″
1	continuously rated electric motor, approximately ¼ hp
1	plastic bowl approximately 15″–18″ diameter
1	plastic cup
1	single-ended grinding arbor
2	bearings to suit arbor
2	3½″ multi-V pulleys
1	pulley belt to suit motor and V pulleys
1	sheet plywood to support bowl, approximately ½″ thick
	approximately 30 ft of wood 2″ × 2″
	approximately 18 ft of wood 4″ × 2″
	small angle-brackets and screws as required
	nuts and bolts to suit motor fixing slots
	grinding wheels of required diameter and grit size with holes of correct size for arbor
	hard felt wheels
	rubber backing discs
	wet-and-dry abrasive papers
	cerium oxide
	plug, wire, switches, etc

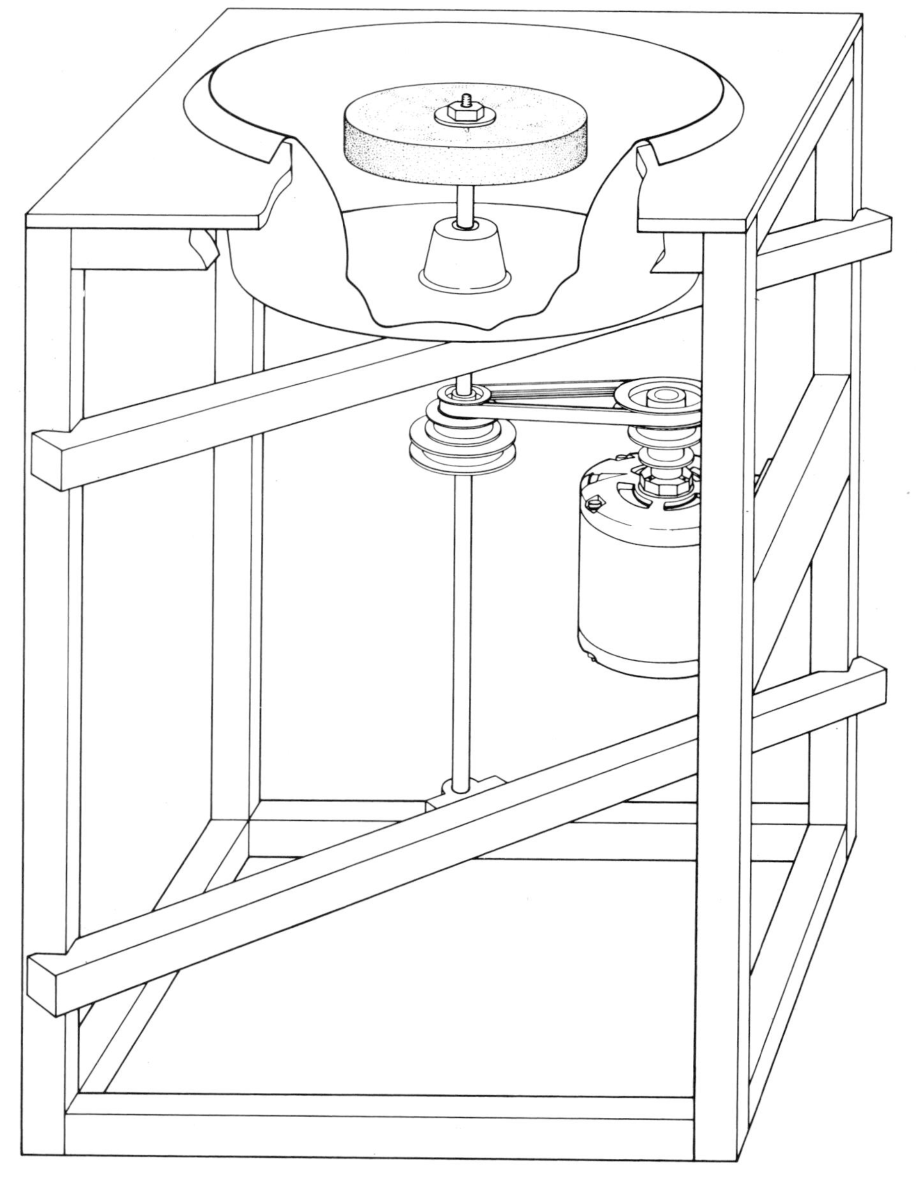

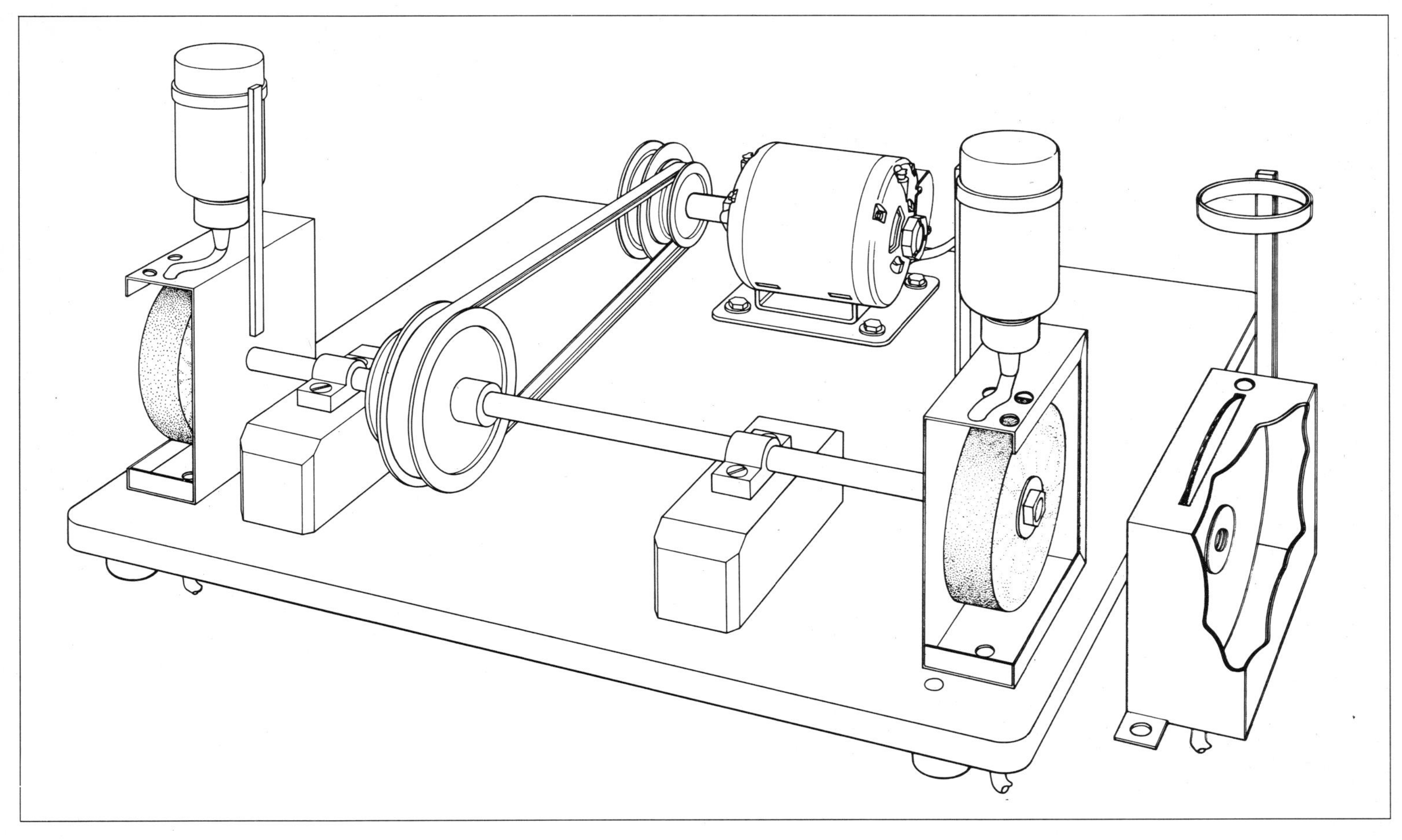

Above: Suggested layout for a home-made vertical combination machine with sawing, grinding, and polishing facilities. Instead of the method of lubrication shown it would be just as easy, for instance, to run a rubber tube straight from a tap. The saw attachment (right) can replace one of the other two units

Left: A home-made grinding machine of the horizontal type. Make sure that the members holding the motor and shaft bearings are strong enough; otherwise the downward pressure on the stone, combined with the vibration, could make the unit collapse

The horizontal grinding machine

Many machines of this type are available ready-made for the amateur, but they are all somewhat expensive. The grinder and polisher described here will allow most normal lapidary operations to be performed with the exception of sawing and perhaps faceting although a little ingenuity could allow even this last to be carried out. The unit itself consists basically of a large plastic kitchen bowl, the centre of which is cut out to receive an inverted plastic cup; this also has a hole in its base through which the shaft passes. The plastic cup prevents any build-up of water and debris from escaping down over the shaft and damaging the bearings.

The whole unit may be let into an existing workbench top or school desk. With a little extra work a wooden supporting framework may be made similar to that shown in the diagram. Whichever way it is used, however, there must be a strong support for the motor and rigid cross-members for the shaft so that the grinding wheel does not wobble excessively.

Before starting to make the supporting framework measure the length of the arbor so that an approximate position may be determined for the diagonal cross-members that will hold the arbor bearings and for the member that will support the motor. If it is possible to obtain an end thrust bearing, it would be preferable to the lower bearing block; for extra strength all three bearings may be used together.

When assembling the plastic cup and the bowl it is important that a good seal is formed. The hole should therefore provide a tight fit for the cup which should be finally sealed with a suitable plastic adhesive.

When the unit is finally assembled you may wish to fill the sides in with thin sheeting. If so, remember to make a panel removable, so that the speed may be altered by means of the multi-ratio pulleys.

Care of the grinding wheels. When handled carefully the grinding wheels should give lengthy service, but it is inevitable that some surface irregularities will form. These should be removed before the wheel is so out of balance that it becomes dangerous. This may be carried out by dressing the wheel with a special diamond tool, obtainable from a local engineering supplier or a lapidary grinding-machine manufacturer.

Felt wheels should not be used for both polishing and prepolishing. It is a good idea to store wheels in plastic bags when not in use.

Conserving and Displaying Minerals

Left: A display of fluorescent minerals. They glow in colours that in most cases are quite different from those that appear in ordinary light (inset). From left to right and top to bottom: zinc blende, scapolite and fluorite; calcite, sodalite and again calcite (showing that a mineral's fluorescence is variable from specimen to specimen); gem quality topaz (showing distinct zoning of its fluorescence); autunite; two specimens of fluorite, glowing blue and white respectively; and at the bottom, further examples of scapolite and sodalite

While it is true that minerals obtained in the field usually need some kind of treatment if their beauty is to be fully visible, the best treatment for a specimen may be to leave it alone. Some minerals are water-soluble and others alter on exposure to light. Only a few minerals have the hardness and durability of gemstones and even these will not always stand up to modern methods of cleaning. The ultrasonic cleaner is especially likely to damage stones with an easy cleavage.

Care of a mineral specimen must begin as soon as it leaves the ground. If a large number of specimens are to be carried they should be individually wrapped in paper; on no account should they be allowed to jostle together. Specimens placed in a cardboard box with no wrapping will be very badly affected by a journey in a car boot; tiny crystals will snap off from their matrix and larger ones may cleave. In addition, it may be difficult to identify specimens once they have become intermingled. It is therefore wise to label your finds in some way before wrapping them. At home the specimens should be carefully unwrapped and placed somewhere where they are not likely to be damaged. The lids and bases of cardboard specimen boxes are quite adequate for storage.

CLEANING

Soil or clay that is compacted or difficult to reach may be removed by judicious probing with a piece of soft wood; metal should not be used. Some finds will consist of tiny pebbles or gravels; these should be sieved (not into the sink) and detritus should be washed away after screening. Many collectors remove matrix from their finds to improve their appearance; this is very often done if the piece is to be shown or sold. However, this is a risky practice since so many minerals can be affected by the sharp blows that are usually required. It should be borne in mind that pressure applied to large rocks in the wrong direction can damage the minerals that they contain. Cavity specimens are almost always better left alone because of the risk of damage.

Some minerals that are found coating others whose features need to be shown can be removed with gentle abrasion. Mica coatings, for example, come away without much difficulty if a stiff brush is used; there is no need to use any metal tool. A calcite coating can be partly removed if use is made of its cleavage. A residue will be left that will need to be dissolved away with dilute hydrochloric acid, provided the underlying mineral is not also soluble in it.

An ultrasonic cleaner can be used on mineral specimens that are known not to respond unfavourably. The best liquid to use is water with a little household detergent.

Before reaching for the tap it is vital to obtain some idea of the nature of your find. You may know what it is, so that the treatment can be checked in a reference book. Or it may be new to you, in which case it should be examined carefully with the help of a geological guide to the district and a mineralogical text book in order to identify it.

Nitrates and sulphates are especially likely to dissolve in water. Some borates are also soluble. In these cases alcohol may often be used. It should not be used near a naked flame. Acetone, CH_3COCH_3, should be treated with similar care; its uses are similar to those of alcohol. Alcohol is also used for quick drying of specimens that have been immersed in water.

Chemical methods

Many minerals have to be cleaned chemically. Stains are often removable with acids. By far the most dangerous of these is hydrofluoric acid, HF. It cannot be stressed too strongly that this highly reactive liquid should only be handled by very experienced workers, and only then with extreme care. It must be kept in polyethylene containers since it will attack some metals and plastics. In no circumstances should it touch the skin or be inhaled. Its chief use is in the cleaning of those silicates that are not attacked by it. It can be used to remove feldspar coatings from tourmaline and beryl crystals. Specimens should not be left in the acid since crystal faces may become etched. If a particular part of the specimen needs to be cleaned with HF while the rest needs to be protected from it, the latter part can be covered with any substance that will resist the acid. Beeswax is suitable and can be dissolved away later with acetone. After treatment with HF specimens must be rinsed for several hours at least, so that all traces of the acid are removed.

Important reminders

1. Don't inhale the fumes that HF gives off.
2. Don't let HF touch the skin; if it does, wash it off immediately with plenty of water.
3. Don't pour HF from one container to another.
4. Don't leave the acid in places where others might be able to tamper with it.
5. Neutralize used acid by placing calcite in it and wash it away with plenty of water.

Ammonium bifluoride (NH_4HF_2) will also be found useful in the cleaning of silicates. It acts less quickly than HF and is not quite so dangerous. However, HF is formed when it mixes with warm water, so great care should be taken. As with HF, surfaces not requiring treatment should be protected.

Some metal specimens, notably gold, silver and copper, may be cleaned with sodium cyanide. This is also a very dangerous poison and should not be inhaled or allowed to touch the skin. In fact, all work with HF and cyanides should be carried out with the aid of a fume cupboard or, failing that, in the open air. If you work outside, ensure that fumes do not blow towards you.

Left: Breithauptite (bronze-coloured) and ullmannite (silver-coloured), their forms largely obscured by a coating of unwanted white calcite

Left: A blende crystal in a matrix of the ubiquitous calcite

Right: Beautiful twinned crystals of prismatic quartz. This would be a rewarding specimen to clean and display

Concentrated hydrochloric, nitric and sulphuric acids are only slightly less dangerous. Normally, however, these acids can be used in dilute form. An important rule must always be borne in mind: *never* add water to an acid—always add the acid, in a stream, to the water. This is especially important with sulphuric acid.

Hydrochloric acid, HCl, is very useful for cleaning off carbonates and will dissolve the calcite that is so commonly found as a coating. A dilute solution of about 5 to 10 per cent is effective. Hydrochloric acid should only be kept in plastic or glass containers, never in metal ones. Keep a careful eye on plastic containers since the acid may percolate slowly through the sides.

Sulphuric acid, H_2SO_4, should be kept in glass containers. It is not very often used for cleaning purposes; it will attack some hydrous minerals and remove water from them.

Nitric acid, HNO_3, must be stored in glass containers. They should be of dark glass; otherwise they will need to be kept in the dark, since the acid decomposes on exposure to light with the release of dangerous fumes of nitrogen dioxide. Like sulphuric acid, it is not often used for cleaning.

Nitric acid mixed with hydrochloric acid in the proportion of one part to three, respectively, will dissolve gold. This mixture is called aqua regia. A mixture of two parts nitric acid and nine parts hydrochloric acid will dissolve platinum.

Acetic acid, CH_3COOH, may be used to dissolve calcite encrustations. It is dangerous in the concentrated form and its fumes should never be inhaled.

Oxalic acid, $(COOH)_2.2H_2O$, is very useful in the removal of iron stains, which are usually composed of the mineral limonite and appear as yellow or brown markings. Oxalic acid is most dangerous in the powder form and should not be swallowed. It is usually quicker to remove staining caused by iron compounds with warm dilute hydrochloric acid.

Before commencing any cleaning involving acids ensure that the reaction will not be disastrous by testing the acid on an inconspicuous part of the specimen first. As might be expected, specimens with no cracks or fissures will be attacked less speedily than much-flawed material. It should also be remembered that in a specimen consisting of more than one mineral, treatment with an acid may attack one part quickly and another part more slowly. Hydrofluoric acid, for example, will dissolve clay minerals from beryl before the latter is damaged.

The mineral classes may be listed according to their solubilities.

soluble in water :

- nitrates;
- some hydroxides;
- chlorides (but not lead, silver and mercuric chlorides);
- carbonates (ammonium, potassium and sodium carbonates only);
- borates (those containing water of crystallization);
- sulphates (not barium, lead or calcium sulphates);
- phosphates (ammonium, potassium and sodium phosphates only);
- some arsenates;

soluble in acids :

- metals (gold, platinum in aqua regia);
- sulphides;
- some oxides;

hydroxides;
fluorides;
carbonates;
borates;
silicates (mostly soluble with difficulty in boiling hydrofluoric acid; but zeolites generally soluble in hydrochloric acid).
sulphates;
phosphates;
arsenates;
vanadates;
molybdates;
nitrates.

Left: A fine encrustation of grossular garnet crystals from Monte Rosso di Verra in Italy

It may be useful to give some examples of the treatment best suited to some minerals commonly encountered.

Albite. Remove iron staining with oxalic or hydrochloric acid; clean with distilled water. Albite sometimes shows black staining due to organic matter, which can be removed with sulphuric acid.

Blende. Clean away any calcite with hydrochloric acid before alteration of bright crystal faces takes place.

Fluorite. Remove calcite coating with hydrochloric acid.

Garnets. Remove mica encrustations with stiff wire brush or needle; remove iron stains with oxalic acid. Some demantoid garnet, notably that from Italy, is found with chrysotile (asbestos) fibres, which can be removed by hand.

Gold. Iron stains will be removed by almost any acid. Hydrofluoric acid will remove encrustations of quartz. Care should be taken not to scratch the specimen.

Gypsum. Adhering clay should be washed off with water. Gypsum is very soft, so care in handling is needed.

Halite. Clean with alcohol, not water.

Malachite. Wash in distilled water containing a little ammonia; then soak in clean water; finally immerse in acetone. (This treatment is unsuitable for azurmalachite.)

Marcasite. Smells of sulphur if decomposition has set in. Wash in warm distilled water, later adding ammonia; rinse in clean water and soak in acetone to dry.

Mica family. Water tends to separate the individual leaves; acids in general may be used for cleaning.

Microcline feldspar. Remove iron stains with oxalic acid and hematite coating with warm concentrated hydrochloric acid.

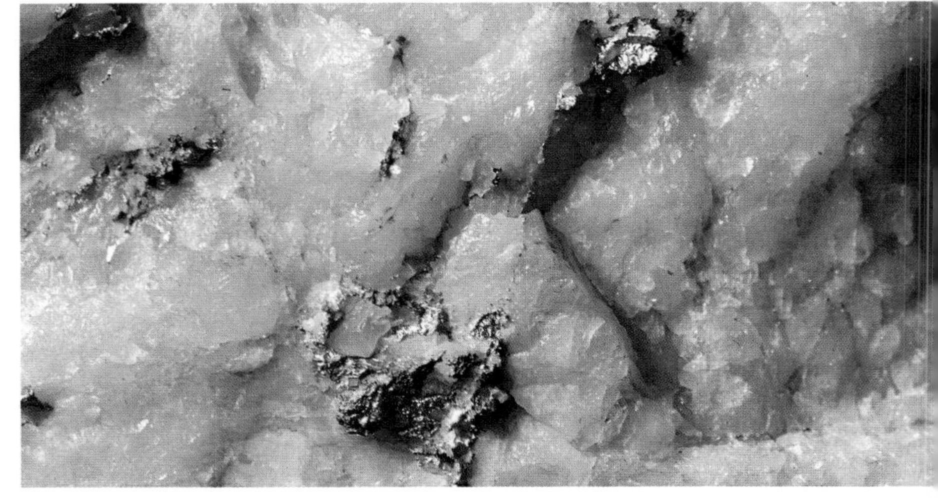

Above: Native gold in quartz, from Val d'Ayas, Italy. Hydrofluoric acid will remove quartz, but gold makes a more interesting and attractive specimen if some matrix is retained

Orpiment. Calcite encrustations are best removed with acetic acid.

Prehnite. Although hydrochloric acid tends to render surfaces powdery it can be used for the quick removal of unwanted mineral encrustations. If the iron stains sometimes encountered penetrate between the fibres they cannot be removed.

Proustite. Surface alteration on exposure to light can be removed by rubbing followed by ultrasonic cleaning.

Pyrite. Iron stains removed with oxalic acid, quartz coatings with hydrofluoric acid.

Quartz. Iron stains removed with oxalic acid; calcite removed with hydrochloric acid. Avoid rapid cooling after immersion in warm cleaning agents.

Realgar. Alters to orpiment and crumbles if kept in light.

Rutile. Remove other silicates with hydrofluoric acid.

Silver. The black tarnish may be removed by a number of methods including immersion in a solution of potassium or sodium cyanide and then in distilled water. Store away from materials containing sulphur.

Spodumene. Adhering clay minerals, often containing iron, can be removed with hydrofluoric acid.

Topaz. Remove iron stains with oxalic acid and clays with hydrofluoric acid.

Tourmaline. Remove adhering clay minerals with hydrofluoric acid; do this quickly to avoid loss of brightness on crystal faces.

Turquoise. Clean with distilled water; avoid oils and greases.

Vanadinite. Calcite removable with organic acids.

Vivianite. Store in the dark and if possible in a moist atmosphere, which may help to prevent alteration of colour and possible disintegration of the specimen.

Willemite. Calcite encrustations should be picked away; acids should not be used.

STORING MINERALS

Minerals displayed under normal conditions are subjected to relatively bright light, to varying temperatures and to varying atmospheric humidity. Specimens must be selected that will stand up to these stresses.

It is wise to keep temperatures as constant as possible. Sulphur is a good example of a mineral that can crack if temperatures change greatly.

Below: The gem display at the National Museum of Natural History, in the Smithsonian Institution, Washington DC. The wall cases contain fine examples of jewellery, while the flat-topped cases contain rough and cut gemstones

Minerals that are liable to dry out include autunite, torbernite and some types of opal, particularly specimens from Virgin Valley, Nevada. Ulexite, sylvine and halite are among those that can be spoiled in the opposite way—by absorbing moisture from the air.

Many minerals develop a tarnish. This may be desirable, but rarely in metallic specimens such as silver and copper. Such minerals should be stored away from sulphurous fumes and in a reasonably dry atmosphere.

Some minerals naturally change when exposed to light; realgar, for example, changes to orpiment. But the effects of light are more frequently noticeable in materials that have been treated to improve their appearance. Gemstones of a pale colour, such as light blue beryl and topaz, are often irradiated to darken them. The colour of some of these stones fades remarkably quickly. In general pinks seem prone to fading, and some colourless stones such as heat-treated zircon may discolour over time. Such discoloration can sometimes be removed by careful heating.

Some practices undertaken to improve the appearance of a specimen may eventually prove to be mistaken. From time to time emeralds appear on the market that have been oiled—that is, immersed in a clear oil, which coats the stone and disguises inclusions. This oil will ooze out under warm conditions, and the appearance of the stone will be quite changed.

Some gemstones—especially, it seems, green and blue materials such as lapis lazuli and jadeite—are occasionally dyed. The dye will eventually fade or react unfavourably with the air to leave the specimen much less attractive.

Recording data

A visitor to a mineral gallery, even some of the most celebrated, will soon discover specimens labelled 'provenance unknown'. Naturally some very old collections contain material whose original source is unknown, but there is no excuse for such ignorance on the part of anyone starting a new collection. When a specimen is collected in the field and individually wrapped, the wrapping should be marked with a reference number corresponding with a detailed entry in a notebook. Many collectors and museums affix a tiny label on which the number can be written. Some mark the pieces with Indian ink in an inconspicuous place.

As soon as identification is complete every possible detail of the specimen should be recorded in a card index or notebook. The following information should be included:

- reference number;
- name of mineral;
- size;
- locality of origin (map references are useful);
- notes on variety, any unusual colour or markings, inclusions, associated minerals and so on;
- good reference in a standard textbook;
- date of acquisition;
- how acquired (finding, purchase, exchange, gift, and so on);
- price paid or estimated value.

DISPLAYING MINERALS

There are as many ways of housing specimens as there are collectors. Many collectors of cut gemstones keep them wrapped in folded papers. The advantage is that information about the contents can be written on the outside and this can be very convenient if trading is expected. They are also easily portable in large quantities. They do, however, wear quite quickly.

Larger rough crystals can be kept in various types of box; those made of plastic with a perspex top are quite satisfactory. These can be labelled underneath and should be excellent for display purposes since they can be stored together in a cabinet. The interiors of cabinets or boxes should be light-coloured to ensure that the maximum light reaches the specimen when on display. The sets of small shallow drawers offered for sale in office equipment shops are excellent for the storage of small specimens, while larger ones can be accommodated in deeper drawers. These can also hold jars in which water-soluble minerals and others needing special storage conditions can be housed. Some mineral-dealers will sell specially constructed cabinets; these often have no advantages apart from an impressive appearance. If a number of specimens are fluorescent it may be a good plan to keep these in a separate display where ultra-violet light can be positioned to illuminate them.

The display of gemstones raises special questions of security. They should never be kept in cabinets without a good lock (the usual filing-cabinet or furniture lock is useless). This applies both in the home (to discourage children) and in particular at shows, where experience has shown that such desirable items as gemstones are frequently the victims of the light-fingered. Many collectors of gemstones become blasé and forget that even a collection of faceted smoky quartz may be desirable to someone or thought to have substantial value.

It is preferable that collections of gemstones and other specimens of obvious value be kept in a safe-deposit or a bank. Those who insist that their stones be visible all the time are simply not being sensible. However, when gemstones are displayed, even only temporarily, certain things must be borne in mind. The first is that most gemstones are very small. They need to be firmly fastened to whatever surface is chosen for their display so that their best aspect faces the viewer. Secondly, the maximum possible light must reach them and special optical effects must be catered for, such as asterism and chatoyancy.

The owner of the collection will have to decide whether to try for an attractive arrangement or whether to group the stones by species. Some museum gemstone collections are mounted on specially constructed plastic stands which can be made at very little cost. Sometimes the plastic can be engraved with the details of the stone although this is expensive and may limit the further usefulness of the stand if a better specimen is subsequently acquired.

The information contained on museum labels is very varied. In general it is better to avoid handwriting since yours will inevitably displease someone; have them typed on an electric typewriter if possible. You should give the stone's crystal system as well as its mineral species and colloquial name; its place of origin, including the mine if possible; any particularly interesting features such as inclusions or associated minerals,

Right: An excellent means of storing specimens, provided by a steel cabinet containing sliding plastic drawers. The drawers are removable, so specimens can be arranged in any desired order. The cabinet was made for housing small items such as screws

Left: A display of small specimens in plastic boxes with transparent lids. Boxes can be obtained with magnifying lenses built into the lids, for viewing 'micromount' mineral specimens

or the collection it was acquired from, if this is noteworthy. For gemstones the weight in carats should also be included.

Undoubtedly the most vexed question in mineral and gemstone displays is that of lighting. Fluorescent tubes (mercury-discharge lamps) should certainly be avoided; they give too diffuse a light and many mineral specimens show their least interesting colour under them. Furthermore they will not enable the viewer to see asterism or chatoyancy, which require a small light-source. One justly celebrated gemstone collection has used car spotlights for its cat's-eye stones and starstones with great success. Many shop windows today are illuminated by small spotlights with tungsten-filament bulbs and these, placed at the correct angles, are very effective in mineral displays.

BUYING SPECIMENS

One of the most difficult things to learn about the world of minerals and gemstones is how they are bought and sold. Many collectors obtain their specimens by visiting likely localities themselves. Others who are members of mineral societies exchange specimens with other members. Those who cannot pursue their collecting activities in either of these ways must visit a mineral dealer. There are relatively few such dealers if one discounts the do-it-yourself retailers of small jewellery findings and miscellaneous agates whose advertisements appear in rockhound magazines. Most of the latter have only rare access to any very interesting material and some, regrettably, have very little knowledge.

Many dealers advertise by post. Those that are situated in well-known mining localities are worth following up. The size and experience of an establishment are good guides in the mineral world, as elsewhere in commerce. The collector will soon discover from experience which are the most useful sources.

Many collectors visit mine dumps and these can still be a fruitful source of good material. It is wrong to suppose that since working is abandoned there can be no danger, and mine owners unwilling to face possible court claims for injuries often restrict access.

Most gem dealers will sell either cut stones or diamonds and only the very largest firms make a practice of dealing regularly in both. Naturally a dealer in coloured stones will be pleased to obtain a diamond for a customer if asked to, but the serious collector would normally choose a dealer who specializes in the stone he wants. In most countries there is an established gem centre. The world's most important are London (Hatton Garden), New York, Idar-Oberstein (West Germany), Tel-Aviv, Antwerp and Amsterdam. The last three specialize in diamonds. In all these places dealers in individual species can be found. Some dealers will be pleased to handle rare gemstones for the museum and collectors' market. Others specialize in smaller *calibré* stones (these are small stones, usually of ruby, sapphire or emerald, which are cut to fixed sizes). Stones are sold by the carat. If a number of stones are placed together in a parcel by the dealer he will normally quote one price per carat for the whole parcel and another price, somewhat higher, for 'picking'. There is no standard scale of prices for gems (or for minerals). Advertisements in the appropriate journals will give some idea of the market at any particular time. Diamond, however, does have a scale of price since the supply of rough is so closely controlled.

Gemstone prices rarely go down in the long term, but the appearance of a large supply of material on the market may depress the price for a while. Buying for investment is not recommended unless one is able to put down a very large sum. It is only worth while for the very finest of the 'classic' gemstones—diamond, ruby, emerald and sapphire, probably in that order. No other kind of stone can appreciate so much as to form a true investment. However, a well balanced and documented collection will always bring much more money to the seller than could the individual items in it.

Selecting material

When gem rough is bought, certain tell-tale signs should be looked for. Any rainbow-like colours in a crystal may indicate incipient cleavage which will ruin the stone if faceting is attempted. Colour may be distributed in bands that are impossible to show to advantage however the stone is fashioned. There may be signs of staining (the dye collects in cracks and hollows and can be seen as a patch of concentrated colour). Inclusions that break the surface may cause the whole stone to fracture when you try to cut it. Much gem material is opaque when recovered from the mine and it may be necessary for a lapidary to polish a 'window' in it to see whether the interior is clear or flawed. If you have rough stones on approval never do this without the permission of the owner.

Obviously rough material that can be cut into good quality finished stones will be expensive. The purchase of faceted stones, especially if they are commercially important, needs a great deal of care. The following points especially should be examined.

Type and depth of colour. For example, many Australian blue sapphires are almost black compared to the better-quality Siam stones. Pale or very dark rubies are much less valuable than the best crimson kinds, which traditionally come from Burma. The most desirable colour for emerald is a bluish-green, best shown in stones from Colombia. Diamonds inclining to yellow are less highly regarded than white ones, but those of a deep, bright yellow are rare and desirable. Other colours, such as blue, green or (very rarely) red are prized. For the more usual white diamonds a number of colour-grading schemes have been proposed, but none has been generally adopted.

Clarity. This is especially important for diamond; grading systems usually take it into account along with colour. A certain number of inclusions does not detract from the value of emerald, since totally clean stones are

Above: Rough material of gem quality. The finished stones take up a far smaller space, since up to half of the volume is lost in cutting

almost never encountered in nature.

Style of cutting. Stones cut in the country of origin used to be 'lumpy'; this was especially the case with Ceylon sapphires and some Indian material. This is now much less true because of the progress of cutting techniques in these countries.

Chatoyancy and asterism. Eyes or stars should be as close to the centre of their stones as possible. If they are a great deal off-centre the value of the stone is greatly reduced.

Synthetic crystals have been produced, often in large quantities, to imitate most of the well known and expensive gemstones. A gemmologist will be able to identify most of these without too much trouble. Some that are not of obvious use as gems may be offered for sale as mineral specimens. They look too 'finished' to be natural, but could deceive the inexperienced. If you are offered anything suspiciously bright in colour, especially if it is a bright green, have it checked by an expert.

The Mineral Kingdom

Left: Characteristic rosettes of wavellite from Barnstaple, England, showing a fibrous radiating structure

The following pages contain the essential information on over one thousand minerals. These are nearly half of all those known to science, and they include all those that the vast majority of mineralogists will ever be able to examine, whether they be enthusiastic collectors or professionals.

The ordering of so many entries presents difficulties; there are many different systems of classification of minerals, and all are arbitrary to some extent. In large part these entries follow the order of the Chemical Index of Minerals (CIM) published by the British Museum (Natural History) in London. The first section of CIM consists of the elements and their alloys; further minerals are entered according to traditional subdivisions into oxides, sulphates, silicates, phosphates and so on. They are further grouped by their metals, according to their atomic weight.

The ordering of the entries here departs from CIM to make clearer some of the important relationships among minerals. Members of the main mineral families are grouped together. Each such family begins with a general introduction describing some of its chemical properties. Within each group the minerals are listed according to CIM; and minerals not belonging to the main families are listed according to CIM too.

To find a particular mineral in this section, consult the Index. Under each mineral name in the Index there are often several page numbers—numbers in bold type refer to this reference section.

If you have a sample of a mineral of whose identity you are unsure, you should consult the Identification Tables in the following section. When you think you know the name of the mineral you can find its entry by consulting the Index.

Details of cutting and polishing minerals are included here whenever the mineral is suitable for fashioning and is likely to be available to the ordinary lapidary. Information has not been provided when the mineral is very rare, too difficult to work or too easily damaged to be worth fashioning.

The information here has been compiled from many detailed works of reference, which are listed in the Bibliography of this Encyclopedia. Of necessity they are expressed in a condensed form. The technical terms that have been used are explained fully in the appropriate sections of this book. Brief definitions of them will be found in the Glossary. Most are easy to understand and are necessary to anyone with a serious interest in minerals.

The information in the entries is listed in the same order throughout: name of mineral; chemical composition; general properties; mode of occurrence; localities in which it is found; mode of treatment; fashioning details.

Name. Usually this is the name given in CIM. Chemical designations have been used rather than more fanciful ones—rutile (synthetic) in preference to titania, for example. The spellings of names follow CIM; thus pyrite, baryte and sylvine replace the older pyrites, barytes and sylvite.

Composition. In a few cases, such as psilomelane and limonite, the exact composition of a mineral is uncertain. In these cases, it has been described in words; otherwise it would have been necessary to give several formulae.

General properties. Here the mineral's *crystal system*—the basic arrangement of its atoms—is specified. Six systems are used, rather than seven; this means that crystals that some authors describe as belonging to the trigonal system are here classified under the hexagonal system. This usage is the more modern one. The usual shape of the mineral's crystals, its *habit*, is described too, together with its *lustre* and *colour*. *Twinning* is mentioned when it is a conspicuous feature—crystals are described as twinned when they appear in two parts, which are mirror images of each other, or which interpenetrate.

The *cleavage*, or tendency to splitting along certain directions, is mentioned here. It is an important way of identifying minerals, and it is important to a collector to know about cleavage before he attempts to separate a sample from other minerals. It is even more important to the cutter of stones, since a stone with a cleavage can shatter when it is polished in certain directions. The *hardness* of the mineral is specified here, according to Mohs' scale, which runs from 1 (talc) to 10 (diamond). It is an important guide to the identity of a mineral, and is comparatively easy to test. *Specific gravity*, SG, is the density of a mineral, its mass per unit volume; it is given here in grams per cubic centimetre, gm/cm^3. A careful series of measurements is needed to make accurate determinations of it, but it is again a good method of getting at a mineral's identity.

Occurrence. Each mineral occurs in association with characteristic companion minerals and in characteristic geological circumstances; atacamite, for example, is found in the oxidized zones of copper deposits, often with cuprite and malachite.

Localities. Very often a mineral is known from only a few sites, and then the discussion of its mode of occurrence merges with the description of these localities. In general, the localities have been included in which outstanding specimens are to be found, or the particular place that has given its name to the mineral. Sometimes the places mentioned are the only known places of occurrence of the mineral; more often, they are a selection from localities too numerous to list. All place names have been modernized where necessary, and as far as possible their spellings follow the usage of the Universal Postal Union. To find a particular site the collector must consult the mineralogical literature of the area he is interested in.

Treatment. In most cases the best treatment for a mineral species is to leave it alone. Do not attempt to clean any of the minerals listed here unless a treatment procedure is described. In certain cases fragility or unfavourable reaction to certain acids used in fashioning has been mentioned. Where some cleaning agent has been found to be particularly useful, such as alcohol for the cleaning of fibrous material, it has been mentioned.

Fashioning. Details of fashioning stones that are of interest to lapidaries are given here (provided by Colin Winter). They include the kinds of shapes that the mineral can be cut into, whether faceted or cabochon (rounded); how it splits, and how readily; what angles facets should be cut at so that the light rays follow paths that give the maximum fire and colour to the stone; and the mineral's sensitivity to heat—an important consideration, since stones are heated during fashioning, both deliberately, when mounted with wax, and unavoidably, owing to friction when being ground and polished.

At the end of the section, a few materials of organic origin, such as jet and amber, are listed for their aesthetic appeal.

Wherever possible, minerals that are especially important, interesting or attractive, or simply very common, have been illustrated with photographs. No amount of verbal description of the appearance of a mineral can replace one good picture—to the mineral enthusiast, the next-best thing to holding a sample in his hands.

Right: Colourless prismatic crystals of celestine with yellow crystals of sulphur

Above: Copper from Lake Superior, Michigan. This crystal shows the enlarged faces of the rhombic dodecahedron, in this example out of proportion

Above right: This copper specimen shows dendritic (plant-like) habit and is from Houghton, Michigan

Elements and Alloys

Copper Cu

Native copper occurs in small amounts only, the mineral being almost always secondary in origin. It crystallizes in the cubic system but crystals are rare; when found they are cubic or rhombdodecahedral. Dendritic or filiform habit is far more common. The colour is copper-red with a brown tarnish, the lustre metallic, and the streak copper-red. It is ductile and malleable with a hackly fracture. H 2.5–3.0; SG 8.9.

Occurrence. Copper may occur in basaltic lavas, sandstones and conglomerates, and is recovered by the reduction of copper minerals or solutions. It is found in veins and beds in association with chalcopyrite, chalcocite, cuprite, malachite and azurite. It often occurs in the vicinity of igneous rocks.

Localities. Occurs in dendritic forms in limestone at Turnisk, USSR; as crystals in Cornwall; at Wallaroo, South Australia and Broken Hill, New South Wales. Found in sandstone at Corocoro, Bolivia, and at Cananea, Mexico. In the USA copper is found in abundance in the Keweenaw area of north Michigan; at the Copper Queen mine, Bisbee, Arizona and at Georgetown, New Mexico.

Treatment. Copper specimens should be cleaned with water as they are readily attacked by acids, especially nitric. Black deposits of copper oxide may be removed by the use of a solution of one part by weight sodium hydroxide, three parts potassium sodium tartrate and 20 parts distilled water.

Silver Ag

Silver is a member of the cubic crystal system but crystals are rare, the mineral occurring more commonly as dendritic wires or as scales. The colour is silver-white, tarnishing quickly to black. The streak is silver-white and the lustre metallic. It is malleable and ductile with a hackly fracture. H 2.5–3.0; SG 10.1–11.1

Occurrence. Found in hydrothermal veins, where it may have been formed by the action of hot waters on silver sulphides, or of metallic sulphides or arsenides on silver chloride. It occurs disseminated in various metallic sulphides.

Localities. Silver is found at Kongsberg, Norway, sometimes as fine, large specimens; at Freiberg, E. Germany; Příbram, Czechoslovakia; Broken Hill, New South Wales; Chanarcillo, Atacama, Chile; and in the states of Guanajuato and Chihuahua, Mexico. In the USA silver occurs at Keweenaw, Michigan, in association with copper ores, and at Butte, Montana, as well as Colorado, Arizona and Idaho. In Canada it occurs at Cobalt, Ontario.

Treatment. Dissolves readily in nitric acid, hot sulphuric acid, and aqua regia. Tarnish may be removed electrolytically or with sodium cyanide or potassium cyanide. These compounds should be used with the utmost care, since they are dangerous. The tendency of silver to tarnish is greatly increased when it is stored in the neighbourhood of sulphur-bearing agents.

Right: This silver specimen from Secci, Sarrabus, Sardinia, is called filiform from its threadlike appearance

Gold Au

Gold crystals are rare; they belong to the cubic system and when found may be octahedra, cubes or rhombdodecahedra. Gold is more commonly found as grains, in dendritic forms or as nuggets, which are irregular lumps with rounded surfaces. Twinning is common. The colour is golden yellow, lighter when there is a high silver content; the streak is a similar colour. The lustre is metallic. Gold is ductile and is the only highly malleable yellow mineral. Gold is usually alloyed with varying amounts of silver and may also contain some copper. Electrum contains up to 20% silver and rhodite contains some rhodium. Porpezite contains palladium. H 2.5–3.0; SG of pure gold 19.3; SG of alloyed gold 15.6–19.3

Occurrence. Gold is quite widely distributed in igneous, sedimentary and metamorphic rock, and in sea water. Auriferous quartz is a major source for gold and occurs in hydrothermal veins. In examining this type of occurrence care should be taken not to confuse gold with pyrite, which is superficially similar. Alluvial deposits, which contain gold released from rocks by the action of weathering, are good areas for the prospector. Consolidated deposits such as those in the Witwatersrand, South Africa, are the major economic gold producers. The Republic of South Africa is in fact the world's largest producer of gold, being responsible for over half the total annual production.

Localities. Apart from South Africa, gold is found in Rhodesia, Egypt and Ghana. Also found in Wales and sparingly in some other parts of Great Britain; in the Ural Mountains, in Czechoslovakia and Romania; in the Alps of South Island, New Zealand; from Charters Towers, Queensland, from Bendigo and Ballarat, Victoria and from Kalgoorlie in Western Australia. In the USA gold is often obtained by individual prospectors by placer mining. It is found along the mountain ranges of the western states, including California, Colorado, Nevada and Alaska. In Canada deposits occur at the Klondike, Yukon Territory, in Nova Scotia and Ontario.

Treatment. Gold is dissolved only by aqua regia and iron stains may be removed by any convenient acid. Quartz may be removed by strong hydrofluoric acid.

Above: Gold on quartz from California. Such a specimen would command very high prices, owing to the large amount of gold visible

Left: Another fine example of gold on quartz, also from California

Left: Silver showing dendritic form; strictly speaking it is reticular dendritic since the 'branches' form a mesh. From the Harz Mountains, Germany

Right: Native mercury from Lucca, Italy. This form is extremely rare as mercury is normally found in the combined state

Mercury Hg

Mercury is rare in the metallic state, usually being of secondary origin and obtained from cinnabar. It crystallizes at −40°C and shows a hexagonal structure. It is tin-white with a brilliant metallic lustre and is found in small fluid globules. SG 13.6

Occurrence. Mercury is usually found as small isolated drops although it may sometimes occur as large fluid masses in rock cavities, usually in regions of volcanic activity. Deposits from hot springs occasionally contain mercury as well as cinnabar.

Localities. Mercury is found at Idrija and Mount Avala, Yugoslavia and near Landsberg, W. Germany. It also occurs in the United States, in California and Texas; and in Spain and Italy.

Treatment. Occurs naturally in such small drops that cleaning would not be attempted. In large quantities it is highly poisonous and all activity associated with it should be carried out in very well-ventilated conditions.

Amalgam

Amalgam is the alpha-phase of the system Ag–Hg. The gamma-phase is moschellandsbergite. Amalgam is a member of the cubic crystal system and is found as dodecahedra and in a massive form. The colour and streak are silver-white; the lustre is brilliant metallic. H 3.0–3.5; SG 13.7–14.1

Occurrence. Amalgam is rare and found in mercury or silver deposits as grains or scattered crystals.

Localities. Fine crystals from Landsberg, W. Germany; and Coquimbo, Chile.

Diamond C

The transparent, crystalline form of carbon, diamond is polymorphous with graphite. It crystallizes in the cubic system and occurs as octahedra, dodecahedra or icositetrahedra; twinning is common but cube forms are rare. There is a perfect octahedral cleavage and the fracture is conchoidal. Many stones are found as worn pebbles. Some stones may display anomalous birefringence

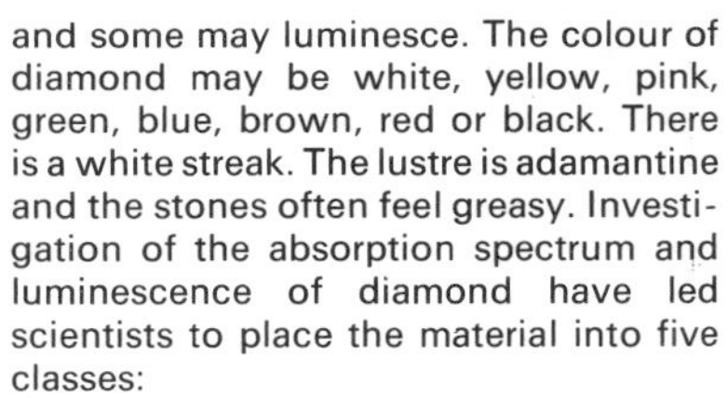

and some may luminesce. The colour of diamond may be white, yellow, pink, green, blue, brown, red or black. There is a white streak. The lustre is adamantine and the stones often feel greasy. Investigation of the absorption spectrum and luminescence of diamond have led scientists to place the material into five classes:

- *Type 1.* Nitrogen contained in platelet form; the majority of all natural diamonds.
- *Type 1b.* Nitrogen contained in dispersed form.
- *Type 2a.* Stones with no significant nitrogen.
- *Type 2b.* Diamonds with no nitrogen but containing aluminium; all natural blue stones are of this type.
- *Type 3.* The meteoric diamond with a hexagonal structure, named lonsdaleite.

Diamonds are graded both when rough and after cutting, different systems being used for each type of appraisal. Many

Above right: Crystals of diamond, some displaying the characteristic forms (particularly the octahedron) of the cubic system. Note the high lustre

Far right: Diamond in its parent rock of kimberlite from Kimberley, South Africa

Right: A single crystal of diamond in its matrix. The faces are curved

shades of white are recognised for the cut stones, though the term blue-white has fallen into disuse; the best quality stones are called 'finest white' or 'river' according to the country where the grading takes place. Cape stones are yellow in appearance – not a bright yellow but a perceptible off-white. H 10; SG 3.52; RI 2.42, dispersion 0.044
Occurrence. Diamond occurs in alluvial gravel deposits associated with quartz, corundum, platinum, zircon and other minerals. The other main type of occurrence is in the 'pipes'; these are composed of kimberlite, in which the most prominent mineral is serpentinized olivine. Diamond is randomly distributed in the pipes.
Localities. Diamonds are found in India, the celebrated Golconda mines being in the Madras area; in Brazil, from Diamantina, Minas Gerais; from South Africa in a number of areas including Kimberley, Pretoria (the Premier mine is in this area), and Jagersfontein in the Orange Free State. Also found in Tanzania, Sierra Leone, Ghana, Zaire, Botswana, South-West Africa, Angola and Guinea. Other sources are Australia, Borneo, Guyana, Venezuela and Siberia, where the Mir pipe is one of the largest in the world, and parts of the US.
Treatment. Diamond is unaffected by acids and crystals may be cleaned by boiling in concentrated hydrofluoric acid.
Fashioning. Uses: faceting, impure or badly coloured material used as abrasive; *cleavage*: perfect, octahedral; brittle; *cutting angles*: crown 34° 30', pavilion 40° 45'; *heat sensitivity*: very low.

Left: Graphite, one of the forms of carbon, from Siberia. The other form is diamond

Graphite C

Graphite, sometimes called black lead, crystallizes in the hexagonal system as flat six-sided crystals, but is usually found massive. There is a perfect basal cleavage. Both graphite and diamond are pure carbon; the phenomenon of widely differing forms in substances of identical chemical composition is called *polymorphism*. Graphite has a typically greasy feel and will mark paper; it is black and gives a black streak. The lustre is dull metallic. It is sectile and is easily scratched. H 1–2; SG 2.1–2.3
Occurrence. Graphite occurs in flakes in metamorphic rocks that themselves are derived from rocks containing carbon. It also occurs in veins and pegmatites.
Localities. Found in the Irkutsk area of Siberia, and in Sri Lanka, where it occurs in veins. Graphite crystals occur in limestone at Pargas, Finland; a compact form has been mined at Borrowdale, Cumbria; graphite suitable for pencils comes from Sonora, Mexico. In the USA a major locality is at Ticonderoga, New York State, where it is found in quartzites and with quartz in small veins running through gneiss.

Lead Pb

Native lead occurs only rarely, and sometimes contains some silver or antimony. It is a member of the cubic crystal system and forms thin plates and small globular masses of a grey colour. It is soft enough to be scratched with the fingernail, is very malleable and somewhat ductile. The lustre is metallic. H 1.5; SG 11.4
Occurrence. Very rare as native metal. Occurs in compact limestone with hematite, and in gold placers.
Localities. Lead is found in Sweden at the Harstig mine, Pajsberg; from the gold placers of the Ural Mountains, USSR; from Mexico near Veracruz; and from Franklin, New Jersey.
Treatment. Any acid apart from nitric can be used as a cleaning agent.

Arsenic As

Arsenic is found as light grey reniform or botryoidal masses; it may also be stalactitic. Crystals of the hexagonal system with rhombohedral form are occasionally found. There is a perfect cleavage. The lustre is metallic and the streak light grey. It may smell of garlic when struck with a hammer or on heating. It tarnishes quickly to a darker grey. H 3.5; SG 5.7
Occurrence. Arsenic occurs in hydrothermal veins and often contains some silver, iron, gold or bismuth as traces; it may also contain antimony. It is sometimes accompanied by ores of nickel and cobalt.
Localities. Fine reniform masses occur at St Andreasberg in the Harz Mountains. Also from the silver mines at Ste. Marie-aux-Mines in Alsace, France, and from Fukui Prefecture, Japan, where globular masses of rhombohedral crystals are found. It occurs in Arizona, and in the Montreal area of Canada.
Treatment. Arsenic is attacked by acids except hydrofluoric, which may therefore be used to remove adhering material. It is not possible to remove the tarnish.

Domeykite Cu_3As

Domeykite is a copper arsenide closely related to mohawkite, which contains Ni and Co. It is a member of the cubic crystal system; reniform and botryoidal forms are found. White to grey in colour, it tarnishes easily. The lustre is metallic. H 3.0–3.5; SG 7.9–8.1
Occurrence. With copper ores.

Left: Native arsenic in massive form

Right: Hopper crystals of bismuth. Natural bismuth rarely shows its crystal form as well as does this synthetic specimen

Far right: Sulphur crystal from Perticara, Italy

Localities. Domeykite has been found in several localities in Chile, and from the Mohawk mine, Keweenaw County, Michigan; also from Lake Superior, Ontario.
Treatment. Brown coating will eventually form; impossible to remove permanently.

Algodonite
Some workers do not consider this copper compound as a separate mineral as there is a close resemblance to domeykite. The formula Cu_6As has been proposed. It is silver-white to steel-grey in colour, tarnishing quickly to brown on exposure to light. H 4; SG 8.3
Localities. Algodonite has been found in Chile and from the area of Lake Superior.

Antimony Sb
Antimony belongs to the hexagonal system, but is commonly found massive and reniform, with granular forms also occurring. Polysynthetic twinning may also be displayed. There is a perfect cleavage. It is light-grey with a metallic lustre and grey streak. H 3.0–3.5; SG 6.7
Occurrence. Antimony occurs in hydrothermal metal-bearing veins and is frequently associated with silver and arsenic. It is often accompanied by stibnite and sphalerite.
Localities. Near Sala, Sweden; St. Andreasberg in the Harz Mountains, Germany; Příbram, Czechoslovakia; Chile and Borneo. Occurs in the USA in California, and in Canada at South Ham, Quebec and Prince William, New Brunswick.
Treatment. Any acid apart from sulphuric may be used as a cleaning agent.

Above right: Native antimony. It is normally found massive like this. Well-formed crystals are rare

Horsfordite
Horsfordite, an antimonide of copper, has been given the formula Cu_6Sb. An alternative is Cu_5Sb. SG 8.8
Localities. Near Mytilene, Greece.

Bismuth Bi
Bismuth, a silver-white mineral, which becomes reddish with tarnish, crystallizes in the hexagonal crystal system. Crystals are rare and the massive forms are those usually encountered. It may be found granular or arborescent. The lustre is metallic and the streak silver-white. Cleavage is perfect. H 2.5; SG 9.8.
Occurrence. Bismuth is comparatively rare, the metal being commonly obtained from the smelting of gold and silver. It occurs naturally in veins in granite and gneiss, often accompanying ores of cobalt and silver; it is deposited hydrothermally.
Localities. Major deposits of economic importance are at San Baldomero, Bolivia and in other areas in the vicinity of La Paz; also found in E. Germany. There are important deposits in North Queensland and New South Wales. Some bismuth occurs in Devon and Cornwall, California, South Dakota and Colorado.

Sulphur S
The commonest form of sulphur, alpha-sulphur, crystallizes in the orthorhombic crystal system; but the mineral is polymorphous and beta-sulphur and gamma-sulphur or rosickyite, both monoclinic, are found. The orthorhombic form commonly occurs as acute pyramidal crystals but may also be tabular; massive forms are found and also granular aggregates. The lustre is resinous and there is a white streak. The crystals are coloured a bright yellow and are slightly sectile and rather brittle. H 1.5–2.5; SG 2.0–2.1; RI 1.958, 2.038, 2.245, with a strong birefringence.
Occurrence. Sulphur is associated with volcanic activity and occurs in gases given off at fumaroles. It is also formed by the decomposition of hydrogen sulphide in hot springs, which may result from the action of acid water on metallic sulphates or from the reduction of sulphates such as gypsum. It is most commonly found in sedimentary rocks of Tertiary age, associated most fre-

Right: Sulphur crystals from Sicily showing a group of bipyramids

quently with gypsum and limestone.
Localities. The finest crystals come from Girgenti, Sicily and adjoining localities; it is also common in volcanic regions of Japan, Iceland, Hawaii, Mexico and the Andes. There are large deposits of economic importance in Louisiana and Texas, where it is associated with evaporate deposits, particularly salt domes.
Treatment. Sulphur is extremely sensitive to heat, so that the warmth of the hand may cause it to crack. It should not be exposed to bright sunlight for the same reason. It is not advisable to attempt to clean this mineral.

Tellurium Te
Tellurium occurs as prismatic crystals of the hexagonal system and is also found massive. It has a perfect prismatic cleavage, is tin-white in colour and has a similar streak. The lustre is metallic. It gives a red solution in warm concentrated sulphuric acid. H 2.0–2.5; SG 6.2
Occurrence. Occurs in hydrothermal vein deposits.
Localities. Tellurium has been found in Romania, Western Australia, the United States, particularly Colorado, and Mexico.

Iron Fe
Native iron is rare. It is found either as a terrestrial form, in which it occurs as grains and masses in rocks; or as meteoritic iron, forming the entire mass or a matrix in which silicate minerals are embedded, or as grains or scales throughout the meteorite. Kamacite is a form of meteoritic iron and is the cubic alpha-phase of the Fe–Ni system with up to 6% Ni. Metakamacite is an unstable phase of similar composition, while taenite is the gamma-phase, with upwards of 33% Fe. The terrestrial form, awaruite, is Ni, Fe, with 60% or more Ni. Iron is grey to black, with a metallic lustre; it is strongly magnetic and malleable. H 4.5; SG 7.3–7.8
Localities. Terrestrial iron is found in basalt at Disco Island, West Greenland and in small grains with pyrrhotite near Kassel, W. Germany. Awaruite is found in Awarua Bay in South Island, New Zealand, where it is associated with gold and platinum.
Treatment. Iron dissolves rapidly in acids, especially nitric. This reaction gives off dangerous fumes of nitrogen oxides. Native iron may oxidize in moist conditions and should be stored with a dehydrating agent such as silica gel.

Iridosmine Os, Ir
This is the osmiridium-rich member of the system Os–Ir, in which the osmium content exceeds 35%; a synonym is iridium-osmine. It crystallizes in the hexagonal system and occurs as flattened grains of a metallic white colour. It has a metallic lustre, is slightly malleable and somewhat brittle. A variety with osmium content from 35% to 50% is nevyanskite and another, with osmium content from 50% to 80% is sysertskite. H 6–7; SG 19.3–21.1
Occurrence. Iridosmine occurs with platinum and contains this and such metals as rhodium and ruthenium.
Localities. Chocó area of Colombia and in the Ural mountains, USSR; in Australia at Platina, New South Wales, and in Oregon, California and British Columbia; with gold at Witwatersrand, South Africa.
Treatment. Any dilute acid may be used as a cleaning agent.

Osmiridium Ir, Os
This is the cubic, iridium-rich phase of the system Os–Ir, with osmium less than 35%. Data as for iridosmine, above.

Platinum Pt
Platinum, a precious metal, is rare in its native form, being alloyed with iron, iridium, rhodium, palladium, osmium and other metals. It crystallizes in the cubic system but is most commonly found as grains. It has a metallic lustre and a steel-grey to whitish streak. It is malleable and ductile. It is sometimes magnetic and occasionally shows polarity. H 4–4.5; SG 14–19 (native mineral); 21–22 (pure).
Occurrence. Platinum is usually found in granules or nuggets resulting from the erosion of platinum-bearing rocks. Associated minerals are chromite, magnetite, zircon and corundum, and the various platinum metals.
Localities. The most important locality, now said to be nearing exhaustion, is in the Ural Mountains; there the primary sources are ultrabasic igneous rocks, including dunites (olivine-rich rocks). Outside the Ural Mountains, platinum has been found in the Chocó area of Colombia, near the river Pinto. It occurs in the Broken Hill area of New South Wales and in New Zealand and the Transvaal. In the USA platinum has been found in North Carolina and California (in placer deposits) and in Canada from the Kamloops area, British Columbia.
Treatment. Any acid apart from aqua regia can be used as a cleaning agent.

Dyscrasite Ag_3Sb
Dyscrasite is an important silver mineral. It belongs to the orthorhombic crystal system although crystals – often found as pseudo-hexagonal twins – are rare, the mineral normally being found massive. It is silver in colour and shows a similar streak; it has a metallic lustre. It may be sectile. A yellow or black tarnish sometimes appears. H 3.5–4.0; SG 9.4–10.0
Occurrence. Frequently with silver ores.
Localities. An important West German source is at Wolfach in the Black Forest, in West Germany. Crystals associated with calcite are found in the Harz Mountains, Germany. Other localities are Atacama, Chile and Cobalt, Ontario; Broken Hill, New South Wales.
Treatment. Any dilute acid apart from nitric may be used as a cleaning agent.

Carbides

Moissanite SiC
Probably the only natural occurrence of moissanite is that in the Diablo Canyon meteorite in Arizona, where it is associated with tiny diamonds. It occurs in small green hexagonal plates. Data for the artificial material, carborundum, are as follows: H 9.5; SG 3.1; RI 2.6–2.7

The following are found only in meteorites:
osbornite TiN
cohenite Fe_3C
schreibersite $(Fe, Ni)_3P$

Sulphides, Selenides, Tellurides, Arsenides, Antimonides and Bismuthides

Chalcocite Cu_2S
Chalcocite crystallizes in the orthorhombic system, sometimes forming pseudo-hexagonal twins. It is more frequently found massive. A cubic modification occurs at temperatures over

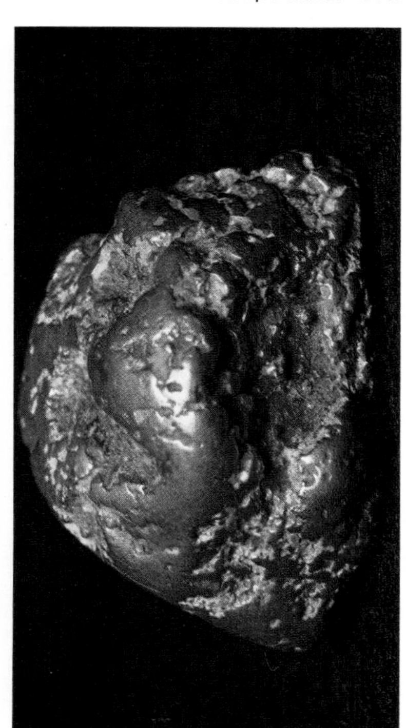

Far left: Meteoritic iron from Henbury, Australia. Most specimens of iron are meteoritic; in terrestrial rocks iron is usually found as grains

Left: Native platinum. This specimen weighs 105 gm and was found in Siberia

Right: Chalcocite from Cornwall

Far right: Bornite from Carn Brea, Cornwall

91°C. The colour is grey, tarnishing to black; the lustre is metallic and the streak black. H 2.5–3.0; SG 5.5–5.8
Occurrence. Chalcocite is found with native copper or with cuprite, usually occurring in secondary sulphide zones.
Localities. Fine crystals of chalcocite occur in the St. Just, St. Ives, Camborne and Redruth areas of Cornwall, England; very fine crystals are found at Bristol, Connecticut; large quantities from Montana, Nevada and California.

Digenite $Cu_{1\cdot5-1\cdot9}S$
Digenite crystallizes in the cubic system in blue or black masses with a submetallic lustre. H 2.5–3; SG 5.6
Occurrence. Occurs in copper ore deposits.
Localities. From Butte, Montana; South-West Africa; Sweden; Kennecott, Alaska.

Covelline CuS
Covelline crystallizes in the hexagonal crystal system, forming hexagonal plates. However, massive forms, often foliated, are more common. There is a perfect basal cleavage; the colour is dark blue with a purple iridescent tarnish. If the plates are sufficiently thin they may be translucent. The lustre is metallic and the streak dark grey to black. H 1.5–2.0; SG 4.6–4.8
Occurrence. Covelline occurs in copper veins, associated with chalcopyrite, bornite (from which it may be distinguished by its perfect cleavage), and chalcocite.
Localities. Fine crystals from the Calabona mine, Alghero, Sardinia; other localities are Chile, Bolivia, Argentine and Peru. It is also found at Butte, Montana.

Berzelianite Cu_2Se
Berzelianite crystallizes in the cubic system although a non-cubic phase appears to exist. It is found in thin dendritic crusts, silver-white in colour, with a metallic lustre, tarnishing to grey. H 2; SG 6.71
Major Localities. Found with copper minerals at the Skrikerum mine, Kalmar, Sweden and from the Harz Mountains, Germany.

Umangite Cu_3Se_2
Umangite crystallizes in the orthorhombic system as masses or grains coloured blue to black with a tinge of red. The lustre is metallic and the streak black. H 3; SG 6.4
Localities. Occurs in a pink calcite as veins, in association with berzelianite and other minerals, at Lake Athabasca, Saskatchewan; from Sierra de Umango, La Rioja, Argentina; Harz Mountains, Germany; from the copper mine at Skrikerum, Sweden.

Klockmannite CuSe
Klockmannite crystallizes in the hexagonal system as granular aggregates coloured grey to black. There is a perfect cleavage. H about 2; SG 6.0
Localities. Occurs in uranium ores at Lake Athabasca, Saskatchewan; Skrikerum, Sweden; Harz Mountains, W. Germany; etc.

Rickardite
Rickardite, a copper telluride of uncertain composition, crystallizes in the tetragonal system as compact purple masses, which lose brightness on exposure to light. The streak is red. H 3.5; SG 7.5
Occurrence. Occurs with tellurides in veins bearing gold and quartz.
Localities. From Gunnison Co., Colorado; Arizona; USSR; San Salvador; South Africa; Japan; etc.

Lautite CuAsS
Lautite crystallizes in the orthorhombic system as tabular crystals or as radiating granular masses coloured grey or black with a tinge of red. H 3.5; SG 4.9
Occurrence. Occurs with other sulphides in veins with native arsenic.
Localities. From Lauta, Germany; Ste. Marie-aux-Mines, Alsace, France.

Bornite Cu_5FeS_4
Bornite crystallizes in the cubic system and is usually massive, though some cubic and rhombdodecahedral crystals are found; twins may occur. The colour is reddish-brown but an iridescent tarnish quickly develops which is purple in colour; the name 'peacock ore' is derived from this tarnish. The lustre is metallic and the streak a pale grey-black. H 3; SG 4.9–5.4
Occurrence. Bornite is found in hydrothermal veins with chalcopyrite and chalcocite. It may also occur as a primary mineral in igneous rocks and pegmatite veins.
Localities. Fine crystals are found in the Redruth area of Cornwall, England and in the copper ores of Montecatini, Tuscany, Italy. Fine-coloured specimens are found at Androta, Malagasy Republic. It is the principal ore of some Chilean copper mines and also occurs at Butte, Montana.
Treatment. No steps are normally taken to remove the tarnish, which is considered desirable; clean with water.

Far right: This specimen of covelline from Sardinia shows the lamellar (thin, platy) structure typical of the mineral

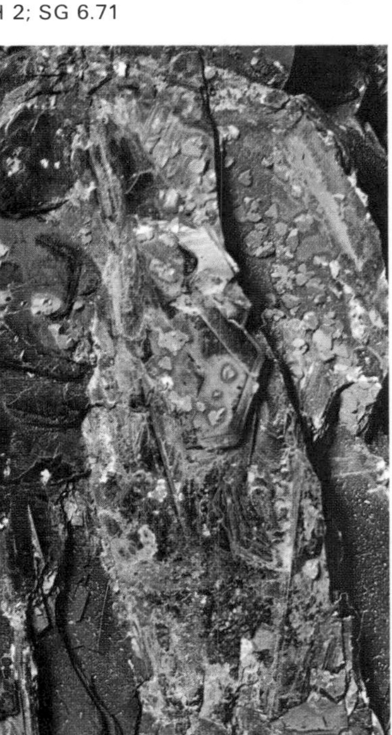

Right: The copper sulphide covelline. It is rarely found as single crystals, and is usually encountered in masses like this one from Sardinia

Left: Crystals of chalcopyrite, a copper-iron sulphide. They appear tetrahedral but their true form is bisphenoidal. They come from Cornwall

Below left: Pyrite. These crystals are pentagonal dodecahedra

Below right: Acanthite from Saxony. This is the stable form of silver sulphide at normal temperatures

Chalcopyrite $CuFeS_2$
The most important ore of copper, chalcopyrite crystallizes in the tetragonal system, often forming crystals resembling tetrahedra. Massive forms are common and various types of twinning occur. An alternative name is copper pyrite. The colour is brass-

yellow with an iridescent tarnish; the lustre is metallic and the streak greenish-black. It is less hard and is a deeper yellow than iron pyrite, and it is harder than gold. Fracture uneven, sometimes conchoidal. H 3.5–4.0; SG 4.1–4.3
Occurrence. Occurs as a primary mineral in igneous rocks and in hydrothermal vein deposits with pyrite, sphalerite and other minerals. It also occurs in pegmatites and contact metamorphic deposits.
Localities. Chalcopyrite deposits occur in many localities. Important sulphide copper deposits in which chalcopyrite is the chief mineral are found in the USA in the states of Arizona, Montana, Nevada, New Mexico and California. Very fine crystals are found in Chester Co., Pennsylvania. In Europe good crystals occur in Westphalia and at Ste. Marie-aux-Mines, Alsace. Chalcopyrite may also be found at St. Agnes, Cornwall and in Devon.
Treatment. Cyanide solutions may remove the black coating often found. Hydrochloric acid may sometimes induce an iridescent coating. Distilled water should be used for cleaning.

Cubanite $CuFe_2S_3$
Sometimes known as chalmersite, cubanite crystallizes in the orthorhombic crystal system forming thin elongated prisms, vertically striated. Twinning is common. The colour is brass-yellow and the mineral is strongly magnetic. There is a metallic lustre. H 3.5; SG 4.1
Occurrence. Cubanite is found with copper ores and with gold.
Localities. From Barracanao, Cuba and from the Morro Velho gold mine, Minas Gerais, Brazil.
Treatment. As for chalcopyrite.

Acanthite and **Argentite Ag_2S**
Acanthite is the low-temperature monoclinic modification of Ag_2S; most natural silver sulphide specimens are acanthite pseudomorphous after argentite, which is the high-temperature cubic modification. Acanthite forms slender prismatic crystals which are sectile. The colour is iron-black. SG 7.2–7.3
Argentite forms octahedra or cubes, frequently in arborescent or reticulated groups; it is found massive and as a coating. The colour and streak are grey to black; the lustre is metallic and shining, and there is a frequent alteration of the surface to a black sulphide. H 2.0–2.5; SG 7.20–7.36
Localities. Acanthite is found at Jachymov, Czechoslovakia, and at Freiberg, E. Germany. It also occurs at Georgetown, Colorado and in Chihuahua and Zacatecas, Mexico. Argentite is the most important primary mineral of silver and occurs in hydrothermal veins with pyrargyrite, proustite and native silver. It may also occur as a product of silver sulphides that have been attacked by weather. It occurs at Jachymov, Czechoslovakia; Freiberg, E. Germany; Andreasberg, Harz Mountains, Germany; Liskeard, Cornwall; it is common in the silver mines of Mexico, especially in Guanajuato and Zacatecas and at Arizpe, Sonora. It is found with copper ores at Butte, Montana and in the silver districts of Colorado and Nevada.
Treatment. Both minerals darken when exposed to bright light. This coating may be removed ultrasonically. Clean with water.

Aguilarite Ag_4SSe
Aguilarite is polymorphous, though the high- and low-temperature forms are not distinguished by separate names. The high-temperature form crystallizes in the cubic system, forming skeleton dodecahedral crystals which are sectile. The colour is black, and lustre metallic. H 2.5; SG 7.58
Localities. Aguilarite is found with silver at Guanajuato, Mexico.
Treatment. Clean with water.

Hessite Ag_2Te
Hessite is polymorphous though the high- and low-temperature forms do not possess distinguishing names. It crystallizes in the monoclinic system and is usually found massive. The colour is grey and lustre metallic; the material may be sectile. When gold is contained the mineral graduates towards petzite. H 2.5–3.0; SG 8.3–8.4
Occurrence. Found with silver and gold.
Localities. Altai Mountains, USSR; Coquimbo, Chile; Colorado; San Sebastian and Jalisco, Mexico.
Treatment. Should be cleaned with water or alternatively with dilute hydrochloric acid.

Empressite
Empressite, a silver telluride of varying composition, crystallizes in the orthorhombic system, although some authorities have noted hexagonal forms. It occurs as yellow granular masses with a grey to black streak. H 3.5; SG 7.6
Localities. Occurs with galena and tellurium at the Empress Josephine mine, Kerber Creek, Colorado.

Stromeyerite AgCuS
Stromeyerite crystallizes in the orthorhombic system as pseudohexagonal prisms or as compact masses coloured dark grey with a blue tarnish; the streak is steel grey. H 2.5–3; SG 6.2
Localities. Occurs in Colorado as a major ore of silver; also from British Columbia; Mexico; Chile; Germany; USSR; Tasmania; and also at some other localities.

Above: Petzite from Kalgoorlie, Australia

Above right: Calaverite from Cripple Creek, Colorado

Jalpaite Ag_3CuS_2
Jalpaite crystallizes in the tetragonal system as foliated masses coloured grey but tarnishing to dark grey or yellow. The streak is black. H 2.5; SG 6.8
Localities. Occurs with chalcopyrite at Jalpa, Mexico; also from Příbram, Czechoslovakia; and Colorado.

Eucairite AgCuSe
Eucairite crystallizes in the orthorhombic system as granular masses or as films on calcite. The colour is silver-white or grey tarnishing to light yellow. H 2.5; SG 7.6–7.8
Localities. Occurs in basalt at Lake Athabasca, Saskatchewan; in quartz and calcite at Skrikerum, Sweden; and in a copper ore at Sierra de Umango, La Rioja, Argentina.

Sternbergite $AgFe_2S_3$
Sometimes known as flexible silver ore, sternbergite crystallizes in the orthorhombic crystal system forming fan-like aggregates; twinning may also occur. The individual crystals take the form of thin laminae which are flexible. There is a perfect cleavage. The colour is brown with a black streak and a metallic lustre. H 1.0–1.5; SG 4.2
Frieseite is a mixture of sternbergite and pyrite; and it is thought that *argyropyrite* may be identical in composition with sternbergite or argentopyrite.
Occurrence. Sternbergite occurs in silver mines with pyrargyrite and stephanite.
Localities. From Jachymov, Czechoslovakia; Freiberg, E. Germany; Colorado and California.
Treatment. Clean with water.

Argentopyrite $AgFe_2S_3$
Argentopyrite crystallizes in the orthorhombic system as pseudohexagonal crystals coloured white or grey tarnishing to brown, blue, green or purple. H 4; SG 4.2
Localities. Occurs with arsenic and proustite at Freiberg, E. Germany; Jachymov, Czechoslovakia.

Calaverite $AuTe_2$
Calaverite contains some silver and crystallizes in the monoclinic system, forming thin prisms striated parallel to their length. Massive forms also occur. The colour is white, often tinged with yellow; the lustre is metallic. Calaverite may be distinguished from sylvanite by its lower silver content, but this can only be established by chemical tests. H 2.5; SG 9.0
Occurrence. In vein deposits, with native gold and other minerals.
Localities. Found in Western Australia at Kalgoorlie; in the Cripple Creek district, Teller Co., Colorado and in Calaveras Co., California.
Treatment. Clean with water.

Krennerite $AuTe_2$
Krennerite crystallizes in the orthorhombic system as prismatic crystals with a perfect cleavage. The colour is pale yellow or white with a metallic lustre. H 3; SG 8.6
Occurrence. Occurs in low-temperature veins with calaverite at Cripple Creek, Teller Co., Colorado; also from Kalgoorlie, Western Australia; and from other localities.

Petzite Ag_3AuTe_2
Petzite contains three parts of silver to one of gold and crystallizes in the cubic crystal system, in which it forms granular masses. It is grey to black and may tarnish to the latter colour. Its lustre is metallic. Petzite graduates toward hessite (q.v.) which is silver telluride, Ag_2Te. It is brittle and somewhat sectile. H 2.5–3; SG 8.7–9.0
Occurrence. In vein deposits with tellurides.
Localities. Petzite is found at Kalgoorlie, Western Australia and in Calaveras County, California.
Treatment. It may be cleaned in dilute acids avoiding nitric acid.

Sylvanite $AgAuTe_4$
An ore of gold, sylvanite is a telluride of gold and silver. It crystallizes in the monoclinic system, commonly as twins, which are arborescent and resemble writing; also as bladed and columnar crystals and as granular masses. It is steel-grey in colour and streak and has a brilliant metallic lustre and perfect cleavage. H 1.5–2.0; SG 7.9–8.3
Occurrence. Sylvanite occurs in gold-bearing veins and is usually associated with igneous rocks.
Localities. Found in Romania and at Kalgoorlie in Western Australia. Also found in the mines at Carson Hill, Calaveras County, California.
Treatment. Surface alteration takes place on exposure to strong light. Any acid apart from nitric acid or aqua regia can be used as a cleaning agent.

Right: Sylvanite crystals. Sylvanite is a telluride of gold and silver

Above: Red blende (sphalerite) from Wales

Left: Blende from Switzerland. These crystals show 'spinel twinning'; that is, crystals interpenetrate each other, the whole resembling a spinel crystal

Blende ZnS

Blende, also called sphalerite, is an important ore of zinc and has also been used, despite its softness and easy cleavage, as a gemstone. The mineral crystallizes in the cubic system, common forms being tetrahedra, dodecahedra and cubes. Polysynthetic twinning is common. There is a perfect dodecahedral cleavage and a nearly adamantine lustre. The colour is usually yellow to yellow-brown or black, but red and colourless examples may also be found. The streak is brownish to yellow. Some specimens may exhibit triboluminescence. Blende is the stable form of ZnS below 1020°C; wurtzite (q.v.) the stable form at higher temperatures. H 3.5–4; SG 3.9–4.1; RI 2.37–2.42; dispersion 0.156

Varieties. Marmatite contains up to 20% iron; pribramite contains some cadmium. Blende forms veins in limestone or occurs as concretions.

Occurrence. Blende is found in areas of contact metamorphism where intrusive igneous material has introduced sulphides into adjacent rocks. Limestones or dolomites frequently contain ores formed by replacement. Blende is found in veins and is frequently found associated with galena, pyrite, calcite and dolomite.

Localities. Fine gem-quality material is found at Picos de Europa, Santander, Spain and also from Joplin, Missouri; the St. Agnes area of Cornwall provides material; the most important economic deposits are in Missouri, Oklahoma and Kansas, and from the Bou-Beker area in Morocco.

Treatment. Calcite encrustation may be removed with dilute hydrochloric acid though prolonged application may dull bright surfaces.

Fashioning. Uses: faceting or cabochons; *cleavage*: perfect, dodecahedral; brittle; *cutting angles*: crown 30°–40°, pavilion 37°–40°; *heat sensitivity*: low.

Right: This crystal of cinnabar from Almaden, Spain, shows pinacoidal (top) and rhombohedral (side) faces

Below right: Cinnabar in its commoner form of a granular mass

Wurtzite ZnS
Wurtzite crystallizes in the hexagonal system as hemimorphic pyramidal crystals or as crusts or fibrous masses. The colour is brown or orange to black with a brown streak. May fluoresce orange under ultra-violet light. H 3.5–4; SG 4.0–4.1
Occurrence. Occurs in hydrothermal veins with blende; sometimes in sedimentary deposits.
Localities. From England; Peru; USA; Czechoslovakia; etc.

Greenockite CdS
Greenockite is an ore of cadmium. It crystallizes in the hexagonal crystal system in hemimorphic pyramidal crystals with a distinct cleavage. The colour is honey or orange-yellow and the streak may be the same colour, or a brick-red. The lustre is resinous. H 3.0–3.5; SG 4.9–5.0; RI (when transparent) 2.50, 2.52
Occurrence. Greenockite is associated, often as a coating, with zinc minerals, and particularly blende.
Localities. Good crystals may be found at Bishopton, Renfrewshire, Scotland; at Příbram, Czechoslovakia; at Franklin, New Jersey and in Missouri.
Treatment. May be cleaned with any acid apart from hydrochloric.

Cinnabar HgS
Cinnabar is the only common ore of mercury and crystallizes in the hexagonal system, forming rhombohedral or thick tabular crystals; penetration twins are common. Most cinnabar, however, is found as compact granular masses. There is a perfect prismatic cleavage. It is cochineal-red with a tendency to brown; the lustre is close to metallic and the streak is vermilion. H 2–2.5; SG 8.0–8.2; RI 2.91–3.27
Occurrence. Cinnabar is found in fractures, often in sedimentary rocks. In many cases it occurs in areas associated with volcanic activity, and also in the vicinity of hot springs.
Localities. From Mt. Avala near Belgrade, Yugoslavia. Good crystals are found at Moschellandsberg (Landsberg) in W. Germany. The most important economic deposit is at Almaden, Ciudad Real, Spain. In California cinnabar is found in the Coast Range. Some large transparent red crystals come from China.
Treatment. Cinnabar may tarnish when exposed to strong light. Can be cleaned with cold concentrated citric acid.
Fashioning. Uses: faceting or cabochons; *cleavage:* perfect prismatic; sectile; *cutting angles:* crown 30°–40°, pavilion 37°–40°; *heat sensitivity:* high.

Right: The cadmium sulphide greenockite is usually found as a coating as in this specimen

Metacinnabarite HgS
An ore of mercury, metacinnabarite belongs to the cubic crystal system. It forms tetrahedral crystals, although these are rare; it is more usually found massive. The colour and streak are black, and the lustre is metallic. H 3; SG 7.7
Occurrence. Found in the upper portions of mercury deposits.
Localities. From Idria, Italy; the Bedington mine near Knoxville, California; and at other places in the same state.
Treatment. May be cleaned with dilute acids.

Tiemannite HgSe
Tiemannite crystallizes in the cubic system as tetrahedra or as masses coloured grey with a tinge of purple. The streak is black. H 2.5; SG 8.2–8.3
Localities. Occurs with baryte and calcite with manganese in a limestone at Piute Co., Utah; Harz Mountains, Germany.

Coloradoite HgTe
Coloradoite crystallizes in the cubic system as black granular masses. H 2.5; SG 8.1–8.6
Localities. Occurs in telluride ores in Colorado; also from California and Kalgoorlie, Western Australia.

Galena PbS
The most important ore of lead, also containing silver, galena crystallizes in well-shaped crystals of the cubic system, most commonly in cubes, though octahedra are also found. It may also form skeleton crystals and contact and interpenetrant twins. There is a perfect cleavage. The colour is lead grey and the

Above: Crystals of galena occurring as cubes from Joplin, Missouri. Octahedra are also found

streak similar; the lustre is metallic. H 2.5–2.75; SG 7.4–7.6
Occurrence. Galena is formed by hydrothermal reaction and is found in beds and veins, usually in eruptive rocks and in association with sphalerite, silver ores, quartz and calcite. It also occurs in contact metamorphic deposits, and particularly as replacement deposits in dolomites or limestones.
Localities. Galena is found with silver in Czechoslovakia; from Freiberg and the Harz Mountains, Germany; Truro and Liskeard in Cornwall, and from Missouri and Wisconsin.
Treatment. Calcite covering may be removed with dilute acetic acid.

Clausthalite PbSe
Clausthalite crystallizes in the cubic system as masses coloured lead-grey with a tinge of blue; some surfaces may show black with red or brown spots. H 2.5–3; SG 8.1–8.2
Localities. Occurs with selenides at Lake Athabasca, Saskatchewan; Harz Mountains, Germany; Falun, Sweden.

Altaite PbTe
Altaite crystallizes in the cubic system, as cubes or octahedra, but is usually found massive. It is white, with a metallic lustre, but possesses a yellowish tinge which tarnishes to a bronze colour. There is a perfect cubic cleavage and a black streak. H 2.0–3.0; SG 8.16
Occurrence. Found with hessite.
Localities. Altai Mountains, Siberia, USSR; Coquimbo, Chile; California and other areas of the United States.
Treatment. May be cleaned with any acid apart from nitric.

Realgar AsS
Realgar crystallizes in the monoclinic crystal system and is found as red or orange-yellow short prismatic crystals with vertical striations and a resinous lustre. The streak is orange to orange-red. H 1.5–2.0; SG 3.56
Occurrence. Realgar is found with ores of silver and lead.
Localities. Fine crystals occur in dolomite in Valais, Switzerland; also in the USA; and in Japan.
Treatment. Realgar alters to orpiment and finally crumbles to powder on prolonged exposure to light. It should therefore be kept in a container impervious to light. May be cleaned with water.

Left: The arsenic trisulphide orpiment usually occurs massive. This specimen is from Pakistan

Orpiment As_2S_3
Orpiment, which is also known as yellow arsenic, is used as a pigment. It crystallizes in the monoclinic system. Crystals are small and the material is more commonly found massive, with a lemon-yellow colour and a pearly to resinous lustre. There is a perfect cleavage and the streak is pale yellow. H 1.5–2.0; SG 3.4–3.5
Occurrence. Frequently found with realgar. Results from the alteration of arsenic and some silver minerals.
Localities. Fibrous masses are found at Moldava, Romania and also in Kurdistan and near Salonika, Greece. Good quality crystals are found at Mercur, Utah and the mineral is also found as a

Above: Prismatic crystals of stibnite, the commonest antimony mineral. From Victoria, Australia

hot-spring deposit at Steamboat Springs, Nevada. Large workable deposits are found in Turkey, Iran and in Georgia, USSR.

Treatment. Dilute acetic acid is probably the most satisfactory medium for removing calcite encrustations, since orpiment is readily attacked by most acids. Its softness makes the use of sharp tools unwise.

Stibnite Sb_2S_3

Stibnite is the most important ore of antimony. It crystallizes in the orthorhombic system as prismatic crystals with vertical striations and, in some cases, with numerous well-developed faces. It may also be found massive or in radiating groups of columnar crystals. It is lead-grey in colour and in streak; there is a perfect cleavage and a metallic lustre which is very brilliant on freshly broken surfaces. Crystals may tarnish to black and may contain silver or gold. H 2.0; SG 4.52–4.62

Occurrence. Stibnite is most commonly found in veins with quartz, frequently in granitic rocks. It may also be found as beds in schists and in hot-spring deposits.

Localities. Fine crystals have been found at the Ichinokawa mines, Shikoku, Japan, and at Wolfsburg, Germany. Major economic deposits are in Honan Province, China, in Algeria and in Mexico.

Treatment. Best cleaning agent is water. Soaps and detergents should be avoided. Lustre dulls after exposure to strong light.

Metastibnite Sb_2S_3

This is the amorphous form of *stibnite,* with the same chemical composition. It is brick-red in colour with a sub-metallic lustre and has been found with cinnabar and arsenic sulphide on siliceous sinter at Steamboat Springs, Nevada.

Treatment. As for stibnite, the best cleaning agent is water, without the addition of soaps or detergents.

Bismuthinite Bi_2S_3

An ore of bismuth, bismuthinite sometimes contains added copper and iron. It is lead-grey in colour and streak with a metallic lustre, and crystallizes in the orthorhombic system as acicular crystals, though the material is far more commonly found massive. It has a perfect cleavage and may be sectile. H 2.0; SG 6.7–6.8

Occurrence. Most commonly found associated with igneous rocks and in

Right: These acicular (needle-shaped) crystals of bismuthinite are rare; the mineral is more commonly found in masses. From Carn Brea, Cornwall

tourmaline-bearing copper deposits and tourmaline-quartz veins.
Localities. Found in Bolivia, Cornwall, and with chrysoberyl at Haddam, Connecticut.
Treatment. It is fairly resistant to acids, but concentrated nitric should be avoided.

Guanajuatite $Bi_2(Se, S)_3$
Guanajuatite is the orthorhombic phase of bismuth selenide; the trigonal phase is paraguanajuatite. It forms bluish-grey acicular crystals and is also found massive. The lustre is metallic. H 2.5–3.5; SG 6.25–6.98
Occurrence. Found in the mines at Guanajuato, Mexico, associated with bismuth and pyrite; and also in the Harz Mountains, Germany, in calcite veins.
Treatment. May be cleaned with dilute acids.

Tetradymite Bi_2Te_2S
Tetradymite crystallizes in the hexagonal crystal system, forming small crystals sometimes bladed but more commonly massive. The colour is pale grey and the lustre metallic. There is a perfect cleavage. H 1.5–2.0; SG 7.2–7.6
Occurrence. Commonly found in gold-bearing quartz veins and in surface deposits formed by the action of hot water on igneous rocks.
Localities. Found in Czechoslovakia and Norway; in Bolivia; in California and Colorado in the United States; and in Canada.

Joseite $Bi_3Te(Se, S)$
Grunlingite and oruetite appear to be similar in composition to joseite.
Localities. From Minas Gerais, Brazil.

Molybdenite MoS_2
An important ore of molybdenum, molybdenite crystallizes in the hexagonal crystal system as tabular hexagonal prisms which taper and show horizontal striations. Granular and massive forms may also be encountered. The colour is lead-grey and the streak bluish-grey on paper and greenish-grey on porcelain. It is sectile and has a metallic lustre. It may feel greasy; it differs from graphite in its streak on paper. H 1.0–1.5; SG 4.7–5.0
Occurrence. Molybdenite is found in pneumatolytic contact deposits and is associated with cassiterite, scheelite, and wolframite; it may also be found in pegmatites and in quartz veins in association with granite.
Localities. Good quality specimens are found at Raade, Norway, at Blue Hill, Maine and in Colorado. There are several Canadian localities.
Treatment. Iron stains removable with oxalic acid.

Tungstenite WS_2
Tungstenite crystallizes in the hexagonal system as masses or scales coloured grey. H 2.5; SG 7.7
Localities. Occurs in limestone at the Emma mine, Salt Lake Co., Utah.

Alabandite MnS
Alabandite crystallizes in the cubic system and is usually found granular massive. The colour is black and the streak green; it has a metallic lustre. There is a perfect cleavage. H 3.5–4.0; SG 3.93–4.04
Occurrence. In ore veins with sphalerite and other sulphides.
Localities. Alabanda, Asia Minor; Morococha, Peru; Tombstone, Arizona.
Treatment. If necessary may be cleaned with water. It is liable to oxidize if stored in a moist atmosphere and may undergo surface alteration on exposure to strong light.

Hauerite MnS_2
Related to pyrite, hauerite is manganese disulphide and crystallizes in the cubic crystal system; although it is most commonly found massive, octahedral or pyritohedral crystals may occur. It is brown in colour with a metallic lustre and a brownish-red streak. H 4.0; SG 3.46
Occurrence. Found in association with gypsum and sulphur, probably deposited from manganese and sulphur-bearing water.
Localities. It is found in clay at the Destricello mine near Raddusa in Italy.
Treatment. Use any dilute acid apart from hydrochloric as a cleaning agent.

Left: A rosette of pyrrhotine, an iron sulphide. From Chiuzbaia, Romania

Below left: This tapering hexagonal prism of molybdenite from Pontiac, Canada, shows a metallic lustre

Pyrrhotine FeS
An alternative name for this ferrous sulphide is magnetic pyrite although its magnetic powers vary in intensity. It has a varying sulphur content. Crystallizing in the hexagonal crystal system, it is usually found massive, although tabular crystals or pyramidal ones with horizontal striations may occur. Its colour is bronze-yellow to copper-red and the streak is dark-grey to black. Crystals tarnish easily and have a metallic lustre. As pyrrhotine often includes nickel, its main importance is as an ore of that metal. H 3.5–4.5; SG 4.58–4.64
Occurrence. Associated with basic igneous rocks such as gabbro and often found with pyrite and magnetite. It may also be found in contact metamorphic

Below: A crystal of hauerite displaying a combination of forms, in this case the cube and octahedron. From Sicily

Right: Pyrite crystal from Traversella, Italy

Far right: Pyrite from Blackdene, Durham, showing globular form

Below right: Arsenopyrite from Panasqueira, Portugal

deposits. Some metamorphic limestones include pyrrhotine.
Localities. Large crystal groups are found in Carinthia, Austria; in Trentino, Italy; and in Norway and Sweden. The nickel-bearing variety is found in large workable quantities at Sudbury, Ontario.
Treatment. Iron stains may be removed with oxalic acid, and calcite by judicious use of hydrochloric acid. Careful washing must follow to ensure that all traces of acid are removed.

Pyrite FeS_2
Pyrite is very often confused with gold, which it somewhat resembles. Gold and copper are found in pyrite and the mineral is mined both as a source of these metals and also for its sulphur. It crystallizes in the cubic system as cubes, which are striated on each face, the striations being at right angles to each other on adjacent faces, due to oscillation between this form and that of the pyritohedron. Pyrite is also found massive. The streak is greenish-black. It displays a metallic lustre and may be distinguished from gold by its far lower specific gravity and from chalcopyrite by a paler colour and greater hardness. It may display triboluminescence. H 6.0–6.5; SG 4.95–5.10
Occurrence. Pyrite is a common constituent of many rock-types and is sometimes associated with coal. It usually occurs as small individual crystals but may display a radiating structure when found in concretions.
Localities. Pyrite is of world-wide occurrence; fine crystals are found in Germany and the St. Gotthard region of Switzerland. Pyritohedra are found in the island of Elba. In England good crystals have been found in the Liskeard, St. Just and St. Ives areas of Cornwall. Peru, Bolivia, Chile, Brazil, Japan and Mexico also provide excellent material.
Treatment. Remove iron stains with oxalic acid.
Fashioning. Uses: cabochons or faceting; *cleavage*: indistinct to cubic and octahedral faces; brittle and sensitive to shock; *cutting angles*: variable, to suit requirements; *heat sensitivity*: fairly high.

Marcasite FeS_2
It should be noted that what is known as marcasite in the jewellery trade is in fact pyrite (q.v.). Marcasite has an identical chemical composition, but crystallizes in the orthorhombic crystal system. The crystals are commonly tabular or pyramidal; it is also found massive and may display a wide variety of shapes such as reniform and globular. The colour is a bronze-yellow and this may deepen on exposure to light. The streak is greyish or brownish-black and the lustre metallic. It may be distinguished from pyrite by its lower SG and by its differing crystal forms; the colour is also paler. H 6.0–6.5; SG 4.85–4.90
Occurrence. Marcasite is associated with galena, sphalerite, calcite and dolomite in replacement deposits in limestones; also in druses of ore veins.
Localities. The Karlovy Vary area of Czechoslovakia supplies crystals of a spear-like shape. It is also found in the Kentish chalk marl between Folkestone and Dover. A stalactitic form is found at Galena, Illinois.
Treatment. Marcasite is a fairly unstable mineral and is liable to disintegrate with the formation of ferrous sulphate and sulphuric acid. It should not be cleaned.
Fashioning. Uses: cabochons or faceting; *cleavage*: poor, prismatic; brittle; *cutting angles*: variable, to suit requirements; *heat sensitivity*: fairly high.

Löllingite $FeAs_2$
Löllingite (or loellingite) is a diarsenide of iron, but passes into Fe_3As_4, which is leucopyrite; it is also close to arsenopyrite and safflorite $CoAs_2$. It is a member of the orthorhombic crystal system and is normally found massive, coloured silver-white to steel-grey to black. H 5.0–5.5; SG 7.0–7.4
Occurrence. Löllingite occurs in veins and is associated with calcite, silver, cobalt and gold ores. It is deposited from solution at medium temperatures.
Localities. The classic locality from which the name is taken is Lölling, near Hüttenberg, Carinthia, Austria; other localities are the Harz Mountains, Saxony and Silesia in Germany, and Poland.
Treatment. Hydrochloric acid may be used to remove calcite covering that is sometimes found.

Arsenopyrite FeAsS
An ore of arsenic, arsenopyrite is sometimes known as mispickel. Sometimes

Above right: The iron disulphide marcasite from Cornwall. The name derives from the Arabic word used for pyrite

Right: Cobaltite on chalcopyrite, from Sweden

part of the iron is replaced by cobalt (up to 12%); the name given to the mineral of this composition is danaite. Arsenopyrite crystallizes in the monoclinic crystal system and forms twins, which may be trillings or cruciform; the crystals themselves are prismatic, sometimes vertically flattened. Granular forms are also found. The colour is white inclining to grey and the streak greyish-black. There is a cleavage and the lustre is metallic. H 5.5–6.0; SG 5.9–6.2

Occurrence. Arsenopyrite is often associated with gold and also with tin and tungsten in pneumatolytic deposits; it is also associated with silver ores and with quartz in veins deposited by hot waters. It may occur in a disseminated form in limestones and serpentinites.

Localities. Arsenopyrite is found in the silver mines at Freiberg, E. Germany; in crystals from the Binnental, Valais, Switzerland; fine crystals from Franconia, New Hampshire. It is also found at Franklin, New Jersey. Large masses found in quartz veins at Deloro, Hastings County, Ontario.

Treatment. Unaffected by hydrochloric acid, which may be used to remove adhering minerals.

Above: Acicular millerite crystals from Westwald, Germany. The crystals can also be found in radiating groups

Left: Skutterudite in calcite from Morocco. Skutterudite is part of a series which also includes smaltite and chloanthite

Gudmundite FeSbS

Gudmundite crystallizes in the monoclinic system as prismatic crystals, often twinned and coloured grey. H about 6; SG 6.7

Occurrence. Occurs in sulphide deposits.

Localities. From Sweden; Norway; Yellowknife, Northwest Territories; etc.

Linnaeite Co_3S_4

Linnaeite crystallizes in the cubic system as octahedra or as granular masses. The colour is grey, tarnishing to red or violet. H 4.5–5.5; SG 4.8–5.0

Occurrence. Occurs in hydrothermal veins with other sulphides.

Localities. From Siegen, Germany; Katanga, Zaire; Maryland, California; and other localities.

Safflorite (Co, Fe)As_2

Safflorite crystallizes in the orthorhombic and monoclinic systems as fibrous masses coloured white tarnishing to grey. The streak is grey to black. H 4.5–5; SG 7.2

Occurrence. Occurs with minerals containing cobalt or nickel in vein deposits.

Localities. From Great Bear Lake, Ontario; Sweden; Germany; etc.

Smaltite (Co, Ni) As_{2-3}

Smaltite is closely related to skutterudite. It is an ore of cobalt, crystallizing in the cubic crystal system as massive granular forms. It is tin-white inclining to grey, with a metallic lustre and a greyish-black streak. It is not possible to sharply differentiate between the species skutterudite and chloanthite since they grade into one another with the increase/decrease of cobalt/nickel. H 5.5–6.0; SG 5.7–6.8

Occurrence. In veins with ores of silver and cobalt; associated minerals include sphalerite, galena and sometimes bismuth.

Localities. From Czechoslovakia, where it is found embedded in calcite and from copper-bearing schists at Hesse-Nassau, W. Germany; a large deposit is located at Cobalt, Ontario.

Treatment. May undergo alteration in moist atmosphere. Adhering minerals may be removed with the aid of hydrochloric acid.

Skutterudite (Co, Ni) As_3

An ore of cobalt, skutterudite is close to smaltite, $CoAs_2$, and to chloanthite, $NiAs_2$. It crystallizes in the cubic crystal system, usually in massive form. It is white inclining to grey with a metallic lustre, and may sometimes be iridescent. The streak is greyish-black. H 6.0; SG 6.5–6.9

Occurrence. Skutterudite usually occurs in veins with other cobalt and nickel minerals.

Localities. Found in New Jersey at Franklin. A large deposit is located at Cobalt, Ontario. The name comes from the classic locality at Skutterud, Norway.

Treatment. Adhering minerals removable with hydrochloric acid.

Cobaltite CoAsS

Cobaltite crystallizes in the cubic system as cubes, octahedra, pyritohedra and combinations of cube and pyritohedron. Faces are striated and there is a perfect cleavage. The colour is white with a brilliant metallic lustre and a grey streak. H 5.5; SG 6.3

Occurrence. Occurs in hydrothermal veins.

Localities. From Boulder, Colorado; Idaho; Nevada; California; Cobalt, Ontario; Sonora, Mexico; Sweden; England; etc.

Carrollite $CuCo_2S_4$

Carrollite crystallizes in the cubic system as octahedra and is also found as granular masses. The colour is grey tarnishing to red or violet. H 4.5–5.5; SG 4.5–4.8

Occurrence. Occurs in hydrothermal veins with other sulphides.

Localities. From Carroll Co., Maryland; Sweden.

Glaucodot (Co, Fe) AsS

A sulpharsenide of cobalt and iron, glaucodot crystallizes in the orthorhombic system in rhombic crystals or in massive form. The colour is grey to white and the lustre metallic. H 5.0; SG 5.90–6.01

Occurrence. With other cobalt minerals; frequently found intergrown with cobaltite.

Localities. Good crystals from Hakansbö, Vastmanland, Sweden; from Skutterud, Norway; and from the silver veins at Cobalt, Ontario.

Treatment. Clean with water.

Millerite NiS

Millerite is an ore of nickel. It is a member of the trigonal crystal system and is found as very slender crystals often in radiating groups; it may also occur as

Right: Niccolite in its usual form of a crystalline aggregate

Far right, above: Typical form of the nickel arsenate rammelsbergite from Schneeberg, East Germany

Below right: Breithauptite with ullmannite in calcite

Far right, below: Another example of ullmannite

tufted coatings. There is a perfect cleavage and the colour is brass-yellow with a greenish-black streak and metallic lustre. The crystals may develop a grey iridescent tarnish. H 3.0–3.5; SG 5.3–5.6
Occurrence. Millerite occurs with other nickel minerals and sometimes with coal; it has also been found associated with serpentinites and in meteoritic iron.
Localities. Millerite has been found with nickel, cobalt and silver ores in Czechoslovakia; with iron ores in Westphalia, W. Germany; very fine hair-like crystals from Merthyr Tydfil, Mid Glamorgan, Wales; with a green chrome-bearing garnet at Orford, Ontario.
Treatment. The thin crystals should only be cleaned, if at all, in alcohol. They are very easily matted and may also be attacked by acids.

Polydymite Ni_3S_4
Polydymite crystallizes in the cubic system as octahedra or granular masses. The colour is grey tarnishing to red or violet. H 4.5–5.5; SG 4.5–4.8
Occurrence. Occurs in hydrothermal veins.
Localities. From Siegen, Westphalia, Germany.

Melonite $NiTe_2$
Melonite crystallizes in the hexagonal system as lamellae with a perfect cleavage and is reddish-brown in colour with a dark grey streak. H 1–1.5; SG 7.7
Occurrence. Occurs in hydrothermal veins with other tellurides, also with gold and quartz.
Localities. From Calaveras Co., California; Canada; South Australia; USSR.

Maucherite $Ni_{11}As_8$
Maucherite crystallizes in the tetragonal system as tabular crystals or as fibrous masses. The colour is grey tinged with red with a dark grey streak. H 5; SG 7.9
Localities. With nickel minerals at Sudbury, Ontario; also from Germany and Malaga, Spain.

Niccolite NiAs
An ore of nickel, niccolite is sometimes known as copper nickel. It is a member of the hexagonal crystal system although crystals are rare; it is usually found massive. Reniform or columnar shapes are sometimes found. It is pale copper-red with a pale-brown to black streak. It has a metallic lustre. Some iron and cobalt is found in niccolite and in some cases the arsenic is partly replaced by antimony; this brings it closer to breithauptite, NiSb. H 5–5.5; SG 7.8
Occurrence. Niccolite is found associated with native silver and silver-arsenic minerals.
Localities. It is found at Franklin, New Jersey; Styria, Austria; and Saxony and Thuringia, E. Germany. It is found with silver and cobalt ores at Cobalt, Ontario.
Treatment. Dilute hydrochloric acid may be used to remove calcite encrustations.

Rammelsbergite $NiAs_2$
Rammelsbergite is a member of the orthorhombic crystal system; it is usually found massive. The colour is white with a reddish tinge; the lustre is metallic. H 5.5–6.0; SG 6.9–7.2
Occurrence. Often found associated with quartz and pyrite and with other nickel minerals.
Localities. Lölling, Carinthia, Austria; in the silver veins, Cobalt, Ontario.
Treatment. A coating of green annabergite may develop in a moist atmosphere.

Gersdorffite NiAsS
Gersdorffite crystallizes in the cubic system; often massive. Iron and cobalt may replace some of the nickel. The colour is white to grey, and the lustre metallic. H 5.5; SG 5.6–6.2
Occurrence. Often in veins with siderite and other iron minerals.
Localities. Lobenstein, E. Germany; Sudbury, Ontario.
Treatment. If necessary, any dilute acid apart from nitric may be used as a cleaning agent.

Breithauptite NiSb
Breithauptite is a mineral that crystallizes in the hexagonal system as prismatic crystals but is more commonly found massive. The colour is light red with a metallic lustre; the streak is reddish-brown. H 5.5; SG 7.5–8.2
Occurrence. Associated with native silver and silver ores, also with silver-arsenic minerals (cf niccolite).
Localities. St. Andreasberg, Harz Mountains, Germany; Cobalt, Ontario.
Treatment. Encrustations removable with hydrochloric acid.

Ullmannite NiSbS
Ullmannite, a member of the cobaltite family, usually contains some arsenic. It crystallizes in the cubic crystal system; usually massive. When crystals do occur they may be pyritohedra or tetrahedra. The colour is grey to white with a metallic lustre. H 5.0–5.5; SG 6.2–6.7
Occurrence. Occurs in veins, often with iron minerals, and sometimes with sphalerite and galena.
Localities. Found at Lölling, near Hüttenberg, Carinthia, Austria; from Sardinia and France; from Brancepeth Colliery, Co. Durham, and from Gunnison Co., Colorado.
Treatment. Cleaned with water.

Pentlandite $(Fe, Ni)_9S_8$
Pentlandite is a sulphide of iron and nickel. It occurs in massive form and is a member of the cubic crystal system. It is a light bronze-yellow in colour, with a metallic lustre, and shows a light

brown streak. H 3.5–4.0; SG 5.0
Occurrence. Usually found intergrown with pyrrhotine and associated with millerite, pyrite, chalcopyrite and niccolite.
Localities. Occurs with chalcopyrite in Norway and with pyrrhotine in the Sudbury district of Ontario; San Diego Co., California.
Treatment. Clean with any dilute acid.

Bravoite (Fe, Ni)S_2
Bravoite is closely related to pyrite but may contain up to 20% nickel. It crystallizes in the cubic system, though it is commonly only found as small fragments or grains. It is pale yellow with a reddish tarnish. H 5.5–6.0; SG 4.6
Localities. Bravoite is found in the vanadium ores at Minas Ragra, Peru; Derbyshire; Spain; Germany; Colorado.
Treatment. Clean with dilute acid.

Kermesite Sb_2S_2O
Alternative names for this oxysulphide of antimony are red antimony, purple blende and antimony blende. Kermesite is a member of the monoclinic or triclinic crystal system. It occurs as tufts of crystals of a cherry-red colour with a perfect cleavage and adamantine lustre; the streak is brownish-red. H 1.0–1.5; SG 4.5–4.6
Occurrence. Occurs as an alteration product of stibnite and is commonly associated with other secondary antimony minerals.
Localities. Kermesite is found in Czechoslovakia, north-west of Bratislava; south-east of Constantine in Algeria; in Wolfe County, Quebec and in Nova Scotia; at Sonora, Mexico.
Treatment. As it loses water in dry atmospheres, store in a closed container.

Sulpho-Salts

Tennantite Cu_3AsS_3 or $Cu_{12}As_4S_{13}$
Tennantite is related to tetrahedrite. It is a member of the cubic crystal system and forms dodecahedral crystals; alternatively it may be found massive. The colour is black to grey, with a metallic lustre, and the streak black, brown or dark red. H 3.0–4.5; SG 4.59–4.75
Occurrence. Often with silver or zinc.
Localities. The variety from Cornwall (Wheal Jewel) contains copper and iron; an argentiferous form, known as fredericite, occurs in Sweden at Falun. The variety binnite is found in the Binnental, Valais, Switzerland; this variety contains up to 14% silver. Sandbergerite is the zinc-bearing variety and is found at Morococha, Peru.

Tetrahedrite $Cu_{12}Sb_4S_{13}$
Tetrahedrite is a member of the cubic crystal system in which it forms tetrahedral crystals, often twinned; also found massive. The colour is grey to black; the streak is similar but may also incline to red or brown. The lustre is metallic. Arsenic is usually present and this brings the mineral close to tennantite. H 3–4.5; SG 4.4–5.1

Above: Tetrahedra of tetrahedrite partly covered by a layer of chalcopyrite. This particular form is called a triakistetrahedron. From the Harz Mountains of Germany

Left: Tennantite crystals. This mineral is associated with tetrahedrite in a series. The well-shaped crystals shown here are from Idaho Springs, Colorado

Above: Pyrargyrite with amethyst from Guanajuato, Mexico

Right: Two proustite crystals from Chanarcillo, Chile. The adamantine lustre is clear

Far right: Tabular pseudo hexagonal crystals of stephanite from East Germany

Occurrence. Commonly found in copper or silver veins or in contact-metamorphic deposits. Common associates include chalcopyrite, bornite, sphalerite, galena and the ruby silvers.
Localities. Very fine crystals have been found in Romania at Cavnic (Kapnikbanya); an argentiferous form has been found at St. Andreasberg in the Harz Mountains, Germany; fine specimens come from the copper and tin mines in Cornwall, in particular the Herodsfoot mine near Liskeard.
Treatment. Clean with water.

Chalcostibite $CuSbS_2$
Chalcostibite crystallizes in the orthorhombic system as prismatic crystals or as granular masses. There is a perfect cleavage. H 3–4; SG 4.8–5.0
Localities. Occurs with quartz and sulphides at Oruro, Bolivia; Harz Mountains, Germany; Granada, Spain.

Emplectite $CuBiS_2$
Emplectite crystallizes in the orthorhombic system as prismatic crystals with a perfect cleavage. The colour is grey or white. H 2; SG 6.3–6.5
Occurrence. Occurs with chalcopyrite in veins.
Localities. From Colorado; Chile; Saxony, Germany; etc.

Wittichenite Cu_3BiS_3
Wittichenite crystallizes in the orthorhombic system as prismatic crystals or as masses. The colour is grey or white, tarnishing to yellow; the streak is black. H 2–3; SG 6.2
Localities. Occurs with enargite at Butte, Montana; Wittichen and Baden, Germany; Peru; Japan.

Smithite $AgAsS_2$
Smithite crystallizes in the monoclinic system as red tabular crystals with a red streak. H 1.5–2.0; SG 4.9
Occurrence. Occurs in dolomite with sphalerite and pyrite.
Localities. From the Binnental, Valais, Switzerland.

Proustite Ag_3AsS_3
An alternative name for this silver-bearing mineral is ruby silver ore (cf pyrargyrite). Belonging to the hexagonal crystal system, it is found as acute rhombohedral or scalenohedral crystals; massive forms also occur. There is a distinct cleavage and the colour is scarlet with a similarly coloured streak. The lustre is adamantine and the mineral suffers surface alteration when exposed to light. H 2.0–2.5; SG 5.57–5.64; RI (when transparent) 3.0, 2.7
Occurrence. Proustite is found in the upper portions of silver veins with galena and sphalerite.
Localities. Jachymov, Czechoslovakia; fine crystals from Ste. Marie-aux-Mines, Alsace, France; from the Dolores mine, Chanarcillo, and from Atacama, Chile; silver-bearing districts in USA; W. Germany.
Treatment. Silvery coating forms on surface on exposure to light; removable by gentle wiping, preferably with cotton.
Fashioning. Uses: faceting or cabochons; *cleavage:* distinct, rhombohedral; brittle; *cutting angles:* crown 30°–40°, pavilion 37°–40°; *heat sensitivity:* high, dop with cold-setting cement.

Xanthoconite Ag_3AsS_3
Xanthoconite crystallizes in the monoclinic system as tabular crystals or as reniform masses. The colour is yellow, red or orange with an orange-yellow streak. H 2–3; SG 5.5
Occurrence. Occurs in hydrothermal veins with other silver minerals.
Localities. From Ste. Marie-aux-Mines, Alsace, France; Chanarcillo, Chile; Freiberg, E. Germany; Czechoslovakia.

Miargyrite $AgSbS_2$
Miargyrite crystallizes in the monoclinic system as complex crystals; also massive. The colour is black to grey, but deep-red in thin section. The streak is cherry-red and the lustre metallic. H 2.0–2.5; SG 5.10–5.30
Occurrence. Associated with silver and antimony ores. Found in hydrothermal vein deposits formed at low temperatures with other silver sulpho-salts.
Localities. From Příbram, Czechoslovakia; Chile, Bolivia; Silver City, Idaho.
Treatment. Surface darkens on exposure to strong light.

Pyrargyrite Ag_3SbS_3
Pyrargyrite crystallizes in prismatic crystals of the hexagonal crystal system and may also be found massive. Multiple twins are common. The colour is black to greyish-black by reflected light and deep red by transmitted light through thin section. The streak is purple-red. There is a distinct cleavage and the lustre is metallic to adamantine. H 2.5; SG 5.77–5.86
Occurrence. Pyrargyrite is found in the upper portions of silver veins; associated minerals include galena and sphalerite.
Localities. It is found with proustite at Atacama, Chile; from Příbram, Czechoslovakia; in the silver-bearing areas of the Rocky Mountains, including the Silver City area of Colorado; Harz Mountains, Germany.
Treatment. As for proustite.

Stephanite Ag_5SbS_4
Sometimes known as brittle silver ore, stephanite crystallizes in the orthorhombic crystal system as short prismatic or tabular crystals; pseudo-hexagonal twins are common and the crystal faces are sometimes obliquely striated. The

colour and streak are black, and the lustre metallic. H 2.0–2.5; SG 6.2–6.3
Occurrence. Occurring in silver deposits, stephanite is associated with other silver minerals and with galena and blende.
Localities. From Příbram, Czechoslovakia; Atacama, Chile; and Cornwall. An economically important deposit is located at Comstock Lode, Nevada; Andreasberg, Harz Mts., Germany.
Treatment. Exposure to strong light causes a dark coating to form which can be removed ultrasonically.

Polybasite $Ag_{16}Sb_2S_{11}$
Polybasite belongs to the monoclinic crystal system, in which it crystallizes as six-sided tabular prisms; twins are also found. The colour and streak is black and the lustre metallic; in thin section the colour is cherry-red. May contain up to 13% copper. H 2.0–3.0; SG 6.0–6.2
Occurrence. Found in silver veins; and is associated with silver and lead sulpho-salts, and with quartz and baryte.
Localities. Příbram, Czechoslovakia; Atacama, Chile; the silver mines of Colorado and Nevada; Las Chiapas, Sonora, Mexico.
Treatment. Dark surface alteration may be removed ultrasonically.

Matildite $AgBiS_2$
Matildite crystallizes in the hexagonal crystal system as slender prismatic crystals. Crystals, however, are rare, the mineral more often being found massive. The colour is grey and the lustre metallic. H 2.5; SG 6.9
Occurrence. Found with silver ores.
Localities. From the Matilda mine, Morococha, Peru; Cobalt, Ontario; Inyo Co., California; Boise Co., Idaho.

Nagyagite $Pb_5Au(Te, Sb)_4S_{5-8}$
A telluride and sulphide of antimony, lead and gold, with Sb_2S_3 replacing some PbS. It crystallizes in the tetragonal system as tabular crystals and is also found as granular masses. The colour and streak are grey to black and

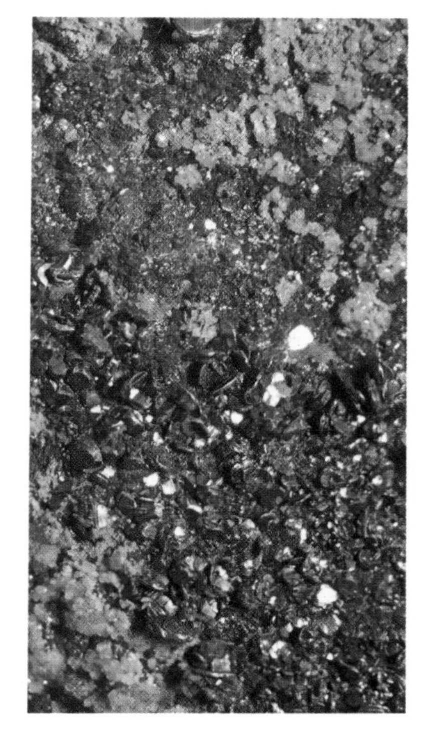

there is a perfect cleavage. The lustre is metallic. H 1.0–1.5; SG 7.4
Occurrence. In hydrothermal vein deposits associated with native gold.
Localities. From Sacarambu (formerly Nagy-Ag), Romania; also from Colorado; Kalgoorlie, Western Australia.

Hutchinsonite $(Tl, Pb)_2(Cu, Ag)As_5S_{10}$
Hutchinsonite is a member of the orthorhombic crystal system, in which it forms flattened rhombic prisms with a good cleavage. The colour is scarlet to red, and the lustre is adamantine. There is a high birefringence. H 1.5–2.0; SG 4.6
Localities. Found in dolomite from the Binnental, Valais, Switzerland; Quiruvilca, Peru.

Sartorite $PbAs_2S_4$
Sartorite crystallizes in the orthorhombic crystal system as slender striated crystals of a dark-grey colour and metallic lustre. H 3.0; SG 5.4
Localities. In dolomite from the Binnental, Valais, Switzerland.

Jordanite $(Pb, Tl)_{13}As_7S_{23}$
Jordanite is isomorphous with geocronite. It is a member of the monoclinic crystal system and forms pseudo-hexagonal twins. It is grey in colour, with a metallic lustre and a black streak. H 3.0; SG 5.5–6.4
Occurrence. In cavities in dolomite.
Localities. Found in the Binnental, Valais, Switzerland.

Zinkenite $Pb_6Sb_{14}S_{27}$
Zinkenite crystallizes in the hexagonal system, as crystals resembling hexagonal forms through twinning. Massive forms are more commonly found. The colour and streak are steel-grey and the lustre metallic. H 3.0–3.5; SG 5.12–5.35
Occurrence. Occurs with quartz and with other sulpho-salts.
Localities. From Wolfsburg in the Harz Mountains, Germany; a silver-bearing variety from Dundas, Tasmania; from Nevada and Colorado; Bolivia; British Columbia.
Treatment. Clean fibrous forms in alcohol.

Boulangerite $Pb_5Sb_4S_{11}$
Boulangerite crystallizes in the monoclinic crystal system, in which it may either form tabular prismatic crystals or occur massive. The colour is bluish lead-grey and the lustre metallic; there are two directions of cleavage and the streak is reddish-brown. H 2.5–3.0; SG 5.7–6.3
Occurrence. Found in ore veins with lead sulpho-salts, galena and pyrite. It may also be found associated with quartz and dolomite.
Localities. From Příbram, Czechoslovakia; fine crystals from Sala, Vastmanland, Sweden; also from Nevada and Washington State.

Geocronite Pb_5SbAsS_8
Isomorphous with jordanite, geocronite is a member of the monoclinic system and is usually found massive granular; it is lead-grey or white and has a metallic lustre. H 2.0–3.0; SG 6.4

Occurrence. Often found as nodules in galena.
Localities. From Sala, Vastmanland, Sweden: Meredo, Asturias, Spain; Val di Castello, Tuscany, Italy; called kilbrickenite, from Co. Clare, Ireland.

Cannizzarite
This is a lead bismuth sulphide, but the exact chemical formula has not yet been determined. It is a bladed mineral with a metallic lustre found in fumaroles on Vulcano, Lipari Islands, Sicily, Italy. H not known; SG 6.7

Meneghinite $Pb_{13}Sb_7S_{23}$
Meneghinite crystallizes in the orthorhombic system as slender prismatic crystals; it is also found massive. The colour is black to grey, and the lustre metallic. H 2.5; SG 6.34–6.43
Occurrence. Associated with galena, sphalerite etc.
Localities. Bottino, Italy; Marble Lake, Frontenac County, Ontario; Santa Cruz, California.

Left: Crystals of nagyagite from Transylvania, Romania

Below left: Crystals of meneghinite from Bottino, Italy

Left: A good example of jordanite from the Binnental, Valais, Switzerland. The light-coloured mineral is dolomite

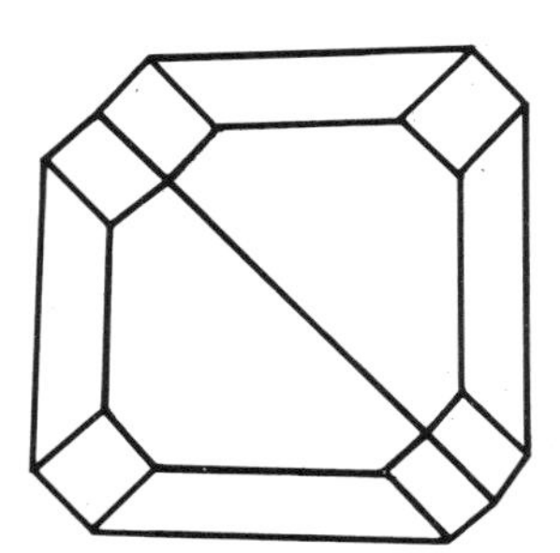

Top: Bournonite from Czechoslovakia. This crystal shows twinning like that depicted in the diagram

Above right: Bournonite crystals on the iron carbonate siderite. From Huttenberg, Austria

Bournonite $CuPbSbS_3$
Sometimes known as cog-wheel ore from its appearance, or endellionite from its original Cornish location, bournonite crystallizes in the orthorhombic crystal system, frequently as twins, which by repetition form crosses and wheel shapes. The colour and streak are grey and there is a brilliant metallic lustre. H 2.5–3.0; SG 5.7–5.9
Occurrence. Bournonite occurs in veins with galena, sphalerite, etc; also with other sulpho-salts.
Localities. Large crystals are found at Neudorf in the Harz Mountains, Germany; from Wheal Boys, Endellion, Cornwall, where it was originally found by Count Bournon. It is also found in Arizona and Nevada.

Dufrenoysite $Pb_2As_2S_5$
Dufrenoysite crystallizes in the monoclinic system as tabular crystals with a perfect cleavage and grey in colour. The streak is reddish-brown. H 3; SG 5.5
Localities. Occurs in dolomite at the Binnental, Valais, Switzerland; also from Dundas, Tasmania; and from some other localities.

Aikinite $CuPbBiS_3$
Aikinite crystallizes in the orthorhombic system as prismatic crystals, or as masses. The colour is black or grey, with a streak of similar colour. H 2.5; SG 7.1
Localities. Occurs with galena, gold and quartz at Beresovsk, USSR; from Utah, Idaho; France; Mexico.

Owyheeite $Ag_2Pb_5Sb_6S_{15}$
An alternative name for this mineral is silver jamesonite. It is a member of the orthorhombic crystal system, in which it may either occur as acicular crystals or massive. There is a cleavage and the colour is steel-grey to silver-white, with a metallic lustre; there is a yellowish tarnish. The streak is reddish-brown. H 2.5; SG 6.2–6.5
Occurrence. In quartz veins.
Localities. Poorman mine, Silver City district, Owyhee County, Idaho.

Freieslebenite $Ag_5Pb_3Sb_5S_{12}$
Freieslebenite is a member of the monoclinic crystal system, in which it forms prismatic crystals. The colour is light-grey to silver-white and it may also be found as a dark grey, resembling lead. The streak is of a similar colour and the lustre is metallic. H 2–2.5; SG 6.2–6.4
Occurrence. Found with silver ores including argentite.
Localities. From Freiberg, E. Germany; near Guadalajara, Spain; Gunnison Co., Colorado.

Berthierite $FeSb_2S_4$
Berthierite crystallizes in the orthorhombic system as prismatic crystals or as masses with a grey colour tarnishing to dark brown. The streak is dark brown to grey. H 3; SG 4.6
Occurrence. Occurs with stibnite and quartz in veins.
Localities. From England; France; Germany; Czechoslovakia; Colorado; New Brunswick.

Jamesonite $Pb_4FeSb_6S_{14}$
Sometimes known as grey antimony, jamesonite is a member of the monoclinic crystal system and forms acicular crystals; it is also found massive with a fibrous texture. The colour is grey and the streak greyish-black. The lustre is metallic and there is a perfect cleavage. H 2.0–3.0; SG 5.5–6.0
Occurrence. Jamesonite is found in ore veins with other lead sulpho-salts, galena, pyrite and calcite.
Localities. Found at Endellion in Cornwall; Mount Bischoff, Tasmania; and at the antimony mines in Arkansas and South Dakota.
Treatment. Gentle washing in distilled water is the best method of cleaning.

Stannite Cu_2FeSnS_4
Sometimes called tin pyrite or bell-

metal ore, stannite is a sulphostannate of copper, iron and sometimes zinc. Traces of germanium may also be found. Bolivianite is a synonym. It crystallizes in the tetragonal crystal system and crystals appear pseudotetrahedral through twinning. Massive forms are also found. It is grey to black with a similar streak; the lustre is metallic. H 3.5; SG 4.3–4.5
Occurrence. Occurs in tin-bearing veins with cassiterite, tetrahedrite and pyrite.
Localities. From Cinvald, Czechoslovakia; Cornwall, at Wheal Rock near St. Agnes; Oruro, Bolivia; Black Hills of South Dakota.

Teallite $PbSnS_2$
Teallite is found in thin flexible folia of the orthorhombic crystal system. It is black to grey with a metallic lustre. There is a black streak and a perfect cleavage. H 1.0–2.0; SG 6.4
Occurrence. With silver minerals and in tin veins.
Localities. Himmelsfürst mine, Freiberg, E. Germany; Santa Rosa, Antequera, Bolivia.

Cylindrite $Pb_3Sn_4Sb_2S_{14}$
Cylindrite crystallizes in the orthorhombic system, occurring in cylindrical forms which separate under pressure into shells or leaves. It is black to grey with a metallic lustre. H 2.5; SG 5.4
Occurrence. Occurs with pyrite and blende.
Localities. From Poopó, Oruro, Bolivia.

Germanite $Cu_3(Ge, Ga, Fe, Zn)(As, S)_4$
Germanite contains 6–10% germanium and up to 2% gallium. It is a member of the cubic crystal system, although it is most commonly found massive. The colour is dark reddish-grey with a metallic lustre. H 4.0; SG 4.46–4.59
Localities. Intergrown with tennantite and pyrite at Tsumeb, South-West Africa.

Argyrodite Ag_8GeS_6
Argyrodite often contains tin, grading towards canfieldite. It is a member of the cubic crystal system and forms octahedra and dodecahedra, although these crystals are often minute. Crystals may be twinned; also occurs massive. The colour is grey but turns to red and violet after exposure to air. H 2.5; SG 6.1–6.3
Occurrence. With silver minerals.
Localities. From the Himmelsfürst mines, Freiberg, E. Germany and from Bolivia.

Enargite Cu_3AsS_4
An ore of copper and arsenic, dimorphous with luzonite, enargite is a member of the orthorhombic crystal system, in which it forms small crystals with vertically striated prismatic faces; star-shaped trillings and massive forms are also found. The colour is grey to black with a similar streak; there is a perfect cleavage and the lustre is metallic. H 3.0; SG 4.43–4.45
Occurrence. An uncommon mineral found in primary deposits and associated with bornite, pyrite, sphalerite, quartz and baryte, among other minerals.
Localities. An important deposit at Bor, Yugoslavia; large masses from Morococha, Peru; at Butte, Montana and from Colorado. Famatinite is closely linked with enargite with the formula Cu_3SbS_4, and is found at the Sierra de Famatina, La Rioja, Argentina; luzonite is found at Luzon Island, Philippines. Also related is epigenite, of probable formula $(Cu, Fe)_5AsS_6$. It is thought to belong to the orthorhombic system and crystallizes in short prisms; found at Wittichen, near Baden, Germany.
Treatment. Dun-black coating is not removable.

Famatinite Cu_3SbS_4
Famatinite crystallizes in the tetragonal system as granular masses coloured deep pink or brown with a black streak. H 3.5; SG 4.6
Occurrence. Occurs in copper deposits.
Localities. From Famatina, Argentina; Nevada; Sonora, Mexico; Japan; etc.

Left: Octahedral crystals of cuprite from Cornwall

Oxides and Hydroxides

Lithiophorite $(Al, Li)MnO_2(OH)_2$
Lithiophorite crystallizes in the monoclinic system as botryoidal masses coloured blue to black and with a perfect cleavage. The streak is dark grey to black. H 3; SG 3.1–3.4
Localities. Occurs at Schneeberg, E. Germany; South Africa; Virginia and Arizona.

Cuprite Cu_2O
An ore of copper, and now sometimes cut as a gemstone, cuprite is sometimes

Above left: Small striated enargite crystals from Mexico

Far left: Stannite with cubic pyrite crystals from Oruro, Bolivia

Left: These crystals of argyrodite from Saxony display mamillary habit

called red copper ore. It is a member of the cubic crystal system, in which it forms octahedra, dodecahedra and cubes; it may also be found massive. It is cochineal-red with a brownish-red streak and an adamantine lustre. H 3.5–4.0; SG 5.8–6.1
Occurrence. Formed by the oxidation and alteration of copper sulphide deposits and associated with native copper, malachite and azurite.
Localities. Ekaterinburg, USSR; St. Day, Cornwall; Broken Hill, New South Wales; South-West Africa (gem quality).
Treatment. Exposure to strong light causes surface film to form.
Fashioning. Uses: faceting; *cleavage*: octahedral, interrupted; brittle; *cutting angles*: no information available; the faceted stones are usually shallow to help reduce the deep colour; *heat sensitivity*: low.

Crednerite $CuMnO_2$
Crednerite crystallizes in the monoclinic system as coatings or as thin black plates with a brownish-black streak. H 4; SG 5.3
Localities. Occurs with cerussite at Higher Pitts, Mendip Hills, England; etc.

Bromellite BeO
Bromellite is a member of the hexagonal crystal system and forms prismatic crystals. It is white in colour. It has been manufactured for commercial use. H 9.0; SG 3.0; RI 1.719, 1.733
Occurrence. With swedenborgite at Långban, Värmland, Sweden.
Treatment. Should be cleaned with water only.

Chrysoberyl $BeAl_2O_4$
A notable gem material, chrysoberyl includes the varieties alexandrite and cymophane (cat's-eye) as well as an attractive yellow-green stone. It crystallizes in the orthorhombic system, often forming pseudo-hexagonal twins or trillings. The lustre is vitreous. The double refraction is 0.009 and there is strong pleochroism; in the alexandrite variety this gives the pleochroic colours red, orange-yellow and green. The presence of chromium, which in alexandrite gives the red and green (so balanced in the absorption spectrum that the colour seen depends on the nature of the incident light—red in tungsten light and green by daylight), can be detected with the spectroscope. The chromium replaces the aluminium, which may also be replaced by ferric iron; this shows in the absorption spectrum as a sharp band at 4,440 Å. The cymophane variety contains a multitude of microscopic channels which reflect light when the stone is cut as a cabochon with the perpendicular to the surface at right angles to them. H 8.5; SG 3.50–3.84; RI 1.747, 1.748, 1.757
Occurrence. In granites, pegmatites and schists; also from sands and gravels.
Localities. Ural Mountains, USSR (alexandrites); Sri Lanka gem gravels (all varieties); Minas Gerais, Brazil (yellow-green and alexandrite); Malagasy Republic (yellow-green); Somabula Forest, Rhodesia (alexandrite); Mogok, Burma (colourless and green varieties).

Right: A superb trilling of chrysoberyl from Brazil. The apparent hexagonal form is due to grouping of orthorhombic crystals. This is called pseudo-hexagonal twinning.

Treatment. Any dilute acid may be used as a cleaning agent.
Fashioning. Uses: faceting or cabochons; *cleavage*: distinct // to the brachydome; brittle; *cutting angles*: crown 40°, pavilion 40°; *pleochroism*: strong in deep colours; *heat sensitivity*: very low.

Right, centre: Typical crystals of psilomelane showing dendritic form. From the Restormel mine, Cornwall

Taaffeite $BeMgAl_4O_8$
Originally thought to be spinel, this pale mauve mineral, with a vitreous lustre, was found to be birefringent by Count Taaffe of Dublin in 1945. It contains a trace of iron. The system was found by X-ray diffraction to be hexagonal and the mineral a member of the trapezohedral class. H 8.0–8.5; SG 3.61; RI 1.721–1.723, and 1.717–1.718
Localities. Sri Lanka gem gravels; also

Right: Octahedra of blue spinel from Franklin, New Jersey

China, from banded sediments in association with fluorite.
Fashioning. Uses: faceting; *cleavage*: not known; *cutting angles*: crown 40°, pavilion 40°; *pleochroism*: weak; *heat sensitivity*: not known. Taaffeite is treated as for spinel by native cutters.

Periclase MgO
Periclase crystallizes in the cubic system as cubes or octahedra; also forms grains. It is colourless, with a vitreous lustre, and has a good cleavage. Periclase has been synthesized and may very occasionally be used as a gem. H 5.5; SG 3.56–3.58; RI 1.74
Occurrence. In contact-metamorphosed limestones.
Localities. Mte. Somma, Mt. Vesuvius, Italy; Nordmark, Värmland, Sweden.
Treatment. Bright surfaces may dull in moist atmosphere.
Fashioning. Uses: faceting or cabochons; *cleavage:* distinct cubic; tough; *cutting angles:* crown 40°, pavilion 40°; *heat sensitivity:* low.

Spinel $MgAl_2O_4$
Spinel is a member of the cubic crystal system, in which it forms octahedra, which occasionally show dodecahedral truncations. Cubes are also found. The colour depends on the variety but for gem use the red is much the most favoured. This variety owes its red colour to chromium; it is distinguished from ruby by the specific gravity. The lustre is vitreous. Blue, violet, black and green colours may also be found. Although a colourless variety is sometimes quoted, it is not found naturally; a seemingly colourless stone is likely to have a tinge of colour, which is not immediately detectable. Spinel has been synthesized by both the Verneuil and flux-melt methods, the products frequently being used as gems. The Verneuil flame-fusion material may be distinguished by its higher SG and refractive index, brought about by an excess of alumina over that found in the natural stones. H 8; SG 3.58 (red variety: other colours may be higher); 3.62 (Verneuil synthetic); RI 1.718 (natural); 1.728 (Verneuil synthetic).
Occurrence. Frequently from gem gravels or embedded in limestone or dolomite.
Major Localities. Mogok, Burma; Sri Lanka; various localities in the USA.
Treatment. Clean with any dilute acid.
Fashioning. Uses: faceting or cabochons; *cleavage*: imperfect octahedral; *cutting angles*: crown 40°, pavilion 40°; *heat sensitivity*: very low.

Psilomelane
A manganese oxide with water and barium, psilomelane is sometimes given the formula $Ba\ Mn^{2+}\ Mn_8{}^{4+}\ O_{16}\ (OH)_4$, but strictly speaking it is a mixture of no fixed composition. It is an ore of manganese and is found massive, botryoidal, reniform and sometimes stalactitic. The colour is black to grey and the streak brownish-black. The lustre is submetallic. H 5.0–7.0; SG 6.4
Occurrence. Associated with pyrolusite often in alternating layers and frequently found in laterite deposits.
Localities. Schneeberg, E. Germany; Lanlivery, Cornwall; Ouro Preto, Minas Gerais, Brazil; in Lake Superior hematite deposits, Michigan.
Treatment. Clean with oxalic acid.
Fashioning. Uses: cabochons, tumbling, faceting, intaglios, etc.; *cleavage*: none; *cutting angles*: variable to suit requirements; *heat sensitivity*: low.

Zincite ZnO
An ore of zinc, zincite is occasionally cut as a gem. It is deep red with a tinge of yellow and an orange-yellow streak. The lustre is sub-adamantine. It crystallizes in the hexagonal system, though it is usually found massive or as grains. There is a perfect cleavage. The red colour is caused by the manganese content. H 4.0–4.5; SG 5.43–5.70; RI 2.013; 2.029
Localities. With franklinite and willemite at Franklin, Sussex County, New Jersey.

Franklinite
$(Zn, Mn, Fe^{2+})\ (Fe^{3+}, Mn^{3+})_2O_4$
An ore of zinc, franklinite is predominantly $ZnFe_2O_4$. It forms octahedra in the cubic system or may be found massive and granular. The colour is black and the streak reddish-brown or black. H 5.5–6.5; SG 5.07–5.22; RI 2.36 (in very thin section).
Occurrence. From ores in Precambrian limestones; associated with willemite, zincite and calcite.
Localities. Franklin, Sussex County, New Jersey.
Treatment. Calcite coating removable with HCl.

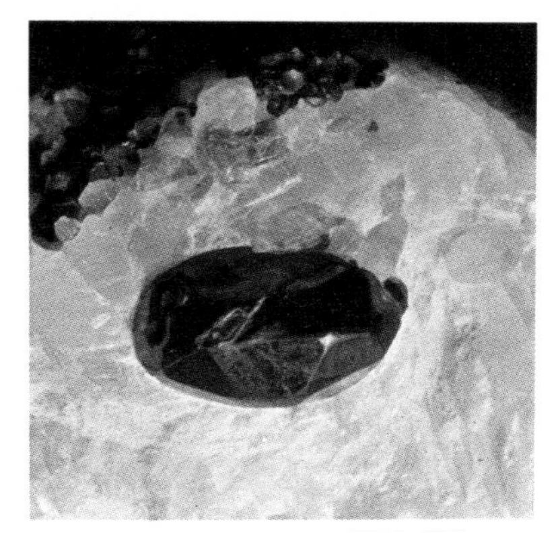

Left: A crystal of franklinite, named after its locality, Franklin, New Jersey

Corundum Al_2O_3
Celebrated for its varieties ruby and blue sapphire, corundum is aluminium oxide, with Cr_2O_3 substituting for the alumina in ruby, and Fe_2O_3 substituting similarly in sapphire of various colours; in blue sapphires titanium is present in addition to iron. The mineral is a member of the hexagonal crystal system, forming tabular prisms with rhombohedra in the case of ruby, and hexagonal bipyramids in the case of sapphire,

Above left: A crystal of ruby, a variety of corundum, in green zoisite from Tanzania

Left: A rare mineral only known from Franklin, New Jersey, zincite. It has sometimes been faceted

Right: In this crystal of quartz from Italy the apparent pyramid faces at the top are really two rhombohedra. Note also the horizontal striation of the prism faces

Below: Left-handed and right-handed forms of quartz. These are sometimes combined in a single twinned crystal

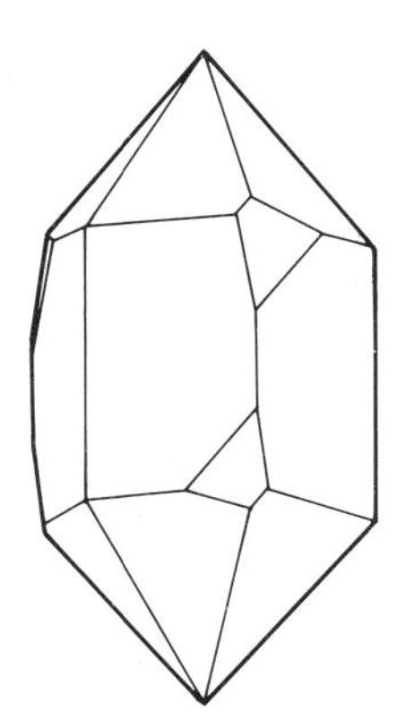

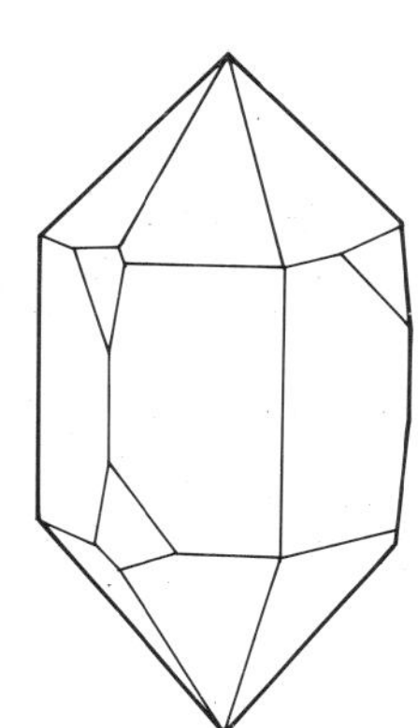

particularly the Sri Lanka specimens; much material is found water-worn. Lamellar twinning is common and parting may take place parallel to the lamellae; there is no cleavage. The lustre is vitreous. Dichroism is marked, ruby showing crimson and orange-red, blue sapphire showing dark and light blue. Asterism is quite common and is due to reflection of light from crystals of rutile arranged at right angles to the vertical crystal axis. The presence of chromium may be detected by the spectroscope; the absorption spectrum differs from that shown by red spinel in that two or three lines are present in the blue. H 9; SG 3.99; RI 1.76; 1.77, with a double refraction of 0.008.

Occurrence. As an accessory mineral in dolomites, limestones, gneisses or mica schists. Associated minerals include spinel and tourmaline as well as other aluminium minerals.

Localities. The finest rubies come from Mogok, Upper Burma; this area also produces good quality sapphires. The Sri Lanka gem gravels produce rubies which tend to pink and verge into pink sapphire; parti-coloured stones also found in this region. Good quality rubies and sapphires come from Cambodia and Thailand; the rubies, owing to their iron content, are rather darker than the Burmese stones. Also from Montana (sapphire); from Central and East Africa (ruby and sapphire), particularly Tanzania, where blue sapphire is of good quality. Fine yellows come from Australia and Sri Lanka; greens from Australia, especially from the Anakie area, Queensland.

Treatment. Any acid can be used as a cleaning agent.

Fashioning. Uses: faceting, cabochons, intaglios, carvings, etc.; *cleavage*: none; planes of parting exist: 1) perfect // to basal pinacoid, 2) prominent, pyramidal; brittle; *cutting angles*: crown 40°, pavilion 40°; *dichroism*: strong in deeply coloured stones; *heat sensitivity*: low, but overheating during faceting and polishing may induce cracks along facet edges.

Diaspore AlO.OH

Diaspore is the alpha-phase of AlO.OH and forms flattened prismatic crystals of the orthorhombic crystal system; it may also be found granular massive or stalactitic. The colour is white or greenish-grey, with a brilliant lustre. When observed, pleochroism is dark violet and faint yellow. The gamma phase of this material is boehmite. H 6.5–7.0; SG 3.3–3.5

Occurrence. As for corundum (see preceding entry).

Localities. In a granular limestone near Ekaterinburg, Ural Mountains, USSR; in dolomite at Campolungo, St. Gotthard, Switzerland; mangandiaspore, the dark-red manganese-bearing variety, from Postmasburg, South Africa; in the emery mines, Chester, Massachusetts.

Gibbsite $Al(OH)_3$

Gibbsite is also known by the alternative name of hydrargillite. It is a member of the monoclinic crystal system, in which it forms tabular hexagonal crystals; it is also found as concretions or stalactitic. There is a perfect cleavage and the colour varies from white to reddish or greenish-white. The lustre is pearly. When breathed on the mineral may give the smell of wet clay. Some specimens may be transparent. H 2.5–3.5; SG 2.3–2.4

Occurrence. Gibbsite is formed from the alteration of aluminium-bearing minerals. It is frequently present together with bauxite.

Localities. Crystalline gibbsite from the Ural Mountains, USSR; with natrolite in the Langesundfiord, Norway; Ouro Preto, Minas Gerais, Brazil; the stalactitic form is found at Richmond, Massachusetts.

Quartz SiO_2

There are a number of polymorphs of silica, that which is known as quartz being thermally stable below 573°C. High quartz is stable between 573°–870°C. Above this temperature the stable polymorph is tridymite and this retains stability up to 1470°C. Cristobalite is the polymorph stable above this temperature up to its melting point at 1723°C. Liquid silica, when cooled below 1720°C, can form a glass (silica-glass, lechatelierite), which, if heated over 1000°C, devitrifies and forms cristobalite as a stable or metastable phase. Keatite is a metastable form of silica, which has been synthesized at temperatures between 380°C and 585°C; coesite

and stishovite are high-pressure polymorphs, which can exist metastably at ordinary temperatures and pressures.

Quartz is one of the most commonly occurring minerals and crystallizes in the hexagonal system; it is a member of the trigonal trapezohedral class. This class possesses the property of enantiomorphism; that is, the atoms of silica are arranged helically along the axes of reference. Crystals are described as being left- or right-handed. If a crystal is right-handed, there is a trigonal pyramid in the upper right corner of the prism face below the positive rhombohedron, and if there are any striations they slope to the right. Left-handed crystals show opposite relations.

The habit of quartz is most commonly the prism with positive and negative rhombohedra. Twinning is virtually universal although it is not usually observable in the external form of the crystal. Quartz is piezoelectric; that is, when it is subjected to mechanical stress it develops electric charges on the surface. Conversely, when it is subjected to an electric charge it shows mechanical strain. It is also pyroelectric; that is, it develops an electric charge on the surface as a result of a change of temperature. Quartz rotates the plane of polarization of light travelling parallel to the c-axis either to the right or left. H 7; SG 2.65; RI 1.544, 1.553

Varieties. Varieties of quartz include citrine (yellow); amethyst (violet to purple); rock crystal (colourless); smoky quartz (smoky brown to black); rose quartz (pink), usually but not invariably found massive; blue quartz, which is found as grains in some metamorphic and igneous rocks. Finer-grained varieties include chalcedony, which may be varied in colour; sard, light brown to reddish-brown; carnelian, uniform reddish-brown; moss agate, grey or bluish chalcedony with dark dendritic inclusions, often of an iron compound; agate, a chalcedonic variety in which successive layers differ in colour and translucency; chrysoprase, an apple-green nickel-coloured variety of chalcedony;

Left: Quartz from Bolzano, Italy

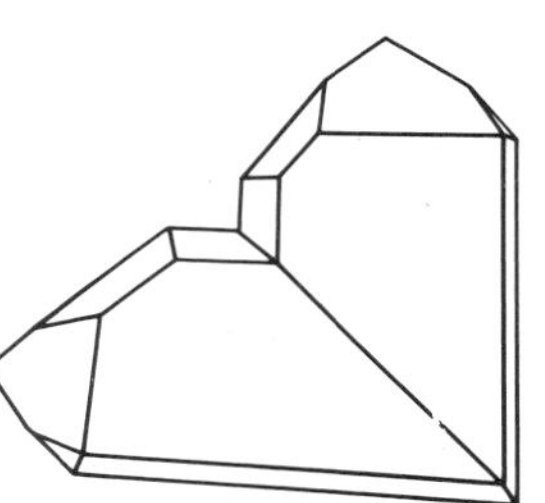

Below: The so-called Japan twinning of quartz in which two crystals join at nearly 90°. The striation of the prism faces is prominent

plasma, a green microgranular quartz; bloodstone, which is greenish with red flecks of iron oxide or red jasper; flint and chert, the distinction between the two being indefinite; jasper, a massive fine-grained quartz with large amounts of additional material, in particular iron oxide, variable in colour but most commonly dark-red to yellow. Inclusions of rutile and tourmaline are common and specimens are known as rutilated or tourmalinated quartz. Star-stones and cat's-eyes are fairly common.

Occurrence. Silicon constitutes 27.7% by weight of the Earth's crust and is the second most abundant element after oxygen. Quartz is an important rock-forming mineral and occurs in many ways, though most of the best-developed crystals occur in granite pegmatites and may be deposited by hydrothermal action in granular form as a replacement of pre-existing rocks.

Localities. Very fine quartz crystals come from the St. Gotthard area of Switzerland and from Bourg d'Oisans, Isère, France; this locality is notable for twin crystals. In England, Cleator Moor in Cumbria furnishes good crystals; fine agates come from Brazil and formerly from Idar-Oberstein, Germany; small well-formed crystals occur in Herkimer County, New York State.

Fine amethyst is found in the Sverdlovsk area of the Ural Mountains, USSR; at Idar-Oberstein, Germany; Brazil;

Right: Smoky quartz from the St. Gotthard, Switzerland

Uruguay; India, where geodes are found in the Deccan. Rock crystal is widely distributed but good specimens come from the St. Gotthard area of Switzerland and from Brazil; rose quartz is found in Rio Grande do Norte at the Alto Feio; in the Malagasy Republic and India and in South-West Africa and South Africa; tiger's-eye comes from South Africa; citrine from Brazil; agates and the non-transparent quartzes particularly from Brazil, South Africa and India.

Treatment. Quartz may be cleaned in warm hydrochloric or other acids; hydrofluoric acid should be avoided as should rapid cooling, which may crack the material. Some quartz may be heated to give different colours. Much citrine is obtained by heating amethyst and some green transparent quartz is also obtained in this way.

Fashioning. Uses: faceting, cabochons, tumbling, carving, intaglios, etc; *cleavage*: indistinct, conchoidal fracture; brittle in crystalline varieties and tough in some cryptocrystalline varieties; *cutting angles*: crown 40°–50°, pavilion 43°; *dichroism*: strong in deeply coloured varieties of crystalline material; *heat sensitivity*: very low.

Opal $SiO_2.nH_2O$

Opal is amorphous and contains up to 10% water. It is found massive, sometimes reniform or stalactitic; frequently pseudomorphous after other minerals, shells or wood. It has a vitreous lustre.

Right: Amethyst crystals from Bolzano, Italy, displaying parallel growth

Far left: Fire opal from Zimapan, Mexico

Left: Opal from Queensland, Australia

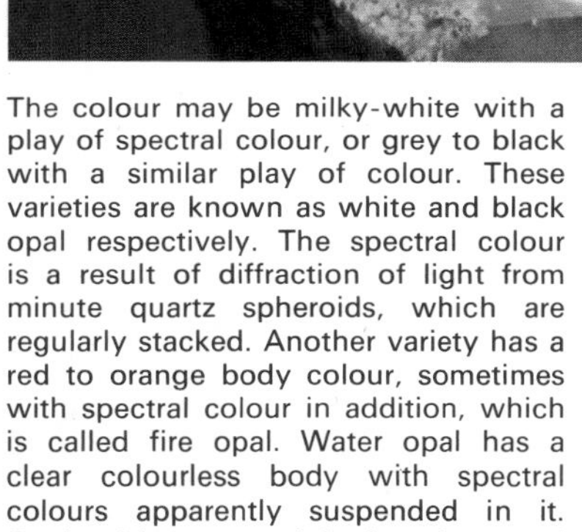

The colour may be milky-white with a play of spectral colour, or grey to black with a similar play of colour. These varieties are known as white and black opal respectively. The spectral colour is a result of diffraction of light from minute quartz spheroids, which are regularly stacked. Another variety has a red to orange body colour, sometimes with spectral colour in addition, which is called fire opal. Water opal has a clear colourless body with spectral colours apparently suspended in it. Opal without a red body-colour and without a play of colour is called common opal or potch. Most common opal is grey or black, although a green form, perhaps coloured by nickel, is also found. Boulder opal is brown and may sometimes display spectral colour.

H 5.5–6.5; SG 1.9–2.3; RI 1.43, 1.45

Occurrence. Opal is deposited at low temperatures from water containing silica and occurs with many types of rock. It fills seams and fissures of igneous rock, occurs in mineral veins, and may be deposited from hot springs. Boulder opal is found in sandstone as veins, or as concretions made of shells of sandstone and a hard siliceous clay, with opal between them or filling the centre.

Localities. Opal has been found in the area of Cervenica, Czechoslovakia; the fire opal occurs in a trachyte porphyry at Zimapan, Hidalgo, Mexico and the water opal comes from San Luis Potosi in the same country. Some opal is found in Honduras and Guatemala. The finest opals come from Australia where the black opals are found at Lightning Ridge, north-west New South Wales; fine white opals come from the fields at Coober Pedy, and Andamooka in South Australia; boulder opal is found in Queensland. In the USA opal, often pseudomorphous after wood, is found in the Virgin Valley area of Nevada. Recently white opal has been found in the state of Piaui, Brazil.

Treatment. Clean with water, which must not contain any dirt or colouring matter.

Fashioning. Uses: faceting or cabochons; *cleavage*: none, conchoidal fracture; *cutting angles*: crown 40°–50°, pavilion 43°, table facet often slightly

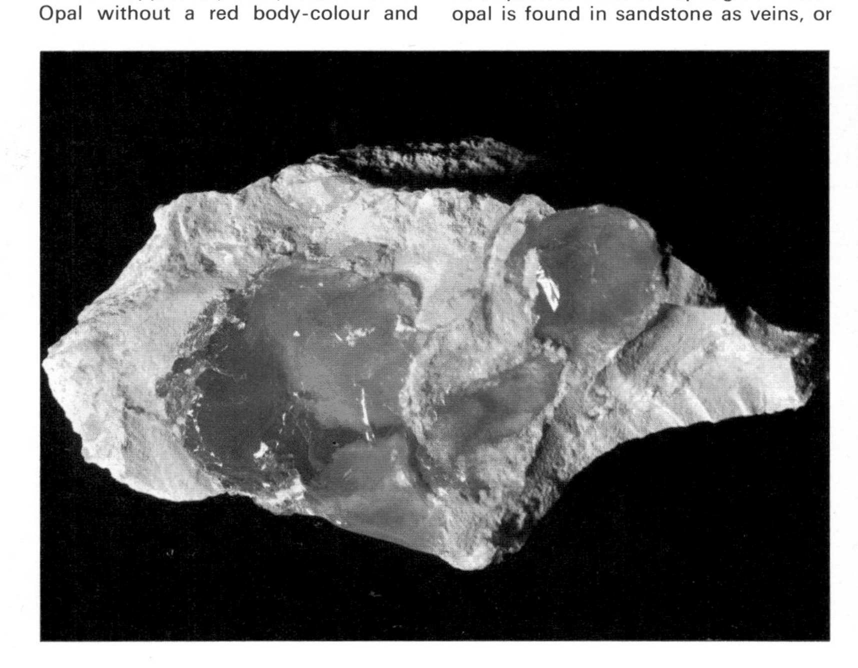

Far left: Clear masses of fire opal from Mexico

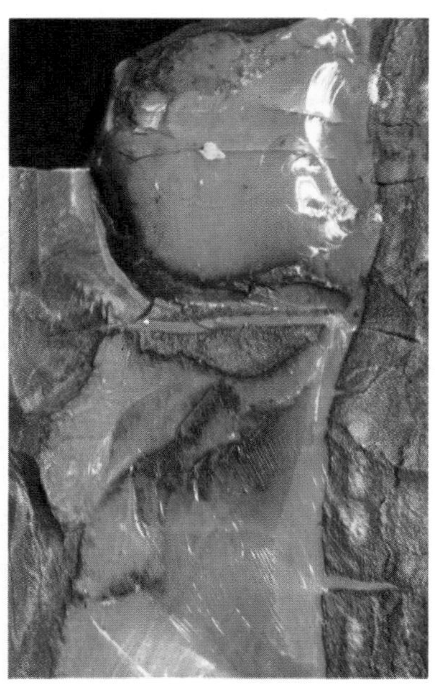

Left: Queensland opal

Right: Fine geniculate (knee-shaped) rutile crystal from Brazil

Far right: Another habit of rutile. These are acicular crystals in quartz from Brazil. Rutile completely enclosed by quartz is sometimes called Venus hair stone

domed; *heat sensitivity*: high, care should be taken when dopping and polishing.

Rutile TiO_2

Rutile, anatase and brookite are polymorphs of TiO_2. Rutile crystallizes in the tetragonal system, forming prismatic crystals terminated by bipyramids; twins are found which are sometimes geniculate, or knee-shaped. The colour is reddish-brown to black, the streak pale brown and the lustre adamantine. There is an uneven fracture. H 6–6.5; SG 4.2–4.4; RI 2.62, 2.90 (synthetic rutile). The birefringence is 0.287 and the dispersion about 0.3.

Occurrence. Rutile is found as an accessory mineral in igneous rocks and in gneiss, mica schist, granular limestone and dolomite. It may be secondary in origin, occurring as an alteration of mica or sphene. It is found as acicular crystals penetrating quartz (rutilated quartz).

Localities. Rutile is found south-west of Graz, Austria; in the St. Gotthard area, Switzerland; from Minas Gerais, Brazil; Waterbury, Vermont; North Carolina; and Graves' Mountain, Georgia. Fine crystals of rutilated quartz from Val Travetsch, Switzerland and from Alexander County, North Carolina.

Treatment. Silicate coatings may be removed with hydrofluoric acid, and iron stains with oxalic acid.

Fashioning. Uses: faceting; *cleavage*: distinct, prismatic (1st and 2nd order); *cutting angles*: crown 30°–40°, pavilion 37°–40°; *heat sensitivity*: low.

Anatase TiO_2

Sometimes known as octahedrite, anatase, with rutile and brookite, is a polymorph of TiO_2. It forms bipyramidal crystals of the tetragonal system and is also found tabular. It may be yellow, blue, brown or black in colour, with a white streak. The lustre is adamantine to metallic, the latter when the mineral is black and opaque. The fracture is subconchoidal and there is a perfect cleavage. H 5.5–6.0; SG 3.8–4.0

Occurrence. Anatase is found as an accessory mineral in igneous and meta-

Far right: Prismatic anatase from Binnental, Switzerland. Anatase and rutile, which are tetragonal, form a series with brookite, which is orthorhombic

Right: This crystal of anatase from Bourg d'Oisans, France, at first sight seems to be an octahedron but is in fact a bipyramid

morphic rocks, and is derived from the alteration of other titanium minerals. It is associated with quartz, hematite, apatite, sphene and other minerals.
Localities. Anatase occurs as fine crystals in the area of Bourg d'Oisans, Isère, France; in the Binnental, Valais, Switzerland; and in chlorite near Tavistock, Devon. It is also found at Somerville, Massachusetts, and as blue crystals in Beaver Creek, Colorado.
Treatment. Iron stains may be removed with oxalic acid.

Brookite TiO_2
Brookite, rutile and anatase are polymorphs of TiO_2. Brookite crystallizes in the orthorhombic system, in platy or tabular crystals. The colour is reddish-brown to black, the streak white and the lustre metallic to adamantine. The cleavage is poor and the fracture uneven. H 5.5–6.0; SG 3.9–4.2
Occurrence. Brookite is found as an accessory mineral in igneous and metamorphic rocks and in hydrothermal veins.
Localities. Brookite occurs in the gold placers of the Urals, USSR, and at St. Gotthard, Switzerland. It is found on quartz with chalcopyrite and galena at Ellenville, Ulster County, New York and in thick black crystals at Magnet Cove, near Hot Springs, Arkansas.
Treatment. Clean with water or dilute acids.

Perovskite $CaTiO_3$
Perovskite crystallizes in pseudocubic forms of the orthorhombic system, forming cubes with irregularly distributed faces; these are striated parallel to the edges and are probably penetration twins. It may also be found in reniform masses showing small cubes. The colour ranges from pale yellow, through orange, to greyish-black; the streak is colourless to grey. There is an imperfect cleavage and the lustre is metallic-adamantine. H 5.5; SG 4.0
Occurrence. Perovskite is found in chlorite, talc or serpentinous rocks, or as an accessory constituent of melilite, nepheline or leucite basalts.
Localities. Found in a chlorite slate in the Urals, USSR; in the neighbourhood of the Findelen glacier, Zermatt, Switzerland; Val d'Aosta, Piedmont, Italy.
Treatment. Perovskite should be cleaned with dilute acids or water.

Left: Brookite on albite from Switzerland

Far left: Brown anatase from the Binnental, Switzerland

Below left: The calcium titanate perovskite. It often looks cubic but the crystal system is orthorhombic. This specimen is from Achmatovsk, USSR

Ilmenite $FeTiO_3$
Ilmenite contains up to about 6% Fe_2O_3 and crystallizes in the hexagonal system in thick tabular crystals. Massive compact forms are common and twinning on the basal pinacoid is frequently found. The colour is black, the lustre metallic, and the streak black to brownish-red. There is a basal parting; the fracture is conchoidal. H 5–6; SG 4.5–5.0
Occurrence. Ilmenite occurs as an accessory mineral in igneous rocks, including gabbro and diorite. It may occur in quartz veins and in pegmatites in association with hematite and chalcopyrite. It is found concentrated in alluvial sands with magnetite, rutile and monazite.

Far left: Another crystal of perovskite showing interpenetration

Left: Brown perovskite from Val Malenco, Italy

Right: Minium from Broken Hill, New South Wales, showing the scaly habit

Below right: Prismatic cassiterite from La Villeder, France. It shows the mineral's characteristic high lustre

Localities. Found in the Ilmen mountains in the Urals, USSR; in the Binnental, Switzerland; and as small crystals near Bourg d'Oisans, Isère, France. Fine crystals are found in New York State at Warwick, Amity and Monroe; also with magnetite deposits in the Adirondack region of the same state. Ilmenite in the form of sand is found at Menaccan, Cornwall.

Baddeleyite ZrO_2
Baddeleyite crystallizes in the monoclinic system forming tabular crystals with a nearly perfect cleavage. It is yellow, brown to black or colourless, and has a greasy lustre. Baddeleyite contains some hafnium. H 6.5; SG 5.5–6.0
Localities. Found in the diamond sands at Minas Gerais, Brazil, as rolled pebbles. From Sri Lanka and from Mte. Somma, Mt. Vesuvius, Italy; also near Bozeman, Montana.
Treatment. Clean with dilute acids.

Cassiterite SnO_2
Cassiterite or tinstone crystallizes in the tetragonal system, forming pyramidal or short prismatic crystals. It is also found massive and granular. The colour is reddish-brown to black; it may also be yellow. The streak is white or grey and the lustre adamantine. The fracture is uneven. H 6–7; SG 6.8–7.1; RI 1.9, 2.1; dispersion 0.071, birefringence 0.096
Occurrence. Cassiterite is the principal ore of tin and occurs in high-temperature hydrothermal veins and pegmatites in or near granites. It is associated with wolframite, topaz, quartz, tourmaline and other minerals. 'Stream tin', weathered cassiterite, is found in alluvial deposits. It may range from a fine powder through sand-sized grains to rounded pebbles. 'Wood tin' has a fibrous structure, displaying concentric bands and resembling wood.
Localities. Large veins of tin ores existed in Cornwall but the most accessible have now been worked out. The classic localities for specimens are St. Agnes, Callington, Camborne, St. Just and St. Austell. At Wheal Coates, near St. Agnes, pseudomorphs after feldspar are found. Other localities include northwest Czechoslovakia; Villeder, France, where fine twin crystals are found, and the Lake Ladoga area of Finland. There are important deposits in Malaysia and in New South Wales; in Bolivia near La Paz, Oruro and Potosi, and in Mexico from the states of Durango, Guanajuato and Jalisco. Small deposits occur in the USA, in the Santa Ana Mountains, California. Good material comes from Arandis, South-West Africa.
Treatment. Bolivian crystals are sometimes encrusted with a clay-like material, which can be removed by scraping, or with a sharp knife. Clean with dilute acid.
Fashioning. Uses: faceting or cabochons; *cleavage*: imperfect, prismatic (2nd order); *cutting angles*: crown 30°–40°, pavilion 37°–40°; *heat sensitivity*: very low.

Massicot and **Litharge PbO**
Massicot is the orthorhombic and litharge the tetragonal phase of PbO; massicot is yellow and litharge yellow-orange. The material is found earthy or scaly massive. Study of some crystals shows that the centre of crystal plates is massicot while the borders are litharge. Massicot is optically positive and litharge negative. Massicot: H 2; SG 9.6; litharge: H 2; SG 9.3
Occurrence. Both minerals are of secondary origin, associated with galena and other minerals.
Localities. Massicot and litharge are found at Freiberg, E. Germany; Sardinia and Colorado.
Treatment. Both minerals are attacked by acids and should be cleaned with water only.

Minium Pb_3O_4
Minium, sometimes called red lead, crystallizes in the tetragonal system and is found as crystalline scales of a vivid red mixed with yellow, with a greasy lustre. The streak is orange-yellow. More commonly seen as an artificial mineral made by heating litharge or massicot in air. H 2–3; SG 8.9–9.2
Occurrence. Minium is of secondary origin and occurs from the alteration of galena or cerussite.
Localities. Minium is found in the Altai Mountains, USSR; from the Eifel region of Germany; from Broken Hill, New South Wales and with native lead at the Jay Gould mine, Idaho.
Treatment. Clean with water.

Plattnerite PbO_2
Plattnerite crystallizes in the tetragonal system, and is usually found massive, though prismatic crystals sometimes occur. It is iron black with a chestnut-brown streak and a submetallic lustre. H 5–5.5; SG 9.40
Localities. Plattnerite is found with lead minerals at Leadhills, Lanarkshire, and at Wanlockhead, Dumfries and Galloway; and at Shoshone County, Idaho.
Treatment. Easily attacked by acids; clean with water.

Quenselite $PbMnO_2.OH$
Crystallizing in the monoclinic crystal system, quenselite has a perfect basal cleavage. It is pitch-black with a metallic lustre and a brownish-grey streak. H 2.5; SG 6.8
Localities. Found at Långban, Värmland, Sweden.

Cesarolite $PbMn_3O_7.H_2O$
Cesarolite forms cellular masses, steel-grey in colour. H 4.5; SG 5.2
Localities. Found at Sidi-Amer-ben-Salen, Tunisia.

Magnetoplumbite $(Pb, Mn)_2Fe_6O_{11}$
Magnetoplumbite crystallizes in the hexagonal system, forming pyramidal crystals. It is black with a dark-brown streak and possesses a good basal cleavage. It is strongly magnetic. H 6; SG 5.5
Localities. Magnetoplumbite is found associated with manganophyllite and

Left: Chromite nodules from the USSR

kentrolite at Långban, Värmland, Sweden.
Treatment. Clean with water.

Navajoite $V_2O_5 . 3H_2O$
Navajoite probably crystallizes in the monoclinic system. It is found in fibrous aggregates, dark-brown in colour.
Localities. Found with vanadium minerals in Utah, Arizona and Colorado.

Arsenolite As_2O_3
Arsenolite is the cubic phase of As_2O_3 and forms octahedra; it is sometimes found as crusts. There is an octahedral cleavage. The colour is white, with a vitreous lustre, and the mineral is in fact sometimes called white arsenic. H 1.5; SG 3.7
Occurrence. Arsenolite is a mineral of secondary origin associated with other arsenic minerals.
Localities. Found at Jachymov, Czechoslovakia; at the Amargosa mine, California, and as crystals with enargite in the Alps.
Treatment. As the mineral is slightly water-soluble and is affected by acids it has to be handled with care.

Claudetite As_2O_3
Claudetite is the monoclinic phase of As_2O_3 and forms thin plates with a perfect cleavage. The colour is white. The lustre is vitreous to pearly. H 2.5; SG 3.8–4.1
Occurrence. Claudetite is of secondary origin and is usually a sublimation product of mine fires.
Localities. Claudetite is found at Schmöllnitz, Czechoslovakia and at San Domingos, Algarve, Portugal; Decazeville, France.

Senarmontite Sb_2O_3
Senarmontite is the cubic phase of Sb_2O_3. It is found as octahedra or as crusts, but more commonly massive. It is colourless to grey, and has a resinous lustre. H 2; SG 5.3
Occurrence. Senarmontite is formed by the oxidation of stibnite and other antimony minerals.
Localities. Found in octahedral crystals and massive at the mine of Djebel-Haminate, Constantine, Algeria; Wolfe Co., Quebec.
Treatment. Clean with water.

Valentinite Sb_2O_3
Valentinite is the orthorhombic phase of Sb_2O_3 and occurs as prismatic crystals with a perfect cleavage. The colour is white and the lustre adamantine. H 2.5–3.0; SG 5.7
Occurrence. Valentinite is an oxidation product of antimony minerals.
Localities. Found at Příbram, Czechoslovakia; near Freiberg, E. Germany; at the Sensa mine, near Constantine, Algeria, and from South Ham, Wolfe Co., Quebec.
Treatment. Clean with water.

Stibiconite $Sb_3O_6(OH)$
Stibiconite is usually amorphous and massive in form. Its crystals are sometimes assigned to the cubic system. Some antimony is replaced by calcium. The colour is pale yellow to yellowish-white. It is partly isotropic and partly birefringent. H 3–7; SG 3.3–5.5
Occurrence. Stibiconite is an alteration product of antimony minerals.
Localities. Found at Goldkronach, in the Fichtelgebirge, W. Germany; Altar, Sonora, Mexico; the Empire district, Nevada; Australia.
Treatment. Avoid acids; clean ultrasonically; a white coating sometimes encountered is also stibiconite and is not removable.

Bismite Bi_2O_3
Bismite is the monoclinic alpha-phase of Bi_2O_3 and is straw-yellow in colour; its lustre is subadamantine. The cubic phase of Bi_2O_3 is sillenite. H 4.5; SG 8.64–9.22
Occurrence. Bismite results from the oxidation of other bismuth compounds.
Localities. Found with gold at Beresovsk, near Ekaterinburg, Ural Mountains, USSR; Schneeberg, E. Germany; Bolivia; California; New Mexico.

Chromite $FeCr_2O_4$
Chromite is the only ore of chromium and often includes some magnesium and aluminium. It crystallizes in the cubic system, forming octahedra, but massive forms are much more common. The colour is black to brownish-black and the streak dark brown. The lustre is metallic and the fracture uneven. H 5.5; SG 4.1–5.1
Occurrence. Chromite is an accessory mineral in igneous rocks such as peridotite and serpentinite.
Localities. Found in the Ural Mountains, USSR; from Styria, Austria and Zabkowke, Poland; as crystals from the Isle of Unst, Shetlands; from Selukwe, Rhodesia; and from North Carolina and California.
Treatment. Clean with dilute acids.

Tungstite $WO_3 . H_2O$
Found as powdery coatings, tungstite crystallizes in the orthorhombic system. The colour is yellow to yellowish-green. H 1–2; SG 5.5
Occurrence. Tungstite is a secondary mineral found with wolframite.
Localities. Found at Salmo, British

Below left: Tungstite in massive form from Bolivia

Below: The octahedral form of senarmontite is clearly shown in this group of crystals from Algeria

Right: Becquerelite from Zaire. Many other uranium-bearing minerals are bright yellow

Columbia; from Cornwall; Connecticut; and Bolivia.
Treatment. May be cleaned with dilute acids.

Hydrotungstite $H_2WO_4.H_2O$
This mineral, crystallizing in the monoclinic system, is dark green to yellowish-grey. It is found in Cornwall, New Zealand, Bolivia, etc., as an alteration product of wolframite or scheelite.

Uraninite UO_2
Uraninite is an important ore of uranium and the source of radium. Well-known by its alternative name of pitchblende, it crystallizes in the cubic system, though crystals are rare. The mineral is usually found massive botryoidal and is brown to black with a brownish-black to grey streak. The lustre is submetallic and greasy, resembling pitch and the fracture is conchoidal to uneven. It is radioactive. The presence of various oxidation and decay products makes the composition very variable. The name uraninite is normally applied to the crystals; pitchblende to the massive material. H 5–6; SG 6.5–8.5 (massive material); 7.5–10.0 (crystals).
Occurrence. Crystalline material occurs in granite pegmatites associated with tourmaline, monazite and zircon. The massive forms are usually found as crusts in high-temperature hydrothermal veins in association with cassiterite, pyrite, chalcopyrite and galena. It may appear in alluvial deposits.
Localities. From Přibram and Jachymov, Czechoslovakia; various localities in Cornwall; Tanzania; feldspar quarries at Middletown, Connecticut; mica mines at Mitchell, North Carolina; Great Bear Lake, Canada; Shaba, Zaire; Rum Jungle, Northern Territory, Australia.
Treatment. Clean in dilute HCl.

Right: Uraninite from Zaire. This mineral is radioactive

Becquerelite $CaU_6O_{19}.11H_2O$
Becquerelite crystallizes in minute crystals of the orthorhombic system; twinning is common. The colour is brownish-yellow and there is a perfect cleavage and a resinous lustre. It is radioactive. H 2.5; SG 5.1
Localities. Found with curite, soddyite, anglesite and uraninite at Kasolo, Zaire. The variety schoepite is from the same locality ($UO_3.2H_2O$), as is ianthinite, at present regarded as $UO_2.5UO_3.10H_2O$. It crystallizes in the orthorhombic system and forms acicular crystals, coloured violet-black with alteration to yellow on the edges. The streak is brown-violet.

Davidite
$(Fe^{2+}, Ce, U)_2(Ti, Fe^{3+}, V, Cr)_5O_{12}$
Davidite crystallizes in the hexagonal system but is usually found massive; it is black with a metallic lustre. It is radioactive. H 6; SG 4.5
Localities. Davidite is found at Radium Hill, Australia; and Tete, Mozambique.
Treatment. Alteration coatings cannot be removed. Clean with dilute acid.

Brannerite (U, Ca, Ce) (Ti, Fe)O_{16}
Brannerite crystallizes in the monoclinic system as prismatic crystals or as granular masses. The colour is black and the streak dark greenish-brown. H 4.5; SG 4.5–5.4
Occurrence. Found chiefly as a primary mineral in pegmatites; also found in gold placers.
Localities. From Stanley Basin, Idaho; Bou-Azzer, Morocco; Cordoba, Spain.
Treatment. Clean with water.

Fourmarierite $PbU_4O_{13}.4H_2O$
Fourmarierite crystallizes in the orthorhombic crystal system, forming tabular crystals. The colour is red and the lustre adamantine. H 3–4; SG 6.0
Occurrence. Occurs as an alteration product of uraninite.
Localities. Found at Shinkolobwe, Shaba, Zaire.
Treatment. Clean with water.

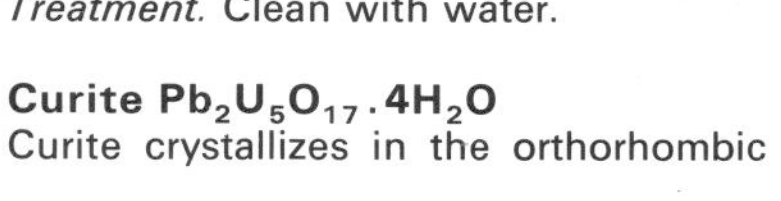

Curite $Pb_2U_5O_{17}.4H_2O$
Curite crystallizes in the orthorhombic

system, often forming minute brownish needles. The colour is reddish-brown to deep yellow by transmitted light and the lustre is adamantine. The streak is orange. It is strongly radioactive. H 4–5; SG 7.1
Localities. Found with other uranium minerals at Kasolo, Shaba, Zaire; Puy-de-Dôme, France.

Clarkeite $(Na, K, Pb)_2\ U_2O_7 . nH_2O$
Clarkeite forms massive coatings, dark brown in colour, surrounding uraninite, of which it is an alteration product. Found at Spruce Pine, North Carolina. H 4.0–4.5; SG 6.3

Gummite
A general term for secondary uranium oxide. Much material described as gummite is curite.

Uranosphaerite $Bi_2U_2O_9 . 3H_2O$
Uranosphaerite crystallizes in the orthorhombic system forming semi-globular aggregates. The colour is orange-yellow to brick-red, and the lustre greasy. There is a distinct cleavage. H 2–3; SG 6.3
Localities. From Neustädtel, south of Schneeberg, E. Germany.

Tellurite TeO_2
Tellurite crystallizes in the orthorhombic system, forming slender prismatic crystals. The colour is white to yellow and the lustre subadamantine; there is a perfect cleavage. H 2; SG 5.9
Occurrence. Tellurite is an oxidation product of tellurides and of native tellurium.
Localities. Found at Sacarambu, Romania; Boulder County, Colorado; Nye County, Nevada.
Treatment. Readily attacked by most acids; clean with water.

Manganosite MnO
Manganosite crystallizes in the cubic system forming octahedra with a cubic cleavage. The colour is emerald-green, becoming black on exposure to light. The streak is brown and lustre vitreous. H 5–6; SG 5.1
Localities. Found at Långban, Värmland, Sweden; Franklin, New Jersey.
Treatment. Clean with water.

Above: Bright yellow prismatic tellurite crystals

Left, centre: Pyrolusite crystals from Restormel, Cornwall. This example shows the rather rare individual crystals

Pyrochroite $Mn(OH)_2$
Pyrochroite crystallizes in the hexagonal system and is usually found foliated. The colour is white, with a pearly lustre, but darkens on exposure to light. There is a perfect basal cleavage. H 2.5; SG 3.2
Occurrence. Found associated with haussmannite.
Localities. Långban, Värmland, Sweden; Franklin, New Jersey.
Treatment. Clean with water.

Haussmannite Mn_3O_4
Haussmannite may contain up to 7% FeO and up to 8% ZnO. It crystallizes in the tetragonal system, forming bipyramids; twinning is frequent. Massive forms with granular structure are often found. There is a cleavage; the colour is brownish-black, the streak chestnut-brown and the lustre is submetallic. H 5–5.5; SG 4.8
Occurrence. Found in veins associated with acid igneous rocks.
Localities. From Långban, Värmland, Sweden; Ouro Preto, Minas Gerais, Brazil; with psilomelane in the Batesville district, Arkansas.
Treatment. Clean with water.

Pyrolusite MnO_2
Pyrolusite crystallizes in the tetragonal system, sometimes forming dendritic coatings along joints or bedding planes in sedimentary rocks. It is found in many forms including divergent fibrous masses, although crystals are rare. The colour is black to bluish steel-grey and the streak is black. The lustre is metallic to dull, and the fracture is uneven or splintery. H 1–2 (massive material), 6–6.5 (crystals); SG 4.5–7.9
Occurrence. As a secondary mineral from the alteration of manganese-bearing minerals.
Locations. From Czechoslovakia; Lanlivery, Cornwall; central provinces of India; important deposits associated

Left: Pyrochroite from Gwynedd, Wales

Right: Manganite. The radiating habit of manganite aggregates is well shown by this example from the Harz Mountains of Germany

Above: Octahedra of magnetite from the Binnental, Switzerland

with psilomelane and brucite from Minas Gerais, Brazil, where there are high-grade manganese deposits; with the hematite ores of the Lake Superior region and from Colorado; from Hants County, Nova Scotia.

Ramsdellite MnO_2
Ramsdellite crystallizes in the orthorhombic system forming black tubular crystals, but these are less commonly found than the granular massive forms. The colour is black and the lustre metallic. H 3; SG 4.4
Occurrence. Found with other manganese minerals including pyrolusite and psilomelane.
Localities. Ramsdellite occurs at Horni Blatna, Czechoslovakia; in Minnesota, California and Montana.
Treatment. Clean with water.

Manganite MnO(OH)
Manganite crystallizes in the monoclinic system and is pseudo-orthorhombic. The crystals are prismatic and striated, often occurring as radiating aggregates. Twinning is common. The colour is dark-grey to black and the streak reddish-brown to black. The lustre is submetallic and there is a perfect cleavage. H 4; SG 4.2–4.4
Occurrence. Manganite, an ore of manganese, occurs in deposits precipitated from water under oxidizing conditions and is associated with pyrolusite and baryte. It may also occur in low-temperature hydrothermal veins in association with granitic rocks.
Localities. Manganite is found crystallized at the Botallack mine, St. Just, Cornwall; at Egremont, Cumbria, and at Exeter, Devon. It is found in the Lake Superior iron district, USA and in Nova Scotia. It is also found in the Harz Mountains, Germany.
Treatment. Black coating may be removed ultrasonically. Clean with water.

Wad
Wad may consist of one or more hydrous manganese oxides occurring in the oxidized zone of ore deposits. Minerals included in the mixture, not separately described, may include lithiophorite, lampadite, lubeckite, wackenrodite, rabdionite, asbolane, kakochlor. Wad often consists merely of pyrolusite and/or psilomelane. It is really a field term. Wad is amorphous and is stalactitic or massive reniform, often in earthy masses. The colour is black to brownish-black and the streak black. The lustre is earthy. SG 2.8–4.4

Bixbyite $(Mn, Fe)_2O_3$
Bixbyite is the cubic phase of manganese-iron oxide. It forms black crystals with a metallic lustre, which are often cubes modified by the octahedron. H 6–6.5; SG 4.9
Occurrence. Occurs with topaz in cavities in rhyolite.
Localities. From the Thomas Range, Utah; Ribas, Gerona, Spain; Valle de las Plumas, northern Patagonia, Argentina.
Treatment. May be cleaned with dilute acids.

Jacobsite $(Mn, Fe, Mg) (Fe, Mn)_2O_4$
Jacobsite contains MgO up to 10%, $Fe^{2+}:Mn^{2+}$ up to 1% and $Mn^{3+}:Fe^{3+}$ up to 0.45%. It crystallizes in the cubic system forming distorted octahedra. The fracture is conchoidal and the colour black with a reddish-brown streak. It is magnetic. H 6; SG 4.7
Localities. Found in Värmland, Sweden; and in Bulgaria.
Treatment. Clean with water.

Magnetite Fe_3O_4
Magnetite crystallizes in the cubic system and forms octahedra, rhombdodecahedra and granular masses. Twinning on the octahedron is common. It is strongly magnetic. The colour and streak are black and the lustre metallic. The fracture is uneven. H 5.5–6.5; SG 5.2
Occurrence. Magnetite is found in igne-

ous rocks as an accessory mineral, also in contact and regionally metamorphosed rocks, and occurs in high-temperature mineral veins. It is frequently associated with corundum in emery deposits.
Localities. The name may be taken from Magnesia, Greece; the largest deposits known are at Norrbotten, northern Sweden; other deposits are in Värmland and Vastmanland; Ural Mountains, USSR; as twinned octahedra in the Zillertal, Austria and the Binnental, Valais, Switzerland; fine crystals are found at Port Henry, New York and from Fiormeza, Cuba.
Treatment. Clean with dilute acid or water.

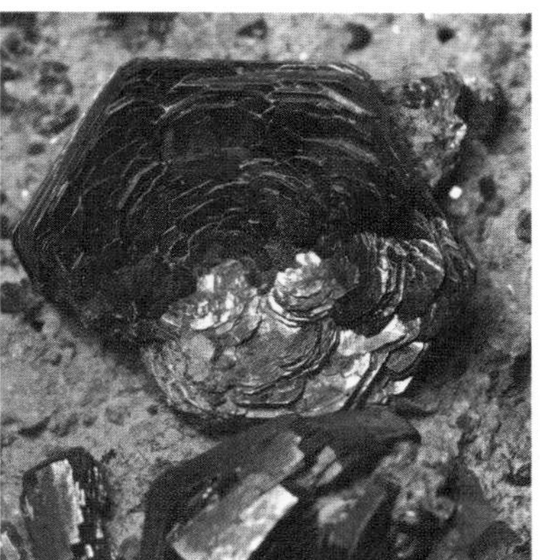

Hematite Fe_2O_3
Hematite crystallizes in the hexagonal system forming tabular or rhombohedral crystals, sometimes showing curved or striated rhombohedral faces. It may also be found columnar, laminated or massive in mamillary or botryoidal forms. Penetration twinning is found on the basal pinacoid. The colour is grey to black; for massive and earthy material dull brownish red. The streak is red to reddish-brown. The fracture is uneven and the lustre metallic. H 5–6; SG 4.9–5.3
Occurrence. Hematite is the most important ore of iron. It occurs as an accessory mineral in igneous rocks and hydrothermal veins, and also in sedimentary rocks where it may be of primary origin. It may occur as a secondary mineral, being precipitated from iron-bearing water.
Localities. Hematite is very widely distributed. The most important locations are along the shores of Lake Superior, USA; fine crystals are found at St. Gotthard, Switzerland; Mt. Vesuvius Italy; the Isle of Elba; Cleator Moor, Cumbria; and Minas Gerais, Brazil.
Treatment. Clean in dilute hydrochloric acid.

Fashioning. Uses: faceting, cabochons, beads, intaglios, etc; *cleavage:* none, but basal and prismatic parting due to twinning, splintery; *cutting angles:* variable, to suit requirements; *heat sensitivity:* very low.

Goethite FeO.(OH)
Goethite is the alpha-phase of FeO.OH and crystallizes in the orthorhombic system. It occurs as vertically striated prisms, often flattened into scales, and also forms fibrous masses. Reniform and stalactitic forms with a radiated concentric structure are found. There is a perfect cleavage. The colour is yellow through red to blackish-brown and the streak brownish-yellow. The fracture is uneven and the lustre sub-adamantine. H 5–5.5; SG 3.3–4.3
Occurrence. Goethite is found with quartz and limonite. It is an alteration product of iron-bearing minerals.
Localities. Good crystals come from the

Above: Fine banded hematite crystals. These are less common than the massive forms. This example is from the island of Elba

Left: Hematite. This rosette formation is called 'iron rose'

Far left: A fine single crystal of hematite

Left: Another good example of an octahedron of magnetite from the Binnental

Right: Fine crystals of goethite displaying stalagmitic habit. From Nassau, Germany

Below right: Halite crystal. This bluish-white colour is typical of the mineral

Botallack mine, St. Just, Cornwall, and from Lostwithiel and Lanlivery in the same county. It is common in the Lake Superior hematite deposits and in the Pike's Peak region of Colorado.
Treatment. Clean with dilute hydrochloric acid.

Lepidocrocite FeO . OH
Lepidocrocite is the gamma-phase of FeO . OH and crystallizes in the orthorhombic system, forming scaly or fibrous aggregates, red to reddish-brown in colour. There is a perfect cleavage and an orange streak. H 5; SG 4.1
Occurrence. Like goethite, it is found with quartz and limonite.
Treatment. Clean with dilute hydrochloric acid.

Limonite
Limonite is a generic term for mixed oxides or hydroxides of iron. It has a yellowish-brown streak and a vitreous lustre. The colour and lack of cleavage distinguish it from goethite. Much so-called ferric oxide described as limonite is in fact goethite.

Magnesioferrite (Mg, Fe)Fe_2O_4
Magnesioferrite crystallizes in the cubic system commonly forming octahedra. The colour and streak are black and the lustre is metallic. It is strongly magnetic. H 6–6.5; SG 4.5–4.6
Occurrence. Magnesioferrite results from the action of heated water-bearing vapours on magnesium and ferric chlorides and occurs in the vicinity of volcanic fumaroles.
Localities. Found on Mt. Vesuvius, Italy; Mt. Stromboli, Lipari Islands, Italy.
Treatment. Clean with dilute hydrochloric acid.

Hercynite $FeAl_2O_4$
A member of the spinel series, hercynite often contains some Mg or Fe^{3+} or both. It crystallizes in the cubic system and is normally massive with a granular texture. The colour is black and the lustre vitreous. H 7.5–8; SG 3.9–4.2
Occurrence. Hercynite occurs in igneous rocks, often with corundum.
Localities. Found in Czechoslovakia at Pobezovice; from a gabbro at Lago di Poschiavo, Grisons, Switzerland; from tin drifts near Morrina, Tasmania; from near Erode, Coimbatore district, India.
Treatment. Clean with dilute acid.

Heterogenite CoO . OH
Heterogenite is hexagonal, occurring as globular or reniform masses with a vitreous lustre. Varieties closely linked are mindigite with 7–10% CuO, and trieuite with up to 20% CuO.
Occurrence. Found at Schneeberg, E. Germany and in northern Chile.

Bunsenite NiO
Bunsenite crystallizes in the cubic system, forming green octahedra with a vitreous lustre. H 5.5; SG 6.4–6.8
Localities. Found at Johanngeorgenstadt, E. Germany, with nickel and bismuth.
Treatment. Clean with water.

Trevorite $NiFe_2O_4$
Though it crystallizes in the cubic system, trevorite is found as black grains with a greenish tint. It gives a black streak. It is strongly magnetic. H 5; SG 5.1
Localities. Found in a talcose rock in the Transvaal, South Africa.

Halides

Villiaumite NaF
Villiaumite crystallizes in the cubic system forming small carmine-coloured grains. The birefringence and pleochroism are exceptionally strong. H 2–2.5; SG 2.8
Localities. Found in a nepheline syenite from the Islands of Los, Guinea; Kola Peninsula, USSR; Mont St. Hilaire, Quebec.

Halite NaCl
Halite crystallizes in the cubic system, and is usually found as cubes sometimes displaying concave faces; also massive, granular to compact, to which form the term rock salt is applied. The mineral is colourless but may display shades of yellow, red or blue. The streak is white and the lustre vitreous. There is a perfect cleavage and the fracture is conchoidal. Halite is water-soluble and tastes of salt. H 2.5; SG 2.1–2.2 (pure material 2.16).
Occurrence. Halite is found in sedimentary rocks and has a world-wide distribution. Salt deposits are formed by the drying up of brines, or sea water that has been enclosed. Halite beds are associated with gypsum, sylvine and anhydrite.
Localities. Large deposits are found in southern USSR; at Stassfurt, E. Germany; Agrigento, Sicily, Italy; Northwich, Cheshire; Algeria; Ethiopia; Punjab, India; Colombia; China.
Treatment. Clean with alcohol.

Sylvine KCl
Sylvine crystallizes in the cubic system and may show cubes or a combination of cube and octahedron; it is also found massive and compact. It is colourless or may display shades of yellow, blue or red; the streak is white and the lustre vitreous. The fracture is uneven. Has a more bitter taste than halite. H 2; SG 2.0
Occurrence. Sylvine occurs in beds with other evaporites.
Localities. Stassfurt, E. Germany; Kalush, USSR.
Treatment. Easily soluble in both water and acids. Should be stored in a sealed container. It cannot be cleaned.

Nantokite CuCl
Nantokite crystallizes in the cubic system in granular masses. The colour is white to grey and the lustre adamantine. H 2–2.5; SG 3.9–4.1
Localities. From Nantoko, Chile; Broken Hill, New South Wales, Australia.

Atacamite $Cu_2Cl(OH)_3$
Atacamite crystallizes in the orthorhombic system and forms slender prismatic crystals showing striations; tabular forms are also found. Also occurs as fibrous or granular masses. The colour is bright to dark green; the lustre is adamantine to vitreous and the streak apple-green. There is also a perfect cleavage. Paratacamite is the rhombohedral phase of $Cu_2Cl(OH)_3$. H 3–3.5; SG 3.8
Occurrence. Atacamite is found in the oxidized zones of copper deposits, associated with cuprite and malachite.
Localities. Found at Atacama, Chile; Wallaroo, South Australia; Vesuvius.
Treatment. Clean with distilled water.

Marshite CuI
Marshite crystallizes in the cubic system and may form tetrahedra. There is a dodecahedral cleavage. It has an adamantine lustre, and is colourless to pale yellow when fresh, becoming flesh pink to dark brownish-red on exposure. It fluoresces red in ultraviolet light. H 2.5; SG 5.6
Localities. From Broken Hill, New South Wales, Australia; Chile.
Treatment. Clean with water.

Chlorargyrite AgCl
Chlorargyrite, also known as cerargyrite, crystallizes in the cubic system, forming cubes. These, however, are rare, the mineral more commonly being found massive. There is a conchoidal fracture. It is colourless, becoming violet-brown on exposure to light; the lustre is resinous. It is sectile. H 1.5–2.5; SG 5.5–5.6
Occurrence. Chlorargyrite occurs as a secondary mineral in the oxidized zone of silver deposits, with native silver and cerussite.
Localities. Atacama, Chile; Broken Hill, New South Wales, Australia; Colorado, Idaho, Utah.

Embolite Ag (Cl, Br)
Embolite crystallizes in the cubic system and is usually found massive. The colour is greyish-green to yellowish-green and yellow; the lustre is resinous. H 2.5; SG 5.7
Localities. From Chañarcillo, Chile; Broken Hill, New South Wales.

Miersite (Ag, Cu)I
Miersite has more silver than copper, approaching a proportion of 4:1, which is the stability limit for the cubic phase at normal temperatures. It forms tetrahedra with a dodecahedral cleavage and is bright yellow. It is thought that cuproiodargyrite is a variety of miersite. H 2–3; SG 5.6
Localities. From Broken Hill, New South Wales; cuproiodargyrite from Huantajaya, Peru.

Sellaite MgF_2
Sellaite crystallizes in the tetragonal system and forms prismatic crystals with a vitreous lustre. There are two directions of perfect cleavage. Sellaite is colourless. H 5–6; SG 2.9–3.1
Localities. Found on moraines of the Gebroulaz glacier near Moutiers, Savoie, France; Mt. Vesuvius, Italy.

Bischofite $MgCl_2 . 6H_2O$
Bischofite crystallizes in the monoclinic system in granular or fibrous forms. It is colourless, with a vitreous lustre. H 1.5; SG 1.56
Localities. Found at Stassfurt, E. Germany.
Treatment. Decomposes in air; keep in sealed container.

Above: Marshite crystals from Broken Hill, New South Wales

Left: Crystals of miersite from Broken Hill

Far left: Chlorargyrite crystals showing cubic form from Saxony, East Germany

Left: Slender crystals of atacamite from Sierra Gorda, Chile

Right: Yellow cubes of fluorite on galena. From Wheal Mary Ann, Cornwall

Below right: Crystals of fluorite displaying interpenetrant twinning, very characteristic of this mineral

Above right: Octahedral fluorite from Piedmont, Italy

Right: Another example of interpenetrant fluorite twins

Carnallite $KMgCl_3.6H_2O$
Carnallite crystallizes in the orthorhombic system, forming pseudo-hexagonal crystals. It is more often found granular massive. The colour is white to yellow or reddish; the lustre is greasy and the fracture conchoidal. Carnallite has a bitter taste and is deliquescent. H 2.5; SG 1.6
Occurrence. Carnallite occurs in evaporite deposits with halite and sylvine.
Localities. From Stassfurt, E. Germany; and Beienrode, W. Germany.
Treatment. Store in sealed container.

Tachhydrite $CaMg_2Cl_6.12H_2O$
Tachhydrite crystallizes in the hexagonal system with a rhombohedral cleavage. The colour is honey to yellow and the lustre is vitreous. The mineral is found massive. H 2; SG 1.6
Localities. Found in salt deposits of northern Germany including Stassfurt, E. Germany.
Treatment. Deliquescent; store in a sealed container.

Fluorite CaF_2
Fluorite crystallizes in the cubic system forming cubes and, rarely, octahedra or rhombdodecahedra. There is a perfect octahedral cleavage. There is a wide variety of colour including pink, blue, green, purple, red, black and colourless. Individual specimens are often colour-zoned. Penetration twins are common. The streak is white and the lustre vitreous. The fracture is subconchoidal. Fluoresces under long-wave ultra-violet light. H 4; SG 3.2; RI 1.43
Occurrence. Fluorite is found in mineral veins, alone or as a gangue mineral with metallic ores. Associated minerals include quartz, baryte, calcite, celestine, dolomite, galena, sphalerite and topaz. Fluorite is worked for its properties as a flux in the smelting of iron.
Yttrian fluorite contains 15–20% YtF_3 and crystallizes in the cubic system, being found massive and earthy. The colour is violet through grey to reddish-brown and there is a cleavage. H 4–5; SG 3–4
Localities. Found in the lead veins of Northumberland; Cleator Moor, Cumbria; Weardale, Durham; Derbyshire, Cornwall and Devon; from the mining districts of E. Germany; as pink octahedra from St. Gotthard, Switzerland; from Hardin County, Illinois and Crystal Peak, Colorado. Yttrian fluorite found near Falun, Sweden and Sussex County, New York State.
Treatment. Calcite coatings removed with dilute HCl and iron stains with oxalic acid. It is heat-sensitive. Do not attempt to clean yttrian fluorite.
Fashioning. Uses: faceting, cabochons, carvings, intaglios, etc; *cleavage:* perfect octahedral; brittle; *cutting angles:* crown 40°–50°, pavilion 43°; *heat sensitivity:* very high.

Chlorocalcite $KCaCl_3$
Chlorocalcite crystallizes in the orthorhombic system with twinned pseudo-cubic crystals. The colour is white. H 2.5–3.0; SG 2.1
Occurrence. Found on Mt. Vesuvius as a sublimation product, in the form of white cubes or as an encrustation.
Treatment. Deliquescent. Store in sealed container.

Jarlite $NaSr_3Al_3F_{16}$
Jarlite crystallizes in the monoclinic system as very small crystals; the mineral is more commonly found massive. The colour is white, with a vitreous lustre. Meta-jarlite is high in Ca and Ba. H 4–4.5; SG 3.8–3.9
Localities. Found with cryolite at Ivigtut, Greenland.

Calomel Hg_2Cl_2
Calomel crystallizes in the tetragonal system, giving tabular or pyramidal crystals, which are often highly complex. The colour ranges from white through yellow to brown, and lustre is adamantine. The streak is pale yellowish-white. It fluoresces dark red under ultraviolet light. Found as a coating on other minerals, usually being deposited from hot solutions. It also occurs as a secondary mineral. It is associated with cinnabar. H 1–2; SG 6.4
Localities. Calomel is found at Moschellandsberg, W. Germany; near Příbram, Czechoslovakia; Almaden, Ciudad Real, Spain; Terlingua, Texas.
Treatment. May be cleaned with hydrochloric acid.

Left: Matlockite. It takes its name from one of its localities, Matlock in Derbyshire

Far left: Crystals of a fragile mineral, thomsenolite from Ivigtut, Greenland

Cryolite Na_3AlF_6
Cryolite crystallizes in the monoclinic system, being rarely found as pseudo-cubic crystals. Massive granular forms are more common. There is no cleavage, but a basal and prismatic parting; twinning is common. It is colourless to brown or reddish and the streak is white. The fracture is uneven. The lustre is greasy. H 2.5; SG 3
Occurrence. Cryolite occurs in a pegmatite associated with siderite, quartz, galena, sphalerite and fluorite.
Localities. Cryolite is found in Greenland, at Ivigtut.
Treatment. Clean in water.

Cryolithionite $Li_3Na_3Al_2F_{12}$
The crystal structure of cryolithionite is identical with that of the garnets; that is, it is a member of the cubic system. It forms colourless dodecahedral crystals with a vitreous lustre and a dodecahedral cleavage. H 2.5; SG 2.7
Localities. Cryolithionite occurs with cryolite in the Ilmen mountains in the Soviet Union.
Treatment. As for cryolite.

Chiolite $Na_5Al_3F_{14}$
Chiolite crystallizes in the tetragonal system and forms small pyramidal crystals; it is also found granular and massive. There is a perfect cleavage and the colour is white, with a vitreous lustre. H 3.5–4; SG 2.8–2.9
Occurrence. Found with cryolite in the Greenland deposit at Ivigtut.
Treatment. As for cryolite.

Ralstonite $NaMgAl(F, OH)_6 . H_2O$
Ralstonite crystallizes in the cubic crystal system forming octahedra, which are white with a vitreous lustre. There is an octahedral cleavage. H 4.5; SG 2.5
Occurrence. Found with cryolite at Ivigtut, Greenland.

Thomsenolite $NaCaAlF_6 . H_2O$
Thomsenolite crystallizes in the monoclinic system and the crystals may resemble cubes, although prismatic forms are also found. It is colourless with a pearly lustre. There is a perfect cleavage. H 2; SG 2.9
Occurrence. Occurs with cryolite at Ivigtut, Greenland.

Pachnolite $NaCaAlF_6 . H_2O$
Pachnolite crystallizes in the monoclinic system forming prismatic crystals and pseudo-orthorhombic crystals through twinning. It is colourless, with a vitreous lustre. H 3; SG 2.9
Occurrence. Found in Greenland as an alteration product of cryolite.

Prosopite $CaAl_2(F, OH)_8$
Prosopite crystallizes in the monoclinic system and is commonly found granular massive. The colour is white to grey, and the lustre is vitreous to dull. H 4.5; SG 2.8
Localities. Found in Altenberg, E. Germany; Pike's Peak, Colorado.

Fluocerite $(Ce, La)F_3$
Fluocerite crystallizes in the hexagonal system, forming thick hexagonal prisms. It is also found massive. There is a distinct cleavage. The colour is pale yellow, changing to yellowish or reddish brown; the lustre is vitreous. H 4.5–5; SG 6.1
Occurrence. Found in the granite at Pike's Peak, Colorado; from Falun, Sweden.

Matlockite PbFCl
Matlockite crystallizes in the tetragonal system forming tabular crystals. It is yellow to green, with an adamantine lustre. H 2–3; SG 7.2
Localities. Found with percylite at Challacolla, Chile and at Laurion, Greece.
Treatment. Clean with distilled water.

Below: Another Greenland mineral, pachnolite. Here it shows the pseudo-orthorhombic crystals that arise through twinning

Right: Cotunnite aggregates from Vesuvius. The background material is lava

Cotunnite $PbCl_2$
Cotunnite crystallizes in the orthorhombic system as acicular crystals and semicrystalline masses. There is a perfect cleavage. The lustre is adamantine to pearly, and the colour is white to yellow. H 2.5; SG 5.3–5.8
Localities. Found in the area of Mt. Vesuvius, Italy; Tarapaca, Chile.

Laurionite PbClOH
Laurionite crystallizes in the orthorhombic system, forming minute colourless prismatic crystals with a cleavage. H 2–3; SG 6.2
Occurrence. Found in lead slags with paralaurionite, penfieldite, matlockite, phosgenite, etc.
Localities. Laurion, Greece; Wheal Rose, Cornwall.

Penfieldite Pb_2Cl_3OH
Penfieldite crystallizes in the hexagonal system and forms hexagonal prismatic crystals with a cleavage. It is white with an adamantine lustre.
Occurrence. Occurs with fiedlerite in lead slags.
Localities. Laurion, Greece.

Mendipite $Pb_3Cl_2O_2$
Mendipite crystallizes in the orthorhombic system, forming fibrous or columnar masses, often radiated. Cleavage is perfect and the colour is white. The lustre is pearly to adamantine. H 2.5–3; SG 7–7.2
Localities. Found in the Mendip Hills, Somerset.
Treatment. Clean with distilled water.

Lorettoite $Pb_7Cl_2O_6$
Lorettoite crystallizes in the orthorhombic system and is found as masses of coarse fibres or blades. It has a perfect basal cleavage. The colour is honey-yellow and the lustre adamantine. H 3; SG 7.6
Localities. Loretto, Tennessee; a reddish form previously considered a separate mineral and called chubutite is found at Chubut, Argentina.

Fiedlerite $Pb_3Cl_4(OH)_2$
Fiedlerite crystallizes in the monoclinic system and forms tabular crystals with a cleavage. The colour is white, and the lustre adamantine. H 3.5; SG 5.8
Occurrence. From the lead slags at Laurion, Greece.

Cumengéite $Pb_4Cu_4Cl_8(OH)_8 \cdot H_2O$
Cumengéite crystallizes in the tetragonal system and occurs intimately associated with boléite, from which it is distinguished by its lighter shade of indigo-blue. H 2.5; SG 4.6
Localities. Found at Boléo, near Santa Rosalia, Baja California, Mexico.
Treatment. May be cleaned with water.

Boléite $Pb_9Cu_8Ag_3Cl_{21}(OH)_{16} \cdot H_2O$
Boléite crystallizes in the tetragonal system and forms pseudocubes through twinning. There is a perfect cleavage. The colour is indigo-blue and the lustre vitreous to pearly. H 3–3.5; SG 5.0
Localities. Occurs with copper and lead minerals, including cumengéite and pseudoboléite, at Boléo, Baja California, Mexico; Broken Hill, New South Wales.
Pseudoboléite is $Pb_5Cu_4Cl_{10}(OH)_8 \cdot 2H_2O$ and is found only in parallel growth with boléite and cumengéite.
Diaboléite is $Pb_2CuCl_2(OH)_4$ and is found with chloroxiphite embedded in mendipite from Higher Pitts, Somerset. It is a bright sky-blue and crystallizes in the tetragonal system; H 2.5; SG 5.4
Treatment. May be cleaned with water.

Chloroxiphite $Pb_3CuCl_2O_2(OH)_2$
Chloroxiphite crystallizes in the monoclinic system and occurs as elongated crystals with a perfect cleavage. The colour is dull olive-green and the lustre is resinous to adamantine. Strong pleochroism shown by cleavage plates, the colours being bright emerald-green and yellowish-brown. H 2.5; SG 6.7
Localities. Found embedded in mendipite in the Mendip Hills, Somerset.
Treatment. Clean with water.

Hematophanite $Pb_4Fe_4O_9(OH, Cl)_2$
Hematophanite crystallizes in the tetragonal system forming lamellar aggregates of thin plates. The colour is dark reddish-brown and the lustre submetallic. There is a cleavage parallel to the base. The streak is yellowish-red. H 2–3; SG 7.7
Occurrence. Occurs with plumboferrite, etc, at Jakobsberg, Värmland, Sweden.

Bismoclite BiOCl
Bismoclite occurs in white to yellowish-brown earthy masses.

Below: A group of penfieldite crystals in a cavity. The mineral's adamantine lustre can be seen

Left: The pseudocubic form taken by boleite crystals can be clearly seen in these examples from Boleo, Mexico

Localities. It is found in Australia, Bolivia, Nevada and Utah; in Italy with bismuthinite at Rio Marina, Elba.

Scacchite $MnCl_2$
An extremely rare mineral isomorphous with chloromagnesite and lawrencite. SG 2.9
Occurrence. Scacchite is found on Mt. Vesuvius as pale rose or colourless encrustations, associated with sylvine, hydrophyllite and lawrencite.

Kempite $Mn_2Cl(OH)_3$
Kempite crystallizes in the orthorhombic system forming minute prismatic crystals coloured emerald-green. H 3.5; SG 2.9
Occurrence. Found with pyrochroite, hausmannite and rhodochrosite in a manganese ore boulder, formerly existing near San José, California.

Chloromanganokalite K_4MnCl_6
Chloromanganokalite crystallizes in the hexagonal system, forming rhombohedral crystals in parallel aggregates. It is isomorphous with rinneite and is coloured yellow. H 2.5; SG 2.3
Localities. Found around the fumaroles of Mt. Vesuvius.
Treatment. Decomposes very easily; do not attempt to clean.

Lawrencite $FeCl_2$
Lawrencite is isostructural with chloromagnesite, and forms compact masses. SG 3.1
Occurrence. It is found in meteoritic iron at Ouifak, Greenland.

Douglasite $K_2FeCl_4.2H_2O$
Douglasite crystallizes in the monoclinic system and is found as green granular masses. This colour alters to a reddish-brown on exposure to air.
Occurrence. Douglasite is found at Douglashall, near Stassfurt, E. Germany.

Erythrosiderite $K_2FeCl_5.H_2O$
Erythrosiderite crystallizes in the orthorhombic system and forms tabular or pseudo-octahedral crystals. The colour is red and the lustre vitreous. SG 2.3
Localities. Found on Mt. Vesuvius and Mt. Etna; also from Stassfurt in East Germany.
Treatment. Deliquescent. Do not attempt to clean.

Rinneite $K_3NaFeCl_6$
Rinneite crystallizes in the hexagonal system, forming coarse granular masses. It is colourless or rose-coloured, violet or yellow when fresh, altering to brown on exposure to air; its lustre is silky. H 3; SG 2.3
Localities. Found at Wolkramshausen, near Nordhausen, E. Germany.

Ferruccite $NaBF_4$
A rare mineral from the Mt. Vesuvius area, associated with avogadrite, malladrite and hieratite. SG 2.4

Left: Kernite from Kern Co., California

Avogadrite $(K, Cs)BF_4$
Avogadrite crystallizes in the orthorhombic system in 8-sided tabular crystals or as granular masses. The colour is snow-white. Hardness and cleavage have not been determined: SG 3.
Occurrence. Found as a sublimate around fumaroles on Vesuvius.

Malladrite Na_2SiF_6
Malladrite crystallizes in the hexagonal system, forming minute prisms sometimes pyramidally terminated. It is the low-temperature phase of Na_2SiF_6.
Occurrence. Associated with avogadrite and hieratite on Mt. Vesuvius.

Hieratite K_2SiF_6
Hieratite is the cubic high-temperature phase of K_2SiF_6.
Localities. It forms stalactitic concretions from the fumaroles of Mt. Vesuvius and from Mt. Vulcano, Lipari Islands.

Camermanite K_2SiF_6
Camermanite is the hexagonal low-temperature phase of K_2SiF_6, forming masses.
Localities. Mt. Vesuvius.

Cryptohalite $(NH_4)_2SiF_6$
Cryptohalite is the cubic, high-temperature phase and forms whitish-yellow encrustations. H 2.5
Localities. The crater of Mt. Vesuvius.

Bararite $(NH_4)_2SiF_6$
Bararite is the low-temperature hexagonal phase.
Localities. Mt. Vesuvius.

Borates

Sassolite $B(OH)_3$
Sassolite crystallizes in the triclinic system, forming pseudo-hexagonal plates or, more commonly, small pearly scales. The colour is white and there is a perfect basal cleavage. H 1; SG 1.48
Occurrence. Found deposited from the gases of fumaroles and in hot springs.
Localities. Some lagoons in Tuscany, Italy; the crater of Mt. Vulcano, Lipari Islands.

Kernite $Na_2B_4O_7.4H_2O$
Kernite crystallizes in the monoclinic system with perfect cleavage. The colour is white and the lustre vitreous to pearly. H 3; SG 1.95
Occurrence. Found associated with ulexite in south-eastern Kern Co., California.
Treatment. Store in sealed container.

Borax $Na_2B_4O_7.10H_2O$
Borax crystallizes in the monoclinic system forming short prismatic crystals; it is also found massive. The colour is white, sometimes tinged with grey or blue; the streak is white. There is a perfect cleavage and the lustre is vitreous to resinous; the fracture is conchoidal. H 2–2.5; SG 1.7
Occurrence. Borax is precipitated by evaporation of salt lakes. Associated minerals include halite, sulphates, carbonates and other borates.
Localities. Borax occurs in the Death Valley area, California; Nevada; and in Kashmir and Tibet.

Right: Hambergite from the Malagasy Republic. This mineral may sometimes be cut as a gemstone when it displays a very high birefringence

Below right: Another mineral from the Malagasy Republic, rhodizite

Treatment. Leave crust; borax alters to tincalconite, $Na_2B_4O_7 . 5H_2O$, on exposure; this forms an opaque white layer, which thickens until the entire crystal disintegrates.

Rhodizite $CsB_{12}Be_4Al_4O_{28}$
Rhodizite crystallizes in the cubic system, forming white dodecahedra with a silky lustre. H 8; SG 3.4; RI 1.69
Occurrence. Rhodizite is found on red tourmaline from Ural Mts, USSR; from Antandrokomby and Manjakandriana in the Malagasy Republic.
Treatment. Usually so integrated with matrix that attempts to free it are useless and may spoil the crystal. May be cleaned with dilute acids.

Hambergite $Be_2BO_3(OH)$
Hambergite crystallizes in the orthorhombic system and forms greyish-white prismatic crystals with a vitreous lustre and a perfect cleavage. H 7.5; SG 2.3
Occurrence. Found in a pegmatite vein near Halgaraen, southern Norway; good crystals from south of Betafo, Imalo and Maharitra, Malagasy Republic.
Treatment. Remove iron stains with oxalic acid.

Szajbelyite $MgBO_2OH$
Szajbelyite crystallizes in the orthorhombic system forming small nodules, which are white outside, with a silky lustre and yellow within. H 3.5; SG 2.60
Occurrence. Found with boracite and sylvine at Aschersleben, Ilmen Mts, E. Germany; Rézbânya, Romania; with ludwigite from Lincoln Co., Nevada.

Pinnoite $MgB_2O_4 . 3H_2O$
Pinnoite crystallizes in the tetragonal system and is found as nodules which are radiating and fibrous. The colour is sulphur-yellow and the lustre vitreous. H 3–4; SG 2.2
Occurrence. Found at Stassfurt, E. Germany, in saline deposits with kainite and boracite.

Kurnakovite $Mg_2B_6O_{11} . 15H_2O$
Kurnakovite crystallizes in the triclinic system forming compact masses which appear cubic and are white in colour, with a vitreous lustre. H 3; SG 1.8
Occurrence. Found with borates at Inder, Kazakhstan, USSR; also from Boron, California.

Inderite $Mg_2B_6O_{11} . 15H_2O$
Inderite crystallizes in the monoclinic system forming nodular masses and, sometimes, striated prismatic crystals, which are white in colour and have a vitreous lustre. H 3; SG 1.8–1.9
Localities. From Inder, Kazakhstan, USSR; Boron, California.

Kaliborite $HKMg_2B_{12}O_{21} . 9H_2O$
Kaliborite crystallizes in the monoclinic system and forms aggregates of crystals with perfect cleavage. The colour is colourless to white. The lustre is vitreous. H 4–5; SG 2.1
Localities. Found at Stassfurt, E. Germany; Inder, Kazakhstan, in the Soviet Union.

Colemanite $Ca_2B_6O_{11} . 5H_2O$
Colemanite crystallizes in the monoclinic system forming short prismatic crystals or granular compact masses. There is a perfect cleavage and the lustre is vitreous to adamantine. The fracture is uneven to conchoidal. The colour is milky white, yellow or grey. H 4–4.5; SG 2.4
Localities. Found in Death Valley and Calico, California.
Fashioning. Uses: faceting or cabochons; *cleavage:* perfect clinopinacoidal; very brittle; *cutting angles:* crown 40°, pavilion 40°; *heat sensitivity:* low.

Meyerhofferite $Ca_2B_6O_{11} . 7H_2O$
Meyerhofferite crystallizes in the triclinic system and forms prismatic tabular crystals and fibrous masses. The colour is white and the lustre vitreous or fibrous. There is a cleavage. Inyoite alters to meyerhofferite. H 2; SG 2.1
Occurrence. Found with inyoite in Death Valley, California.

Inyoite $Ca_2B_6O_{11} . 13H_2O$
Inyoite crystallizes in the monoclinic system forming tabular crystals with a cleavage. The colour is white, and the lustre vitreous. Alters to meyerhofferite. H 2; SG 1.8
Occurrence. From Death Valley, Inyo Co., California in association with colemanite.

Right: Colemanite from San Bernardino Co., California, showing typical short prismatic crystals

Priceite $Ca_4B_{10}O_{19}.7H_2O$
Priceite probably crystallizes in the triclinic system, forming microscopic rhomboidal plates with perfect cleavage or compact masses. The colour is snow-white. H 3–3.5; SG 2.4
Localities. Found in Curry Co., Oregon and Inyo Co., California.

Ginorite $Ca_2B_{14}O_{23}.8H_2O$
Ginorite is found as white fibrous masses. H 3.5; SG 2.0
Localities. Sasso Pisano, Italy.

Ulexite $NaCaB_5O_9.8H_2O$
Ulexite crystallizes in the triclinic system forming rounded masses of fine fibrous crystals, or parallel fibrous aggregates. The colour is white and the lustre silky. The mineral is transparent and print can sometimes be read along the length of the fibres, from which property the name 'television stone' is taken. H 2.5; SG 1.9–2
Occurrence. Ulexite is an evaporite mineral found in sedimentary rocks in borax deposits. It is accompanied by colemanite.
Localities. In the dry plains at Tarapaca, Chile; from Salinas de la Puna, Argentina; Esmeralda, Nevada; Death Valley, Desert Wells, Kern Co., and Los Angeles Co., California.
Treatment. Store in sealed containers with silica gel.

Probertite $NaCaB_5O_9.5H_2O$
Probertite crystallizes in the monoclinic system and forms radiating prismatic crystals. They are colourless and have a vitreous lustre. They possess a perfect cleavage. H 3.5; SG 2.1
Occurrence. Occurs in clay, borax or kernite in the Kramer district, Kern Co., California.

Hydroboracite $CaMgB_6O_{11}.6H_2O$
Hydroboracite crystallizes in the monoclinic system and forms fibrous foliated masses, which resemble gypsum. There is a perfect cleavage and the colour is white, with a vitreous to silky lustre. H 2; SG 2
Localities. Found at Stassfurt, E. Germany; and from the Ryan area, Inyo Co., California.

Jeremejevite $Al_6B_5O_{15}(OH)_3$
Jeremejevite, sometimes spelt jeremeyevite, crystallizes in the hexagonal system and forms prismatic pseudohexagonal crystals. These show both uniaxial and biaxial parts and are colourless to pale yellowish-brown, with a vitreous lustre. H 6.5; SG 3.2
Localities. Found at Mt. Soktuj, in the Adun-Chilon mountains, in Transbaikalia, USSR; also from South-West Africa. Quantities of facetable material have recently been discovered in this area.
Treatment. Clean with water or dilute acids.

Sinhalite $MgAlBO_4$
Sinhalite was discovered to be a separate mineral in 1951/52; it was long thought to be a brown variety of peridot. Sinhalite crystallizes in the orthorhombic system but little is yet known of the crystal morphology since much of the material was first noted as cut stones. The colour is dark greenish-brown, sometimes a golden yellow, with strong pleochroism. It has a vitreous lustre. It may be distinguished from peridot by an extra band in the absorption spectrum at 4630Å. H 6.5; SG 3.47–3.50
Localities. Found in Ceylon gem gravels, etc.
Treatment. Iron stains may be removed with oxalic acid but prolonged immersion should be avoided.
Fashioning. Uses: faceting or cabochons; *cleavage:* probably distinct macro- and brachypinacoidal; *cutting angles:* crown 40°, pavilion 40°; *pleochroism:* distinct; *heat sensitivity:* low.

Left: Fragments of jeremejevite. The vitreous lustre is very apparent

Left: Ulexite. Its fibrous nature is seen here in a specimen from Boron, California

Warwickite $(Mg, Fe)_3TiB_2O_8$
Warwickite, which may contain more magnesium than iron, crystallizes in the orthorhombic system as elongated prismatic crystals. The colour is dark brown to black, with a pearly lustre, and there is a cleavage. H 3–4; SG 3.3
Localities. Found at Warwick, Orange Co., New York State.

Nordenskiöldine $CaSn_2O_6$
Nordenskiöldine crystallizes in the hexagonal system, forming thin or thick tabular crystals or lens-like crystals; it is also found as parallel growths with calcite and other minerals. The colour is yellow or colourless; there is a perfect cleavage and the lustre is vitreous. H 5.5–6.0; SG 4.2
Occurrence. Found in an alkaline pegmatite with feldspar, zircon and other minerals on Arö island, Norway; with tourmaline, stannite and sulphides in marble at Arandis, South-West Africa.

Sussexite $(Mn, Mg)BO_2(OH)$
Sussexite contains more manganese than magnesium and may contain up to 3% zinc. It crystallizes in the orthorhombic system and forms veins or masses, which have a fibrous structure. The colour is white, pink or straw-yellow with a white streak and silky to earthy lustre. H 3–3.5; SG 3.3
Occurrence. Sussexite occurs with pyrochroite and other minerals at Franklin, New Jersey, where it is found in hydrothermal veins in franklinite; it is also found as veins in hematite at the Chicagon mine, Iron Co., Michigan and the Gonzen mine, Switzerland.

Pinakiolite $(Mg, Mn^{2+})_2Mn^{3+}BO_5$
Pinakiolite contains magnesium and manganese in the proportion 3:1 and crystallizes in the monoclinic system, forming thin tabular crystals; twinning is common. There is a good cleavage and the mineral is very brittle. The colour is black with a brilliant metallic lustre and the streak is brownish-grey. H 6; SG 3.8
Occurrence. Pinakiolite is found in a granular dolomite at Långban, Sweden.

Above: Ludwigite in the form of radiating fibres, from Brosso, Italy

Ludwigite $(Mg, Fe^{2+})_2Fe^{3+}BO_5$
Ludwigite crystallizes in the orthorhombic system forming fibrous granular masses or aggregates of needle-like crystals; prismatic crystals with rhombic cross-section are rare. There is a perfect cleavage and the colour is dark-green to black with a silky lustre. H 5; SG 3.8
Occurrence. Ludwigite is found as a high-temperature mineral in contact metamorphic deposits. It occurs in cassiterite ores and iron deposits.
Localities. Found with cassiterite in Kern Co., California; in magnetite skarn deposits in Norway; from Suan, Korea; and from other localities.

Fluoborite $Mg_3BO_3(F, OH)_3$
Fluoborite (nocerite) crystallizes in the hexagonal system and occurs as prismatic crystals, as compact fibrous masses or as divergent aggregates of prismatic crystals. The cqlour is white with a vitreous lustre; fibrous aggregates are silky. H 3.5; SG 2.9
Occurrence. Fluoborite is found as crystals in thermally metamorphosed impure limestones. It also occurs as veins in franklinite.
Localities. In limestones at Crestmore and with calcite near Ludlow, California; from Sterling Hill, New Jersey; from Broadford, Skye, etc.

Boracite $Mg_3B_7O_{13}Cl$
Alpha-boracite, the high-temperature form, is stable above 265° and crystallizes in the cubic system. Beta-boracite crystallizes in the orthorhombic system, and may be colourless, grey, yellow, green or bluish-green; crystals are pseudocubic and occur as cubes, dodecahedra, or octahedra. Fine-grained or fibrous masses also occur. The fracture is conchoidal and the lustre vitreous. H 7–7.5; SG 2.9
Occurrence. Boracite occurs in beds of gypsum, anhydrite and rock salt.
Localities. Found at Aislaby, Yorkshire; Luneville, France; the Choctaw salt dome, Iberville Parish, Louisiana; Otis, California; etc.
Treatment. Clean quickly in distilled water.

Hilgardite $Ca_2B_5ClO_8(OH)_2$
Hilgardite is dimorphous with parahilgardite and crystallizes in the monoclinic system, forming tabular crystals with hemimorphic habit. There is a perfect cleavage. The crystals are colourless with a vitreous lustre and are transparent. H 5; SG 2.7
Occurrence. Hilgardite occurs in salt domes, particularly the Choctaw dome, Iberville Parish, Louisiana.

Lüneburgite $Mg_3(PO_4)_2B_2O_3 . 8H_2O$
Lüneburgite crystallizes in the monoclinic system, forming minute pseudohexagonal tablets; it is also found as flattened masses with a fibrous structure. The colour is white to brown or green. H about 2; SG 2.0
Occurrence. Lüneburgite is found in clay with halite, sylvine and polyhalite. It has also been found in a guano deposit.
Localities. From Carlsbad, New Mexico and from parts of Texas; in Marl at Lüneburg, W. Germany; and in guano at Mejillones, Peru.

Cahnite $Ca_2BAsO_4(OH)_4$
Cahnite crystallizes in the tetragonal system and forms minute penetration twins with etched and rounded crystal faces. There is a perfect cleavage. The colour is white and the lustre vitreous. H 3; SG 3.1
Occurrence. Found with franklinite at Franklin, New Jersey and at some other localities.

Seamanite $Mn_3PO_4BO_3 . 3H_2O$
Seamanite crystallizes in the orthorhombic system and forms small acicular crystals which are pale yellow in colour. H 4; SG 3.1
Occurrence. Found in fractures cutting siliceous rock associated with sussexite and calcite at the Chicagon mine, Michigan.

Sulphoborite $Mg_3SO_4B_2O_4(OH)_2 . 4H_2O$
Sulphoborite crystallizes in the orthorhombic system and forms short to long colourless prismatic crystals. H 4–4.5; SG 2.4
Occurrence. Sulphoborite is found in salt deposits.
Localities. From Wittmar, Brunswick, W. Germany, and Westeregeln, E. Germany; the Inder Lake area, western Kazakhstan, USSR.

Carbonates

Thermonatrite $Na_2CO_3 . H_2O$
Thermonatrite crystallizes in the orthorhombic system and forms crusts or efflorescences; it may also occur as clusters of platy needles. It is colourless to grey or yellow. Thermonatrite is sectile and the lustre is vitreous. H 1–1.5; SG 2.2
Occurrence. Found as a deposit from saline lakes, in association with fumaroles, and as an efflorescence on the soil in desert areas.
Localities. From Death Valley, California as an efflorescence; from Mt. Vesuvius; as a bedded deposit at the Gorodki oil field in the Kama region of the USSR; from the Sudan desert, etc.
Treatment. Store in a sealed container.

Natron $Na_2CO_3 . 10H_2O$
Natron crystallizes in the monoclinic system forming efflorescences and crusts; synthetic crystals have a tabular form. The crystals are colourless to yellow and are brittle with a vitreous lustre. H 1–1.5; SG 1.4
Occurrence. Natron is found in solution in soda lakes or as an efflorescence on lava.
Localities. From Wadi Natrum, Egypt;

Right: Pirssonite from Searles Lake, California

Left: Malachite in botryoidal concretionary form from Zaire

Mt. Vesuvius; Owens Lake, California.
Treatment. Store in a sealed container.

Trona $Na_3H(CO_3)_2 . 2H_2O$
Trona crystallizes in the monoclinic system and is commonly found massive, fibrous or columnar; crystals are elongated and flattened. The colour is greyish-white to pale yellow or brown; there is a perfect cleavage. The lustre is vitreous and glistening. H 2.5–3; SG 2.1
Occurrence. Trona occurs as a deposit in saline lakes or as an efflorescence on soil in dry, hot regions.
Localities. Occurs with borax at Searles Lake, and at Death Valley, California; also from Tibet, Mongolia and Iran, etc.
Treatment. Store in a sealed container.

Nahcolite $NaHCO_3$
Nahcolite crystallizes in the monoclinic system, forming concretionary masses which may be several feet in diameter; it may also occur as crystal aggregates or as elongated prismatic crystals. There is a perfect cleavage. The colour is white, but may be discoloured by impurities to grey or brown; it is transparent to translucent and the lustre is vitreous to resinous. H 2.5; SG 2.2
Occurrence. Nahcolite occurs in unweathered oil-shale of the Green River formation, north-west Colorado; in Searles Lake, California; from Egypt; and from Lake Magadi, Kenya; etc.
Treatment. Store in sealed container.

Pirssonite $Na_2Ca(CO_3)_2 . 2H_2O$
Pirssonite crystallizes in the orthorhombic system and forms short transparent prismatic crystals which may be colourless or grey. The lustre is vitreous. H 3–3.5; SG 2.3
Occurrence. Found in mud layers as euhedral crystals at Searles Lake, California; at Borax Lake, and as an efflorescence at Deep Spring Lake, California, and other localities.
Treatment. Store in a sealed container.

Gaylussite $Na_2Ca(CO_3)_2 . 5H_2O$
Gaylussite crystallizes in the monoclinic system and forms flattened, wedge-shaped elongated crystals with a perfect cleavage. The colour is white to yellow; the lustre is vitreous and the fracture conchoidal. H 2.5–3; SG 1.9
Occurrence. Gaylussite is found in lake deposits, particularly at Searles Lake and Deep Spring Lake, California; also from the eastern Gobi Desert, Mongolia; etc.
Treatment. Store in a sealed container.

Malachite $Cu_2CO_3(OH)_2$
Malachite crystallizes in the monoclinic system and is usually found massive as thick compact crusts with a botryoidal or mamillary surface and a fibrous and banded structure. It may also be stalactitic. When crystals are found they are small, acicular or prismatic with wedge-shaped terminations. There is a perfect cleavage. The colour is bright to dark green with a vitreous lustre. The streak is pale green. H 3.5–4; SG 4.0
Occurrence. Malachite occurs as a secondary mineral in the oxidation zone of copper deposits. It is associated with azurite and a variety combining the two is called azur-malachite.

Right: Prismatic azurite crystals on a limonitic matrix. From Sardinia

Opposite page, upper: Prisms of calcite from Andreasberg in the Harz Mountains, West Germany

Opposite page, lower: Parallel growth of calcite in scalenohedral form. From Korsnäs, Finland

Below right: An encrustation of acicular artinite crystals from Val Malenco, Sondrio, Italy

Below: A fine prismatic crystal of azurite from Tsumeb, South-West Africa

Localities. Malachite is found in the Ural Mountains; in Zaire; at Tsumeb, South-West Africa; at Broken Hill, New South Wales; at Bisbee, Arizona; and at Liskeard, Cornwall.
Treatment. Wash with distilled water, especially if cracked. Add ammonia until a blue precipitate forms and then add more water. The specimen can be allowed to soak for one hour and then should be soaked in distilled water. The water can be removed after several hours by soaking in wood alcohol and allowing to dry. Do not carry out this process if azurite is present.
Fashioning. Uses: cabochons, carvings, tumbling, beads, etc; *cleavage:* perfect pinacoidal in crystals (rare), parting easily in fibre directions in massive material; ***heat sensitivity:*** high, so the mineral should be treated with care.

Azurite $Cu_3(CO_3)_2(OH)_2$
Sometimes called blue malachite, azurite crystallizes in the monoclinic system forming massive specimens, nodular concretions, films and stains and, more rarely, tabular or short prismatic crystals. The colour is light to dark blue with a vitreous lustre and a blue streak; commonly associated with malachite to form the variety azur-malachite. H 3.5–4; SG 3.7
Occurrence. Found as a secondary mineral in the oxidation zone of copper deposits associated with malachite.
Locations. As for malachite; particularly fine crystals are found near Tsumeb, South-West Africa.
Treatment. Clean with distilled water.
Fashioning. Uses: faceting or cabochons; *cleavage:* perfect interrupted // to the clinodome; *cutting angles:* crown 40°, pavilion 40°; *heat sensitivity:* high in transparent faceting-quality crystals.

Magnesite $MgCO_3$
Magnesite crystallizes in the hexagonal system and is commonly found massive, compact, coarse- to fine-grained, lamellar or fibrous. Crystals, which are rare, may be rhombohedral, prismatic, tabular or scalenohedral. There is a perfect cleavage and the fracture is conchoidal. The colour is white to grey and may be yellowish-brown; the lustre is vitreous. The streak is white. Some specimens may be transparent. H 3.5–4.5; SG 3.0–3.1
Occurrence. Magnesite occurs as an alteration product of magnesium-rich rocks or as beds in metamorphic rocks. It may also occur as a gangue mineral in hydrothermal ore veins or in sedimentary deposits.
Locations. Found in Oberdorf, Austria, in fine crystals and as very fine crystals at Serra das Eguas, Bahia, Brazil. Large deposits occur in California and in Nevada. Cuttable crystals have been found at Bom Jesus dos Meiras, Brazil, and many other localities.
Treatment. Liquids best avoided, because of porosity.

Artinite $Mg_2CO_3(OH)_2 . 3H_2O$
Artinite crystallizes in the monoclinic system and forms crusts of acicular crystals or fibrous aggregates. There is a perfect cleavage and a high degree of brittleness. The colour is white and the lustre is vitreous or silky; the crystals may be transparent. H 2.5; SG 2.0
Occurrence. Found with hydromagnesite and other magnesium-bearing minerals coating surfaces in serpentinized ultrabasic rocks.
Localities. Fine specimens from the Aosta Valley, Italy; radiating fibrous aggregates from San Benito Co., California; sprays of acicular crystals from Long Island, New York; etc.
Treatment. Soil removed by a quick immersion in alcohol.

Hydromagnesite $Mg_4(CO_3)_3(OH)_2 . 3H_2O$
Hydromagnesite crystallizes in the monoclinic system and forms acicular or bladed crystals; sometimes, in large examples, they are vertically striated. It also occurs in crusts and massive forms. Twinning is common. The colour is white and crystals are transparent. The lustre is vitreous. There is a perfect cleavage. H 3.5; SG 2.2
Occurrence. Hydromagnesite is found as an alteration of serpentinite or other magnesian rocks.
Localities. Found as excellent crystals in the chromite deposits of Iran; from British Columbia; with artinite on Long Island, New York; from California; etc.

Calcite $CaCO_3$
Calcite crystallizes in the hexagonal system and forms a wide variety of crystals. The commonest forms are the scalenohedron and the rhombohedron; it may also be found massive, granular or stalactitic. Twinning is common and there is a perfect cleavage. The colour is white and various shades of grey, yellow, brown, red, green, blue and black may result from impurities. The lustre is vitreous to pearly. It is transparent to translucent with a whitish-grey streak. H 3; SG 2.7; RI 1.48, 1.65; dispersion very high.
Occurrence. Calcite is one of the commonest of minerals and occurs as the chief constituent in limestones. It may also be found as a component of other sedimentary and metamorphic rocks. It may be found in pegmatites and as a gangue mineral in ore deposits. It occurs as stalactites and stalagmites in caves and in geodes and concretions.
Localities. Fine specimens are found in Cumbria, Durham and Cornwall; Tsumeb, South West Africa; with copper on the Keweenaw Peninsula, Michigan; in the lead-zinc regions of Wisconsin and Illinois; etc.
Treatment. Clean if necessary with distilled water.
Fashioning. Uses: faceting, cabochons, tumbling, carvings, etc; *cleavage:* perfect rhombohedral in crystalline varieties; *cutting angles:* crown 40°–50°, pavilion 43°; *heat sensitivity:* high.

Aragonite $CaCO_3$
Dimorphous with calcite, aragonite crystallizes in the orthorhombic system forming acicular crystals elongated along the c-axis; often twinned; also in fibrous, columnar, stalactitic and micro-

Right: Flos ferri, a form of aragonite

Far right: Interpenetrating twins of aragonite

Below right: The pseudo-hexagonal form of aragonite, well shown in these fine crystals from Agrigento, Italy

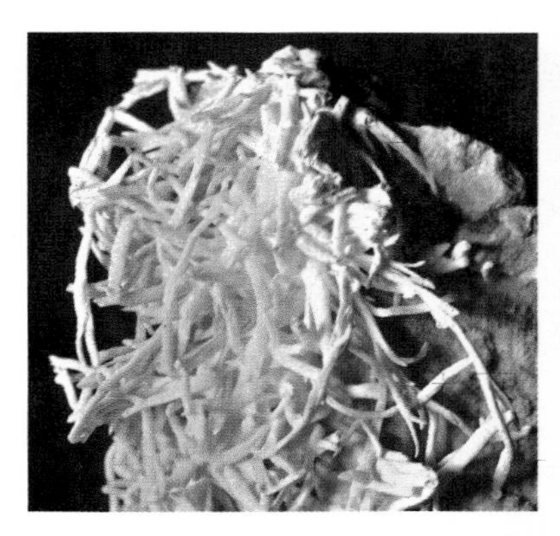

crystalline forms. There is a distinct cleavage. The colour is white, yellowish, green or blue, and there is some degree of transparency. The lustre is vitreous. H 3.5–4; SG 2.9

Occurrence. Aragonite occurs in low-temperature deposits, especially in limestone caves and in the vicinity of hot springs. It may be found in the oxidized zone of ore deposits and in some sedimentary rocks. Secreted by marine organisms.

Localities. Fine specimens from Cumbria, where they occur as aggregates of acicular crystals; from Carinthia, Austria; transparent crystals from Horschenz, Czechoslovakia; lead- and zinc-bearing varieties from Tsumeb, South West Africa; from Wind Cave, Custer Co., South Dakota; from the Magdalena district, Socorro Co., New Mexico; etc.

Treatment. Clean with distilled water.

Fashioning. Uses: faceting, cabochons, carvings; *cleavage:* distinct brachypinacoidal; brittle; *cutting angles:* crown 40°, pavilion 40°; *pleochroism:* weak; *heat sensitivity:* fairly high.

Dolomite $CaMg(CO_3)_2$

Dolomite crystallizes in the hexagonal system and forms simple rhombohedra; it may also be found as prismatic crystals terminated by rhombohedra; as twins; crystal aggregates; and massive, granular. There is a perfect cleavage. The colour is white, grey, through pink to pale brown, and the lustre is vitreous. RI 1.67, 1.50; H 3.5–4; SG 2.8

Occurrence. Dolomite occurs as strata formed in a number of different ways; in hydrothermal vein deposits, in veins in serpentinite and in altered basic igneous rocks containing magnesium.

Localities. Fine crystals from Switzerland and Italy; from Lockport, New York; in quartz geodes at Keokuk, Iowa; St. Eustache, Quebec; Guanajuato, Mexico.

Right: Dolomite crystals with hematite and quartz

Treatment. Clean if necessary with distilled water.

Huntite $CaMg_3(CO_3)_4$

Huntite crystallizes in the hexagonal system and forms compact masses resembling chalk; these are fibrous. The colour is white and the lustre earthy. H very soft; SG 2.6

Occurrence. Found in caverns where the country rock is rich in magnesium.

Localities. From Grotte de la Clamouse, Hérault, France; Dorog, Hungary; along the Trucial Coast; Carlsbad Caverns, New Mexico; etc.

Strontianite $SrCO_3$

Strontianite crystallizes in the orthorhombic system in prismatic crystals, which may be acicular; it is more frequently found massive, fibrous or concretionary. Twinning is frequent. The colour is white, grey, yellowish or brown, greenish or reddish, transparent to translucent. The lustre is vitreous. H 3.5; SG 3.7

Occurrence. Strontianite occurs as a

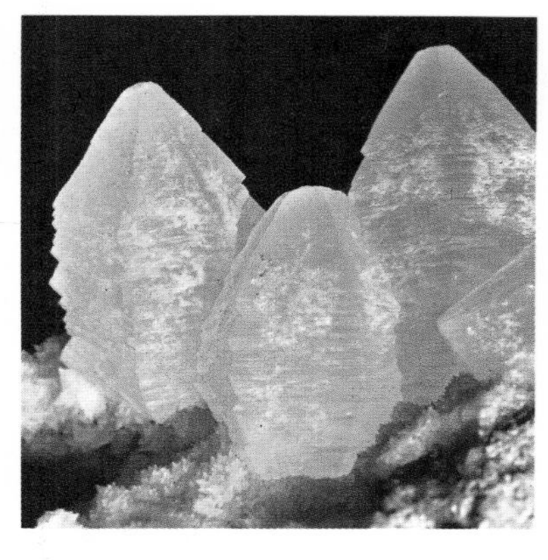

Left: Smithsonite crystals from Berg Aukas, Grootfontein, South-West Africa

Far left: Characteristic dipyramidal crystals of witherite with horizontal striation of the faces clearly shown

Far left, lower: Strontianite from Thuringia

Below left: The rhombohedral form of smithsonite can be seen in these crystals from Chessy, France

low-temperature mineral in veins and geodes in limestone, or in sulphide vein deposits as a gangue mineral.
Localities. Fine crystals from the Austrian Tyrol; Westphalia; Strontian, Strathclyde, Scotland; Strontium Hills, San Bernardino Co., California; etc.
Treatment. Clean with distilled water.

Witherite $BaCO_3$
Witherite crystallizes in the orthorhombic system and crystals are always twinned, forming pseudohexagonal dipyramids; faces are horizontally striated; massive and granular forms are also found. The colour is white to grey, sometimes tinged with yellow, green or brown. The mineral is transparent to translucent with a vitreous lustre. It phosphoresces blue under short-wave UV. H 3–3.5; SG 4.2
Occurrence. Found as a low-temperature mineral in hydrothermal vein deposits.
Localities. Very fine crystals from Northumberland, Cumbria and Durham; from the Minerva mine, Rosiclare, Illinois with fluorite and calcite; etc.
Treatment. Clean with distilled water.

Alstonite $BaCa(CO_3)_2$
Alstonite crystallizes in the orthorhombic system. It is dimorphous with barytocalcite and forms pseudohexagonal prisms. The colour is white, grey, or pink with a vitreous lustre. H 4–4.5; SG 3.6
Occurrence. Occurs as a low-temperature mineral in hydrothermal veins.
Localities. From the Brownley Hill mine, Alston, Cumbria; New Brancepeth, Durham; Hexham, Northumberland.

Barytocalcite $BaCa(CO_3)_2$
Dimorphous with alstonite, barytocalcite crystallizes in the monoclinic system; forms short to long prismatic crystals with striations; massive forms are also found. There is a perfect cleavage. The colour is white, grey, yellowish or greenish with a vitreous lustre. H 4; SG 3.6
Occurrence. Barytocalcite occurs associated with calcite, baryte and fluorite in veins in limestone from Alston Moor, Cumbria; also from Långban, Sweden.
Treatment. Clean with distilled water.

Smithsonite $ZnCO_3$
Smithsonite crystallizes in the hexagonal system and is found as rhombohedral crystals with curved faces or as scalenohedra; commonly botryoidal, reniform or stalactitic, massive, granular or as crusts. The colour may range from white, grey and pale yellow to deep yellow, pale to bright apple green, bluish green and blue, deep pink to purple and brown. The streak is white and the lustre vitreous to pearly. H 4–4.5; SG 4.3–4.4
Occurrence. Smithsonite occurs as a secondary mineral in the oxidized zone of ore deposits; as a replacement of calcareous rocks near ore deposits and sometimes as a secondary mineral derived from sphalerite in pegmatites.
Localities. Found at the Kabwe mine, Zambia, as fine transparent crystals; pink, white and grey crystals from Tsumeb, South West Africa; botryoidal encrustations from Kelly, Socorro Co., New Mexico; in pegmatite with hemimorphite and other minerals from Tin Mountain mine, South Dakota; as yellow banded crusts in lead mines at Yellville, Marion Co., Arkansas; Laurion, Greece; etc.
Treatment. Clean with distilled water.
Fashioning. Uses: faceting (rare), cabochons, beads, carvings, etc; *cleavage.* perfect rhombohedral, massive types separate easily between growth layers; brittle; *cutting angles:* crown 40°, pavilion 40°; *heat sensitivity:* low.

Above: Concretionary form of hydrozincite from Powys, Wales

Above right: Dawsonite crystals from Monte Amiata, Italy

Hydrozincite $Zn_5(CO_3)_2(OH)_6$
Hydrozincite crystallizes in the monoclinic system, and is found as compact masses, or as crusts, stalactites and reniform masses. Crystals are flattened and elongated, tapering to a sharp point. The colour is white, grey and pale pink, yellow and brown. The lustre is pearly if crystalline, dull to silky when massive. There is a perfect cleavage. H 2–2.5; SG 3.5–3.8
Occurrence. Hydrozincite is found as a secondary mineral formed by the alteration of sphalerite in the oxidation zone of ore deposits. It may also be found in pegmatites.
Localities. Found at Santander, Spain; Constantine, Algeria; Goodsprings, Nevada; Tin Mountain mine, Custer Co., South Dakota; Franklin, New Jersey; blade-like crystals from Mapimi, Mexico; and many other localities.
Treatment. Clean with distilled water.

Rosasite $(Cu, Zn)_2CO_3(OH)_2$
Rosasite crystallizes in the monoclinic system as botryoidal or mamillary crusts with a fibrous structure; crusts of microscopic crystals are sometimes found. The Cu to Zn ratio is near 3:2. The colour is bluish-green, green or sky-blue. H about 4.5; SG 4.0–4.2
Occurrence. Rosasite occurs as a secondary mineral in the oxidation zone of zinc-copper-lead deposits.
Localities. Found at the Rosas mine, Sulcis, Sardinia; with smithsonite and hemimorphite at Cerro Gordo, Inyo Co., California; with aurichalcite at Kelly, Socorro Co., New Mexico.
Treatment. Clean with distilled water.

Aurichalcite $(Zn, Cu)_5(CO_3)_2(OH)_6$
Aurichalcite has a Cu:Zn ratio of about 2:5. It crystallizes in the orthorhombic system and forms acicular crystals; more commonly tufted aggregates or encrustations. It may also be found granular, columnar or laminated. There is a perfect cleavage and the mineral is very brittle. The colour is pale green to greenish-blue, transparent. The lustre is silky to pearly. H 1–2; SG 3.9
Occurrence. Aurichalcite is found as a secondary mineral in oxidized zinc-copper deposits and sometimes in pegmatites.
Localities. Found at Laurion, Greece; Sardinia; Tin Mountain mine, South Dakota; very fine specimens from Mapimi, Durango, Mexico; etc.
Treatment. Clean with distilled water.

Loseyite $(Zn, Mn)_7(CO_3)_2(OH)_{10}$
Loseyite crystallizes in the monoclinic system forming radiating bundles of crystals coloured bluish-white to brown. H about 3; SG 3.2
Occurrence. Loseyite is found in small veins in massive ore with calcite, at Franklin, New Jersey.
Treatment. Clean with water.

Otavite $CdCO_3$
Otavite crystallizes in the hexagonal system, forming crusts of minute rhombohedral crystals coloured white to yellow-brown or reddish, with an adamantine lustre. H not determined; SG 4.9
Occurrence. Otavite occurs as a secondary mineral associated with smithsonite, azurite and malachite at Tsumeb, South West Africa.
Treatment. Do not attempt to clean.

Dawsonite $NaAlCO_3(OH)_2$
Dawsonite crystallizes in the orthorhombic system and is found as radial encrustations of acicular crystals. There is a perfect cleavage. The colour is white, transparent. The lustre is vitreous or silky. H 3; SG 2.4
Occurrence. Found as a low-temperature mineral in shale of the Green River formation, western Colorado; fine microscopic crystals from St. Michel, Quebec; in saline soils at Olduvai Gorge, Tanzania; etc.
Treatment. Do not attempt to clean.

Hydrotalcite $Mg_6Al_2CO_3(OH)_{16}.4H_2O$
Hydrotalcite crystallizes in the hexagonal system and is dimorphous with manasseite. It is found massive, foliated, or lamellar with a perfect cleavage. It is greasy to the touch. The colour is white, transparent; white streak. H 2; SG 2.0

Far right: Aurichalcite from Yanga-Koubanza, Zaire, showing the acicular form of the individual crystals

Right: Rosasite as an encrustation, from Sardinia

Localities. Found with serpentine at Nordmark and Snarum, Norway; Vernon, New Jersey; etc.

Manasseite
$Mg_6Al_2CO_3(OH)_{16}.4H_2O$
Dimorphous with hydrotalcite, manasseite crystallizes in the hexagonal system, forming massive foliated bodies. It has a perfect cleavage and a greasy feel. The colour may be white, grey or pale blue with a pearly lustre. H 2; SG 2
Localities. Associated with serpentine at Amity, New York; Nordmark and Snarum, Norway.

Dundasite $PbAl_2(CO_3)_2(OH)_4.2H_2O$
Dundasite crystallizes in the orthorhombic system forming small round aggregates of radiating crystals and matted crusts. There is a perfect cleavage. The colour is white transparent and the lustre is vitreous to silky. H 2; SG 3.4
Occurrence. Dundasite is found as a secondary mineral in the oxidation zone of lead deposits.
Localities. Found with crocoite in the Adelaide Proprietary mine, Dundas, Tasmania, from Wensley, Derbyshire and Trefriw, Gwynedd, Wales; etc.

Lanthanite $(La, Ce)_2(CO_3)_3.8H_2O$
Lanthanite crystallizes in the orthorhombic system and is found as scales, plates or thick tabular crystals. Twinning occurs. The colour is white, yellowish, pinkish, transparent. The lustre is pearly. H 2.5–3; SG 2.6–2.7
Occurrence. Found in oxidized zinc ores at Bethlehem, Lehigh Co., Pennsylvania; on cerite at Bastnaes, Sweden; etc.

Calkinsite $(La, Ce)_2(CO_3)_3.4H_2O$
Calkinsite crystallizes in the orthorhombic system as yellow platy crystals with a perfect cleavage. Twinning is common. H about 2.5; SG 3.2
Localities. Occurs with barite and goethite in burbankite, Hill Co., Montana.

Tengerite
Tengerite is perhaps two distinct species: $BeYtCO_3(OH)_3.H_2O$ with the Yt including some Ce, etc; or $CaYt_3(CO_3)_4(OH)_3.3H_2O$.
Occurrence. It occurs as crystalline powdery coatings and as fibrous mamillary crusts, a dull white in colour. H not measured; SG 3.1
Localities. Found on gadolinite at Clear Creek Canyon, Jefferson Co., Colorado; on melanocerite at the Cardiff uranium mine, Wilberforce, Ontario; etc.

Cerussite $PbCO_3$
Cerussite crystallizes in the orthorhombic system and forms crystals of varying habit; single crystals are usually tabular and elongated along the c-axis; clusters of crystals, reticulated masses and aggregates are also found. It may occur massive, granular or stalactitic. The colour ranges from transparent white to smoky and blue-green; the lustre is adamantine or submetallic and the streak colourless to white. H 3–3.5; SG 6.5
Occurrence. Cerussite is a secondary mineral in the oxidation zone of ore deposits and is associated with other secondary lead minerals.

Left: Hydrotalcite from Snarum, Norway

Localities. From Leadhills, Strathclyde; Tsumeb, South West Africa; Broken Hill, New South Wales; chatoyant material from the Mammoth mine, Pinal County, Arizona; etc.
Fashioning. Uses: faceting or cabochons; *cleavage:* distinct prismatic; very brittle; *cutting angles:* crown 40°, pavilion 40°; *heat sensitivity:* high.

Hydrocerussite $Pb_3(CO_3)_2(OH)_2$
Hydrocerussite crystallizes in the hexagonal system forming thick to thin tabular crystals with a hexagonal outline; steep pyramidal crystals are also found. There is a perfect cleavage. The colour is white or grey, transparent. The lustre is adamantine. H 3.5; SG 6.8
Occurrence. Found as a secondary mineral in oxidized ores, associated with cerussite, dioptase and boléite at the Mammoth mine, Tiger, Arizona; from the Mendip Hills, Somerset; Laurion, Greece; Altai Mts., USSR.

Bismutite $(BiO)_2CO_3$
Bismutite crystallizes in the tetragonal (orthorhombic) system, forming massive

Below: Prismatic cerussite crystals from Tsumeb, South-West Africa

Right: The compact form of stichtite from Bou Azzer, Morocco

Below right: An exceptionally well formed crystal of liebigite from Utah

bodies or fibrous crusts; lamellar aggregates are also found. The colour is yellow to brown, or greenish, grey or black. Some green or blue comes from associated copper minerals. The lustre is vitreous to pearly. H 2.5–3.5; SG 6.1–7.7 (a synthetic form was 8.1).
Occurrence. Found as a secondary mineral in the oxidized zone of veins and pegmatites which contain primary bismuth minerals.
Localities. From St. Just and Redruth, Cornwall; Colorado, Arizona and New Mexico; Bolivia, Australia; Malagasy Republic; etc.

Beyerite $Ca(BiO)_2(CO_3)_2$
Sometimes with lead replacing some calcium, beyerite crystallizes in the tetragonal system as thin rectangular flattened plates or as compact earthy masses. The colour is white to bright yellow, or greyish-green with a vitreous lustre. H 2–3; SG 6.5
Localities. Found as a secondary mineral with bismutite; as crystals in bismutite cavities near Wickenburg, Arizona; also from Schneeberg, E. Germany; etc.

Stichtite $Mg_6Cr_2CO_3(OH)_{16}.4H_2O$
Stichtite crystallizes in the hexagonal system and is dimorphous with barbertonite. It is found lamellar and foliated or in micaceous scales; the colour is lilac to rose-pink. There is a perfect cleavage. The streak is white to lilac and the lustre waxy. H 1.5–2; SG 2.1
Occurrence. Found in serpentinite rocks with chromite a common associate.
Localities. From Dundas, Tasmania; Barberton district, Transvaal; Black Lake, Quebec.

Barbertonite $Mg_6Cr_2CO_3(OH)_{16}.4H_2O$
Dimorphous with stichtite, barbertonite crystallizes in the hexagonal system, forming twisted masses of fibres, or as flattened plates with a perfect cleavage. The colour is lilac to rose-pink, translucent with a pearly lustre. The streak is white to pale lilac. H 1.5–2; SG 2.1
Occurrence. Found in serpentinite rock with stichtite and chromite.
Localities. From Cunningsburgh, Scotland; Dundas, Tasmania; Barberton, Transvaal.

Rutherfordine UO_2CO_3
Rutherfordine crystallizes in the orthorhombic system forming thin single crystals or microscopic fibrous aggregates; there is a perfect cleavage. The colour is white, pale yellow to orange or brown; yellowish-green colours also occur. H not measured; SG 5.7
Localities. Found with becquerelite and masuyite from Shinkolobwe, Zaire; in a pegmatite resulting from the alteration of uraninite at Morogoro, Tanzania; from the Beryl Mountain pegmatite, New Hampshire; and from Newry, Maine; etc.

Sharpite $(UO_2)(CO_3).H_2O$
Sharpite apparently crystallizes in the orthorhombic system, though this is not certain; it is found as crusts of radiating fibres, greenish-yellow in colour. H about 2.5; SG at least 3.3
Localities. Found with uranophane, curite, becquerelite and masuyite at Shinkolobwe, Zaire; from Krüth, Haut-Rhin, France, with uraninite.

Bayleyite $Mg_2UO_2(CO_3)_3.18H_2O$
Bayleyite crystallizes in the monoclinic system forming short prismatic crystals or crusts. The colour is yellow becoming yellowish-white on exposure. Lustre vitreous. H not determined; SG 2.0
Localities. Bayleyite is found as a secondary mineral, which is water-soluble, at the Hillside mine, Bagdad, Arizona; as an efflorescence with epsomite and gypsum at Azegour, Morocco.

Liebigite $Ca_2U(CO_3)_4.10H_2O$
Liebigite crystallizes in the orthorhombic system and is usually found as granular or scaly crusts; crystals are short prismatic with rounded edges. The colour is green to yellowish-green, transparent, with a vitreous lustre. H 2.5–3; SG 2.4
Localities. Found at Wheal Basset, Cornwall; Jachymov, Czechoslovakia, as an efflorescence on mine walls; and in the oxidized zone of uraninite veins; from Utah and New Mexico.

Andersonite $Na_2CaUO_2(CO_3)_3.6H_2O$
Andersonite crystallizes in the hexagonal system forming thick crusts; crystals may be pseudocubic. The colour is bright yellow-green, transparent to translucent with a vitreous lustre. H 2.5; SG 2.8
Localities. Occurs as a secondary mineral associated with bayleyite, gypsum and other minerals on mine walls at Bagdad, Arizona; as an efflorescence with liebigite from Jim Thorpe, Pennsylvania.

Swartzite $CaMgUO_2(CO_3)_3.12H_2O$
Swartzite crystallizes in the monoclinic system and forms crusts of bright green prismatic crystals which turn yellowish-white on dehydration. H not measured; SG 2.3
Localities. Found as an efflorescence on mine walls at the Hillside mine, Yavapai Co., Arizona, with bayleyite and other minerals.

Rhodochrosite $MnCO_3$
Rhodochrosite crystallizes in the hexagonal system and is commonly found massive, granular; it may also occur stalactitic or as rhombohedral crystals. There is a perfect cleavage. The colour is pale pink to deep red and may be orange or brown; the lustre is vitreous and the material may be transparent. H 3.5–4; SG 3.7
Occurrence. Occurs as a gangue mineral in hydrothermal ore veins or in high-temperature contact metasomatic deposits and as a secondary mineral in manganese deposits; occasionally in pegmatites.
Localities. Fine scalenohedra from Hotazel, Kuruman, South Africa; from Catamarca province, Argentina, where stalactitic aggregates are found; pink groups from Butte, Montana; from the Harz Mountains, Germany, and from Magdalena, Sonora, Mexico.
Treatment. May be cleaned with dilute hydrochloric acid; white coating formed after weathering cannot be removed.
Fashioning. Uses: faceting, cabochons, tumbling, beads, etc; *cleavage:* perfect rhombohedral, massive types may sep-

arate between growth layers; brittle; *cutting angles;* crown 40°, pavilion 40°; *heat sensitivity:* moderate, sensitive to shock.

Manganosiderite (Mn, Fe)CO_3
Manganosiderite crystallizes in the hexagonal system and is isomorphous with rhodochrosite. It is considered to be a variety of that mineral.

Siderite $FeCO_3$
Siderite crystallizes in the hexagonal system and is commonly found massive, granular. Crystals are rhombohedral, scalenohedral or tabular. Botryoidal and globular forms also occur. There is a perfect cleavage. The colour is pale yellow, grey, through browns, greens, reds and sometimes nearly colourless. The lustre is vitreous and the streak white. H 3.5–4.5; SG 3.9
Occurrence. Siderite is found in bedded sedimentary deposits and as a gangue mineral in hydrothermal ore veins; it occurs in basaltic igneous rocks and sometimes in pegmatites.
Localities. Found at Bodmin, St. Austell, Camborne, Lanlivery and Redruth, Cornwall; Příbram, Czechoslovakia; fine crystals from Val Taevetsch, Grisons, Switzerland; Broken Hill, New South Wales; Morro Velho gold mine, Ouro Preto, Minas Gerais, Brazil; etc.

Pyroaurite
$Mg_6Fe_2CO_3(OH)_{16}.4H_2O$
Pyroaurite crystallizes in the hexagonal system and is dimorphous with sjögrenite. Crystals are thin to thick tabular; fibrous forms are also found. There is a perfect cleavage. The colour is yellow to brownish-white, silver-white, greenish or colourless; the lustre is vitreous to pearly. H 2.5; SG 2.1
Occurrence. Found in serpentinite or as a low-temperature hydrothermal vein mineral.
Localities. From a dolomitic rock at Rutherglen, Ontario; also from Half-Grunay, north of Fetlar, in the Shetland Islands.

Left: Pyroaurite on calcite from Långban, Sweden

Sjögrenite
$Mg_6Fe_2CO_3(OH)_{16}.4H_2O$
Sjögrenite crystallizes in the hexagonal system and is dimorphous with pyroaurite. Crystals are thin plates with a perfect cleavage, white, yellow or brownish in colour. The lustre is vitreous or waxy, transparent or translucent. H 2.5; SG 2.1
Localities. Occurs as a low-temperature hydrothermal mineral with calcite and pyroaurite at Långban, Sweden.

Brugnatellite
$Mg_6FeCO_3(OH)_{13}.4H_2O$
Brugnatellite crystallizes in the hexagonal system forming foliated masses of small flakes with a perfect cleavage. The colour is flesh-pink to yellowish or brownish-white. The lustre is pearly and the streak white. H about 2; SG 2.1
Localities. Occurs as crusts in hydrothermally altered serpentinites and as an alteration product of melilite at Iron Hill, Gunnison County, Colorado; found also in Piedmont, Italy.

Ankerite Ca(Mg, Fe)$(CO_3)_2$
Ankerite has over 10% iron and often contains some manganese. It crystallizes in the hexagonal system forming simple rhombohedral or massive, granular forms. There is a perfect cleavage. The colour is white, grey, yellow to brown with a vitreous to pearly lustre. H 3.5–4; SG 2.9
Localities. Occurs at Oldham, near Manchester; at Erzberg, Austria; as a gangue mineral at the Homestake gold mine, Lead, South Dakota; in sulphide veins in the Cour d'Alene region, Idaho; etc.

Cobaltocalcite $CoCO_3$
Cobaltocalcite crystallizes in the hexagonal system forming small spherical masses with radiated structure, or crusts. Crystals rare. The colour is rose-red, altering to grey-brown or black. The lustre is vitreous to waxy. H 4; SG 4.1
Localities. Occurs in the cobalt-nickel veins at Schneeberg, E. Germany; fine specimens from Shaba, Zaire; and from other localities.

Left, above: Rhodochrosite from Cavnic, Romania

Left: Rhombohedral crystals of rhodochrosite from the Sweet Home mine, Colorado

Right: Octahedron of northupite from California

Zaratite $Ni_3CO_3(OH)_4 . 4H_2O$
Zaratite, sometimes called emerald nickel, crystallizes in the cubic system forming compact masses and mamillary encrustations coloured emerald green. The lustre is greasy. H 3.5; SG 2.5–2.6
Occurrence. Found as a secondary mineral in serpentinites and basic igneous rocks, often associated with chromite and sometimes as an alteration of millerite or meteoritic iron.
Localities. From the Shetland Islands; India; Australia; the Coast Range counties of California; etc.

Northupite $Na_3Mg(CO_3)_2Cl$
Northupite crystallizes in the cubic system forming octahedral crystals, which are colourless, yellow or grey to brown. They are transparent with a vitreous lustre. H 3.5–4; SG 2.3
Localities. Found associated with galena, galeite and other minerals at Searles Lake, San Bernardino Co., California (this locality yields fine octahedra); and at other localities.

Parisite $Ca(Ce, La)_2(CO_3)_3F_2$
Parisite crystallizes in the hexagonal system and forms double hexagonal pyramids and sometimes rhombohedra. The colour is brown or brownish-yellow, transparent. The lustre is vitreous or resinous. H 4.5; SG 4.3
Localities. Occurs in carbonaceous shale beds in the Muzo emerald deposit, Colombia; in pegmatite pipes in riebeckite-aegirine-granite at Quincy, Massachusetts; and also from a number of other localities.

Phosgenite $Pb_2CO_3Cl_2$
Phosgenite crystallizes in the tetragonal system forming long to short prismatic crystals; massive granular forms are also found. The colour is colourless, white, yellow, brown and sometimes greenish or pinkish; the lustre is adamantine and the streak white. H 2–3; SG 6.1
Occurrence. Phosgenite occurs as a secondary mineral in the oxidation zone of lead deposits and is commonly associated with other secondary lead minerals.
Localities. From Matlock, Derbyshire; Tarnow, Poland; Monteponi, Sicily, as fine crystals; as acicular crystals in quartz at the Silver Sprout mine, Inyo Co., California; as large masses at the

Right: Phosgenite crystal from Monteponi, Sardinia. Prism and pyramid forms can be clearly seen

Terrible mine, Custer Co., Colorado; and at other localities.

Caledonite $Cu_2Pb_5(SO_4)_3CO_3(OH)_6$
Caledonite crystallizes in the orthorhombic system forming small elongated prismatic crystals usually in divergent groups; also found as coatings and massive. There is a perfect cleavage. The colour is dark blue to bluish-green, transparent, with a greenish-blue or bluish-white streak. The lustre is vitreous to resinous. H 2.5–3; SG 5.6
Occurrence. Found in the oxidation zone of copper-lead deposits.
Localities. From Leadhills, Strathclyde, Scotland; very fine blue crystals from the Mammoth mine, Tiger, Arizona; with linarite at the Wonder prospect, California; etc.

Left: Caledonite from the Mammoth mine, Tiger Co., Arizona

Leadhillite $Pb_4SO_4(CO_3)_2(OH)_2$
Leadhillite crystallizes in the monoclinic system and is dimorphous with susannite. Crystals are pseudohexagonal and massive; granular forms also occur; contact twins are found. There is a perfect cleavage. The colour ranges from colourless to white, grey, yellowish, pale green to blue; transparent to translucent. The lustre is resinous to adamantine. H 2.5–3; SG 6.5
Occurrence. Found in the oxidized zone of lead deposits as a secondary mineral associated with cerussite, lanarkite, galena, etc.
Localities. From Leadhills, Strathclyde, Scotland; Wanlockhead, Dumfries and Galloway also in Scotland; Tsumeb, South West Africa; the Mammoth mine, Tiger, Arizona; etc.

Left: Phosgenite, showing several prism faces and a pinacoid (the upper horizontal face). From Monteponi, Sardinia

Susannite $Pb_4SO_4(CO_3)_2(OH)_2$
Dimorphous with leadhillite, susannite is the high-temperature variety. It crystallizes in the hexagonal system forming rhombohedral crystals; there is a perfect cleavage. It is colourless to greenish or yellowish and the lustre is adamantine to resinous. H 2.5–3; SG 6.5
Occurrence. Occurs as a secondary mineral in the oxidation zones of lead deposits.
Localities. From Leadhills, Strathclyde; Mammoth mine, Tiger, Arizona; etc.

Wherryite
$Pb_4Cu(SO_4)_2CO_3(Cl, OH)_2O$
Wherryite crystallizes in the monoclinic system in massive granular forms; fibrous aggregates of acicular crystals are also found. The colour is light green, yellow, or bright yellowish-green and the lustre is vitreous. H not measured; SG 6.4
Localities. Found in a vug at the Mammoth mine, Tiger, Arizona with chrysocolla, leadhillite and other minerals.

Schrökingerite
$NaCa_3UO_2SO_4(CO_3)_3F.10H_2O$
Schrökingerite crystallizes in the orthorhombic system as pseudohexagonal plates or as crusts or globular aggregates. There is a perfect cleavage and the colour is between green and yellow. The lustre is vitreous to pearly. H 2.5; SG 2.5
Occurrence. Found as a secondary mineral often of recent formation.
Localities. From gypsum-bearing clays in Lost Creek, Sweetwater Co., Wyoming; Jachymov, Czechoslovakia; La Soberania mine, San Isidro, Mendoza, Argentina.

Hanksite $Na_{22}K(SO_4)_9(CO_3)_2Cl$
Hanksite crystallizes in the hexagonal system as tabular to prismatic crystals often showing interpenetrant twinning. It is colourless to grey, yellow or almost black, with a vitreous lustre. The streak is white. Saline taste. H 3–3.5; SG 2.5
Occurrence. Found with halite and other minerals in saline beds.
Localities. From the borax areas of Death Valley, California and the saline beds of Searles Lake in the same state.

Nitrates

Nitratine $NaNO_3$
Often called soda-nitre, nitratine crystallizes in the hexagonal system forming crusts. Twinning is common and there is a perfect cleavage. The colour is colourless or white, discoloured grey or yellowish, transparent. The lustre is vitreous. H 1.5–2; SG 2.2
Occurrence. Occurs as an efflorescence in hot dry regions.
Localities. From Nevada, New Mexico, California; Peru; Bolivia; etc.

Nitre KNO_3
Nitre crystallizes in the orthorhombic system and occurs as thin crusts; also massive granular. It is colourless or white, discoloured by impurities. The lustre is vitreous. H 2; SG 2.1
Occurrence. Occurs as an efflorescence on soil or in caves.
Localities. From northern Chile; Bolivia; Italy; USSR; Kentucky, New Mexico; etc.

Gerhardtite $Cu_2NO_3(OH)_3$
Gerhardtite crystallizes in the orthorhombic system forming thick tabular crystals with a perfect cleavage. The colour is emerald green and the streak light green. H 2; SG 3.4
Occurrence. Occurs as a secondary mineral in the oxidation zone of copper deposits.
Localities. From Shaba, Zaire; with cuprite at the United Verde mine, Jerome, Arizona.

Nitrocalcite $Ca(NO_3)_2.4H_2O$
Nitrocalcite crystallizes in the monoclinic system. Its colour is white or grey. H soft; SG 1.9
Occurrence. It occurs as an efflorescence on soil or in limestone caverns and on calcareous rocks.
Localities. From France, Spain and Hungary; Arizona; California; etc.

Above right: Nitrocalcite from the McDonnell Range, Australia

Right: Nitratine from Tarapaca, Chile

Buttgenbachite
$Cu_{19}Cl_4(NO_3)_2(OH)_{32}.2$ or $3H_2O$
Buttgenbachite crystallizes in the hexagonal system as acicular crystals, radiated sprays and felt-like aggregates. The colour is deep blue and the lustre vitreous. H 3; SG 3.4
Localities. Found as a secondary mineral at Likasi, Shaba, Zaire.

Darapskite $Na_3NO_3SO_4.H_2O$
Darapskite crystallizes in the monoclinic system, forming long prismatic crystals; occurs also as platy or granular material. Twinning is common. The crystals are colourless, transparent and the lustre is vitreous. H about 2.5; SG 2.2
Localities. From the nitrate deposits of Atacama, Chile, and other places.

Silicates

OLIVINE GROUP
A general name for the series from forsterite (Mg_2SiO_4) to fayalite (Fe_2SiO_4). The gem variety is peridot.

Forsterite Mg_2SiO_4
Forsterite crystallizes in the orthorhombic system forming thick tabular crystals with vertical striations and wedge-shaped terminations. Massive forms are also found. The colour is green, pale yellow or white and the lustre is vitreous. The streak is colourless. H 7; SG 3.2
Occurrence. Forsterite occurs in basic and ultrabasic igneous rocks and thermally metamorphosed impure dolomitic limestones.
Localities. From Germany; Sweden; Norway; USSR; San Bernardino Co., California; etc.
Treatment. Remove iron stains with oxalic acid.

Hortonolite $(Fe, Mg)_2SiO_4$
Hortonolite crystallizes in the orthorhombic system and is usually massive, granular. Crystals are thick tabular. The colour is light green to brown with a greasy lustre and colourless streak. H 7; SG 3.8
Localities. Found in dolerite and gabbro; Monroe, Orange Co., New York; Greenland; Transvaal; and from some other localities.
Treatment. Clean with distilled water.

Peridot $(Mg, Fe)_2SiO_4$
Peridot forms prismatic crystals, often flattened; masses or grains. There is a distinct cleavage. The colour is green to brown with a colourless or yellowish streak and vitreous lustre. H 6.5–7; SG 3.2–3.3; RI 1.63, 1.69; double refraction 0.038
Occurrence. Occurs in igneous rocks such as basalt or gabbro, or as a metamorphic product of sedimentary rocks containing silica and magnesium, such as impure dolomites. May occur as crystals in meteoritic iron.
Localities. Found on the Island of Zabarjad in the Red Sea; Burma; Hawaii (this variety may contain some chromium); Arizona; the Eifel region of W. Germany; Minas Gerais, Brazil.
Fashioning. Uses: faceting, cabochons, tumbling, beads, etc; *cleavage*: distinct brachypinacoidal; *cutting angles*:

crown 40°, pavilion 40°; *pleochroism*: distinct; *heat sensitivity*: low.

Fayalite Fe_2SiO_4
Fayalite crystallizes in the orthorhombic system as thick tabular crystals, sometimes with wedge-shaped terminations; usually found massive. The colour is greenish-yellow, yellow or brown with a colourless streak; the lustre is vitreous. H 7; SG 4.3
Occurrence. Occurs in acid and alkaline volcanic and plutonic rocks; metamorphosed iron-rich sediments.
Localities. From Sweden; France; Salt Lake Crater, Oahu, Hawaii; and numerous other localities.
Treatment. Remove iron stains with oxalic acid.

Monticellite $MgCaSiO_4$
Monticellite crystallizes in the orthorhombic system forming prismatic crystals; it is more commonly found massive granular or as disseminated grains. The colour is colourless, grey or greenish with a vitreous lustre. H 5.5; SG 3.0–3.2
Occurrence. Occurs as a metamorphic or metasomatic mineral; also from ultrabasic rocks.
Localities. From Crestmore, Riverside Co., California; Italy; USSR; Zaire; and other localities.

Tephroite Mn_2SiO_4
Tephroite crystallizes in the orthorhombic system, forming short prismatic crystals or compact masses. The colour is grey, olive-green, flesh red or reddish-brown. The lustre is greasy. H 6; SG 4.1
Occurrence. Occurs in iron-manganese ore deposits; sometimes in metamorphic rocks derived from sediments rich in manganese.
Localities. From Franklin, New Jersey; Treburland mine, Cornwall; Långban, Sweden; and from a variety of other localities.
Treatment. Clean with distilled water.

Glaucochroite $CaMnSiO_4$
Glaucochroite crystallizes in the orthorhombic system as long prismatic crystals, sometimes in aggregates. The colour is bluish-grey, pink or white. The lustre is vitreous. Twinning is common. H 6; SG 3.4
Localities. Found with willemite and franklinite at Franklin, New Jersey.
Treatment. Clean with distilled water.

HUMITE GROUP
The members of the humite group have the same ratio of unit cell dimensions b:c, but the a:c and a:b ratios alter from norbergite to clinohumite. The composition of the group can be given as $Mg(OH, F)_2.nMg_2SiO_4$

Norbergite $Mg_3SiO_4(F, OH)_2$
Norbergite crystallizes in the orthorhombic system, often as grains. The colour is orange-yellow to brown, the lustre is vitreous. H 6–6.5; SG 3.2
Occurrence. Occurs in contact zones in limestone or dolomite.
Localities. From Franklin, New Jersey; Norberg, Västmanland, Sweden; etc.
Treatment. Clean with distilled water.

Chondrodite $Mg_5Si_2O_8(F, OH)_2$
Chondrodite crystallizes in the monoclinic system as crystals of varied habit or in masses, often with lamellar twinning. The colour is yellow, red or brown, with a vitreous lustre. H 6–6.5; SG 3.1–3.2
Occurrence. Occurs in contact zones in limestone or dolomite.
Localities. Pargas, Finland; Tilly Foster iron mine, Brewster, New York State; Riverside Co., California; etc.
Treatment. Clean with distilled water.

Humite $Mg_7Si_3O_{12}(F, OH)_2$
Humite crystallizes in the orthorhombic system as small crystals, which may be white, yellow, dark orange or brown; they have a vitreous lustre. H 6; SG 3.2–3.3
Occurrence. Occurs in contact zones in limestone or dolomite or in veins.
Localities. From Finland; Värmland, Sweden; Tilly Foster iron mine, Brewster, New York State; and other places.
Treatment. Clean with distilled water.

Left: Darapskite with nitratine from Chile

Below left: Fine crystals of chondrodite from a classic locality, the Tilly Foster mine, New York

Below: Crystals of humite with brown biotite from Mt Somma, Italy

Right: Five small almandine garnets. Their trapezohedral form can just be discerned

Clinohumite $Mg_9Si_4O_{16}(F, OH)_2$
Clinohumite crystallizes in the monoclinic system, forming crystals which often show lamellar twinning. The colour is white, yellow or brown, with a vitreous lustre. H 6; SG 3.2–3.3
Occurrence. Occurs in contact zones in dolomite, in veins or talc schist.
Localities. From Pargas, Finland: Llanos de Juanar, Malaga, Spain; Tilly Foster iron mine, Brewster, New York State; and elsewhere.
Treatment. Clean with distilled water.

HELVINE GROUP

Three minerals with the composition $R_4Be_3Si_3O_{12}S$, R being manganese in helvine, iron in danalite and zinc in genthelvite.

Helvine $Mn_4Be_3Si_3O_{12}S$
Helvine crystallizes in the cubic system as tetrahedra or rounded aggregates. The colour is reddish-brown, yellow or green, with a vitreous lustre. H 6; SG 3.1–3.3

Right: A pyrope garnet embedded in matrix

Occurrence. Occurs in granite pegmatites, skarns, hydrothermal veins, etc.
Localities. From Pala, San Diego Co., California; Mont St. Hilaire, Quebec, Canada; Brazil; Norway; and other localities.
Treatment. Clean with distilled water.

Danalite $Fe_4Be_3Si_3O_{12}S$
Danalite crystallizes in the cubic system as octahedra or dodecahedra and in granular masses. The lustre is vitreous. The colour is grey, lemon yellow, red or brown. H 5.5–6; SG 3.3–3.4
Occurrence. Occurs in granite pegmatites, contact-metasomatic deposits and hydrothermal veins.
Localities. From Redruth, Cornwall; Sweden; USSR; Carroll Co., New Hampshire; etc.
Treatment. Clean with distilled water.

Genthelvite $Zn_4Be_3Si_3O_{12}S$
Genthelvite crystallizes in the cubic system as tetrahedra or as rounded aggregates. The colour is white, yellow, emerald-green, purple, pink or brown to black, with a vitreous lustre. H 6–6.5; SG 3.6
Localities. Occurs with analcime in carbonatite at Mont St. Hilaire, Quebec; Kola Peninsula, USSR; etc.
Treatment. Clean with distilled water.

GARNET GROUP

The garnets can be described by the general formula $X_3Y_2Si_3O_{12}$, in which X may be Ca, Mg, Mn, or Fe^{2+}, and Y may be Al, Fe^{3+} or Cr. There is a considerable amount of atomic substitution in the garnets; they are therefore often described in terms of pure end members, of which the principal ones are as follows:

pyrope	$Mg_3Al_2Si_3O_{12}$
almandine	$Fe_3Al_2Si_3O_{12}$
spessartine	$Mn_3Al_2Si_3O_{12}$
grossular	$Ca_3Al_2Si_3O_{12}$
uvarovite	$Ca_3Cr_2Si_3O_{12}$
andradite	$Ca_3Fe_2Si_3O_{12}$

The first three and the second three form two separate groups with almost continuous substitution within each, but with very little substitution between them. All garnets crystallize in the cubic system. Although sometimes found in igneous rocks they are common in metamorphic rocks.

Pyrope $Mg_3Al_2Si_3O_{12}$
Pyrope crystallizes in the cubic system, forming dodecahedra or trapezohedra or combinations; more frequently found as pebbles or grains. It has a vitreous lustre; its colour is pink to purplish-red or crimson; some specimens may appear almost black. Chromium may be present and such stones are bright red. H 7–7.5; SG 3.5–3.8; RI 1.73–1.76
Occurrence. Occurs in peridotites or in serpentinites. An associate of diamond in the peridoitites of South Africa.
Localities. From South Africa; Queensland and New South Wales; Czechoslovakia; Madras, India.
Treatment. Clean with dilute acid.
Fashioning. Uses: faceting, cabochons, beads; *cleavage*: none, but a plane of parting exists // to dodecahedron; brittle; *cutting angles*: crown 40°, pavilion 40°, angles reduced in very deeply coloured material; *heat sensitivity*: fairly high, but unaffected by normal dopping temperatures.

Almandine $Fe_3Al_2Si_3O_{12}$
Almandine crystallizes in the cubic system forming dodecahedra or trapezohedra. Compact or granular masses are also found. The colour is deep red or brownish-black. The lustre is resinous and the streak white. H 7–7.5; SG 4.1–4.3; RI 1.83
Occurrence. Occurs in schists, gneiss and other metamorphic rocks. Also contact zones and alluvial deposits.
Localities. From India; Sri Lanka; Tanzania; Australia; Brazil; Malagasy Republic; North Creek, New York State. A star garnet is found in Idaho and elsewhere.
Treatment. Clean with dilute acid.
Fashioning. Uses: faceting, cabochons, beads, tumbling, etc; *cleavage*: none, but a plane of parting exists // to the dodecahedron; brittle; *cutting angles*: crown 40°, pavilion 40°; *heat sensitivity*: fairly high, but unaffected by normal dopping temperatures.

Spessartine $Mn_3Al_2Si_3O_{12}$
Spessartine crystallizes in the cubic system as dodecahedra or trapezohedra or in combination. Faces are often striated. Massive and granular forms also occur. The colour may be orange, orange-red, reddish-brown or brown. The streak is white and the lustre vitreous. H 7–7.5; SG 3.8–4.2; RI 1.8
Occurrence. Occurs in granite pegmatites, schists or quartzites.
Localities. Fine crystals from San Diego Co., California; Rutherford Mine, Amelia, Virginia; Malagasy Republic; etc.
Treatment. Clean with dilute acid.
Fashioning. Uses: faceting or cabochons; *cleavage*: none, but a plane of parting exists // to the dodecahedron; brittle; *cutting angles*: crown 40°, pavilion 40°; *heat sensitivity*: fairly high, but unaffected by normal dopping temperatures.

Left: The manganese aluminium garnet spessartine, from San Diego Co., California. The white mineral is feldspar

Grossular $Ca_3Al_2Si_3O_{12}$
Grossular crystallizes in the cubic system, forming dodecahedra or trapezohedra, or combinations. Massive and granular forms are found. The colour may be bright emerald-green, lime green, yellow, red, reddish-brown, orange, white, grey or black. Some varieties are transparent. The streak is white and the lustre resinous. H 6.5–7; SG 3.4–3.6; RI 1.73, 1.75
Occurrence. Occurs in metamorphosed impure calcareous rocks, particularly in contact zones; associated minerals include idocrase, calcite and diopside; it may also occur in schists.
Localities. A vanadium-bearing variety from Kenya has been named tsavorite; an opaque green variety is sometimes called Transvaal jade; hessonite from Sri Lanka; Brazil; Tanzania; some green stones from Pakistan; yellow material from Elba; Zermatt, Switzerland; etc.
Treatment. Clean with dilute acid.
Fashioning. Uses: faceting, cabochons, beads, carvings, tumbling, etc; *cleavage*: none, but a plane of parting exists // to the dodecahedron; brittle, except in the massive hydrogrossular or 'Transvaal jade' type which is very tough; *cutting angles*: crown 40°, pavilion 40°; *heat sensitivity*: fairly high, but unaffected by normal dopping temperatures.

Uvarovite $Ca_3Cr_2Si_3O_{12}$
Uvarovite crystallizes in the cubic system as dodecahedra or trapezohedra, or in combination; dodecahedral faces often striated; massive, coarse granular forms found. The colour is emerald green with a white streak and a vitreous lustre. H 6.5–7; SG 3.4–3.8; RI 1.86
Occurrence. Occurs with chromite in serpentinite, in skarn deposits and metamorphosed limestones.
Localities. Found with chromite at Bisert, Urals, USSR; fine crystals from Outokumpu, Finland; with diopside at Orford, Quebec; chromite deposits in northern California; etc.
Treatment. Remove calcite with dilute hydrochloric acid.

Andradite $Ca_3Fe_2(SiO_4)_3$
Andradite crystallizes in the cubic system, forming dodecahedra or trapezohedra or combinations. Massive forms also found. The colour may be yellow (topazolite); emerald-green (demantoid); black (melanite) and various shades of the above. The lustre is vitreous to resinous. H 6.5–7; SG 3.7–4.1; RI 1.89
Occurrence. Demantoid and topazolite occur in chlorite schist and serpentinite; melanite in alkaline igneous rocks. Also from metamorphosed limestones or contact zones.
Localities. Demantoid from the Ala Valley, Piedmont, Italy; the Ural Mountains, USSR; melanite from San Benito Co., California; Franklin, New Jersey; etc.
Treatment. Clean with dilute acid.
Fashioning. Uses: faceting or cabochons; *cleavage*: none, but may have a plane of parting // to the dodecahedron; brittle; *cutting angles*: crown 40°, pavilion 40°; *heat sensitivity*: fairly high, but unaffected by normal dopping.

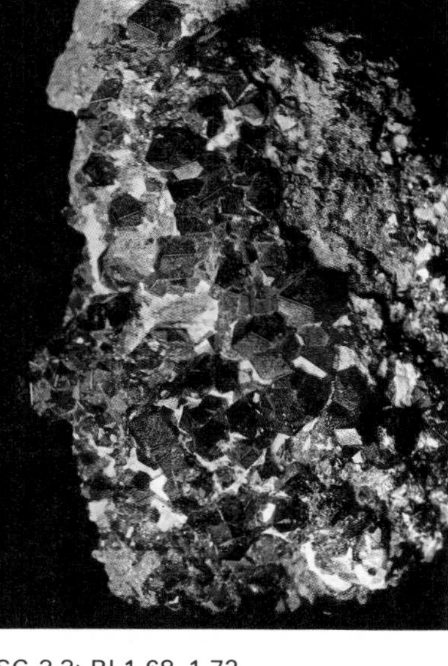

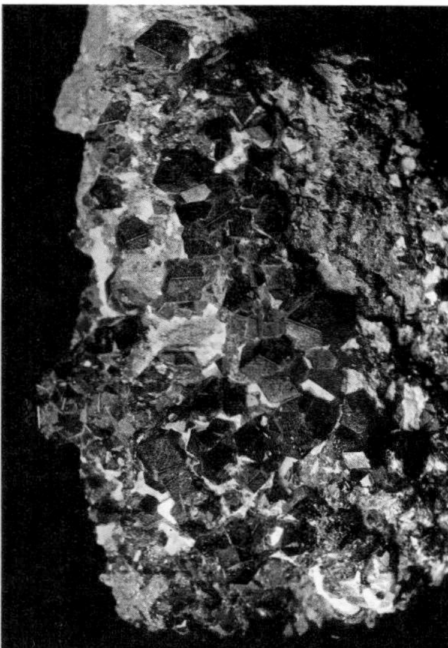

Left: A dark variety of andradite from Oravita, Romania

Far left: Uvarovite, the calcium-chromium garnet. It is normally found only in small crystals like these from the Ural Mountains of the USSR

EPIDOTE GROUP
Minerals in this group are silicates of calcium, aluminium and ferric iron with some replacement by ferrous iron and magnesium.

Zoisite $Ca_2Al_3Si_3O_{12}(OH)$
Zoisite crystallizes in the orthorhombic system forming striated prismatic crystals or massive granular forms. There is a perfect cleavage. The colour is grey, white, pink (thulite); colourless to blue or purple (tanzanite); brown or green. There is a vitreous lustre. H 6.5–7; SG 3.3; RI 1.68, 1.72
Occurrence. Occurs in regionally metamorphosed rocks including calcareous schists, basic igneous rocks and metamorphosed impure limestones and dolomites. Also from quartz veins and pegmatites.
Localities. From Tanzania (tanzanite); Nevada and California; east Greenland (thulite); Italy; Mexico; Finland; etc. A green zoisite with embedded ruby crystals, from Tanzania, has been named anyolite.
Treatment. Clean with dilute acid.

Left: The demantoid variety of andradite. It owes its colour to chromium. These crystals, with chrysotile asbestos, are from Val Malenco, Italy

Opposite page: The hessonite variety of grossular garnet from the Ala Valley, Piedmont, Italy. Here the two forms of rhombdodecahedron and icositetrahedron are combined. The green crystals are diopside and the bluish lamellae are chlorite

Above: Allanite crystals from Ontario, Canada

Fashioning. Uses: faceting (tanzanite), cabochons, beads, carvings, tumbling, etc (thulite); *cleavage*: perfect, brachypinacoidal; brittle; *cutting angles*: crown 40°, pavilion 40°; *pleochroism*: strong; *heat sensitivity*: fairly low, unaffected by normal dopping temperatures, but excessive heat may cause tanzanite to change colour; *mechanical sensitivity*: high; tanzanite should never be cleaned in an ultrasonic cleaner.

Clinozoisite $Ca_2Al_3Si_3O_{12}OH$
Clinozoisite crystallizes in the monoclinic system forming prismatic crystals, often deeply striated and acicular; more commonly found as granular or fibrous masses. There is a perfect cleavage. It is colourless, yellow, green or pink; there is a vitreous lustre. The streak is colourless. H 6.5; SG 3.2
Occurrence. Occurs in regionally metamorphosed igneous and sedimentary rocks or as an alteration product of plagioclase feldspars.
Localities. From Sonora, Mexico; Switzerland; Austria; Italy; California; and other localities.
Treatment. Clinozoisite should be cleaned with dilute acid.

Allanite
$(Ca, Fe^{2+})_2(R, Al, Fe^{3+})_3Si_3O_{12}OH$
Allanite crystallizes in the monoclinic system as compact masses or grains; crystals are tabular but may be long prismatic. Polysynthetic twinning is common. The colour is brown to black and there is a resinous lustre. H 5–6; SG 3.9–4.0
Occurrence. Occurs in granites, granite pegmatites or gneiss.
Localities. From Canada; Norway; Germany; USSR; California; and elsewhere.

Hancockite
$(Pb, Ca, Sr)_2(Al, Fe)_3Si_3O_{12}OH$
Hancockite crystallizes in the monoclinic system as aggregates or compact masses. There is a perfect cleavage. The colour is yellowish-brown, and the lustre is vitreous. H 6–7; SG 4.0
Localities. Occurs with garnet and biotite at Franklin, New Jersey.

Piemontite
$Ca_2(Al, Fe, Mn^{3+})_3Si_3O_{12}OH$
Piemontite crystallizes in the monoclinic system, commonly forming coarse-grained masses; crystals are prismatic, acicular. There is a perfect cleavage. The colour is reddish-brown to black or may be a lighter red or yellow; there is a vitreous lustre. H 6; SG 3.4–3.5
Occurrence. Occurs in schists, or manganese ore deposits.
Localities. From a sericite schist at Piemonte, Italy; Nordmark, Värmland, Sweden; in a quartz schist at Black Peak, northwest Otago, New Zealand; Glen Coe, Highland, Scotland; Riverside Co., California; Annapolis, Missouri; etc.
Treatment. Clean with dilute acid.

Epidote $Ca_2(Al, Fe)_3Si_3O_{12}OH$
Epidote crystallizes in the monoclinic system as prismatic crystals deeply striated or as granular masses. There is a perfect cleavage. The colour is yellowish to brownish-green or black, and the lustre is vitreous. The streak is grey. H 6–7; SG 3.3–3.5; RI 1.74–1.76

Right: Clinozoisite from Amborompotsy, Malagasy Republic. Clinozoisite is sometimes faceted though the material shown is opaque

Occurrence. Occurs in regionally metamorphosed igneous and sedimentary rocks and as an alteration of plagioclase feldspars.
Localities. From the Ala Valley, Piedmont, Italy; fine crystals from the Untersulzbachtal, Salzburg, Austria; Malagasy Republic; USSR; Japan; Australia; Riverside and San Diego Counties, California; Prince of Wales Island, Alaska; Brazil; a chromium-bearing variety, tawmawite, occurs at Tawmaw, Burma, and at Outokumpu, Finland. The variety withamite contains some manganese; fouqueite contains up to 6% iron and up to 4% ferrous iron.
Treatment. Clean with dilute acid.
Fashioning. Uses: faceting or cabochons; *cleavage*: perfect // to basal pinacoid, imperfect // to orthopinacoid; brittle; *cutting angles*: crown 40°, pavilion 40°; *pleochroism*: strong; *heat sensitivity*: fairly low.

MELILITE GROUP

The melilite group has the end-members gehlenite and akermanite.

Gehlenite $Ca_2Al_2SiO_7$

Gehlenite crystallizes in the tetragonal system, forming short prismatic crystals but being more commonly found massive and granular. The colour is grey, green to brown, yellow or colourless, with a resinous lustre. H 5–6; SG 3.0
Occurrence. Occurs in calcium-rich basic eruptive rocks, thermally metamorphosed impure limestones and furnace slags.
Localities. From Durango, Mexico; USSR; Italy; Riverside Co., California; and other localities.
Treatment. Clean with dilute acid.

Melilite $(Ca, Na)_2[(Mg, Fe^{+2}, Al, Si)_3O_7]$

Melilite crystallizes in the tetragonal system as thin tabular crystals or lamellar aggregates with a distinct cleavage. It is colourless, green, brown or yellow. H 5–6; SG 3.0
Occurrence. Occurs in certain basic igneous rocks, and in thermally metamorphosed limestones.
Localities. From Julianehaab, Greenland; Langesundfjord, Norway; and many other places.
Treatment. Clean with dilute acid.

Akermanite $Ca_2(MgSi_2O_7)$

Akermanite, a member of the melilite group, crystallizes in the tetragonal system as prismatic crystals or as

Above: Fine monoclinic prisms of epidote from Traversella, Italy

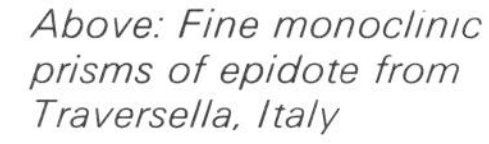

Left: Gehlenite from Val di Fassa, Italy

Right: Fine diopside crystal from the Ala Valley, Piedmont, Italy. The red crystals are hessonite

granular masses. It is grey, brown, green or colourless with a vitreous or resinous lustre. H 5–6; SG 2.9
Occurrence. From rocks rich in calcium and in thermally metamorphosed limestones; also from slags.
Localities. As for gehlenite.

PYROXENE GROUP
The pyroxenes are an important group of rock-forming minerals found in many igneous and metamorphic rocks. They conform to the general formula $X_2Si_2O_6$ in which X is usually Ca, Mg, Fe, Li, Ti, Al or Na. There are two principal groups. The ortho-pyroxenes crystallize in the orthorhombic crystal system and contain very little calcium; the commonest are hypersthene and bronzite. The clino-pyroxenes are monoclinic and contain Ca, Na, Al, Fe^{3+}, or Li; they include augite, pigeonite, diopside, jadeite, aegirine and spodumene. The pyroxenes are characterised by having two cleavages intersecting almost at 90°.

Enstatite $MgSiO_3$
Enstatite crystallizes in the orthorhombic system, usually being found massive, sometimes fibrous or lamellar; crystals are prismatic. It is colourless, yellowish, green, emerald-green (a chrome-bearing variety); and brown, with a vitreous lustre. The streak is grey. H 5–6; SG 3.2–3.4
Occurrence. Enstatite occurs in basic or ultrabasic igneous rocks; in thermally metamorphosed rocks and in meteorites.
Localities. From the Harz Mountains and the Eifel region of Germany; the chrome variety from South Africa, where it is associated with diamond; from Colorado, Montana and other states of the USA; etc.
Treatment. Clean with dilute acid.
Fashioning. Uses: faceting or cabochons; *cleavage*: easy prismatic; *cutting angles*; crown 40°, pavilion 40°; *pleochroism*: distinct; *heat sensitivity*: low.

Below: Enstatite. It may be found sufficiently transparent to facet. This opaque crystal is from Odegardens, Norway

Clinoenstatite $MgSiO_3$
Dimorphous with enstatite, clinoenstatite crystallizes in the monoclinic system, with short prismatic or tabular crystals; massive, lamellar forms are also found. It may be colourless, yellowish to brown or greenish with a vitreous lustre. H 5–6; SG 3.1
Localities. Found in some igneous rocks and in meteorites, and also in a porphyritic volcanic rock from Vogelkop, West Irian, Indonesia.
Treatment. Clinoenstatite should be cleaned with dilute acid.

Diopside $MgCaSi_2O_6$
Diopside crystallizes in the monoclinic system forming short prismatic crystals; massive, lamellar or granular forms are also found. The colour is colourless, white, grey, pale to dark green, yellowish to reddish-brown. The lustre is vitreous and the streak white or grey. Some ferrous oxide is included in addition to the composition given. H 5.5–6.5; SG 3.2–3.3; RI 1.66, 1.72
Occurrence. Occurs in calcium-rich metamorphic rocks, also in some basic and ultrabasic rocks, sometimes in meteorites.
Localities. From Sweden; Switzerland; Italy; USSR; South Africa, Malagasy Republic; Burma; California; Montana; etc.
Treatment. Clean with dilute acid.
Fashioning. Uses: faceting or cabochons; *cleavage*: perfect prismatic; *cutting angles*: crown 40°, pavilion 40°; *pleochroism*: distinct; *heat sensitivity*: very low.

Aegirine (Acmite) $NaFe^{3+}Si_2O_6$
Aegirine crystallizes in the monoclinic system as long prismatic crystals with vertical striations, as tufts or aggregates of minute fibres. Twinning is common. The colour is dark green to black (the aegirine variety); reddish-brown (acmite). Lustre resinous; streak pale yellowish-grey. H 6; SG 3.5–3.6
Occurrence. Occurs in alkaline rocks such as syenites and also in metamorphic rocks.
Localities. From Mont St. Hilaire, Quebec; Greenland; Norway; Nigeria; Kenya; Scotland; etc.

Spodumene $LiAlSi_2O_6$
Spodumene crystallizes in the monoclinic system as prismatic crystals, flattened and vertically striated; cleavable masses also found and cleavage is perfect. It may be colourless, yellow, green, emerald-green (hiddenite), pink (kunzite). The lustre is vitreous and the streak white. May fluoresce under ultraviolet light. H 6.5–7.5; SG 3.0–3.2; RI 1.65, 1.68
Occurrence. Occurs in granite pegmatites with quartz, feldspar and muscovite.
Localities. Minas Gerais, Brazil; hiddenite, coloured by chromium, from Stony Point, North Carolina; kunzite from California; Afghanistan; Malagasy Republic; etc.
Treatment. Clean with dilute acid.
Fashioning. Uses: faceting or cabochons; *cleavage*: perfect prismatic; brittle; *cutting angles*: crown 40°, pavilion 40°; *pleochroism*: strong in deep colours; *heat sensitivity*: fairly high.

Left: The lilac-pink kunzite variety of spodumene is frequently faceted. This crystal is from Pala, San Diego Co., California

Far left: Aegirine crystal from Magnet Cove, Arkansas

Hypersthene (Mg, Fe)SiO_3
Hypersthene crystallizes in the orthorhombic system rarely forming short prismatic crystals; lamellar masses are more commonly found. The colour is brownish-green to black with a pearly or silky lustre. The streak is brownish-grey. Varieties include: bronzite, with 10–30 mols per cent of $FeSiO_3$; eulite, with 70–90 mols of $FeSiO_3$; and ferrohypersthene, with 10–30 mols per cent of $MgSiO_3$. H 5–6; SG 3.4–3.8
Occurrence. Occurs in basic igneous rocks and some metamorphic rocks; also from meteorites.
Localities. From Bodenmais, and the Eifel region, W. Germany; with labradorite on the Island of St. Paul, Labrador, Canada; California, Arizona, Colorado; etc.
Treatment. Clean with dilute acid.

Clinohypersthene (Mg, Fe)SiO_3
Clinohypersthene has 20–50 mols per cent $FeSiO_3$ and crystallizes in the monoclinic system. The habit is massive and the colour green to brown; H 5–6; SG 3.2–3.5
Localities. Occurs mainly in meteorites; from Broken Hill, New South Wales; Bon Accord quarry, Transvaal, South Africa.
Treatment. Clean with dilute acid.

Hedenbergite $CaFeSi_2O_6$
Hedenbergite crystallizes in the monoclinic system and is usually found massive; lamellar; crystals are short prismatic. Twinning is common. The colour is brownish to greyish-green or black with a white or grey streak. There is a vitreous lustre. Ferrohedenbergite has more iron than calcium. H 6; SG 3.5
Occurrence. Occurs in limestone contact zones, iron-rich metamorphic rocks and in granites.
Localities. From California; Arizona; Colorado; Sweden; etc.
Treatment. Clean with dilute acid.

Pigeonite
$(Mg, Fe^{2+}, Ca)(Mg, Fe^{2+})[Si_2O_6]$
Pigeonite crystallizes in the monoclinic system, usually occuring as grains or micro-phenocrysts. Crystals are short prismatic. The colour is brown to light purple or greenish to black. H 6; SG 3.3–3.4
Occurrence. Occurs in volcanic rocks, particularly in andesite and dacite.
Localities. From Pigeon Point, Minnesota, and numerous other localities.
Treatment. Clean with dilute acid.

Jadeite $NaAlSi_2O_6$
Jadeite crystallizes in the monoclinic system, and, with nephrite, is one of the minerals to which the generic name jade

Left: Radiating fibrous crystals of hedenbergite from Tuscany, Italy

Right: Tremolite from Bagat, Ontario

Below right: Bustamite from Broken Hill, New South Wales

is given. It forms fine to coarse granular masses or alluvial boulders; crystals are rare but may be small elongated prismatic. The fracture is splintery and the mineral is very tough. The colour may vary from a translucent emerald-green (Imperial jade) to pale green, mauve, white, brown, yellow or grey. The lustre is vitreous, and the streak is colourless. Chloromelanite is a dark-green variety with black veining; maw-sit-sit is of similar appearance. H 6.5–7; SG 3.3; RI 1.66

Occurrence. Occurs in serpentinites derived from olivine rock in Burma; from cherts in California (San Benito and Contra Costa counties); Tibet; Mexico; New Zealand; perhaps from Yunnan, China; etc.

Treatment. Clean with dilute acid.

Fashioning. Uses: cabochons, carvings, beads, etc, very rarely faceted; *cleavage*: poor prismatic; tough; *heat sensitivity*: low.

Augite
$(Ca, Na)(Mg, Fe, Al)(Si, Al)_2O_6$
Augite crystallizes in the monoclinic system as short prismatic crystals or as compact masses or grains. Twinning is common. The colour is brown, purple or black with a greyish-green streak. The lustre is vitreous. Varieties include titanaugite and fassaite, rich in calcium oxide. H 5.5–6; SG 3.2–3.5

Occurrence. Of widespread occurrence in basic and ultrabasic igneous rocks and in some metamorphic rocks.

Above right: Augite from Ariccia, Italy

Right: Rhodonite. These crystals are too opaque to facet although the material does occur in transparent specimens

Localities. From Trentino, Italy; Czechoslovakia; Germany; South-West Africa; California; Montana; etc.

Treatment. Remove iron stains with oxalic acid.

PYROXENOID GROUP

Certain minerals that are chemically similar to the pyroxenes but have a slightly different atomic structure are classified as 'pyroxenoids'.

Rhodonite $MnSiO_3$
Rhodonite crystallizes in the triclinic system, commonly being found massive; crystals are tabular with a perfect cleavage. The colour is pink to rose red, veined by black alteration products; the lustre is vitreous. H 5.5–6.5; SG 3.5–3.7; RI 1.71, 1.75

Occurrence. Occurs in manganese-bearing ore bodies formed by hydrothermal processes.

Localities. Fine crystals from Franklin, New Jersey; good crystals in galena from Broken Hill, New South Wales; also from USSR; South Africa; India; Japan; etc.

Treatment. Rhodonite should be cleaned with distilled water.

Fashioning. Uses: faceting, cabochons, tumbling, carving, beads, etc; *cleavage*: perfect prismatic, massive types tough; *cutting angles*: crown 40°, pavilion 40°; *pleochroism*: distinct in transparent material; *heat sensitivity*: low, crystalline material very sensitive to shock.

Bustamite $(Ca, Mn)Si_2O_6$
Bustamite crystallizes in the triclinic system, commonly being found massive; crystals are tabular with rounded edges. There is a perfect cleavage. The colour is pale pink to brownish-red; the lustre is vitreous. H 5.5–6.5; SG 3.3–3.4

Occurrence. Occurs in manganese-bearing ore bodies usually formed by metasomatic processes.

Localities. From Franklin, New Jersey.

Treatment. Clean with dilute acetic acid.

AMPHIBOLE GROUP

The amphiboles are a group of hydrous silicates of considerable chemical com-

Left: Fibrous crystals of tremolite from St. Gotthard, Switzerland. The large white crystals are feldspar

Below left: Anthophyllite from Moravia

plexity because of the extensive atomic substitution that takes place. They are characterized by having two cleavages, which intersect at 120°. They are widely distributed in igneous and metamorphic rocks. The commoner amphiboles include anthophyllite, cummingtonite, grunerite, tremolite, actinolite, hornblende, glaucophane, riebeckite and arfvedsonite.

Tremolite $Ca_2Mg_5Si_8O_{22}(OH)_2$
Tremolite crystallizes in the monoclinic system forming long, bladed crystals; more often found as fibrous aggregates, often radiated. Twinning is common. The colour is colourless, white, grey, pale green, pink or brown. The lustre is vitreous. H 5–6; SG 2.9–3.2
Occurrence. Occurs in contact and regionally metamorphosed dolomites or low-grade ultrabasic rocks.
Localities. Italy; Switzerland; Ontario; California; Arizona; and in various other localities.
Treatment. Tremolite should be cleaned with dilute acid.

Actinolite
$Ca_2(Mg, Fe)_5Si_8O_{22}(OH)_2$
Actinolite crystallizes in the monoclinic system, usually forming fibrous or columnar aggregates, often radiated; crystals are long, bladed. The lustre is vitreous. The colour is green and a variety of shades are known; black is also found. H 5–6; SG 3.0–3.4
Occurrence. Occurs in contact and regionally metamorphosed dolomites, magnesian limestones and ultrabasic rocks.
Localities. The variety nephrite (one form of jade) is found in Alaska; British Columbia; New Zealand; Siberia; and California. The variety byssolite is found in Pennsylvania and Virginia, amongst other places.
Fashioning. Uses: faceting or cabochons; *cleavage*: highly perfect, prismatic; very brittle; *cutting angles*: crown 40°, pavilion 40°; *pleochroism*: distinct, increasing with iron content; *heat sensitivity*: fairly high, but unaffected by normal dopping temperatures; *mechanical sensitivity*: very high, treat with care.

Richterite
$(Na, K)_2(Mg, Mn, Ca)_6Si_8O_{22}(OH)_2$
Richterite crystallizes in the monoclinic system forming long prismatic crystals with lamellar twinning and a perfect cleavage. The colour is brown, yellow, dark brownish-red or pale to dark green; there is a vitreous lustre. H 5–6; SG 2.9–3.1
Occurrence. From contact-metasomatic deposits, some alkaline igneous rocks and thermally metamorphosed limestones.
Localities. Leucite Hills, Wyoming; Burma; Malagasy Republic; etc.
Treatment. Clean with distilled water.

Grunerite $(Fe, Mg)_7Si_8O_{22}(OH)_2$
Grunerite crystallizes in the monoclinic system in fibrous lamellar radiating forms, commonly twinned. The colour is grey, dark green or brown. The lustre is silky. H 5–6; SG 3.4–3.6
Occurrence. In contact or regionally metamorphosed iron-rich rocks.
Localities. From Scotland; Arizona; South Africa; etc.
Treatment. Clean with dilute acid.

Anthophyllite
$(Mg, Fe)_7Si_8O_{22}(OH)_2$
Anthophyllite crystallizes in the orthorhombic system, being found massive, fibrous or lamellar; crystals are prismatic. There is a perfect cleavage. The colour is white to green or brown with a grey or colourless streak, and the lustre is vitreous. H 5.5–6; SG 2.8–3.5

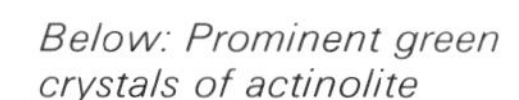

Below: Prominent green crystals of actinolite

Occurrence. Occurs in metamorphic rocks, including schists and gneisses; also metasomatic rocks.
Localities. From California; Arizona; Colorado; Norway; Italy; etc.
Treatment. Iron stains can be removed with oxalic acid.

Cummingtonite
$(Mg, Fe)_7Si_8O_{22}(OH)_2$
Cummingtonite usually contains a ratio of iron to magnesium of 3:2 or more. It crystallizes in the monoclinic system and is fibrous or lamellar with simple or lamellar twinning common. The colour is dark to greyish-green or brown with a silky lustre. H 5–6; SG 3.1–3.4
Occurrence. Occurs in contact or regionally metamorphosed rocks.
Localities. Homestake gold mine, Lead, Lawrence Co., South Dakota; Scotland; Sweden; etc.
Treatment. Clean with dilute acid.

Arfvedsonite
$Na_3(Mg, Fe^{2+})_4Al[Si_8O_{22}](OH, F)_2$
Arfvedsonite crystallizes in the monoclinic system in long prismatic crystals or prismatic aggregates. There is a perfect cleavage and lamellar twinning is common. The colour is greenish-black and the streak dark bluish-grey; the lustre is vitreous. H 5–6; SG 3.3–3.5
Occurrence. Occurs in plutonic alkali igneous rocks, such as nepheline syenite.
Localities. From Julianehaab, Greenland; Norway; Finland; USSR; Boulder Co., Colorado; etc.
Treatment. Clean with dilute acid.

Gedrite
$(Mg, Fe^{2+}, Al)_7(Si, Al)_8O_{22}(OH)_2$
Gedrite crystallizes in the orthorhombic system as fibrous masses; it has a perfect cleavage. The colour is white, greenish, brownish or yellow, with a vitreous or silky lustre. The streak is grey. H 5.5–6; SG 3.1–3.5
Occurrence. Occurs in metamorphic rocks.
Localities. From Greenland; Finland; North Carolina; Japan; Australia; etc.
Treatment. Clean with dilute acid.

Glaucophane
$Na_2(Mg, Fe^{2+})_3Al_2Si_8O_{22}(OH)_2$
Glaucophane crystallizes in the monoclinic system as fibrous masses; individual crystals are prismatic. There is a perfect cleavage and the colour is grey to bluish-black. The streak is bluish-grey, and the lustre is vitreous. H 6; SG 3.0–3.1
Occurrence. Occurs in crystalline schists with minerals such as jadeite, lawsonite and pumpellyite.
Localities. From Zermatt, Switzerland; Coast Range, California; etc.
Treatment. Clean with dilute acid.

Crossite $Na_2(Mg, Fe^{2+})_3(Fe^{3+}, Al)_2Si_8O_{22}(OH)_2$
Crossite crystallizes in the monoclinic system forming fibrous masses; it has a perfect cleavage. The colour is blue-grey. H 6; SG 3.1
Occurrence. Occurs in crystalline schists.
Localities. From the Coast Ranges, California; etc.

Hastingsite $NaCa_2(Fe, Mg, Al)_5(Al_2Si_6)O_{22}(OH)_2$
Hastingsite crystallizes in the monoclinic system as prismatic crystals or compact masses. There is a perfect cleavage. The colour is dark green to black, with a vitreous lustre. H 5–6; SG 3.2–3.6
Occurrence. Of widespread occurrence in igneous and metamorphic rocks.
Localities. From Hastings Co., Ontario; Riverside Co., California; and many other localities.
Treatment. Clean crossite with distilled water.

Riebeckite
$Na_2Fe_3^{2+}Fe_2^{3+}Si_8O_{22}(OH)_2$
Riebeckite crystallizes in the monoclinic system forming long prismatic crystals with striations parallel to the long direction; also massive fibrous (crocidolite), columnar or granular. There is a perfect cleavage. The colour is dark blue to black with a vitreous or silky (crocidolite) lustre. H 5; SG 3.3
Occurrence. Occurs in alkali granites and rhyolites, trachytes, syenites, bedded ironstones and regionally metamorphosed schists.
Localities. From St. Peter's Dome, El Paso Co., Colorado; Cumberland Hill, Rhode Island; Scotland; South Africa; Zambia; Malagasy Republic; and other localities.

Katophorite
$Na_2Ca(Fe^{3+}, Al)_5(AlSi_7)O_{22}(OH)_2$
Katophorite, sometimes known as taramite, crystallizes in the monoclinic system as prismatic crystals with a perfect cleavage. The colour is rose-red to brownish-black, with a vitreous lustre. H 5; SG 3.3–3.5
Occurrence. Occurs in basic alkaline rocks.
Localities. From Oslo, Norway; Ukraine, USSR; and elsewhere.
Treatment. Clean katophorite with dilute acid.

Hornblende $(Ca, Na, K)_{2-3}(Mg, Fe^{2+}, Fe^{3+}, Al)_5(Si, Al)_8O_{22}(OH)_2$
Hornblende crystallizes in the monoclinic system as prismatic crystals or as compact masses. Twinning is common. There is a perfect cleavage. The colour is green, brown to black, with a vitreous lustre. H 5–6; SG 3.0–3.3

Above right: Cummingtonite with biotite mica from Cummington, Massachusetts

Right: Dark crystals of hornblende with oligoclase feldspar from Arendal, Norway

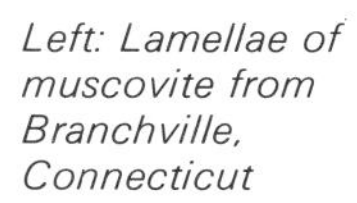

Left: Lamellae of muscovite from Branchville, Connecticut

Occurrence. Occurs in many igneous and metamorphic rocks.
Localities. From many localities, including Mt. Vesuvius, Italy; USSR; Finland; Great Britain; Australia; California; New York State.
Treatment. Hornblende should be cleaned with dilute acid.

Eckermannite $Na_3(Mg, Li)_4(Al, Fe)Si_8O_{22}(OH, F)_2$
Eckermannite crystallizes in the monoclinic system as long prismatic crystals or aggregates with a vitreous lustre. Lamellar twinning is found and there is a perfect cleavage. The colour is dark blue or green. H 5–6; SG 3.0–3.1
Occurrence. Occurs in nepheline syenites as well as in other rocks that are alkali-rich.
Localities. From Tawmaw, Burma; Boulder Co., Colorado; and from other localities.
Treatment. Eckermannite should be cleaned with distilled water.

MICA GROUP

The micas crystallize in the monoclinic system, frequently as pseudohexagonal prisms with a perfect basal cleavage. All the micas contain hydroxyl (OH). Generally those species with a darker colour contain iron and magnesium while lighter species contain aluminium. Lithium, barium, chromium and manganese are sometimes present.
Apart from the minerals described here, the mica group includes anandite, celadonite, ephesite, hendricksite and polylithionite. The vermiculite group consists of altered micas that still show typical micaceous cleavage.

Paragonite $NaAl_3Si_3O_{10}(OH)_2$
Paragonite crystallizes in the monoclinic system in massive compact forms or scales. There is a perfect cleavage and the colour is pale yellow or colourless. The lustre is pearly. H 2.5; SG 2.7–2.9
Occurrence. Occurs in schists, gneisses, quartz veins and sediments.
Localities. From Monte Campione, Switzerland; Piedmont, Italy; Leadville, Colorado.
Treatment. Remove iron stains with oxalic acid.

Muscovite $KAl_3Si_3O_{10}(OH)_2$
Muscovite is one of the commonest of the micas. It crystallizes in the monoclinic system as tabular crystals; more commonly it is lamellar or compact massive. Twinning is common and there is a perfect cleavage. Colours may be white, grey, yellow, green, brown, red or violet, and the lustre is vitreous or pearly. The streak is colourless. Varieties include hydromuscovite with a higher OH and lower K, or K and Al; phengite, which is higher in Si with low H_2O, or with (Fe, Mg) replacing some Al; mariposite, which contains more Si and up to 1 per cent Cr_2O_3: fuchsite, which contains up to 5 per cent Cr_2O_3; and sericite, which is scaly and occurs in fibrous aggregates. H 2.5–4.0; SG 2.7–2.8
Occurrence. Occurs in granites, pegmatites, schists and gneisses or in sediments.
Localities. Widespread occurrence; very large crystals from Custer, South Dakota; also from Sweden; Switzerland; USSR; India; etc.
Treatment. Muscovite should be cleaned with dilute acid.

Illite $K_{1-1.5}Al_4[Si_{7-6.5}Al_{1-1.5}O_{20}](OH)_4$
The name illite has been given to micaceous clays in general and also to a mineral with the composition of muscovite or hydromuscovite, but giving a poor x-ray diffraction pattern.
Treatment. Clean with distilled water.

Margarite $CaAl_4Si_2O_{10}(OH)_2$
Margarite crystallizes in the monoclinic system in platy aggregates with brittle laminae. There is a perfect cleavage. The colour is grey, pink, yellow or green with a colourless streak. The lustre is pearly. H 3.5–4.5; SG 3.0–3.1
Occurrence. Occurs with corundum in metamorphic emery deposits and with staurolite in chlorite and mica schists.
Localities. From Italy; Greece; Austria; California; etc.
Treatment. Clean with dilute acid.

Phlogopite $KMg_3AlSi_3O_{10}(OH)_2$
Phlogopite crystallizes in the monoclinic and hexagonal systems as prismatic crystals, plates or scales. Twinning is common and there is a perfect cleavage. Thin laminae are flexible. The

Left: Margarite from Makares, Greece. The platy nature of the mineral is well shown by these specimens

Above: A fine crystal of phlogopite from a classic US locality, Franklin, New Jersey

Above right: Rose-coloured lepidolite from Moravia

colour is yellow to brown or red to green, or white to colourless. The lustre is pearly. The streak is colourless. Phlogopite may show asterism. Its varieties include: hydrophlogopite; natronphlogopite; bariumphlogopite; manganphlogopite; fluorphlogopite. H 2–2.5; SG 2.7–2.9
Occurrence. Occurs in metamorphic limestones and ultrabasic igneous rocks.
Localities. Sweden; Italy; India; USA; and other localities.
Treatment. Clean with dilute acid.

Clintonite
$Ca(Mg, Al)_3(Al, Si)O_{10}(OH)_2$
Clintonite crystallizes in the monoclinic system as tabular pseudohexagonal crystals; also as foliated or lamellar masses. There is a perfect cleavage. It is colourless, yellowish, greenish, reddish-brown or copper-red, with a vitreous or pearly lustre. The streak may be colourless, yellowish or greenish. H 3.5 and 6, according to direction; SG 3.0–3.1
Occurrence. Occurs with calcite and other minerals in crystalline limestones.
Localities. From Riverside Co., California; the Pargas area of Finland; southern Urals, USSR; etc.
Treatment. Clean with dilute acid.

Roscoelite
$K(V, Al, Mg)(Al, Si_3)O_{10}(OH)_2$
Roscoelite crystallizes in the monoclinic system as minute scales. There is a perfect cleavage and the colour is brown to greenish-brown. There is a pearly lustre. H 2.5; SG 3.0
Localities. Occurs with native gold with which it may be intermingled at Granite Creek, El Dorado Co., California; Kalgoorlie, Western Australia.

Glauconite $(K, Na)(Al, Fe^{3+}, Mg)_2(Al, Si)_4O_{10}(OH)_2$
Glauconite crystallizes in the monoclinic system as aggregates of platelets with a perfect cleavage, and dull lustre. It is yellowish-green in colour. H 2; SG 2.4–2.9
Occurrence. Occurs in greensands, impure limestones and many rocks of marine origin.
Localities. From France; USSR; India; and many other localities.
Treatment. Clean with dilute acid.

Biotite
$K(Mg, Fe)_3(Al, Fe)Si_3O_{10}(OH, F)_2$
Biotite crystallizes in the monoclinic and trigonal systems, forming massive scaly aggregates, or disseminated; it has a perfect cleavage and twinning occurs. The colour is brown, reddish-brown or green, with a metallic to vitreous lustre. The streak is colourless. H 2.5–3 (on cleavage surfaces); SG 2.7–3.4
Occurrence. Occurs in granites, pegmatites, gneisses, schists and other rocks. Varieties include lepidomelane, manganophyllite and siderophyllite.
Localities. From Mt. Vesuvius, Italy; Arendal, Norway; Sweden; USSR; Japan; California; Colorado; and very many other localities.
Treatment. Clean with dilute acid.

Tainiolite $KMg_2LiSi_4O_{10}(OH, F)_2$
Tainiolite crystallizes in the monoclinic system as pseudohexagonal tabular crystals and as scales or aggregates. There is a perfect cleavage. The colour is either brown or colourless. H 2.5–3; SG 2.8–2.9
Occurrence. Occurs in nepheline-syenite pegmatites with natrolite and other minerals.
Localities. From Lovozero Massif, Kola Peninsula, USSR; Greenland; Magnet Cove, Arkansas.
Treatment. Clean with dilute acid.

Lepidolite
$K(Li, Al)_3(Si, Al)_4O_{10}(F, OH)_2$
Lepidolite crystallizes in the monoclinic system as cleavable masses or scaly aggregates; crystals are tabular. The cleavage is a perfect one. The mineral's colour is pink to purple, colourless or white. There is a pearly lustre. The streak is colourless. H 2.5–3.0; SG 2.8–3.3
Occurrence. Occurs in granite pegmatites with spodumene and amblygonite; sometimes from tin veins.
Localities. From Brazil; Mozambique; Malagasy Republic; Pala area, San Diego Co., California; and elsewhere.
Treatment. Clean with dilute acid.
Fashioning. Uses: carvings, cabochons,

Far right: A thin section of biotite shown between crossed polarizers

Right: Biotite from Langesundfjord, Norway

faceting (rare); *cleavage*: perfect basal; *cutting angles*: crown 40°–50°, pavilion 43°; *pleochroism*: weak to distinct; *heat sensitivity*: low.

Zinnwaldite
$K(Li, Al, Fe)_3(Al, Si)_4O_{10}(OH, F)_2$
Zinnwaldite crystallizes in the monoclinic system as prismatic crystals or scales with a vitreous or pearly lustre. There is a perfect cleavage. The colour is grey, green or brown. H 2.5–4; SG 2.9–3.3
Occurrence. Occurs principally in granite pegmatites and high-temperature quartz veins.
Localities. From Cornwall; Saxony, E. Germany; San Diego Co., California and other localities.
Treatment. Clean with dilute acid.

CHLORITE GROUP
Minerals of the chlorite group are frequently green and crystallize in the monoclinic system. There is a basal cleavage similar to that in the micas. Chlorites are aluminium silicates with ferrous iron, magnesium and water. Apart from the minerals described here the chlorites include prochlorite, moravite, cronstedtite, and stilpnomelane.

Pennantite
$(Mn, Al)_6(Si, Al)_4O_{10}(OH)_8$
Pennantite crystallizes in the monoclinic system as masses or tiny flakes; there is a perfect cleavage; it is orange in colour. H 2–3; SG 3.0
Localties. Occurs with spessartine, banalsite, etc, at the manganese mine at Benallt, Gwynedd, Wales.

Daphnite
$(Mg, Fe)_3(Fe, Al)_3(Si, Al)_4O_{10}(OH)_8$
Daphnite crystallizes in the monoclinic system as spherical or botryoidal aggregates. There is a perfect cleavage. The colour is dark green, and there is a pearly lustre. H 2–3; SG 3.2
Localities. With quartz and arsenopyrite at Penzance, Cornwall; etc.
Treatment. Clean with dilute water.

Amesite
$(Mg, Fe^{2+})_4Al_4Si_2O_{10}(OH)_8$
Amesite crystallizes in the hexagonal system forming hexagonal plates in foliated aggregates. There is a perfect cleavage and the colour is pale green, with a pearly lustre. H 2.5–3; SG 2.8
Localities. Occurs with corundophilite and magnetite at Chester, Massachusetts; and elsewhere.
Treatment. Clean with distilled water.

Corundophilite
$(Mg, Fe, Al)_6(Si, Al)_4O_{10}(OH)_8$
Corundophilite crystallizes in the triclinic system as granular or foliated masses; it has a perfect cleavage. The colour is dark green, and the lustre is pearly. H 2–3; SG 2.8
Localities. Occurs with magnetite at Chester, Massachusetts; etc.
Treatment. Clean with distilled water.

Ripidolite
$(Mg, Fe, Al)_6(Si, Al)_4O_{10}(OH)_8$
Ripidolite crystallizes in the monoclinic system as tabular crystals; more commonly as granular or scaly masses. There is a perfect cleavage. The colour is dark green with a colourless or pale green streak. The lustre is pearly. H 2–3; SG 2.9–3.1
Occurrence. Occurs in schists, sometimes in ore veins.
Localities. From the tin veins of Cornwall; Malagasy Republic; New Zealand; and elsewhere.
Treatment. Clean ripidolite with distilled water.

Delessite **$(Mg, Fe^{2+}, Fe^{3+}, Al)_6(Si, Al)_4O_{10}(O, OH)_8$**
Delessite crystallizes in the monoclinic system as foliated or granular masses; it has a perfect cleavage. The colour is black tinged with green. It has a streak that is olive-green in colour. H 2–3; SG 2.7
Localities. Occurs at the diabase quarry at Somerville, Massachusetts.
Treatment. Clean delessite with distilled water.

Left: Amesite crystals from Saranovskoje, USSR

Below left: Crystals of zinnwaldite from the original locality, Zinnwald, Czechoslovakia

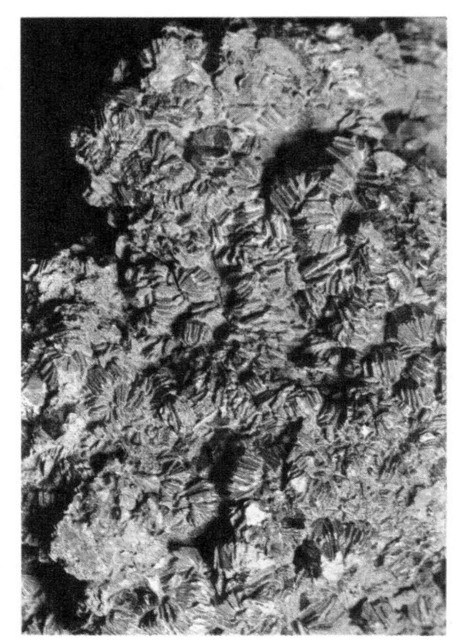

Bottom left: Ripidolite from Rauris, Austria

Right: Clinochlore from the Tilly Foster mine, New York

Chamosite
$(Mg, Fe^{2+})_3Fe_3^{3+}(AlSi_3)O_{10}(OH)_8$
Chamosite crystallizes in the monoclinic and hexagonal systems forming compact masses. The colour is greenish to black. H about 3; SG 3.0–3.4
Occurrence. Found in sedimentary ironstones and in some clay deposits.
Localities. From Chamoson, Valais, Switzerland; Czechoslovakia; etc.
Treatment. Clean with distilled water.

Right: Fibrous chrysotile in an altered serpentine from Silesia, Poland

Below: Penninite in tabular form from Zermatt, Switzerland

Thuringite
$(Fe^{2+}, Fe^{3+}, Mg)_6(Al, Si)_4O_{10}(OH)_8$
Thuringite crystallizes in the monoclinic system as scaly aggregates; it has a perfect cleavage. The colour is olive green to brown, and the lustre is pearly. H 2–3; SG 3.0–3.3
Occurrence. Occurs with iron ores in metamorphic rocks.
Localities. From Thüringer Wald, E. Germany; South Africa; Lake Superior, Michigan; etc.
Treatment. Clean with distilled water.

Clinochlore
$(Mg, Fe^{2+}, Al)_6(Si, Al)_4O_{10}(OH)_8$
Clinochlore crystallizes in the monoclinic system as foliated masses or as tabular crystals with a hexagonal cross-section. There is a perfect cleavage. The colour is white, yellow, colourless, deep green or olive green with a colourless to greenish-white streak. There is a pearly lustre. Varieties include leuchtenbergite, chromeclinochlore and kochubeite. H 2–2.5; SG 2.6–3.0
Occurrence. Occurs in schists and other metamorphic rocks or as an alteration of hydrothermal minerals in igneous rocks.
Localities. From Italy; Austria; Switzerland; California; Colorado; etc.
Treatment. Clean with distilled water.

Brunsvigite
$(Fe^{2+}, Mg, Al)_6(Si, Al)_4O_{10}(OH)_8$
Brunsvigite crystallizes in the monoclinic system as foliated masses or radiating spherical aggregates; crystals are hexagonal tabular and very small. There is a perfect cleavage. The colour is yellow or dark green with a pearly lustre. H about 2; SG 3.0–3.1
Localities. Occurs in veins in granite in the Mourne Mountains, N. Ireland; and elsewhere.
Treatment. Clean with distilled water.

Penninite
$(Mg, Fe, Al)_6(Si, Al)_4O_{10}(OH)_8$
Penninite crystallizes in the monoclinic system as thick tabular crystals, or compact masses. There is a perfect cleavage. The colour is emerald or olive green, white or yellow. There is a vitreous or pearly lustre. H 2–2.5; SG 2.6–2.8
Occurrence. Occurs with serpentine or in schists.
Localities. From the Ala Valley, Piedmont, Italy; Zermatt, Switzerland; etc.
Treatment. Clean with distilled water.

Sudoite
$(Al, Fe, Mg)_{4-5}(Si, Al)_4O_{10}(OH)_8$
Sudoite crystallizes in the monoclinic system as fine-grained masses; it has a perfect cleavage and is white in colour. H not measured; SG 2.6
Localities. Occurs in hematite ore at Negaunee, Michigan; etc.
Treatment. Clean with distilled water.

Diabantite
$(Mg, Fe^{2+}, Al)_6(Si, Al)_4O_{10}(OH)_8$
Diabantite crystallizes in the monoclinic system as compact fibrous masses coloured dark green and with a perfect cleavage. H 2–2.5; SG 2.8–2.9
Localities. Occurs in dolerite at Farmington Hills, Connecticut; Germany; etc.
Treatment. Clean with distilled water.

Nimite
$(Ni, Mg, Fe, Al)_6(AlSi_3)O_{10}(OH)_8$
Nimite crystallizes in the monoclinic system as very small irregular veins in talc. The colour is yellowish-green. H 3; SG 3.2
Localities. From the area of the Scotia talc mine, Barberton Mountain, Transvaal, South Africa.
Treatment. Clean with distilled water.

SERPENTINE GROUP
Closely related to the chlorites are the serpentines, which are fibrous or lamellar. They are usually monoclinic. Other members of the group apart from those described below are marmolite, picrolite, garnierite, iddingsite, etc.

Chrysotile $Mg_3Si_2O_5(OH)_4$
Chrysotile crystallizes in the monoclinic system, forming fibrous masses. The fibres are flexible and easily separated. The colour is white, grey, green to brown with a silky lustre. H 2.5; SG 2.5
Occurrence. Occurs as compact veins in serpentinite rock.
Localities. From Italy; USSR; Rhodesia; in the United States it is found in the Coast Ranges and Sierra Nevada, California; etc.

Left: Crystals of antigorite from Sverdlovsk, USSR

Antigorite $Mg_3Si_2O_5(OH)_4$

Antigorite crystallizes in the monoclinic system and is usually massive and very fine-grained; lamellar and foliated forms are also common; crystals are flaky and very small. There is a perfect cleavage. The colour is white, yellow, green, bluish-green and brownish-red with a white streak and resinous lustre. The term bowenite is used somewhat indiscriminately to describe a translucent greenish serpentine. It is sometimes used as a jade simulant. H 2.5–3.5; SG 2.6

Occurrence. Found mixed with chrysotile in serpentinites.

Localities. Val Antigorio, Italy (this material is not mixed with chrysotile); Switzerland; California; Utah and other states of the USA; etc. Bowenite is found at Smithfield, Rhode Island, the South Island of New Zealand and elsewhere.

Treatment. Clean with dilute acid.

Talc $Mg_3Si_4O_{10}(OH)_2$

Sometimes known as steatite, talc crystallizes in the monoclinic system, usually massive and fine-grained or as foliated or fibrous masses; crystals are thin tabular. There is a perfect cleavage. The colour is pale to dark green, white, grey or brown with a white streak and greasy lustre. H 1; SG 2.5–2.8

Occurrence. Talc is a secondary mineral formed by the hydrothermal alteration of ultrabasic rocks or the thermal metamorphism of siliceous dolomites.

Localities. Found near the Lizard, Cornwall; Shetland Islands; the Zillertal, Austrian Tyrol; Barberton, Transvaal, South Africa; California, Vermont, Pennsylvania and other states; etc.

Treatment. Clean with dilute acid.

Sepiolite $Mg_3Si_4O_{11}.5H_2O$

Sometimes known as meerschaum, sepiolite crystallizes in the orthorhombic system forming fine fibrous masses or compact nodules. The colour may be white or grey, yellowish or blue-green to red. Dry porous masses float on water. H 2–2.5; SG about 2

Occurrence. Sepiolite occurs as an alteration product of serpentine or magnesite.

Localities. Fine nodular masses from Eskişehir, Turkey; also from Spain, Greece, Morocco; California and Arizona; and elsewhere.

Treatment. Clean with distilled water.

KAOLINITE GROUP

These clay minerals, also known as kandites, form scales or plates with angles of 60° and 120°. They usually occur as earthy masses. Apart from kaolinite and dickite, described here, the group includes nacrite and halloysite.

Dickite $Al_2Si_2O_5(OH)_4$

Dickite crystallizes in the monoclinic system, forming thin tabular crystals, sometimes pseudohexagonal, or platelets; clay-like massive forms are the commonest, plastic when moist. There is a perfect cleavage. Usually colourless, may be tinted yellow or brown; it has a dull lustre. H 2–2.5; SG 2.6

Above left: Foliated talc crystals from the USA

Left: Grey sepiolite from Anatolia, Turkey

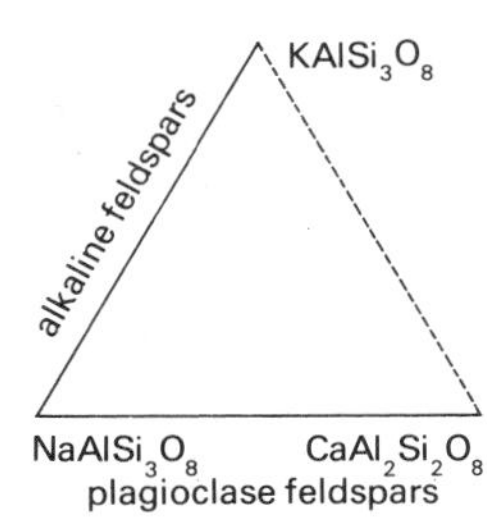

Above: The feldspars

Above right: Microcline just falls within the triclinic crystal system. This crystal is from a classic American locality, Pikes Peak, Colorado

Opposite page: These crystals of orthoclase show Manebach twinning and are accompanied by quartz. From Baveno, Italy

Below: Twinned albite crystals from Schmirn, Austria

Occurrence. Occurs hydrothermally with quartz and sulphides in ore deposits.
Localities. From Anglesey; Boulder Co., Colorado; Shokozan, Japan; etc.
Treatment. Clean with distilled water.

Kaolinite $Al_2Si_2O_5(OH)_4$
Kaolinite crystallizes in the triclinic system as thin hexagonal plates or compact masses. There is a perfect cleavage. The colour is white, colourless, with tints of yellow, blue, red or brown; the lustre is dull. H 2–2.5; SG 2.6
Occurrence. Occurs as a clay mineral formed by the weathering or hydrothermal alteration of feldspars and aluminous silicates.
Localities. From St. Austell, Cornwall; California; Utah; etc.
Treatment. Clean with distilled water.

MONTMORILLONITE GROUP

Also known as smectites, these clay minerals have the general formula $(\frac{1}{2}Ca, Na)_{0.7}(Al, Mg, Fe)_4(Si, Al)_8O_{20}(OH)_4.nH_2O$.

Montmorillonite $(Na, Ca)_{0.33}(Al, Mg)_2Si_4O_{10}(OH)_2.nH_2O$
Montmorillonite crystallizes in the monoclinic system in clay-like massive forms; crystals have a perfect cleavage. The colours are white, grey, yellowish, greenish and pink and the lustre is dull. H 1–2; SG 2–3
Occurrence. Occurs as the principal constituent of bentonite clay deposits; also found in soils, sedimentary and metamorphic rocks.
Localities. France; Italy; USA; and other localities.
Treatment. Clean with distilled water.

Sauconite $Na_{0.33}Zn_3(Si, Al)_4O_{10}(OH)_2.4H_2O$
Sauconite crystallizes in the monoclinic system as very fine-grained masses or micaceous plates forming laminated masses. The colour is reddish-brown or brownish-yellow, with a dull lustre. H 1–2; SG not measured.
Localities. Occurs in the Leadville district, Lake Co., Colorado; Saucon Valley, Pennsylvania; etc.
Treatment. Clean with distilled water.

Hectorite $Na_{0.33}(Mg, Li)_3Si_4O_{10}(F, OH)_2$
Hectorite crystallizes in the monoclinic system as fine-grained masses, white in colour, with a dull lustre. There is a perfect cleavage. H 1–2; SG 2–3
Localities. In bentonite at Hector, San Bernardino Co., California.
Treatment. Clean with distilled water.

Saponite $(Mg, Al, Fe)_3(Al, Si)_4O_{10}(OH)_2$
Saponite crystallizes in the monoclinic system as fine-grained masses with a perfect cleavage. The colour is white or yellowish, grey, greenish, bluish or reddish, with a greasy lustre. H 1–2; SG 2.2–2.3
Occurrence. Occurs in cavities in basic volcanic rocks.
Localities. From Lake Superior; Canada; Transvaal; the variety cathkinite occurs at the Cathkin Hills, Glasgow.
Treatment. Clean with distilled water.

VERMICULITE GROUP

This group, sometimes regarded as belonging to the smectites, has the general formula $(Mg, Fe, Al)_3(Al, Si)_4O_{10}(OH)_2.4H_2O$; varieties include copper vermiculite, jefferisite, vaalite, hydrovermiculite and nickel vermiculite.

FELDSPAR GROUP

These are the most abundant minerals in the Earth's crust and are to be found in a wide variety of rocks. Their chemical composition can be expressed by the general formula $X(Al, Si)_4O_8$ in which X may be Ca, Na, K, or Ba. The barium feldspars are rare. The other feldspars can be considered as belonging to a three-component system; the three components are $KAlSi_3O_8$ (orthoclase, sanidine, or microcline), $NaAlSi_3O_8$ (albite) and $CaAl_2Si_2O_8$ (anorthite). There is a complete series of feldspars from pure potassium to pure sodium feldspar, the 'alkali feldspar series'. Similarly, there is every gradation between pure sodium feldspar and pure calcium feldspar, the 'plagioclase series'. The two series are shown diagramatically in the triangular diagram, which also indicates that there is no potassium-calcium series. The plagioclase series is arbitrarily subdivided according to the proportions of albite component present as follows:

- 'albite': 100–90 per cent albite;
- oligoclase: 90–70 per cent;
- andesine: 70–50 per cent;
- labradorite: 50–30 per cent;
- bytownite: 30–10 per cent;
- anorthite: 10–0 per cent.

Albite $NaAlSi_3O_8$
Albite crystallizes in the triclinic system as tabular or platy crystals. Usually massive, sometimes with curved laminae, twinning is common. There is a perfect cleavage. The colour is white to colourless, bluish, brown or reddish. The streak is white, and the lustre vitreous. H 6–6.5; SG 2.6
Occurrence. Common constituent of pegmatites, granites, andesite, syenite and other alkaline igneous rocks.
Localities. Of very common occurrence, albite is found as fine crystals in the Alps, for example, the St. Gotthard region; Bourg d'Oisans, Isère, France; Morro Velho gold mine, Minas Gerais, Brazil; Arizona; New Mexico; the variety of albite called cleavelandite is found at Amelia, Virginia; peristerite and pericline are other varieties.
Treatment. Clean with dilute acid.
Fashioning. Uses: faceting or cabochons; *cleavage*: perfect // to basal pinacoid, distinct brachypinacoidal; brittle; *cutting angles*: crown 40°–50°, pavilion 43°; *pleochroism*: weak; *heat sensitivity*: fairly high, but unaffected by normal dopping temperatures.

Microcline $KAlSi_3O_8$
Microcline is dimorphous with orthoclase. It crystallizes in the triclinic system in short prismatic crystals; there is a perfect cleavage. It may occur in tabular crystals or massive forms, which are the most common. Carlsbad, Manebach, Baveno and other forms of twinning are very common. The colour is white, grey, yellowish, green. There is a vitreous or pearly lustre. Amazonite and perthite are varieties. H 6–6.5; SG 2.5–2.6
Occurrence. Occurs in plutonic rocks such as granites and in pegmatites. From crystalline schists and sometimes from hydrothermal veins.
Localities. Rhodesia; Brazil; Mexico; Canada; Amelia, Virginia; Black Hills, South Dakota; etc.
Treatment. Clean with dilute acid.
Fashioning. Uses: cabochons, beads, carvings, tumbling; *cleavage*: perfect // to basal pinacoid, distinct brachypinacoidal; brittle; *heat sensitivity*: fairly high, but unaffected by normal dopping temperatures.

Orthoclase $KAlSi_3O_8$
Orthoclase feldspar is dimorphous with microcline and crystallizes in the monoclinic system. It forms short prismatic crystals, but occurs more commonly as masses, which may be lamellar. Carlsbad, Baveno and Manebach twins are

Right: Labradorescence is the name given to the play of colour shown by the feldspar labradorite. This crystal is from Labrador

Below left: Baveno twin of orthoclase from Baveno, Italy

Below right: Prismatic crystals of anorthite

very common and there is a perfect cleavage. The colour is white, colourless, grey or yellowish. The streak is white and the lustre vitreous to pearly. H 6–6.5; SG 2.5–2.6
Occurrence. Occurs in igneous rocks such as pegmatites and granites, or in crystalline schists, ore veins and in sedimentary deposits.
Localities. From the Eifel region of W. Germany (the variety sanidine); Switzerland (adularia or moonstone); yellow transparent material from the pegmatites at Itrongay, Malagasy Republic; California, Oregon, etc; Baveno, Piedmont, Italy; Karlovy Vary, Czechoslovakia; St. Agnes, Cornwall. Moonstones from Sri Lanka, Burma and some other places.
Treatment. Clean with dilute acid.
Fashioning. Uses: cabochons, faceting, beads; *cleavage*: perfect // to basal pinacoid, distinct clinopinacoidal; brittle; *cutting angles*: crown 40°–50°, pavilion 43°; *pleochroism*: weak; *heat sensitivity*: fairly high, but unaffected by normal dopping temperatures.

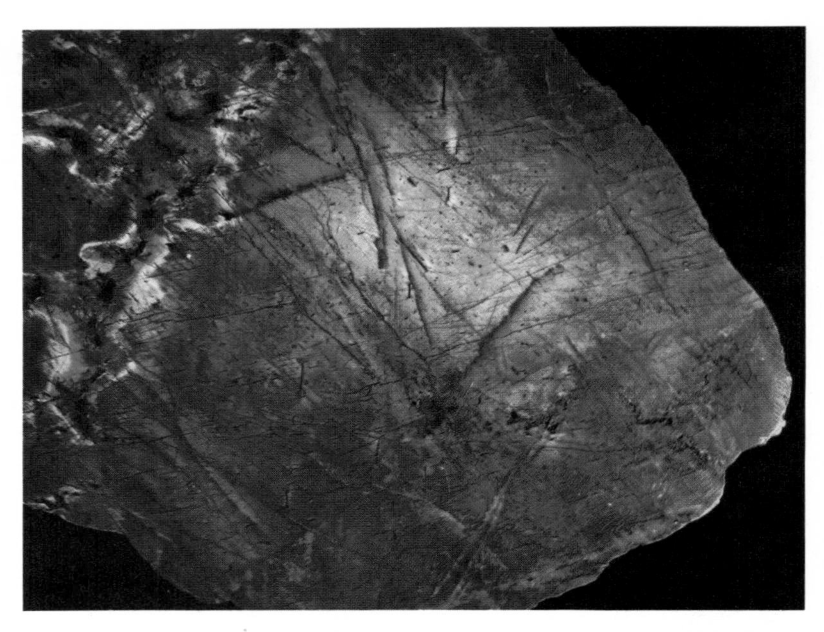

Anorthoclase (Na, K)$AlSi_3O_8$
One of the alkali feldspars, anorthoclase crystallizes in the triclinic system forming short prismatic or tabular crystals, commonly twinned. Massive forms are more common and these may be lamellar. There is a perfect cleavage. The colour is white, colourless to yellowish, brown or red. The lustre is vitreous, and the streak is white. H 6–6.5; SG 2.5–2.6
Occurrence. Mainly in volcanic rocks.
Localities. From Pantelleria, Sicily; Kilimanjaro, E. Africa; Franklin, New Jersey; and many other localities.
Treatment. Clean with dilute acid.

Anorthite $CaAl_2Si_2O_8$
A member of the plagioclase feldspar group, anorthite crystallizes in the triclinic system, forming short prismatic crystals or lamellar, cleavable masses, frequently twinned according to the Manebach, Carlsbad or Baveno laws; albite and pericline twinning are most widespread. The colour is white to grey with a white streak and vitreous lustre. H 6–6.5; SG 2.7
Occurrence. Occurs as a rock-forming mineral in basic volcanic and plutonic rocks and in metamorphic rocks; also from meteorites.
Localities. Widespread occurrence, including Franklin, New Jersey; Mt. Vesuvius, Italy; etc.
Treatment. Clean with distilled water.

Bytownite (Na, Ca)$Al_{1-2}Si_{3-2}O_8$
Bytownite crystallizes in the triclinic system forming tabular crystals or cleavable compact masses. It is colourless, white or grey, with a vitreous lustre. Carlsbad, albite and pericline twinning is common. H 6–6.5; SG 2.7
Occurrence. Occurs as a rock-forming mineral in basic plutonic and volcanic rocks; also from meteorites.
Localities. From Scotland; Transvaal; Montana; etc.
Treatment. Clean with dilute acid.
Fashioning. Uses: faceting or cabochons; *cleavage*: perfect // to basal pinacoid, distinct brachypinacoidal; brittle; *cutting angles*: crown 40°–50°, pavilion 43°; *pleochroism*: weak; *heat sensitivity*: fairly high, but unaffected by normal dopping temperatures.

Labradorite (Na, Ca)$Al_{1-2}Si_{3-2}O_8$
Labradorite crystallizes in the triclinic system as tabular crystals; more commonly found massive, granular with a perfect cleavage and following Carlsbad, albite and pericline twinning laws. The colour is white or grey, often displaying a play of spectrum colour due to interference of light from lamellar twinning. H 6–6.5; SG 2.6–2.7
Occurrence. Occurs as a constituent of basic igneous and metamorphic rocks.
Localities. From eastern Labrador, Canada; Greenland; Sweden; California; South Dakota; etc.
Treatment. Clean with dilute acid.
Fashioning. Uses: faceting, cabochons, carvings, intaglios, tumbling, etc; *cleavage*: perfect // to basal pinacoid, distinct brachypinacoidal; brittle; *cutting angles*: crown 40°–50°, pavilion 43°; *pleochroism*: weak; *heat sensitivity*: fairly high, but unaffected by normal dopping.

Andesine (Na, Ca)$Al_{1-2}Si_{3-2}O_8$
Andesine crystallizes in the triclinic system as tabular crystals or cleavable masses following Carlsbad, albite or pericline twinning laws. The colour is white or grey, with a vitreous lustre. H 6–6.5; SG 2.6
Occurrence. Occurs as a rock-forming mineral in intermediate igneous rocks or in metamorphic rocks.
Localities. Greenland; Norway; etc.
Treatment. Clean with dilute acid.

Left: A rough specimen of sunstone, a variety of oligoclase

Oligoclase $(Na, Ca)Al_{1-2}Si_{3-2}O_8$
Oligoclase crystallizes in the triclinic system as cleavable masses obeying Carlsbad, albite or pericline twinning laws. Crystals are tabular. The colour is grey, greenish, yellow, brown or reddish, with a vitreous lustre. The variety sunstone shows brilliant reflections from inclusions. The streak is white. H 6–6.5; SG 2.6
Occurrence. Occurs in pegmatites, granites and syenites.
Localities. From Canada and Norway (the sunstone variety); California; New York State; Sweden; USSR; etc.
Treatment. Clean with dilute acid.
Fashioning. Uses: faceting, cabochons, carvings, intaglios, etc; *cleavage*: perfect // to basal pinacoid, distinct brachypinacoidal; brittle; *cutting angles*: crown 40°–50°, pavilion 43°; *pleochroism*: weak; *heat sensitivity*: fairly high, but unaffected by normal dopping.

Celsian $BaAl_2Si_2O_8$
Celsian crystallizes in the monoclinic system forming masses or short prismatic crystals with a perfect cleavage. Carlsbad, Manebach and Baveno twinning is common. The colour is colourless, white or yellow, and the lustre is vitreous. H 6–6.5; SG 3.1–3.4
Occurrence. Occurs in the contact zones of manganese deposits.
Localities. From Rhiw, Gwynedd, Wales; Santa Cruz Co., California; Italy; etc.
Treatment. Clean with distilled water.

Paracelsian $BaAl_2Si_2O_8$
Paracelsian crystallizes in the monoclinic system forming large well-formed crystals with multiple twinning. It is colourless and has a vitreous lustre. H 6; SG 3.3
Localities. Occurs in a band in shale and sandstone in a manganese mine at Rhiw, Gwynedd, Wales; etc.
Treatment. Clean with distilled water.

Hyalophane
$(K, Ba)(Al, Si)_2Si_2O_8$
Hyalophane crystallizes in the monoclinic system as prismatic crystals often displaying Carlsbad, Manebach or Baveno twinning. There is a perfect cleavage. The colour is white, colourless or yellow; the lustre is vitreous. H 6–6.5; SG 2.6–2.8
Localities. Occurs in manganese deposits at Otjosondu, South-West Africa; in gneiss at Broken Hill, New South Wales; etc.
Treatment. Clean with distilled water.

FELDSPATHOID GROUP
These are a group of sodium and potassium aluminosilicates that have similar chemical compositions to the alkali feldspars but contain less silica. They occur in alkali-rich igneous rocks that are deficient in silica, and they never occur with quartz. They include leucite, nepheline, cancrinite, sodalite, haüyne, nosean and lazurite.

Sodalite $Na_4Al_3Si_3O_{12}Cl$
Sodalite crystallizes in the cubic system as dodecahedral crystals or as masses. The colour is light to dark blue or colour-

Left: Sodalite from Dungannon, Ontario

Right: Prismatic striated crystal of davyne from Monte Somma, Italy

less; there is a vitreous lustre and a colourless streak. H 5.5–6; SG 2.1–2.4
Occurrence. Occurs in nepheline syenites, other silica-undersaturated igneous rocks and metasomatized calcareous rock.
Localities. From Bancroft, Ontario; Brazil; USSR; Burma; Italy; Montana; New Jersey; and many other localities.
Treatment. Clean with distilled water.
Fashioning. Uses: faceting, cabochons, carvings, beads, tumbling, etc; *cleavage*: distinct, dodecahedral; brittle; *cutting angles*: crown 40°–50°, pavilion 43°; *heat sensitivity*: low.

Lazurite
$(Na, Ca)_8(Al, Si)_{12}O_{24}(S, SO_4)$
Lazurite crystallizes in the cubic system as dodecahedra or as compact masses of a fine dark blue or greenish-blue. There is a dull lustre. The streak is bright blue. H 5–5.5; SG 2.3–2.4
Occurrence. Occurs in limestones as a mineral of contact metamorphism with calcite and pyrite. The mixture lapis lazuli contains lazurite.
Localities. From the Kokcha River valley, Badakshan, Afghanistan; the Andes of Chile; Mt. Vesuvius, Italy; San Bernardino Co., California; etc.

Treatment. Clean with distilled water.
Fashioning. Uses: cabochons, carvings, beads, tumbling, etc; *cleavage*: imperfect, dodecahedral; *heat sensitivity*: low.

Nosean $Na_8Al_6Si_6O_{24}SO_4$
Nosean crystallizes in the cubic system as dodecahedra; more commonly found as colourless, white, blue or grey granular masses. The lustre is vitreous. H 5.5; SG 2.3–2.4
Occurrence. Only in alkali-rich lavas.
Localities. The Laacher See, Germany; Cripple Creek, Colorado; Wolf Rock, Cornwall; etc.
Treatment. Clean with distilled water.

Haüyne
$(Na, Ca)_{4-8}Al_6Si_6O_{24}(SO_4)_{1-2}$
Haüyne crystallizes in the cubic system as dodecahedra or octahedra. They have a vitreous lustre, and the colour is nearly always blue, sometimes white. H 5.5–6; SG 2.4–2.5
Occurrence. In undersaturated lavas.
Localities. From the Laacher See, Germany; Cripple Creek, Colorado; etc.
Treatment. Clean with distilled water.
Fashioning. Uses: cabochons, beads, etc.; *cleavage*: distinct, dodecahedral; *heat sensitivity*: low.

Cancrinite
$Na_6(Al_6Si_6O_{24})CaCO_3.2H_2O$
Cancrinite crystallizes in the hexagonal system as masses or as prismatic crystals with a perfect cleavage. The colour is yellow, orange, pink, white or blue; the lustre is vitreous or pearly. H 5–6; SG 2.4–2.5
Occurrence. Occurs in alkaline igneous rocks and as an alteration product of nepheline.
Localities. From the Fen complex, southern Norway; Finland; India; USSR; Iron Hill, Gunnison Co., Colorado; etc.
Treatment. Clean with distilled water.

Davyne **$(Na, Ca, K)_8Al_6Si_6O_{24}(Cl, SO_4, CO_3)_{2-3}$**
Davyne crystallizes in the hexagonal system as small prismatic crystals with vertical striations and a perfect cleavage. It is white or colourless, and has a vitreous lustre. H 6; SG 2.4–2.5
Localities. Occurs in leucite-rich lava at Mt. Somma, Italy.
Treatment. Clean with distilled water.

Nepheline $NaAlSiO_4$
Usually containing some potassium, nepheline crystallizes in the hexagonal system, forming prismatic crystals or compact masses or grains. Twinning is common. The colour is white, colourless, grey, dark green or brownish-red with a white streak. There is a vitreous or greasy lustre. H 5.5–6; SG 2.5–2.6
Occurrence. Occurs in alkaline plutonic rocks or in pegmatites in association with nepheline-syenites.
Localities. From Langesundfjord, Norway; Julianehaab, Greenland; Zaire; Transvaal, South Africa; etc.
Treatment. Clean with distilled water.

Leucite $KAlSi_2O_6$
Leucite belongs to the tetragonal system and forms pseudocubic crystals, often trapezohedral; the faces may be striated due to twinning. It also occurs as grains and in massive forms. It is white or grey with a colourless streak and a vitreous lustre. H 5.5–6; SG 2.4–2.5
Occurrence. In potassium-rich lavas.
Localities. From Italy (Vesuvius); the Eifel district of W. Germany; Erzgebirge, E. Germany; Leucite Hills, Wyoming; etc.
Treatment. Clean with distilled water.

Right: Lazurite, one of the minerals that together form lapis lazuli. It is shown here with pyrite from Afghanistan

Opposite page: Radiating fibrous crystals of natrolite from Bolzano, Italy

SCAPOLITE GROUP

The theoretical end-members of the scapolite group are marialite,

$Na_4Al_3Si_9O_{24}Cl$,

and meionite,

$Ca_4(Al_6Si_6O_{24})(SO_4, CO_3, Cl_2)$.

These pure end-members do not occur naturally, however. A formula for the whole group could be given as

$(Na, Ca, K)_4Al_3(Al, Si)_3Si_6O_{24}$-$(Cl, F, OH, CO_3, SO_4)$.

Specimens of gemstone quality are pink, white, colourless, violet or yellow; cat's eyes are common. Found in Burma, the Malagasy Republic, Brazil, etc; H 6; SG 2.6–2.7; RI 1.54–1.55, 1.55–1.58

Fashioning. Uses: faceting or cabochons; *cleavage*: distinct prismatic (1st and 2nd order); brittle; *cutting angles*: crown 40°–50°, pavilion 43°; *dichroism*: strong in deep colours; *heat sensitivity*: low.

Marialite $Na_4Al_3Si_9O_{24}Cl$

Marialite crystallizes in the tetragonal system as prismatic crystals or masses. The colour is white, colourless, grey, blue, green, yellow, pink or brown, with a vitreous lustre. The streak is colourless. H 5.5–6; SG 2.5–2.6

Occurrence. Occurs in regionally metamorphosed or altered igneous rocks.

Localities. From Ontario and Quebec; USSR; and other localities.

Meionite $Ca_4(Al_6Si_6O_{24})(SO_4, CO_3, Cl_2)$

Meionite crystallizes in the tetragonal system as prismatic crystals or granular masses. The colour is white, colourless, grey, brown, yellow, pink or violet. The lustre is vitreous and the streak is colourless. H 5.5–6; SG 2.7

Occurrence. Occurs in regionally metamorphosed rocks and contact metamorphic aureoles.

Localities. From Italy, Finland and elsewhere.

ZEOLITE GROUP

Zeolites are silicates of aluminium that contain water that is driven off on heating. The water is removed continuously rather than in definite amounts at certain temperatures. When exposed to water vapour the material once more takes up water. Apart from the minerals described below, the group includes: epistilbite, laumontite, offretite, paulingite, pollucite, scolecite, viséite and wellsite.

Natrolite $Na_2Al_2Si_3O_{10}.2H_2O$

Natrolite crystallizes in the orthorhombic system, forming slender prismatic crystals vertically striated; commonly fibrous and radiating or compact massive. There is a perfect cleavage. It is colourless, white, yellowish or reddish, with a vitreous or greasy lustre. H 5–5.5; SG 2.2

Occurrence. Occurs in cavities in basalts or as an alteration product of nepheline or sodalite in nepheline syenites; also as an alteration product of plagioclase in aplites and dolerites.

Localities. From Mont St. Hilaire, Quebec; Greenland; Bishopton, Strathclyde, Scotland; near Belfast, N. Ireland; San Benito Co., California; etc.

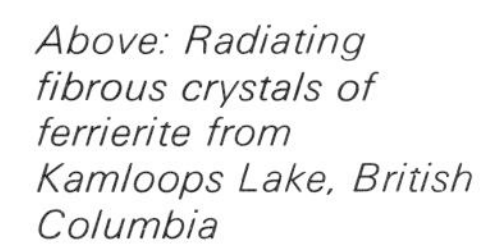

Above: Radiating fibrous crystals of ferrierite from Kamloops Lake, British Columbia

Above right: Radiating crystals of stilbite from Rio Grande do Sul, Brazil

Analcime $NaAlSi_2O_6 . H_2O$
Analcime crystallizes in the cubic system as well formed trapezohedra or modified cubes; massive granular material also found. Lamellar twinning is common. The colour is white, yellowish, pink or greenish, and the lustre is vitreous. H 5–5.5; SG 2.2
Occurrence. Occurs in basalts and other igneous rocks; as an alteration product of nepheline; and in some sandstones.
Localities. From the Isle of Skye; Kilpatrick Hills, Strathclyde; Cowlitz Co., Washington; as well as from a number of other localities.

Ferrierite
$(Na, K)_2MgAl_3Si_{15}O_{36}(OH) . 9H_2O$
Ferrierite crystallizes in the orthorhombic system as thin tabular crystals, often in radiating groups. There is a perfect cleavage. The lustre is vitreous and the colour is white or colourless. H 3–3.5; SG 2.1
Occurrence. Occurs with chalcedony and opal in basalt seams, Kamloops Lake, Canada; etc.

Mesolite
$Na_2Ca_2Al_6Si_9O_{30} . 8H_2O$
Mesolite crystallizes in the monoclinic system forming groups of radiating crystals, which may be acicular. There is a perfect cleavage. The colour is white or colourless. The lustre is silky when fibrous. H 5; SG 2.3
Occurrence. Occurs in cavities in volcanic rocks with other zeolites.
Localities. From the Isle of Skye; Kilmacolm, Strathclyde, Scotland; Giant's Causeway, Moyle, N. Ireland; Kern Co., California; India; Australia; and from some other localities.

Gonnardite
$Na_2CaAl_4Si_6O_{20} . 7H_2O$
Gonnardite crystallizes in the orthorhombic system in fibrous masses. The colour is white with a silky lustre. H 4.5–5; SG 2.3
Occurrence. Occurs with thomsonite in basalt in Sicily; with wollastonite and pyrite at Crestmore, Riverside Co., California; etc.

Right: A sheaf-like aggregate of stilbite crystals

Above: Gismondine, Ballyclare, Northern Ireland

Above left: Gmelinite from New Jersey

Levyne $NaCa_3Al_7Si_{11}O_{36}.15H_2O$
Levyne crystallizes in the hexagonal system as thin tabular crystals or as aggregates, is colourless, grey, white or yellowish, with a vitreous lustre. H 4–4.5; SG 2.1–2.2
Localities. Occurs in basalt as fine crystals at Grant Co., Oregon; Glenarm, N. Ireland; etc.

Gmelinite near
$(Na_2, Ca)Al_2Si_4O_{12}.6H_2O$
Sometimes containing small amounts of potassium, gmelinite crystallizes in the hexagonal system as pyramidal crystals or rhombohedra. Vertical striations and penetration twinning are common. The colour is white, yellowish, reddish or pink, and the lustre is vitreous. H 4.5; SG 2.0–2.1
Occurrence. Occurs in basaltic lavas with other zeolites.
Localities. From Glenarm, N. Ireland; Faroe Islands; New Jersey; and some other localities.

Faujasite
$(Na_2, Ca)Al_2Si_4O_{12}.6H_2O$
Faujasite crystallizes in the cubic system as octahedra, commonly twinned. It is colourless or white and may be stained by impurities, thin sections are transparent. The lustre is vitreous. H 5; SG 1.9
Occurrence. Occurs with other zeolites at the St. Gotthard, Switzerland; with augite in the Kaiserstuhl, Baden, W. Germany.

Stilbite $NaCa_2Al_5Si_{13}O_{36}.14H_2O$
Stilbite crystallizes in the monoclinic system as cruciform penetration twins; as aggregates or in radiating forms. There is a perfect cleavage. It is white, grey, yellowish, pink, orange or brown, with a vitreous lustre. The streak is colourless. H 3.5–4; SG 2.1–2.2
Occurrence. Occurs in cavities in basalt or granite pegmatites; also from some hot spring deposits.
Localities. From the Isle of Skye, Kilmacolm, Strathclyde, and the Kilpatrick Hills, Scotland; Faroe Islands; Switzerland; San Diego Co., California; and from many other places.

Gismondine $CaAl_2Si_2O_8.4H_2O$
Gismondine crystallizes in the monoclinic system as pseudotetragonal bipyramids produced by twinning. The colour is white, colourless, grey or reddish, with a vitreous lustre. H 4.5; SG 2.2
Localities. Occurs in basaltic lavas in Ireland; Iceland; with chlorite in an altered granite from Queensland, Australia; etc.

Chabazite $CaAl_2Si_4O_{12}.6H_2O$
Chabazite crystallizes in the hexagonal system as rhombohedra which may resemble cubes, or as penetration twins. It is white, yellowish, pinkish or greenish with a colourless streak and a vitreous lustre. H 4–5; SG 2.0–2.1
Occurrence. Occurs in cavities in basalt with other zeolites, calcite and quartz. May also be found in schists and crystalline limestones.
Localities. From Kilmacolm, Strathclyde, Scotland; Giant's Causeway, Moyle, Ireland; Riverside Co., California; Bay of Fundy, Nova Scotia; Italy; Germany; USSR; Australia; etc.

Thomsonite $NaCa_2Al_5Si_5O_{20}.6H_2O$
Thomsonite crystallizes in the orthorhombic system as radiating or lamellar aggregates; crystals are acicular prismatic. There is a perfect cleavage. The colour is white, yellowish or pink and the streak is colourless. It has a vitreous lustre. Metathomsonite is a high-temperature polymorph. H 5–5.5; SG 2.2–2.4
Occurrence. Thomsonite occurs in cavities in basalt with other zeolites; also in some schists.
Localities. From the Kilpatrick Hills, Strathclyde and Bishopton, Scotland; Faroe Islands; Italy; Germany; Kern Co., California; etc.

Phillipsite near
$(K_2, Na_2, Ca)Al_2Si_4O_{12}.4\frac{1}{2}H_2O$
Phillipsite crystallizes in the monoclinic system as penetration twins or spherulites. It is white, colourless or reddish, with a vitreous lustre. H 4–4.5; SG 2.2
Occurrence. Occurs in cavities in basalt, saline lake deposits and as deposits from hot springs.

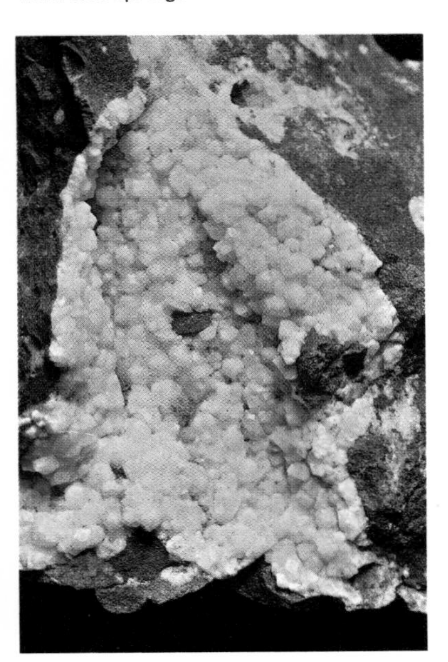

Far left: Pseudocubes of chabazite from Bolzano, Italy

Left: Phillipsite from Iceland

Right: Fibrous aggregates of mordenite from the Val dei Zuccanti, Italy

Localities. From Giant's Causeway, Moyle, N. Ireland; good crystals from Capo di Bove, Italy; from Idar-Oberstein, W. Germany; and from some other localities.

Erionite
$(Ca, Na_2, K_2)_{1-5}Al_9Si_{27}O_{72} . 27H_2O$
Erionite crystallizes in the hexagonal system as fibrous masses resembling wool. The colour is white. H not measured; SG about 2.0
Occurrence. Occurs with opal in rhyolitic tuffs or in basalt.
Localities. From the Faroe Islands; Baker Co., Oregon; Nevada; South Dakota.

Heulandite $(Na, Ca)_{4-6}Al_6$-$(Al, Si)_4Si_{26}O_{72} . 24H_2O$
Heulandite crystallizes in the monoclinic system as granular masses or trapezoidal crystals. There is a perfect cleavage. It is white, colourless, grey, yellow, pink, red or brown with a vitreous lustre and a colourless streak. H 3.5–4; SG 2.1–2.2
Occurrence. Occurs in cavities in basalt with other zeolites; sometimes in gneiss or sandstone.
Localities. From the Isle of Skye, Scotland; northern New Jersey; Switzerland; and elsewhere.

Dachiardite near
$(Ca, K_2, Na_2)_3Al_4Si_{18}O_{45} . 14H_2O$
Dachiardite crystallizes in the monoclinic system forming twinned prismatic crystals with a perfect cleavage. It is colourless, with a vitreous lustre. H 4–4.5; SG 2.1–2.2
Occurrence. Occurs with other zeolites in pegmatite on Elba.

Mordenite
$(Ca, Na_2, K_2)Al_2Si_{10}O_{24} . 7H_2O$
Mordenite crystallizes in the orthorhombic system, often as fibrous aggregates; crystals are prismatic, vertically striated. The colour is white, colourless or yellowish to pink, and the lustre is vitreous. H 4–5; SG 2.1
Occurrence. In cavities in igneous rocks.
Localities. From the Hoodoo Mts., Wyoming; Morden, Nova Scotia; and many other localities.

Brewsterite
$(Sr, Ba, Ca)Al_2Si_6O_{16} . 5H_2O$
Brewsterite crystallizes in the monoclinic system as short prismatic crystals or as fibrous aggregates. Twinning is common and there is a perfect cleavage. The colour is white or colourless, with a vitreous lustre. H 5; SG 2.4
Occurrence. Occurs in cavities in basalt and in schists associated with calcite.
Localities. From Strontian, Highland, Scotland; Mendocino Co., California; etc.

Edingtonite $BaAl_2Si_3O_{10} . 4H_2O$
Edingtonite crystallizes in the orthorhombic system as minute pyramidal crystals or in masses. There is a perfect cleavage. The colour is white, grey or pink, and the lustre is vitreous. H 4; SG 2.8
Localities. Occurs with harmotome and other zeolites in basic igneous rock at Old Kilpatrick, Strathclyde, Scotland; etc.

Harmotome $BaAl_2Si_6O_{16} . 6H_2O$
Harmotome crystallizes in the monoclinic system as penetration twins or as radiating aggregates. The colour is white, colourless, grey, pink, yellow or brown with a white streak. There is a vitreous lustre. H 4.5; SG 2.4–2.5
Occurrence. Occurs in cavities in basalts and with manganese minerals.
Localities. From Old Kilpatrick, Strathclyde and Strontian, Highland, Scotland; Wales; Westchester Co., New York; and elsewhere.

Other Silicates

Dioptase $CuSiO_2(OH)_2$
Dioptase crystallizes in the hexagonal system forming prismatic crystals, crystalline aggregates or massive forms. There is a perfect cleavage. The colour is emerald-green to bluish-green with a vitreous lustre and a pale greenish-blue streak. H 5; SG 3.2
Occurrence. Occurs in the oxidation zone of copper deposits.
Localities. From Altyn-Tube, Khirgiz Steppes, USSR; Shaba, Zaire; with calcite at Tsumeb, South-West Africa; Soda Lake Mountain, San Bernardino Co., California; associated with cerussite, orange wulfenite and other minerals at

Far right: Dioptase from a classic locality, Renéville, Zaire. Crystals are sometimes sufficiently large to facet

Right: Harmotome from Strontian, Scotland

Above: Radiating columnar dioptase crystals from Mindouli, Zaire

the Mammoth mine, Tiger, Arizona.
Treatment. Remove calcite with dilute acetic acid.
Fashioning. Uses: faceting or cabochons; *cleavage*: perfect and rhombohedral; fairly brittle; *cutting angles*: crown 40°, pavilion 40°; *dichroism*: weak; *heat sensitivity*: very low.

Planchéite $Cu_8Si_8O_{22}(OH)_4 . H_2O$
Planchéite crystallizes in the orthorhombic system forming fibrous radial aggregates. The colour is pale to dark blue. H 5.5; SG 3.6–3.8
Occurrence. Occurs as a secondary mineral in the oxidation zone of copper deposits with malachite and other secondary copper minerals.
Localities. From Shaba, Zaire; Table Mountain mine, Klondyke, Arizona.
Treatment. Clean with distilled water.

Chrysocolla $Cu_2H_2Si_2O_5(OH)_4$
Chrysocolla is thought to crystallize in the orthorhombic system; it is found as microscopic acicular crystals in radiating groups or as close-packed aggregates; it is more commonly found cryptocrystalline or botryoidal. The colour is blue to blue-green and may be brown to black. The lustre is vitreous or earthy. H 2–4; SG 2.0–2.4
Occurrence. Found in the oxidation zone of copper deposits.
Localities. From Liskeard, Cornwall; Roughten Gill, Cumbria; Shaba, Zaire; Pennsylvania and other states of the USA; etc.
Treatment. Clean with distilled water.

Phenakite Be_2SiO_4
Phenakite crystallizes in the hexagonal system and forms rhombohedral or prismatic crystals. Granular aggregates are found and radiating fibrous spherulites. Twinning is common. The colour is pale yellow through pink to brown, and colourless; the lustre is vitreous. H 7.5–8; SG 2.9–3.0; RI 1.65, 1.67
Occurrence. Found in pegmatites with topaz; in hydrothermal veins with quartz, cassiterite and beryl; in calcite veins with sulphides and in Alpine veins with adularia and hematite.
Localities. From the Takowaja River, Sverdlovsk, USSR; Minas Gerais, Brazil; France; Switzerland; Pala, San Diego Co., California; Pikes Peak, Colorado; etc.
Treatment. Remove iron stains with oxalic acid.
Fashioning. Uses: faceting or cabochons; *cleavage*: distinct and prismatic; *cutting angles*: crown 40°, pavilion 40°; *heat sensitivity*: very low.

Bertrandite $Be_4Si_2O_7(OH)_2$
Bertrandite crystallizes in the orthorhombic system forming thin tabular or prismatic crystals; it may be pseudomorphous after beryl. There is a perfect cleavage and twinning is common. It is colourless or pale yellow, transparent, with a vitreous lustre. H 6–7; SG 2.6
Occurrence. Occurs in granite pegma-

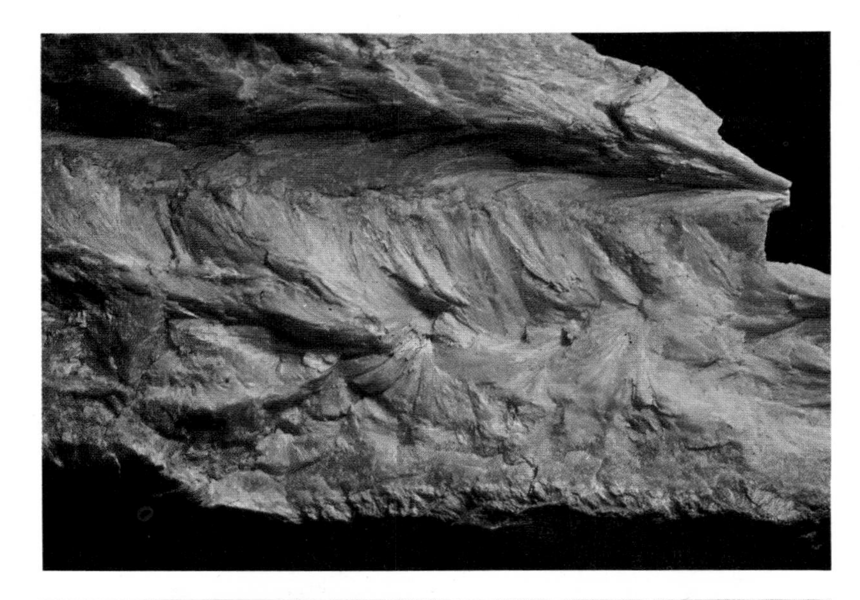

Right: Wollastonite from Meldon, Devon

Below right: Radiating acicular crystals of pectolite. This fine specimen is from Weehawken, New Jersey

Bottom right: Gadolinite from Norway. The greasy lustre can be seen

tites and pneumatolytic-hydrothermal veins with beryl, quartz and calcite.
Localities. From the Pala pegmatites, San Diego Co., California; Cornwall; Mica Creek, Mt. Isa, Queensland; etc.
Treatment. Clean with dilute acid.

Eudidymite $NaBeSi_3O_7OH$
Eudidymite is dimorphous with epididymite. It crystallizes in the monoclinic system forming tabular crystals or spherulites; twinning is common. There is a perfect cleavage. The colour is white or colourless and may be grey, yellow, blue or violet. The lustre is vitreous. H 6–7; SG 2.5
Occurrence. Occurs in nepheline-syenite pegmatites with albite, fluorite and other minerals.
Localities. From Greenland, Langesundfjord, Norway; Kola Peninsula, USSR.
Treatment. Clean with distilled water.

Epididymite $NaBeSi_3O_7OH$
Dimorphous with eudidymite, epididymite crystallizes in the orthorhombic system, forming tabular crystals with a perfect cleavage or spherulites, and with twinning frequent. It is white or colourless and may be shades of grey, blue, yellow or violet. The lustre is vitreous. H 6–7; SG 2.5
Occurrence. Occurs in nepheline-syenite pegmatites with fluorite, albite and other minerals.
Localities. From Greenland; Langesundfjord, Norway; Kola Peninsula, USSR.
Treatment. Clean with distilled water.

Barylite $BaBe_2Si_2O_7$
Barylite crystallizes in the orthorhombic system and forms thin tabular or prismatic crystals; massive forms or disseminated grains are also found. There is a perfect cleavage. The colour is white or bluish, transparent with a vitreous lustre. H 7; SG 4.0
Occurrence. Occurs with willemite and hedyphane in calcite veins at Franklin, New Jersey; fine crystals from Park Co., Colorado; etc.
Treatment. Clean with dilute acid.

Gadolinite $Be_2FeY_2Si_2O_{10}$
Gadolinite crystallizes in the monoclinic system, forming prismatic crystals or masses. The colour is black, greenish-black and sometimes light green. The lustre is vitreous to greasy and the streak greenish-grey. Other rare earths in small quantities may replace yttrium. H 6.5–7; SG 4.0–4.6
Localities. Found in granites and granite pegmatites, commonly with fluorite and allanite. From Sweden; Switzerland; Colorado, Arizona; etc.
Treatment. Clean with dilute acid; do not try to remove brown coating.

Parawollastonite $CaSiO_3$
Parawollastonite crystallizes in the monoclinic system and is the alpha-form of $CaSiO_3$. Crystals are tabular; fibrous masses are more commonly found. There is a perfect cleavage and twinning is common. The colour is white or grey, sometimes yellowish with a vitreous or pearly lustre. H 4.5–5; SG 2.9
Occurrence. Occurs in contacts of igneous intrusions in limestones and in limestone fragments ejected from volcanoes.
Localities. From Mt. Somma, Italy; Csiklova, Romania; Crestmore, Riverside Co., California.
Treatment. Clean with dilute acid.

Wollastonite $CaSiO_3$
The triclinic form of $CaSiO_3$, wollastonite forms tabular crystals but is more commonly found massive and fibrous. There is a perfect cleavage and twinning is common. The colour is white, grey or pale green with a vitreous or pearly lustre. H 4.5–5; SG 2.8–3.0
Occurrence. Occurs in metamorphosed limestones and in some alkaline igneous rocks.
Localities. From Norway; Finland; Inyo Co., California; etc.
Treatment. Clean with dilute acid.

Okenite $CaSi_2O_4(OH)_2.H_2O$
Okenite crystallizes in the triclinic system and forms blade-shaped crystals; more commonly found as fibrous inter-

laced masses. There is a perfect cleavage and twinning occurs. The colour is white, sometimes tinged with yellow or blue. The lustre is vitreous or pearly. H 4.5–5; SG 2.2–2.3
Occurrence. Occurs in amygdales in basalt.
Localities. From Disco Island, Greenland; Faroe Islands; Chile; Syhadree Mts., Bombay.

Gyrolite $Ca_2Si_3O_7(OH)_2 . H_2O$
Gyrolite crystallizes in the hexagonal system forming radiating lamellar masses or concretions. There is a perfect cleavage and the colour is white or colourless. H 3–4; SG 2.3–2.4
Occurrence. Occurs in crevices in rocks as a secondary mineral formed by the alteration of lime silicates.
Localities. From Fort Point, San Francisco Co., California, in association with apophyllite; from Collinward, near Belfast, N. Ireland; Isle of Skye; etc.
Treatment. Clean with distilled water.

Tobermorite
near **$Ca_5Si_6O_{16}(OH)_2 . 4H_2O$**
Tobermorite crystallizes in the orthorhombic system as fibrous or granular masses, or as plates. The colour is white or pinkish-white and the lustre is silky. Small amounts of sodium and potassium may replace calcium in the formula given, and small amounts of aluminium may replace silicon. Three distinct hydrates of similar composition appear to occur, called plombierite, tobermorite and riversideite respectively. H 2.5; SG 2.4
Occurrence. Pseudomorphous after wilkeite in association with calcite, monticellite and vesuvianite at Crestmore, Riverside Co., California; Tobermory, Isle of Mull, Scotland; etc.

Xonotlite $Ca_3Si_3O_8(OH)_2$
Xonotlite crystallizes in the monoclinic system forming acicular crystals; more commonly found massive, compact or fibrous. The colour is chalky white, colourless, pink or grey with a greasy lustre. H 6; SG 2.7
Occurrence. Occurs as veins in serpentine or in contact zones.
Localities. From Tetela de Xonotla, Mexico; Yauco, Puerto Rico; Riverside Co. and San Francisco Co., California.
Treatment. Clean with distilled water.

Afwillite $Ca_3Si_2O_4(OH)_6$
Afwillite crystallizes in the monoclinic system forming tabular crystals with a perfect cleavage and also massive forms. It is white or colourless. H 3; SG 2.6
Localities. Occurs in cracks in blocks of contact rock in Commercial Quarry, Crestmore, California; with calcite in a spurrite rock at Scawt Hill, Antrim, N. Ireland; at Dutoitspan diamond mine, Kimberley, South Africa in a dolerite inclusion in kimberlite.
Treatment. Clean with dilute acid.

Pectolite $NaCa_2Si_3O_8OH$
Pectolite crystallizes in the triclinic system forming aggregates of needle-like crystals, radiating and forming globular masses. There is a perfect cleavage and the colour is white. The lustre is silky. H 4.5–5; SG 2.7–2.8
Occurrence. Occurs in cavities in basaltic rocks, often with zeolites; also in lime-rich metamorphic rocks.
Localities. Paterson and Franklin, New Jersey; Weardale, Durham and Ratho, near Edinburgh; etc.
Treatment. Can be cleaned ultrasonically.

Merwinite $Ca_3Mg[Si_2O_8]$
Merwinite crystallizes in the monoclinic system forming compact granular masses; the crystals have a perfect cleavage. It is colourless to pale green with a greasy lustre. H 6; SG 3.1
Occurrence. Occurs in contact zones with monticellite; spinel, etc.
Localities. From Scawt Hill, Antrim, N. Ireland; limestone quarries, Crestmore, Riverside Co., California.
Treatment. Clean with dilute acid.

Willemite Zn_2SiO_4
Willemite crystallizes in the hexagonal system forming short hexagonal prismatic crystals or fibrous compact masses or grains. It is colourless, white, green, red, yellow, brown or grey with a colourless streak and resinous lustre. Intense yellowish-green fluorescence under ultra-violet light. Strong phosphorescence. H 5.5; SG 3.9–4.2
Occurrence. Occurs as an ore mineral.
Localities. Fine specimens from Franklin and Sterling Hill, New Jersey; also from Greenland; Zaire; Zambia; Tsumeb, South-West Africa; etc.
Treatment. Clean with distilled water.

Hemimorphite $Zn_4Si_2O_7(OH)_2 . H_2O$
Hemimorphite crystallizes in the orthorhombic system forming thin tabular crystals with vertical striations; hemimorphic development shown by doubly-terminated crystals; also found as fan-shaped aggregates and as masses, granular or stalactitic. There is a perfect cleavage. The colour is white or colourless, sometimes pale blue, greenish or grey. The lustre is silky and the streak colourless. H 4.5–5; SG 3.4–3.5
Occurrence. Occurs as a secondary mineral in the oxidized zone of ore deposits or in calcareous rocks; sometimes in pegmatites. Associated with galena, sphalerite, smithsonite etc.
Localities. From Roughten Gill, Cumbria (acicular crystals and mamillary crusts, of a fine green); Matlock, Derbyshire; Mapimi, Durango, Mexico; California; Nevada; etc.

Hardystonite $Ca_2ZnSi_2O_7$
Hardystonite crystallizes in the tetragonal system in granular masses coloured white, pink or brown. H 3–4; SG 3.4
Localities. Found with willemite, etc, at Franklin, New Jersey.

Clinohedrite $CaZnSiO_3(OH)_2$
Clinohedrite crystallizes in the monoclinic system, usually being found massive or as lamellar aggregates; crystals are prismatic, tabular or wedge-shaped. There is a perfect cleavage. It is white or colourless and may be amethystine. Orange fluorescence under short-wave UV. H 5.5; SG 3.2–3.3
Localities. Found with willemite, etc, at Franklin, New Jersey.
Treatment. Clean with distilled water.

Larsenite $PbZnSiO_4$
Larsenite crystallizes in the orthorhombic system forming slender prismatic crystals vertically striated. It is white with an adamantine lustre. H 3; SG 5.9
Localities. Found with willemite, etc, at Franklin, New Jersey.
Treatment. Clean with distilled water.

Hodgkinsonite $MnZn_2SiO_5 . H_2O$
Hodgkinsonite crystallizes in the monoclinic system forming acute pyramidal or stout prismatic crystals; massive forms also found. There is a perfect cleavage and the colour is bright pink to reddish-brown. The lustre is vitreous. H 4.5–5; SG 3.9
Localities. Found with willemite, etc, at Franklin, New Jersey.
Treatment. Clean with distilled water.

Thortveitite $(Sc, Y)_2Si_2O_7$
Thortveitite crystallizes in the monoclinic system forming prismatic crystals showing some distortion and commonly displaying twinning. The colour is grey-

Left: Specimen of willemite fluorescing under short-wave ultraviolet light. The willemite is showing green; the pink mineral is calcite. From Franklin, New Jersey

Far left: The same specimen in normal light

Above: Bright yellow monoclinic crystals of sphene from Bolzano, Italy. The crystals are wedge-shaped and the name sphene derives from the Greek for wedge

ish-green with a streak of the same colour, while the lustre is vitreous. H 6–7; SG 3.5
Occurrence. Found in granite pegmatites rich in rare-earth elements.
Localities. From Iveland, Norway; Befanamo, Malagasy Republic; Ural Mts, USSR; Japan.
Treatment. Clean with distilled water.

Thalenite $Y_2Si_2O_7$
Thalenite crystallizes in the monoclinic system, usually forming compact masses; crystals are tabular or prismatic. The colour is flesh pink, brown or greenish. The lustre is greasy. H 6; SG 4.3–4.6
Occurrence. Found in granite pegmatites with fergusonite, magnetite, etc.
Localities. From Osterby, Sweden; Iizaka, Fukuoka Prefecture, Japan; Siberia; Snowflake feldspar mine, Teller Co., Colorado.
Treatment. Clean with distilled water.

Cerite $(Ca, Mg)_2(RE)_8(SiO_4, FCO_3)_7(OH, H_2O)_3$
(RE in the above formula is any rare earth.) Cerite crystallizes in the hexagonal system forming pseudo-octahedral crystals; more commonly found massive, granular. The colour is brown to cherry-red with a resinous lustre. H about 5.5; SG 4.7
Localities. Found with rare earths in veins at Mountain Pass, California; with fluorite and epidote in the Jamestown district, Boulder Co., Colorado; the Bastnaes mine, Riddarhyttan, Vastmanland, Sweden; Mt. Tenbazan, Korea.
Treatment. Clean with distilled water.

Murmanite $Na_2(Ti, Nb)_2Si_2O_9 . H_2O$
Murmanite crystallizes in the monoclinic system as aggregates of platy crystals with a perfect cleavage. The colour is lilac to bright pink; altered varieties are grey to brown or black. H 2–3; SG 2.7–2.8
Localities. From alkali pegmatites and nepheline-syenite with microcline, eudialyte etc; from Kola Peninsula, USSR.
Treatment. Clean with distilled water.

Sphene $CaTiSiO_5$
Sphene crystallizes in the monoclinic system, forming flattened wedge-shaped crystals; massive, compact and lamellar forms are also found. Lamellar varieties give a distinct parting. Sphene is colourless, brown, green, yellow, rose-red or black; parti-coloured varieties are known. The streak is white and the lustre adamantine to resinous. H 5–5.5; SG 3.4–3.5
Occurrence. Occurs as an accessory mineral in igneous rocks; in schists, gneisses and iron ore; also as a detrital mineral in sedimentary deposits.
Localities. Found at Schwarzenstein and Rotenkopf in the Zillertal, Austria; the Ala Valley, Piedmont, Italy; the Grisons, Switzerland, as fine crystals; Midongy, Malagasy Republic; from the Tilly Foster iron mine, Brewster, New York; a honey-yellow variety from Franklin, New Jersey; from Pino Solo, Baja California Norte, Mexico; etc.
Treatment. Clean with dilute acid.
Fashioning. Uses: faceting or cabochons; *cleavage*: distinct prismatic; *cutting angles*: crown 30°–40°, pavilion 37°–40°; *pleochroism*: strong in deep colours; *heat sensitivity*: very low.

Lamprophyllite $Na_2(Sr, Ba)_2Ti_3(SiO_4)_4(OH, F)_2$
Lamprophyllite crystallizes in the monoclinic and orthorhombic systems, forming tabular elongated crystals or needle-like aggregations. Polysynthetic twinning is found. The colour is golden to dark brown with a vitreous lustre. H 2–3; SG 3.4–3.5
Occurrence. Occurs in nepheline-syenites and associated pegmatites with microcline, eudialyte, etc.
Localities. From Langesundfjord, Norway; Bear Paw Mts., Montana; etc.

Benitoite $BaTiSi_3O_9$
Benitoite crystallizes in the hexagonal system forming pyramidal or tabular crystals somewhat flattened along the c-axis and with a triangular shape. The colour is blue, purple, pink, white or colourless and some crystals are parti-coloured. Blue fluorescence under short-wave ultraviolet light. The streak is colourless and the lustre vitreous. H 6–6.5; SG 3.6; RI 1.7, 1.8
Localities. Found with neptunite, natrolite and joaquinite in serpentine near the headwaters of the San Benito River, San Benito Co., California; from the Eocene sands of south-west Texas; etc.
Treatment. Clean with dilute acid.
Fashioning. Uses: faceting or cabochons; *cutting angles*: crown 40°, pavilion 40°; *dichroism*: strong; *heat sensitivity*: low.

Right: Benitoite, commonly found only in California. The darker crystals are neptunite

Joaquinite $Ba_2NaCe_2Fe(Ti, Nb)_2Si_8O_{26}(OH, F)_2$
Joaquinite crystallizes in the orthorhombic system forming small tabular crystals coloured honey-yellow to brown. The lustre is vitreous. H 5.5; SG 3.8
Localities. Found with neptunite, natrolite and benitoite near the headwaters of the San Benito River, San Benito Co., California.
Treatment. Clean with dilute acid.

Chevkinite $(Ca, Ce)_4(Fe, Mg)_2(Ti, Fe)_3Si_4O_{22}$
Chevkinite crystallizes in the monoclinic system forming slender prisms;

more commonly in irregular masses. The colour is dark reddish-brown to black. Lustre resinous. Varieties are perrierite (an oxidized variety) and thorchevkinite (thorium-bearing). H 5–6; SG 4.3–4.6
Localities. Found in pumice fragments from Cenozoic volcanoes in the western USA; from alkali pegmatites, Kola Peninsula, USSR.
Treatment. Clean with distilled water.

Neptunite
$(Na, K)_2(Fe^{2+}, Mn)TiSi_4O_{12}$
Neptunite crystallizes in the monoclinic system forming prismatic crystals with a square cross-section. There is a perfect cleavage and the colour is black; reddish brown from internal reflection. The streak is reddish-brown and the lustre vitreous. H 5–6; SG 3.1–3.2
Localities. Found with benitoite at the headwaters of the San Benito River, San Benito Co., California. A mangan-neptunite occurs at Mont St. Hilaire, Quebec, and on the Kola Peninsula, USSR.
Treatment. Clean with dilute acid.

Astrophyllite
$(K, Na)_3(Fe, Mn)_7Ti_2Si_8O_{24}(O, OH)_7$
Astrophyllite crystallizes in the triclinic system as bladed crystals in stellate groups with a perfect cleavage. The colour is bronze-yellow or golden-yellow with a submetallic lustre. H 3; SG 3.3–3.4
Localities. Occurs with quartz, feldspar, etc, at St. Peter's Dome, El Paso Co., Colorado; from nepheline-syenite at Mont St. Hilaire, Quebec; etc.
Treatment. Clean with distilled water.

Left: The adamantine lustre of zircon can be seen in this crystal from Slatoust, USSR

Zircon $ZrSiO_4$
Zircon crystallizes in the tetragonal system forming short prismatic crystals; often dipyramidal; metamict varieties with curved faces and edges or as sub-parallel aggregates. The colour may be red, brown, green, yellow or colourless with an adamantine lustre; metamict varieties have a greasy lustre. H 7.5; SG 4.6–4.7 (metamict 4.0); RI 1.92, 2.0; dispersion 0.038
Occurrence. Found as an accessory mineral in igneous rocks and in some metamorphic rocks. Also found as a detrital mineral in sedimentary deposits.
Localities. From Thailand; Sri Lanka (metamict varieties also from this locality); red variety from Expailly-St.-Marcel, France; Burma; Australia; St. Peter's Dome, El Paso Co., Colorado;

Below: A superb group of behitoite crystals showing a well developed pyramidal habit

Right: Acicular crystals of the radioactive mineral sklodowskite from Shinkolobwe, Zaire

Below right: Braunite has a submetallic lustre, which can be seen in these crystals from Thuringia

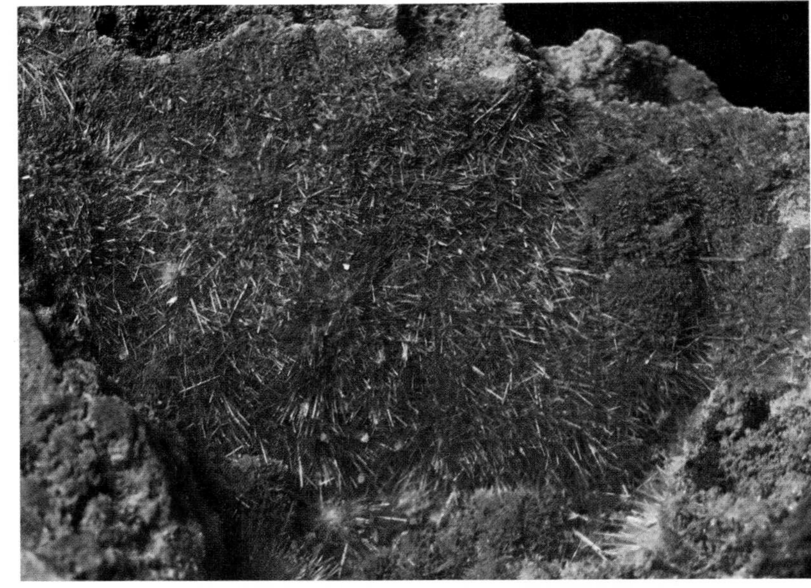

and numerous other localities.
Treatment. Clean with dilute acid.
Fashioning. Uses: faceting or cabochons; *cleavage*: imperfect prismatic; brittle; *cutting angles*: crown 40°, pavilion 40°; *dichroism*: distinct in some colours; *heat sensitivity*: very low.

Catapleiite $(Na_2, Ca)ZrSi_3O_9.2H_2O$
Catapleiite crystallizes in the hexagonal system as thin hexagonal plates or lamellar masses. Twinning is common and there is a perfect cleavage. The colour is light yellow to yellowish-brown or red; sometimes sky blue or colourless. The lustre is vitreous. H 5–6; SG 2.6–2.8
Occurrence. Found in alkaline rocks and pegmatites with feldspars, sphene, etc.
Localities. Mont St. Hilaire, Quebec; Langesundfjord, Norway; Malagasy Republic; Magnet Cove, Arkansas; etc.
Treatment. Clean with distilled water.

Wadeite $K_2CaZrSi_4O_{12}$
Wadeite crystallizes in the hexagonal system, forming prismatic crystals with a hexagonal cross-section. It is colourless, light pink or lilac, with an adamantine lustre. H 5–6; SG 3.1
Occurrence. Occurs in potassium-rich rocks with zeolites, leucite, apatite, etc.
Localities. From Kimberley, South Africa; Australia; Kola Peninsula, USSR.
Treatment. Clean with dilute acid.

Thorite $ThSiO_4$
Thorite crystallizes in the tetragonal system forming short prismatic crystals or reniform masses. The colour is brownish-yellow to orange, black or dark green. Resinous lustre. Varieties include hydrothorite, (a metamict hydrous thorite), calciothorite and freyalite, which contains cerium. H about 4.5; SG 4.1–6.7
Occurrence. Found as a primary mineral in pegmatites or metasomatized zones in impure limestones; hydrothermal veins and detrital deposits.
Localities. From Langesundfjord, Norway; Sicily; Batum, USSR; Henderson Co., North Carolina; etc.
Treatment. Clean with distilled water.

Huttonite $ThSiO_4$
Dimorphous with thorite, huttonite crystallizes in the monoclinic system forming anhedral grains coloured pale cream with an adamantine lustre. SG 7.1
Localities. Associated with scheelite and cassiterite from sands at Gillespie Beach, New Zealand.
Treatment. Clean with distilled water.

Stokesite $CaSnSi_3O_9.2H_2O$
Stokesite crystallizes in the orthorhombic system as pyramidal crystals. There is a perfect cleavage. It is colourless. H 6; SG 3.1
Localities. Occurs with axinite at Roscommon Cliff, St. Just, Cornwall.
Treatment. Clean with distilled water.

Barysilite $MnPb_8(Si_2O_7)_3$
Barysilite crystallizes in the hexagonal system and is found massive with a curved lamellar structure; the colour is white to pink and with a tarnish after exposure. H 3; SG 6.5–6.7
Localities. From Franklin, New Jersey.
Treatment. Clean with distilled water.

Eulytine $Bi_4Si_3O_{12}$
Eulytine crystallizes in the cubic system forming minute tetrahedral crystals or spheres. It is colourless to yellow or grey or dark brown. H 4.5; SG 6.6
Localities. Occurs as crystals on quartz at Johanngeorgenstadt, Schneeberg, E. Germany, in association with native bismuth.
Treatment. Clean with distilled water.

Soddyite $(UO_2)_5Si_2O_9.6H_2O$
Soddyite crystallizes in the orthorhombic system as bipyramidal crystals or as divergent clusters or otherwise as fibrous aggregates. There is a perfect cleavage. The colour is amber in small crystals, larger ones are canary yellow. The streak is pale yellow and the lustre is vitreous to resinous. H 3.5; SG 4.7
Localities. Occurs with uranophane as pseudomorphs after uraninite at the Ruggles pegmatite, Grafton Center, New Hampshire; from pegmatites at Norrabees south of the Orange river in Namaqualand, South Africa; from Shinkolobwe, Shaba, Zaire.
Treatment. Clean with distilled water.

Cuprosklodowskite $Cu(UO_2)_2Si_2O_7.6H_2O$
Cuprosklodowskite crystallizes in the triclinic system forming acicular crystals, stellate aggregates or silky coatings with a radial fibrous structure; they have a vitreous or silky lustre. The colour is pale to emerald green or a darker green. H 4; SG 3.8
Occurrence. Occurs as a secondary mineral from the alteration of primary copper minerals and uraninite.
Localities. From Shaba, Zaire, with sklodowskite, etc; Jachymov, Czechoslovakia; with dioptase from Amelal, Morocco; San Juan Co., Utah with brochantite; etc.

Sklodowskite $Mg(UO_2)_2Si_2O_7.6H_2O$
Sklodowskite crystallizes in the monoclinic system as prismatic acicular crystals or radial fibrous aggregates or granular masses. There is a perfect cleavage. The colour is pale to greenish yellow with a silky lustre. H 2–3; SG 3.6
Localities. Occurs as a secondary mineral with cuprosklodowskite at Kalongwe, Shaba, Zaire; also from New Haven mine, Crook Co., Wyoming; etc.

Uranophane $Ca(UO_2)_2Si_2O_7.6H_2O$
Uranophane crystallizes in the monoclinic system forming acicular to hair-like crystals and as radiated aggregates or tufts. There is a perfect cleavage. The colour is pale to lemon yellow, also amber to brown with a greasy lustre. It fluoresces weakly in ultraviolet light. H about 2.5; SG 3.8
Occurrence. Occurs in pegmatites as pseudomorphs after uraninite and in the oxidized zones of veins of uraninite.
Localities. Fine crystals from Shaba, Zaire; from the Hanosh mine, Grants district, New Mexico; Wölsendorf, E. Germany; and many other localities.

Braunite $Mn^{2+}Mn_6^{3+}SiO_{12}$
Braunite crystallizes in the tetragonal system as pyramidal striated crystals or as masses. There is a perfect cleavage. The colour is black to brown with a submetallic lustre. H 6–6.5; SG 4.7–4.8

Occurrence. Occurs in veins as a secondary mineral formed by weathering and associated with other manganese minerals.
Localities. From Panama; Norway; Sweden; Italy; etc.
Treatment. Clean with distilled water.

Alleghanyite $Mn_5Si_2O_8(OH)_2$
Sometimes containing titanium or fluorine, alleghanyite crystallizes in the monoclinic system forming plate-like crystals deeply striated by twinning; usually massive. The colour is brown or deep pink, and the lustre vitreous. H 5.5; SG 4.0
Localities. Occurs with franklinite and willemite at Franklin, New Jersey, etc.
Treatment. Clean with distilled water.

Leucophoenicite $Mn_7Si_3O_{12}(OH)_2$
Leucophoenicite crystallizes in the monoclinic system, usually found massive, granular. The colour is light to deep pink or purple to brown. The lustre is vitreous. H 5.5–6; SG 3.8
Localities. With willemite in hydrothermal veins at Franklin, New Jersey.
Treatment. Clean with distilled water.

Bementite $Mn_8Si_6O_{15}(OH)_{10}$
Bementite crystallizes in the monoclinic system; it is usually found massive, sometimes radiating or stalactitic. There is a perfect cleavage. The colour is brown to grey and darkens on exposure; the lustre is vitreous. H 6; SG 2.9–3.1
Localities. Occurs in primary manganese ore, notably from Franklin, New Jersey.
Treatment. Clean with distilled water.

Left: Fibrous crystals of johannsenite from Recoaro, Italy

Johannsenite $CaMnSi_2O_6$
Johannsenite crystallizes in the monoclinic system, usually being found massive or as radiating spherulitic aggregates of fibres and prisms; crystals are short prismatic. The colour is grey, green to brown or black, with a vitreous lustre. It is often stained black by manganese oxide. H 6; SG 3.4–3.5
Occurrence. Occurs in metasomatized limestones.
Localities. From Broken Hill, New South Wales, etc.
Treatment. Clean with distilled water.

Babingtonite
$Ca_2Fe^{2+}Fe^{3+}Si_5O_{14}OH$
Babingtonite crystallizes in the triclinic system forming short prismatic or platy crystals with a perfect cleavage. The colour is greenish to brownish-black,

Far left: Babingtonite on orthoclase feldspar. From Baveno, Italy

Left: A triclinic crystal of babingtonite. Here the basal pinacoid can be seen as the vertical face on the right. From Baveno, Italy

Right: The chiastolite variety of andalusite with cross shaped carbonaceous inclusions. From Bimbowrie, Australia

with a vitreous lustre. H 5.5–6; SG 3.3
Occurrence. Occurs in cavities in granite and gneiss.
Localities. From Devon; Baveno, Piedmont, Italy; fine crystals from the Lane trap rock quarry, Westfield, Massachusetts, etc.
Treatment. Clean with dilute acid.

Inesite $Ca_2Mn_7Si_{10}O_{28}(OH)_2 \cdot 5H_2O$
Inesite crystallizes in the triclinic system. It forms prismatic or tabular crystals with a chisel-shape; also acicular radiating aggregates with a perfect cleavage. The colour is pink or orange, darkening to brown on exposure. H 5.5; SG 3.0
Localities. Occurs with rhodochrosite at Hale Creek mine, Trinity Co., California; with apophyllite at New Broken Hill mine, Australia; etc.
Treatment. Clean with distilled water.

Pyroxmangite (Mn, Fe)SiO_3
Pyroxmangite crystallizes in the triclinic system usually in massive forms or grains; crystals are tabular. There is a perfect cleavage. The colour is pale to rose pink or dark reddish-brown due to alteration; there is a vitreous lustre. H 5.5–6; SG 3.6–3.8
Localities. Occurs in metamorphic or metasomatic rocks with rhodochrosite, spessartine, etc; from Broken Hill, New South Wales; Iva, South Carolina; and other localities.
Treatment. Clean with dilute acid.

Cronstedtite $Fe_2^{2+}Fe_2^{3+}SiO_5(OH)_4$
Cronstedtite crystallizes in the monoclinic system forming 3- or 6-sided pyramids or fibrous forms. There is a perfect cleavage. The colour is greenish to brownish-black; thin plates are emerald green by transmitted light. The lustre is vitreous. The streak is dark olive green. H 3.5; SG 3.3–3.4
Localities. Occurs at Ouro Preto, Minas Gerais, Brazil; Truro, Cornwall; Příbram, Czechoslovakia; etc.
Treatment. Clean with distilled water.

Chloropal
$Na_{0.33}Fe_2^{3+}(Al, Si)_4O_{10}(OH)_2 \cdot nH_2O$
Chloropal, sometimes known as nontronite, crystallizes in the monoclinic system, forming fine-grained masses with a perfect cleavage. The colour is yellow to olive green, with a waxy lustre. H1–2; SG 2.3
Occurrence. Occurs as an alteration product of volcanic glasses; in mineral veins with quartz and opal.
Localities. From France; Germany; Malagasy Republic; California; Utah; etc.
Treatment. Clean with distilled water.

Ilvaite $CaFe_2^{2+}Fe^{3+}Si_2O_8(OH)$
Ilvaite crystallizes in the orthorhombic system forming thick prismatic crystals

Below: Ilvaite from Rio Marina, Elba

Left: Radiating pyrophyllite crystals from the Ural Mts, USSR

with diamond-shaped cross section and with vertical striations; also massive, compact or columnar. The colour is grey to black with a black streak (brownish or greenish) and a submetallic lustre. H 5.5–6; SG 3.8–4.1
Occurrence. As a contact-metasomatic mineral, and in iron, zinc and copper ore deposits.
Localities. From Laxey mine, South Mountain, Idaho; Julianehaab, Greenland; Rio Marinha and Capo Calamita, Isle of Elba; etc.
Treatment. Clean with distilled water.

Andalusite Al_2SiO_5
Andalusite crystallizes in the orthorhombic system, forming prismatic crystals with an almost square cross-section. Massive forms and fibrous aggregates also occur. It is trimorphous with sillimanite and kyanite. The colour is pink, reddish-brown or greenish with strong pleochroism. The variety chiastolite contains carbonaceous inclusions, which give a cruciform pattern in cross-section. The streak is colourless and the lustre vitreous. H 6.5–7.5; SG 3.1
Occurrence. Occurs in slates and pelitic schists in contact-metamorphic aureoles, also mica schists; sometimes in pegmatites. Associated minerals include sillimanite and kyanite.
Localities. From Brazil; Sri Lanka; Malagasy Republic; chiastolite from Rhodesia; California; Massachusetts; etc.
Treatment. Clean with dilute acid.
Fashioning. Uses: faceting or cabochons; *cleavage*: distinct to perfect prismatic; brittle; *cutting angles*: crown 40°, pavilion 40°; *pleochroism*: strong; *heat sensitivity*: low.

Kyanite Al_2SiO_5
Kyanite, which is trimorphous with andalusite and sillimanite, crystallizes in the triclinic system, forming long bladed crystals which may be twisted. Massive and fibrous forms also occur; lamellar twinning is frequent and there is a perfect cleavage. The colour is blue with white, green, yellow or pink, sometimes showing in the same crystal; the streak is colourless and the lustre vitreous. H 4–7.5 (varying with crystallographic direction); SG 3.5–3.6
Occurrence. Occurs in schists and gneisses or from granite pegmatites.
Localities. From the Ural Mts, USSR; the Tyrol, and Carinthia, Austria; Kenya; Zaire; Virginia, New York and North Carolina, etc.
Treatment. Remove iron stains with oxalic acid.
Fashioning. Uses: faceting or cabochons; *cleavage*: highly perfect pinacoidal; brittle; *cutting angles*: crown 40°, pavilion 40°; *pleochroism*: distinct; *heat sensitivity*: low.

Sillimanite Al_2SiO_5
Sillimanite, which is trimorphous with andalusite and kyanite, crystallizes in the orthorhombic system forming fibrous crystals, usually felted. Massive forms also occur. There is a perfect cleavage. The colour is colourless, grey to blue or yellowish. The streak is colourless, and there is a vitreous lustre. H 6.5–7.5; SG 3.2
Occurrence. Occurs in schists, gneisses and granites, associated with andalusite or corundum.
Localities. From Mogok, Burma; South Africa; Kenya; Boehls Butte, Idaho; Custer Co., South Dakota; and many other localities.
Treatment. Sillimanite should be cleaned with dilute acid.
Fashioning. Uses: faceting or cabochons; *cleavage*: highly perfect // to the brachypinacoid; *cutting angles*: crown 40°, pavilion 40°; *pleochroism*: weak to distinct; *heat sensitivity*: low.

Mullite $Al_6Si_2O_{13}$
Mullite crystallizes in the orthorhombic system forming colourless or pale pink prismatic crystals, with a vitreous lustre. H 6–7; SG 3.0–3.1
Occurrence. Occurs in fused argillaceous inclusions in Tertiary eruptive rocks on the island of Mull, Scotland; andalusite, sillimanite and kyanite transform to mullite on intense heating.
Treatment. Clean with dilute acid.

Pyrophyllite $Al_2Si_4O_{10}(OH)_2$
Pyrophyllite crystallizes in the monoclinic system in elongated tabular crystals, which are often curved or distorted; usually foliated, radiating, lamellar or fibrous. There is a perfect cleavage; the laminae are flexible. Greasy feel. The colour is white to yellowish-blue to brownish-green. Lustre pearly. H 1–2; SG 2.6–2.9
Occurrence. Occurs in schistose rocks with andalusite, kyanite and sillimanite or in hydrothermal veins with quartz and mica.
Localities. From Minas Gerais, Brazil; San Bernardino Co., California; etc.
Treatment. Remove iron stains with oxalic acid.

Allophane $Al_2SiO_5 . nH_2O$
Crystal system undetermined. Various colours. H 2–3; SG 1.8
Localities. From fissures and cavities in ore veins or coal deposits, western USA; Wheal Hamblyn, Devon; etc.
Treatment. Clean with distilled water.

Below: A fine crystal of kyanite from the Italian Alps. Kyanite is sometimes faceted

Petalite $LiAlSi_4O_{10}$
Petalite crystallizes in the monoclinic system and is usually found massive; polysynthetic twinning common. There is a perfect cleavage. It is colourless, grey, yellow or white. The lustre is vitreous. H 6–6.5; SG 2.3–2.5; RI 1.50, 1.52
Occurrence. Occurs in granite pegmatites with quartz, lepidolite, etc.
Localities. From Elba; Sweden; USSR; Londonderry, Western Austrlia; South-West Africa; etc.
Treatment. Clean with dilute acid.
Fashioning. Uses: faceting or cabochons; *cleavage*: perfect clinopinacoidal; brittle; *cutting angles*: crown 40°–50°, pavilion 43°; *pleochroism*: weak; *heat sensitivity*: low.

Eucryptite $LiAlSiO_4$
Eucryptite crystallizes in the hexagonal system, usually in massive granular aggregates; it forms small, well-formed crystals. Colourless or white. Pink fluorescence under ultra-violet light. H 6.5; SG 2.6
Occurrence. Occurs with albite in New Hampshire and Connecticut; from pegmatite at the Harding mine, Dixon, New Mexico; Bikita, Rhodesia.
Treatment. Clean with distilled water.

Cookeite $LiAl_4(Si, Al)_4O_{10}(OH)_8$
Cookeite crystallizes in the monoclinic system as pseudohexagonal plates or curved radial scales. There is a perfect cleavage. Colour white, greenish, yellowish, pink or brown. Lustre pearly. H 2.5–3.5; SG 2.6–2.7
Occurrence. Occurs in lithium-rich granite pegmatites with tourmaline, albite, quartz, etc.
Localities. From Pala, San Diego Co., California; Londonderry, Western Australia; etc.
Treatment. Clean with distilled water.

Holmquistite near **$(Na, K, Ca)Li(Mg, Fe^{2+})_3Al_2Si_8O_{22}(OH)_2$**
Holmquistite crystallizes in the orthorhombic system, as slender prismatic crystals, often with vertical striations, or massive. There is a perfect cleavage. The colour is dark violet, and the lustre is vitreous. H 5–6; SG 3.1
Occurrence. Occurs where lithium-rich pegmatites come into contact with country rocks of basic composition; associates include quartz, biotite, tourmaline, and plagioclase feldspar.
Localities. From USSR; Utö, Sweden; Edison spodumene mine, Pennington Co., South Dakota; Hiddenite mine, Alexander Co., North Carolina; etc.
Treatment. Clean with distilled water.

Bityite
$CaLiAl_2(AlBeSi_2)O_{10}(OH)_2$
Bityite crystallizes in the monoclinic system as thin tabular crystals, rosettes and micaceous aggregates with a perfect cleavage. It is white to yellow. H 5.5; SG 3.0
Occurrence. Occurs in lithium-rich pegmatites.
Localities. From Maharitra, Mt. Bity, Malagasy Republic in association with pink tourmaline and lepidolite; Londonderry, Western Australia.
Treatment. Clean with distilled water.

Beryl $Be_3Al_2Si_6O_{18}$
Beryl crystallizes in the hexagonal system as prismatic crystals, sometimes vertically striated and sometimes terminated by small pyramid faces. The cleavage is indistinct. The colour may be emerald-green (emerald); golden yellow (heliodor); yellow (yellow beryl); light blue to green (aquamarine); pink (morganite); colourless (goshenite); red (bixbite). The lustre is vitreous. H 7–8; SG 2.6–2.9; RI 1.56, 1.59; birefringence 0.006; synthetics with some exceptions, have lower SG and RI.
Occurrence. Occurs in granite pegmatites, biotite schists and pneumatolytic hydrothermal veins.
Localities. Emerald is found in Colombia, particularly at the mines of Muzo and Chivor; Brazil; India; Pakistan; Rhodesia; Australia; Mozambique; Austria; Zambia; USSR; North Carolina; Transvaal. Heliodor is found in South-West Africa; yellow beryl from Brazil and Malagasy Republic; aquamarine from Brazil, Mourne Mountains of Ireland, Malagasy Republic, USSR. Morganite is found in Brazil, Malagasy Republic, USSR, San Diego Co. in California. Goshenite comes from Goshen, Massachusetts; bixbite from the Thomas Mountains, Utah.
Treatment. Clean with dilute acid.
Fashioning. Uses: faceting, cabochons; carvings, beads, tumbling; *cleavage*: imperfect basal; brittle; *cutting angles*: crown 40°–50°, pavilion 43°; *dichroism*: distinct especially in deep colours; *heat sensitivity*: very low.

Euclase $BeAlSiO_4OH$
Euclase crystallizes in the monoclinic system, forming long prismatic crystals with a perfect cleavage. The colour may be green, blue or colourless; the lustre is vitreous. H 7.5; SG 3.0–3.1
Occurrence. May occur in granite pegmatites, mica schists or placer deposits.
Localities. Fine crystals from Minas Gerais, Brazil; also from Tanzania; the Ural Mts. of the USSR; Bavaria, W. Germany; etc.
Treatment. Clean with dilute acid.
Fashioning. Uses: faceting or cabochons; *cleavage*: perfect // to clinopinacoid; *cutting angles*: crown 40°, pavilion 40°; *pleochroism*: weak; *heat sensitivity*: very low.

Bavenite $Ca_4(Be, Al)_4Si_9(O, OH)_{28}$
Bavenite crystallizes in the orthorhombic system as prismatic crystals or fibrous aggregates with lamellar twinning. There is a perfect cleavage. The colour is white, greenish, pink or brown and the lustre is silky. The streak is white. H 5.5; SG 2.7
Occurrence. Occurs in granite pegmatites with beryl.
Localities. From Pala, San Diego Co., California; Rutherford mine, Amelia, Virginia; Baveno, Italy; etc.
Treatment. Clean with distilled water.

Milarite $KCa_2AlBe_2(Si_{12}O_{30}).H_2O$
Milarite crystallizes in the hexagonal system, forming prismatic crystals. The colour is pale green, yellow or colourless, and the lustre is vitreous. H 5–6;

Left: Globular aggregates of platy bavenite on a background of orthoclase

Below: Milarite crystals from Switzerland. These are hexagonal prisms

SG 2.4–2.6
Occurrence. Occurs in pegmatites and syenites or hydrothermal veins.
Localities. From the Valencia mine, Guanajuato, Mexico; St. Gotthard, Switzerland; Henneberg, East Germany; Kola Peninsula, USSR; Swakopmund, South-West Africa.
Treatment. Milarite should be cleaned with distilled water.

Sapphirine $Mg_{3.5}Al_9Si_{1.5}O_{20}$
Sapphirine crystallizes in the monoclinic system as tabular crystals; more commonly found as disseminated grains. The colour is pale blue or grey to green. H 7.5; SG 3.4–3.5
Occurrence. Occurs in aluminium or magnesium-rich metamorphic rocks with a low silicon content.
Localities. From Greenland; Malagasy Republic; India; Quebec; New York State; etc.
Treatment. Sapphirine should be cleaned with dilute acid.

Sheridanite
$(Mg, Al)_6(Si, Al)_4O_{10}(OH)_8$
Sheridanite crystallizes in the monoclinic system and is found massive or as scaly aggregates; there is a perfect cleavage. The colour is greenish, colourless or yellow and the lustre is pearly. H 2–3; SG 2.6–2.8
Occurrence. Occurs in schists.
Localities. From Sheridan Co., Wyoming; and other localities.
Treatment. Sheridanite should be cleaned with distilled water.

Palygorskite
$(Mg, Al)_2Si_4O_{10}(OH).4H_2O$
Palygorskite crystallizes in the monoclinic and orthorhombic systems, usually as thin flexible sheets composed of interlocking fibres. There is an easy cleavage. The colour is white or grey, and the lustre is dull. H very soft; SG 2.2
Occurrence. Occurs in hydrothermal veins or altered serpentinites or granites.
Localities. From the Shetland Islands; Metaline Falls, Washington; Morocco; etc.
Treatment. Palygorskite should be cleaned with distilled water.

Left: Euclase. Most commonly found in Brazil, this crystal is from Tanzania

Lawsonite
$CaAl_2(Si_2O_7)(OH)_2.H_2O$
Lawsonite crystallizes in the orthorhombic system as prismatic crystals or granular masses. There is a perfect cleavage and the colour is white, grey or pinkish with a white streak. There is a vitreous lustre. H 6; SG 3.0–3.1
Occurrence. Occurs in glaucophane schists.
Localities. From Santa Clara, Cuba; France; Italy; Coast Ranges of California; etc.
Treatment. Clean with dilute acid.

Scolecite $CaAl_2Si_3O_{10}.3H_2O$
Scolecite crystallizes in the monoclinic system as slender prismatic crystals or radiating fibrous masses. There is a perfect cleavage. The colour is white or colourless and the lustre silky. H 5; SG 2.2
Occurrence. Occurs in cavities in basalt or in schists.
Localities. From the Isle of Skye; Faroe Islands; Riverside Co., California; etc.
Treatment. Clean with distilled water.

Opposite page: Hexagonal prism of aquamarine, a variety of beryl, from Brazil

Right: Laumontite with quartz from Berufjordur, Iceland

Below right: Epistilbite from Teigarhorn, Iceland. The radiating habit can easily be seen

Prehnite $Ca_2Al_2Si_3O_{10}(OH)_2$
Prehnite crystallizes in the orthorhombic system, usually as compact granular masses or as botryoidal, stalactitic or reniform forms. Crystals are usually tabular or prismatic. The colour is pale or dark green, yellow or white, and the lustre is vitreous. The streak is colourless. Ferroprehnite is an iron-bearing variety. H 6–6.5; SG 2.9
Occurrence. Occurs as a secondary or hydrothermal mineral in basic igneous rocks or in granites and metamorphosed limestones.
Localities. In Scotland, from the Kilpatrick Hills, Strathclyde, Campsie Hills, Central Region and Corstorphine Hill, Edinburgh; Switzerland; France; Keweenaw Co., Michigan; etc.
Fashioning. Uses: faceting, cabochons, tumbling, beads, etc; *cleavage*: distinct basal, parting easily along fibre directions in massive material; *cutting angles*: crown 40°, pavilion 40°; *pleochroism*: weak; *heat sensitivity*: low.

Laumontite $CaAl_2Si_4O_{12}.4H_2O$
Laumontite crystallizes in the monoclinic system, forming square prisms or fibrous, columnar radiating masses. There is a perfect cleavage. The colour is white, grey, yellowish-pink or brown; there is a vitreous lustre. The mineral may become opaque and powdery on exposure to light. H 3–4; SG 2.2–2.4
Occurrence. Occurs in basalts, decomposed granites or in metallic vein deposits and metamorphic rocks.
Localities. From the Kilpatrick Hills and Kilmacolm, Strathclyde, Scotland; Montecatini, Tuscany, Italy; Pine Creek tungsten mine, Bishop, California; etc.

Epistilbite $CaAl_2Si_6O_{16}.5H_2O$
Epistilbite crystallizes in the monoclinic system, forming twinned prismatic crystals or radiated spherical aggregates. The twinning is interpenetrant cruciform; there is a perfect cleavage. The colour is white or pink, with a vitreous lustre. H 4; SG 2.2
Occurrence. Occurs in pegmatites with beryl at Bedford, New York; in basalt in Hawaii; from the Isle of Skye; etc.
Treatment. Clean with distilled water.

Sarcolite $NaCa_4Al_3Si_5O_{19}$
Sarcolite crystallizes in the tetragonal system in small crystals, flesh-pink in colour. Vitreous lustre. H 6; SG 2.9
Localities. From a volcanic rock on Mt. Vesuvius, Italy.
Treatment. Clean with distilled water.

Cymrite $BaAlSi_3O_8OH$
Cymrite crystallizes in the hexagonal system forming hexagonal prisms or fibrous masses. There is a perfect cleavage. It is colourless, dark green or brown with a satiny lustre. H 2–3; SG 3.4
Occurrence. Occurs at the Benallt manganese mine, Rhiw, Gwynedd, Wales; in copper deposits at Ruby Creek, Alaska; from San Benito Co., California.
Treatment. Clean with distilled water.

Banalsite $Na_2BaAl_4Si_4O_{16}$
Banalsite crystallizes in the orthorhombic system with a massive habit. The colour is white and the lustre vitreous. H 6; SG 3.0
Localities. Occurs in a manganese vein at Rhiw, Gwynedd, Wales.
Treatment. Clean with distilled water.

Armenite $BaCa_2Al_6Si_8O_{28}.2H_2O$
Armenite crystallizes in the hexagonal system as prismatic crystals coloured greyish-green, with a vitreous lustre. H 7.5; SG 2.8
Localities. Occurs with axinite, pyrrhotine and quartz from the silver-bearing calcite veins of the Armen mine, Kongsberg, Norway.
Treatment. Clean with distilled water.

Carpholite $MnAl_2Si_2O_6(OH)_4$
Carpholite crystallizes in the orthorhombic system as radiated tufts coloured yellow. There is a perfect cleavage. H 5–5.5; SG 2.9–3.0
Localities. Occurs near Neuville, Ardennes, Belgium; Harz Mts, Germany; Prilep, Macedonia, Yugoslavia; etc.
Treatment. Clean with dilute acid.

Sursassite $Mn_5Al_4Si_5O_{21}.3H_2O$
Sursassite crystallizes in the monoclinic system as fibrous masses coloured reddish-brown, with a silky lustre. H not measured; SG 3.3
Localities. Occurs with calcite and quartz in veins in iron-manganese ores near Woodstock, New Brunswick, Canada. Also from the manganese deposits at Val d'Err, Grisons, Switzerland.

Ganophyllite $NaMn_3(Si, Al)_4O_{10}(OH)_4$
Ganophyllite crystallizes in the monoclinic system as short prismatic or tabular crystals; rosettes or foliated forms are also found. There is a perfect cleavage. The colour is brown or brownish-yellow. H 4–4.5; SG 2.8–2.9
Occurrence. Occurs with baryte in manganese ores or with rhodonite, etc.
Localities. From manganese ores at Santa Clara Co., California; with rhodonite at Franklin, New Jersey; etc.
Treatment. Clean with distilled water.

Ekmanite $(Fe, Mn, Mg)_6(Si, Al)_8O_{20}(OH)_8.2H_2O$
Ekmanite probably crystallizes in the orthorhombic system as foliated masses or as scales. There is a perfect cleavage.

Left: Fine crystal of cordierite from Orijärvi, Finland. Cordierite is more frequently found as pebbles

The colour is grey to black or green. H 2–2.5; SG 2.8
Localities. Occurs with pyrochroite in cavities in magnetite ore at Brunsjo, Sweden.
Treatment. Clean with distilled water.

Stilpnomelane
$K(Fe, Mg, Al)_3Si_4O_{10}(OH)_2.nH_2O$
Stilpnomelane crystallizes in the triclinic system as thin foliated plates or coatings. There is a perfect cleavage and the colour is black to reddish-brown, sometimes dark green. The lustre is pearly. H 3–4; SG 2.5–2.9
Occurrence. Occurs in schists with chlorite, epidote and lawsonite; also in iron ore deposits.
Localities. From Canada; Greenland; USSR; Switzerland; Minnesota; Michigan; and many other places.
Treatment. Clean with distilled water.

Cordierite $(Mg, Fe^{3+})_2Al_4Si_5O_{18}$
Cordierite, sometimes called iolite, crystallizes in the orthorhombic system forming short prismatic crystals; it is more often found massive. Twinning is common. The colour is blue to brown or yellowish and sometimes strong pleochroism is displayed. There is a vitreous lustre. The streak is colourless. H 7–7.5; SG 2.5–2.8; RI 1.52–1.57
Occurrence. Occurs in thermally altered aluminium-rich rocks, gneisses, schists and pegmatites.
Localities. From Sri Lanka; Malagasy Republic; Guilford, Connecticut; and many other localities.
Treatment. Clean with dilute acid.
Fashioning. Uses: faceting, cabochons, beads, carvings, tumbling, etc.; *cleavage*: distinct brachypinacoidal; brittle; *cutting angles*: crown 40°–50°, pavilion 43°; *pleochroism*: strong; *heat sensitivity*: low.

Staurolite
$(Fe, Mg)_2Al_9Si_4O_{23}(OH)$
Staurolite crystallizes in the monoclinic system forming short prismatic crystals with cruciform twinning common. The colour is dark reddish-brown or yellowish with a resinous lustre. The streak is colourless or grey. H 7–7.5; SG 3.6–3.8
Occurrence. Occurs in mica schists with garnet, muscovite and kyanite.
Localities. Found in the Ticino, Switzerland; Brittany, France; Chesterfield, Hampshire Co., Massachusetts; and many other localities.
Treatment. Clean with dilute acid.
Fashioning. Uses: faceting or cabochons; *cleavage*: distinct brachypinacoidal; brittle; *cutting angles*: crown 40°, pavilion 40°; *pleochroism*: distinct; *heat sensitivity*: low.

Chloritoid
$(Mg, Fe^{2+})_2Al_4Si_2O_{10}(OH)_4$
Chloritoid crystallizes in the triclinic and monoclinic systems as pseudohexagonal tabular crystals; more commonly as foliated masses or as scales. Lamellar twinning is common and there is a perfect cleavage. It has a pearly lustre. The colour is dark grey or greenish to black. Closely related are sismondite

Below left: Twinned crystals of staurolite from Antsirabe, Malagasy Republic. Staurolite is often called cross stone

Below: A staurolite crystal from the Italian Alps

Right: Chloritoid from Pregratten, Austria

and ottrelite, and, reputedly, salmite and venasquite. H 6.5; SG 3.6
Occurrence. Occurs in phyllites, in mica schists or as a hydrothermal alteration product in lavas.
Localities. From the Tyrol, Austria; Deep River, North Carolina; Shetland Islands, Scotland; and many other localities.
Treatment. Clean with distilled water.

Ferrocarpholite
$(Fe, Mg)Al_2Si_2O_6(OH)_4$
Ferrocarpholite crystallizes in the orthorhombic system as prismatic crystals with a perfect cleavage. The colour is dark green. H 5.5; SG 3.0
Localities. Occurs with rutile and zircon in a quartz vein in the eastern Celebes; Calabria, Italy.
Treatment. Clean with distilled water.

Yagiite
$(Na, K)_3Mg_4Al_6(Si, Al)_{12}O_{60}$
Yagiite crystallizes in the hexagonal system and occurs as small grains in the Colomera iron meteorite.

Pumpellyite
$Ca_2MgAl_2(SiO_4)(Si_2O_7)(OH)_2 . H_2O$
Pumpellyite crystallizes in the monoclinic system as fibrous crystals or as flat plates. Twinning is common. The colour is green to brown, with a vitreous lustre. H 6; SG 3.2
Occurrence. Occurs in a wide range of igneous and metamorphic rocks.
Localities. In copper-bearing ores in Michigan; in glaucophane schists in California; from New Jersey; South Africa; Japan; USSR; Haiti; etc.
Treatment. Clean with distilled water.

Idocrase
$Ca_{10}Mg_2Al_4(SiO_4)_5(Si_2O_7)_2(OH)_4$
Sometimes known as vesuvianite, idocrase crystallizes in the tetragonal system as short prismatic crystals or as granular masses. The colour is greenish-yellow to brown, sometimes red, white or blue; the lustre is vitreous or resinous. The streak is white. Varieties include californite, which is translucent white or yellow; cyprine, which is blue and opaque; chrome-idocrase, bright green. H 6–7; SG 3.3–3.4; RI 1.71–1.73
Occurrence. Occurs in metamorphosed calcareous rocks and in veins in ultrabasic rocks.
Localities. From Mt. Vesuvius, Italy; Zermatt, Switzerland; Ala Valley, Piedmont, Italy; Lake Ladoga, Finland; Androscoggin Co., Maine; Franklin, New Jersey; Quebec and Ontario; USSR and elsewhere.
Treatment. Clean with distilled water.
Fashioning. Uses: faceting, cabochons, tumbling, etc; *cleavage*: indistinct prismatic; brittle; *cutting angles*: crown 40°, pavilion 40°; *dichroism*: fairly weak; *heat sensitivity*: low.

Osumilite **$(K, Na)(Mg, Fe^{2+})_2$-$(Al, Fe^{3+})_3(Si, Al)_{12}O_{30} . H_2O$**
Osumilite crystallizes in the hexagonal system as short prismatic crystals, dark blue to black in colour, with a vitreous lustre. H not measured; SG 2.6
Localities. Occurs with feldspars and other minerals in andesite at MacKenzie Pass, Lane Co., Oregon; also from Sakkabira, Kagosima Prefecture, Kyushu, Japan.
Treatment. Clean with distilled water.

Leucophane
$(Ca, Na)_2BeSi_2(O, F, OH)_7$
Leucophane crystallizes in the orthorhombic system as pseudotetragonal crystals with twinning common; sometimes of penetration type. Also as spherules. There is a perfect cleavage and the colour is white to yellowish-green, with a vitreous lustre. H 4; SG 3.0
Localities. Occurs in nepheline–syenite pegmatites at Mont St. Hilaire, Quebec, Canada; Lovozero intrusion, Kola Peninsula, USSR; etc.
Treatment. Clean with dilute acid.

Above right: A crystal of idocrase from Italy, showing prism and pyramid faces

Right: Idocrase crystals from Quetta, Baluchistan, Pakistan

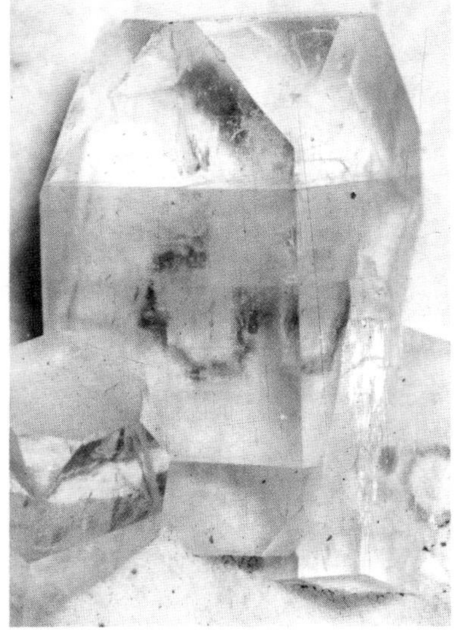

Left: Apophyllite from Poona, India

Far left: Pseudocubic apophyllite. The crystal system is tetragonal. From Rio Grande do Sul, Brazil

Below left: Apophyllite crystals of tabular habit from Bolzano, Italy

Hsianghualite $Ca_3Li_2Be_3(SiO_4)_3F_2$
Hsianghualite crystallizes in the cubic system as dodecahedral crystals or as granular masses. The colour is white, the lustre vitreous. H 6.5; SG 3.0
Localities. Occurs in phlogopite veins in metamorphosed limestone in Hunan Province, China.
Treatment. Clean with distilled water.

Zeophyllite $Ca_4Si_3O_7(OH)_4F_2$
Zeophyllite crystallizes in the triclinic system as plates or in spherical forms with a radiating foliated structure. There is a perfect cleavage. The colour is white. H 3; SG 2.7
Localities. Occurs with apophyllite in basalt at Alten Berg, Czechoslovakia.

Cuspidine $Ca_4Si_2O_7(F, OH)_2$
Cuspidine crystallizes in the monoclinic system as small spear-shaped crystals or as granular masses. Lamellar twinning occurs. The colour is white, grey, greenish or rose red, the lustre is vitreous. H 5–6; SG 2.8–3.0
Localities. In metamorphosed limestone at Crestmore, Riverside Co., California; with nasonite at Franklin, New Jersey; and at other localities.
Treatment. Clean with distilled water.

Bultfonteinite $Ca_2SiO_2(OH, F)_4$
Bultfonteinite crystallizes in the triclinic system as acicular crystals or in spherulitic masses. Twinning is common. The colour is pink or colourless. H 4.5; SG 2.7
Localities. Occurs with calcite and apophyllite at Bultfontein, Dutoitspan and Jagersfontein mines, Kimberley, South Africa.

Apophyllite
$KCa_4Si_8O_{20}(F, OH).8H_2O$
Apophyllite crystallizes in the tetragonal system as pseudocubic crystals or as granular masses. There is a perfect cleavage and the colour is white, grey or pale yellow, greenish or reddish. The lustre is vitreous or pearly and the streak is white. H 4.5–5; SG 2.3–2.4
Occurrence. Occurs in cavities in basalt, also in gneiss or limestone or as a low-temperature mineral in sulphide ores.

Localities. From the Isle of Skye, Scotland; Collinward, N. Ireland; Lake Superior copper mines, Michigan; Mexico; Brazil and elsewhere.
Treatment. Clean with distilled water.
Fashioning. Uses: faceting or cabochons; *cleavage*: highly perfect / / to basal pinacoid, perfect prismatic; brittle; *cutting angles*: crown 40°–50°, pavilion 43°; *pleochroism*: weak; *heat sensitivity*: fairly high, but unaffected by normal dopping temperatures.

Combeite $Na_4Ca_3Si_6O_{16}(OH, F)_2$
Combeite crystallizes in the hexagonal system as hexagonal prisms. They are colourless and have an SG of 2.8
Localities. Occurs with hornblende and other minerals in a nephelinite lava from Mt. Shaheru, Zaire.
Treatment. Clean with distilled water.

Canasite $(Na, K)_5(Ca, Mn, Mg)_4(Si_2O_5)_5(OH, F)_3$
Canasite crystallizes in the monoclinic system as grains with polysynthetic twinning and a perfect cleavage. The colour is greenish-yellow and the streak colourless. The lustre is vitreous. H not measured; SG 2.7
Localities. Occurs with lamprophyllite in pegmatite in the Khibina Tundra alkali massif, USSR.
Treatment. Clean with distilled water.

Mosandrite
$(Na, Ca, Ce)_3Ti(SiO_4)_2F$
Mosandrite crystallizes in the monoclinic system as long prismatic crystals with striations on the prism faces. Massive forms also encountered. Lamellar twinning occurs. The colour is yellowish-green to reddish-brown, with a vitreous lustre. H5; SG 2.9–3.5
Occurrence. Occurs in nepheline-syenites and associated pegmatites.
Localities. From Mont St. Hilaire, Quebec, Canada; Julianehaab, Greenland; Kola Peninsula, USSR; Transvaal, South Africa.
Treatment. Clean with distilled water.

Hiortdahlite
$(Ca, Na)_{13}Zr_3Si_9(O, OH, F)_{33}$
Hiortdahlite crystallizes in the triclinic system as flattened tabular crystals, often with polysynthetic twinning. The colour is light yellow to yellowish-brown, with a vitreous lustre. H 5.5; SG 3.3
Occurrence. Occurs in alkali rocks and pegmatites.
Localities. From Langesundfjord area, Norway; Mt. Somma, Italy; Iles de Los, Guinea; etc.
Treatment. Hiortdahlite should be cleaned with distilled water.

Seidozerite
$Na_2(Zr, Ti, Mn)_2Si_2O_8F_2$
Seidozerite crystallizes in the monoclinic system as radiating crystals or fibrous masses. There is a perfect cleavage. The colour is brownish-red, with a vitreous lustre. H 4–5; SG 3.5
Localities. Occurs in nepheline syenite pegmatites of the Louozero intrusion, Kola Peninsula, USSR.
Treatment. Seidozerite should be cleaned with distilled water.

Rosenbuschite
$(Ca, Na)_3(Zr, Ti)Si_2O_8F$
Rosenbuschite crystallizes in the triclinic system as fibrous masses. Crystals are prismatic, acicular, with a perfect cleavage and a silky lustre. The colour is grey or orange. H 5–6; SG 3.3
Occurrence. Occurs in nepheline-syenite deposits.

Right: Prismatic topaz crystal showing prism and dome faces. From Siberia, USSR

Localities. From the Kola Peninsula, USSR; Langesundfjord area, Norway; Red Hill, Moultonboro, New Hampshire.
Treatment. Clean with distilled water.

Narsarsukite $Na_2(Ti, Fe)Si_4(O, F)_{11}$
Narsarsukite crystallizes in the tetragonal system as prismatic crystals with a perfect cleavage. The colour is yellow, green, brown or colourless; the lustre is vitreous. H 6–7; SG 2.8
Localities. Occurs with quartz and aegirine in north-central Montana and in carbonatite at Mont St. Hilaire, Quebec, Canada; etc.
Treatment. Clean with distilled water.

Sonolite $Mn_9(SiO_4)_4(OH, F)_2$
Sonolite crystallizes in the monoclinic system as prismatic crystals or fine-grained masses. The colour is red or orange to brown. The lustre can be vitreous or dull. H 5.5; SG 3.8
Localities. Occurs with willemite, zincite, etc, at Franklin, New Jersey; and also in Japan.
Treatment. Clean with distilled water.

Right: This is the most prized colour for topaz. The crystal shows prism and dome faces. From Brazil

Topaz $Al_2SiO_4(OH, F)_2$
Topaz crystallizes in the orthorhombic system as prismatic crystals, often with well-developed terminations and a perfect cleavage; often found as water-worn pebbles. The colour is red, golden yellow, blue, brown, colourless or pink; there is a vitreous lustre. The streak is colourless. H 8; SG 3.4–3.5; RI 1.62–1.64
Occurrence. Occurs in pegmatites and quartz veins, contact zones and alluvial deposits.
Localities. From Minas Gerais, Brazil; Nigeria; San Diego Co., California; Rhodesia; Cornwall; Mourne Mts., Ireland; Urals, USSR; Burma; South-West Africa and elsewhere.
Treatment. Clean with dilute acid.
Fashioning. Uses: faceting, cabochons, tumbling, etc; *cleavage*: perfect // to basal pinacoid; brittle; *cutting angles*: crown 40°, pavilion 40°; *pleochroism*: distinct in deep colours; *heat sensitivity*: low.

Leifite
$(NaH_3O)_2(Si, Al, Be, B)_7(O, F, OH)_{14}$
Leifite crystallizes in the hexagonal system as striated colourless acicular crystals with a vitreous lustre. H 6; SG 2.6
Localities. Found in alkali pegmatites at Narsarsuk, Greenland.
Treatment. Clean with distilled water.

Meliphane
$(Ca, Na)_2Be(Si, Al)_2(O, F)_7$
Meliphane crystallizes in the tetragonal system as thin tabular crystals or lamellar aggregates with a vitreous lustre. There is a perfect cleavage. The colour is yellow or colourless. H 5–5.5; SG 3.0
Localities. Found in nepheline-syenite pegmatites at Julianehaab, Greenland; etc.
Treatment. Clean with distilled water.

Götzenite
$(Ca, Na)_7(Ti, Al)_2Si_4O_{15}(F, OH)_3$
Götzenite crystallizes in the triclinic system as colourless prismatic crystals. There is a perfect cleavage. H not measured; SG 3.1
Localities. Occurs in lava at Mt. Shaheru, Zaire.
Treatment. Clean with distilled water.

Barytolamprophyllite
$Na_2(Ba, Sr)_2Ti_3(SiO_4)_4(OH, F)_2$
Barytolamprophyllite crystallizes in the monoclinic system, and has a perfect cleavage. It is found as foliated aggregates and is dark brown. H2–3; SG 3.6
Localities. Occurs with nepheline, cancrinite, etc, in the Lovozero complex, Kola Peninsula, USSR.
Treatment. Clean with distilled water.

Zunyite $Al_{13}Si_5O_{20}(OH, F)_{18}Cl$
Zunyite crystallizes in the cubic system as tetrahedra or twinned octahedra. The colour is white or grey to pink, with a vitreous lustre. H 7; SG 2.8
Occurrence. Occurs in Colorado as a vein mineral or as an alteration product of feldspar, and at other localities.
Treatment. Clean with dilute acid.

Pennaite
Pennaite, a silicate and chloride of sodium, calcium, iron, manganese, titanium and zirconium, crystallizes in the monoclinic system as prismatic crystals with polysynthetic twinning. The colour is yellow to brown. H and SG not known.
Localities. Occurs with nepheline and other minerals on the Pocos de Caldas plateau, Brazil.
Treatment. Clean with distilled water.

Beryllosodalite $Na_4BeAlSi_4O_{12}Cl$
Beryllosodalite crystallizes in the tetragonal system as crypto-crystalline masses with a vitreous lustre. The colour is rose pink, bluish, greenish or white with a strong rose fluorescence under ultraviolet light. H about 4; SG 2.2
Localities. Occurs in pegmatites in the Kola Peninsula, USSR; and in a nepheline-syenite pegmatite at Tugtup agtakôrfia, Ilimaussaq, Greenland.
Treatment. Clean with distilled water.

Tugtupite $Na_4BeAlSi_4O_{12}Cl$
Tugtupite crystallizes in the tetragonal system in compact masses. The colour is pink or red with a bright red fluorescence under ultraviolet light. There is a vitreous lustre. H about 4; SG 2.3
Localities. Occurs in a nepheline-syenite pegmatite at Tugtup agtakôrfia, Ilimaussaq, Greenland.
Treatment. Clean with distilled water.
Fashioning. Uses: faceting or cabochons; *cutting angles*: crown 40°–50°, pavilion 43°; *dichroism*: distinct; *heat sensitivity*: low.

Eudialyte $Na_4(Ca, Fe, Ce, Mn)_2ZrSi_6O_{17}(OH, Cl)_2$
Eudialyte crystallizes in the hexagonal system as tabular crystals of trigonal habit with a vitreous lustre. The colour is yellow, brown or red with a colourless streak. Eucolite is a calcium-rich variety. H 5–5.5; SG 2.7–2.9
Occurrence. Occurs in nepheline-syenite deposits.
Localities. From Mont St. Hilaire, Quebec, Canada; Julianehaab, Greenland; Langesundfjord area, Norway; Kola Peninsula, USSR; Malagasy Republic; etc.
Treatment. Clean with distilled water.

Nasonite $Ca_4Pb_6Si_6O_{21}Cl_2$
Though nasonite crystallizes in the hexagonal system as prismatic crystals,

it more often occurs as granular masses. The colour is white and the lustre greasy. H 4; SG 5.5
Localities. Occurs with datolite, willemite and other minerals at Franklin, New Jersey; and also from Långban in Sweden.
Treatment. Clean with distilled water.

Friedelite
$(Mn, Fe)_8Si_6O_{18}(OH, Cl)_4 . 3H_2O$
Friedelite crystallizes in the hexagonal system as tabular crystals or lamellar aggregates; it has a perfect cleavage. The colour is pale pink to dark red or yellow, the lustre is vitreous. H 4–5; SG 3.0
Localities. Occurs with willemite and other minerals at Franklin, New Jersey; Adervielle, France; and at some other localities.
Treatment. Clean with distilled water.

Left: Tilleyite from California

Pyrosmalite
$(Mn, Fe)_{14}Si_3O_7(OH, Cl)_6$
Pyrosmalite crystallizes in the hexagonal system as hexagonal prisms or foliated masses. There is a perfect cleavage. The colour is dark green to pale brown, with a pearly lustre. H 4–4.5; SG 3.1–3.2
Localities. Occurs with calcite and pyroxene at Broken Hill, New South Wales; Nordmark and Dannemora, Sweden.
Treatment. Clean with distilled water.

Manganpyrosmalite
$(Mn, Fe)_8Si_6O_{15}(OH, Cl)_{10}$
Manganpyrosmalite crystallizes in the hexagonal system as granular masses; it has a perfect cleavage. The colour is brown, the lustre pearly. H 4.5; SG 3.1
Localities. Occurs with franklinite and other minerals at Franklin, New Jersey.
Treatment. Clean with distilled water.

Muirite
$Ba_{10}Ca_2MnTiSi_{10}O_{30}(OH, Cl, F)_{10}$
Muirite crystallizes in the tetragonal system as tetragonal crystals or as grains. The colour is orange, with a vitreous lustre. H 2.5; SG 3.8
Localities. Occurs in metamorphic rock in Fresno Co., California.
Treatment. Clean with distilled water.

Tilleyite $Ca_5Si_2O_7(CO_3)_2$
Tilleyite crystallizes in the monoclinic system as masses or grains with a perfect cleavage. The colour is white or colourless. H not known; SG 2.8
Localities. Occurs in thermally metamorphosed limestones at Carlingford, Ireland; Crestmore, Riverside Co., California.
Treatment. Clean with distilled water.

Spurrite $Ca_5Si_2O_8CO_3$
Spurrite crystallizes in the monoclinic system as granular masses with a vitreous lustre. The colour is grey or pale blue. H 5; SG 3.0
Localities. Occurs as a contact mineral at Scawt Hill, N. Ireland; and other localities.
Treatment. Store in a sealed container.

Scawtite $Ca_7Si_6O_{18}(CO_3) . 2H_2O$
Scawtite crystallizes in the monoclinic system as thin tabular crystals in

Above left: Eudialyte in a specimen from the Kola Peninsula, USSR

Left: Pyrosmalite from Norway. The hexagonal prismatic form is clear

Above: Datolite crystals from New Jersey

divergent groups or as grains. Colourless with a vitreous lustre. H 4.5–5; SG 2.7
Localities. Occurs at Scawt Hill, N. Ireland; and other localities.
Treatment. Clean with distilled water.

Carletonite
$KNa_4Ca_4Si_8O_{18}(CO_3)_4(F, OH) . H_2O$
Carletonite crystallizes in the tetragonal system as masses; it has a perfect cleavage. The colour is pink or blue; the lustre is vitreous to waxy. H 4–4.5; SG 2.4
Localities. Occurs with calcite, fluorite and other minerals at Mont St. Hilaire, Quebec, in a nepheline-syenite.

Kainosite
$Ca_2(Ce, Y)_2(SiO_4)_3(CO_3) . H_2O$
Kainosite (or cenosite) crystallizes in the orthorhombic system as prismatic or pseudotetragonal crystals with a vitreous lustre coloured yellow to brown, or light red to colourless. H 5–6; SG 3.3–3.6
Localities. Occurs in granite pegmatites at Cotopaxi, Colorado; at the uranium mine at Bancroft, Ontario; Val Curnera, Switzerland; and also from some other localities.
Treatment. Clean with distilled water.

Manandonite
$LiAl_4(AlBSi_2O_{10})(OH)_8$
Manandonite crystallizes in the monoclinic system as lamellar aggregates and crusts or plates. It has a perfect cleavage and is colourless. It has a pearly lustre. H about 2.5; SG 2.9
Localities. Occurs with tourmaline and other minerals in pegmatite at Mt. Bity, on the Manandona River in the Malagasy Republic.
Treatment. Clean with distilled water.

Searlesite $NaBSi_2O_6 . H_2O$
Searlesite crystallizes in the monoclinic system as prismatic crystals or radiating fibrous spherulites. There is a perfect cleavage. The colour is white or light brown. H not known, but soft; SG 2.4
Localities. Occurs in clay at Searles Lake, San Bernardino Co., California, and elsewhere in the US.
Treatment. Clean with distilled water.

Reedmergnerite $NaBSi_3O_8$
Reedmergnerite crystallizes in the triclinic system as colourless stubby prisms with a perfect cleavage and a vitreous lustre. H 6–6.5; SG 2.7
Localities. Occurs in unmetamorphosed dolomitic oil shales of the Green River formation in Utah.
Treatment. Clean with distilled water.

Stillwellite (Ce, La, Ca)$BSiO_5$
Stillwellite crystallizes in the hexagonal system as rhombohedral crystals or compact masses. The colour is light blue, brown or yellow, and the lustre is resinous. H not known; SG 4.6
Localities. Occurs with garnet and uraninite in Queensland, Australia: Langesundfjord area, Norway.
Treatment. Clean with distilled water.

Danburite $CaB_2Si_2O_8$
Danburite crystallizes in the orthorhombic system as prismatic crystals with diamond-shaped cross-section, vertically striated; also as granular masses. They are colourless, yellow, gold, pale pink or brown, with a vitreous lustre. The streak is colourless. H 7; SG 3.0; RI 1.63, 1.64
Localities. Occurs with feldspar in dolomite at Danbury, Fairfield Co., Connecticut; from Charcas, San Luis Potosi, Mexico; Uri, Switzerland; fine yellows from Mogok, Burma.
Treatment. Clean with distilled water.
Fashioning. Uses: faceting or cabochons; *cleavage*: none; *cutting angles*: crown 40°, pavilion 40°; *pleochroism*: weak; *heat sensitivity*: very low.

Datolite $CaBSiO_4OH$
Datolite crystallizes in the monoclinic system as short prismatic crystals or as granular masses. The colour is white, colourless, pale yellow, pale green or reddish-brown. There is a vitreous lustre. The streak is colourless. H 5–5.5; SG 2.8–3.0; RI 1.62, 1.67
Occurrence. Occurs in cavities in basic igneous rocks associated with prehnite and calcite; may also occur in metallic veins, in granites and in other rocks.
Localities. From Fife, Scotland; Norway; Italy; New Jersey; the copper area of Keweenaw, Michigan; Riverside and San Bernardino Counties, California; etc.
Treatment. Clean with distilled water.
Fashioning. Uses: faceting or cabochons; *cleavage*: not significant; brittle; *cutting angles*: crown 40°, pavilion 40°; *pleochroism*: weak; *heat sensitivity*: low.

Bakerite
$Ca_4B_4(BO_4)(SiO_4)_3(OH)_3 . H_2O$
Bakerite crystallizes in the monoclinic system as prismatic crystals and occurs in nodules and veins. The crystals are white or colourless, and have a vitreous lustre. H 4.5; SG 2.9
Localities. Occurs in an altered volcanic rock in Upper Baker Canyon, Death Valley, and elsewhere in California.
Treatment. Clean with distilled water.

Howlite $Ca_2B_5SiO_9(OH)_5$
Howlite crystallizes in the monoclinic system as tabular crystals or compact masses. The colour is white, though the mineral is frequently stained to imitate turquoise. It has a vitreous lustre. H 3.5; SG 2.4
Localities. Occurs with colemanite in Los Angeles Co., the Mohave desert, San Bernardino Co., and elsewhere in California; also in Nova Scotia.
Treatment. Clean with distilled water.
Fashioning. Uses: cabochons, tumbling, carving, intaglios, beads, etc; *cleavage*: none, quite tough; *heat sensitivity*: fairly low.

Right: Fine crystals of axinite from Bourg d'Oisans, France. Axinite is strongly pleochroic

Left: Superb tourmaline crystals from Elba

Below left: Fine pink prismatic tourmaline crystals, showing the vertical striation of the prism faces

Harkerite $Ca_{48}Mg_{16}Al_3(BO_3)_{13}$-$(CO_3)_{19}(SiO_4)_{14}(OH)_2Cl_2.2H_2O$
Harkerite crystallizes in the cubic system as colourless octahedra with a vitreous lustre. H not known; SG 2.9
Localities. Occurs at the contact of dolomitic limestone with Tertiary granite at Broadford, Isle of Skye, Scotland.
Treatment. Harkerite should be cleaned with distilled water.

Cappelenite $(Ba, Ca, Ce, Na)_3$-$(Y, Ce, La)_6(BO_3)_6Si_3O_9$
Cappelenite crystallizes in the hexagonal system as prismatic crystals coloured greenish-brown. H 6; SG 4.4
Localities. Occurs in a nepheline-syenite on an island in Langesundfjord, Norway.
Treatment. Cappelenite should be cleaned with distilled water.

Dumortierite $Al_7O_3(BO_3)(SiO_4)_3$
Dumortierite crystallizes in the orthorhombic system as granular or fibrous masses usually coloured blue to violet, or pink to brown. Their lustre is vitreous or dull. The streak is white. H 8.5; SG 3.4
Occurrence. Occurs in aluminium-rich metamorphic rocks and sometimes in pegmatites.
Localities. From the Malagasy Republic; Arizona, Colorado, California; and elsewhere.
Treatment. Remove iron stains with oxalic acid.
Fashioning. Uses: faceting (rare), cabochons, carvings, beads, tumbling, etc; *cleavage*: distinct macropinacoidal; *cutting angles*: crown 40°, pavilion 40°, (estimated); *pleochroism*: strong in transparent material, not significant in massive material; *heat sensitivity*: low.

Holtite $(Al, Sb, Ta)_7(B, Si)_4O_{18}$
Holtite crystallizes in the orthorhombic system as pseudohexagonal crystals; it forms pebbles and twinning occurs. The colour is brown or orange to green with a bright yellow fluorescence under long-wave ultraviolet light. There is a vitreous lustre. H 8.5; SG 3.9
Localities. Occurs with tantalite at Greenbushes, Western Australia.
Treatment. Clean with distilled water.

Painite $Al_{20}Ca_4BSiO_{38}$
Painite crystallizes in the hexagonal system. The only specimen known is in the British Museum (Natural History), and this appears to be pseudo-orthorhombic. It is very dark red and has a vitreous lustre. H about 8; SG 4.0
Localities. From Mogok, Burma.
Treatment. Clean with distilled water.

Leucosphenite $BaNa_4Ti_2B_2Si_{10}O_{30}$
Leucosphenite crystallizes in the monoclinic system as tabular crystals with twinning common. The colour is white, colourless or pale blue. There is a vitreous lustre. H 6.5; SG 3.0
Localities. Occurs with shortite and analcime in the Green River formation, Wyoming and Utah; Mont St. Hilaire, Quebec; etc.
Treatment. Clean with dilute acid.

Axinite
$(Ca, Mn, Fe, Mg)_3Al_2BSi_4O_{15}(OH)$
Axinite's wedge-shaped tabular crystals belong to the triclinic system; the mineral also occurs in lamellar masses. The colour is violet-brown or colourless to yellow, and the lustre is vitreous. The streak is colourless. Varieties include tinzenite, with more manganese than iron, and with calcium up to 1.5%; ferroaxinite, with more iron than manganese, and with calcium above 1.5%; and manganaxinite, with more manganese than iron, and with calcium above 1.5%. H 6.5–7; SG 3.3–3.4; RI 1.68, 1.69
Occurrence. In areas of contact metamorphism or metasomatism with calcite and quartz, etc.
Localities. From Baja California, Mexico; Bourg d'Oisans, Isère, France; Cornwall; tinzenite from Switzerland; fine yellow crystals, though very small, from Franklin, New Jersey; etc.
Treatment. Clean with dilute acid.
Fashioning. Uses: faceting or cabochons; *cleavage*: distinct brachypinacoidal; brittle; *cutting angles*: crown 40°, pavilion 40°; *pleochroism*: strong; *heat sensitivity*: low.

Kornerupine
$Mg_3Al_6(Si, Al, B)_5O_{21}(OH)$
Kornerupine crystallizes in the orthorhombic system as prismatic crystals with a vitreous lustre. They may be dark green, emerald green, pink, yellow, brown, colourless or black. H 6–7 SG 3.3–5.0; RI 1.66, 1.69
Localities. Occurs with quartz, orthoclase feldspar and other minerals at Lac Ste-Marie, Gatineau Co., Quebec; from Itrongay, Malagasy Republic, where it is found with zircon and spinel; a vanadium-coloured (apple-green) variety from Kenya; from the gem gravels of Sri Lanka; etc.
Treatment. Clean with dilute acid.
Fashioning. Uses: faceting or cabochons; *cleavage*: distinct to perfect prismatic; brittle; *cutting angles*: crown 40°, pavilion 40°; *pleochroism*: distinct; *heat sensitivity*: low.

Tourmaline
$(Na, Ca)(Li, Mg, Fe^{2+}, Al)_3$-$(Al, Fe^{3+})_6B_3Si_6O_{27}(O, OH, F)_4$
Tourmaline crystallizes in the hexagonal system as prismatic crystals with a 'rounded triangular' cross-section and vertically striated. They have a vitreous lustre. Colours include red (rubellite), blue (indicolite), yellow, brown, green, black, colourless or parti-coloured. H 7; SG 3.0–3.1; RI 1.62, 1.64; double refraction 0.018
Occurrence. Occurs in granite pegmatites and metamorphic rocks.
Localities. From San Diego Co., California; Sri Lanka; Brazil; Mozambique; Malagasy Republic; Maine; etc.
Treatment. Clean with dilute acid.
Fashioning. Uses: faceting, cabochons, carvings, beads, tumbling, etc; *cleavage*: not significant; brittle; *cutting angles*: crown 40°, pavilion 40°; *dichroism*: strong; *heat sensitivity*: fairly high, keep the stone cool during cutting or polishing.

Homilite $(Ca, Fe)_3B_2Si_2O_{10}$
Homilite crystallizes in the monoclinic system as brown or black tabular crystals. H 5; SG 3.4
Localities. Occurs on islands in the Langesundfjord, Norway.
Treatment. Clean with distilled water.

Serendibite $Ca_4(Mg, Fe, Al)_6$-$(Al, Fe)_9(Si, Al)_6B_3O_{40}$
Serendibite crystallizes in the triclinic system as granular masses coloured

Right: Lomonosovite in a nepheline syenite

various shades of blue and showing polysynthetic twinning. They have a vitreous lustre. H 6.5–7; SG 3.4
Localities. Occurs as a contact mineral in limestone from Crestmore, Riverside Co., California; Tanzania; Sri Lanka; etc.
Treatment. Serendibite should be cleaned with dilute acid.

Grandidierite (Mg, Fe)Al_3BSiO_9
Grandidierite crystallizes in the orthorhombic system as elongated crystals or as masses. There is a perfect cleavage. The colour is greenish-blue, the lustre vitreous. H 7.5; SG 2.9
Localities. Occurs in pegmatites in the Malagasy Republic.
Treatment. Grandidierite should be cleaned with dilute acid.

Hyalotekite
$(Pb, Ca, Ba)_4BSi_6O_{17}(OH, F)$
Hyalotekite probably crystallizes in the orthorhombic system as crystalline masses coloured white or grey, with a vitreous lustre. H 5–5.5; SG 3.8
Localities. Occurs with barylite at Långban, Sweden.
Treatment. Clean with distilled water.

Melanocerite $Ce_4CaBSi_2O_{12}(OH)$
Melanocerite crystallizes in the hexagonal system as rhombohedral crystals or compact masses. The colour is brown or black, the lustre is resinous. H 6; SG 4.1–4.2
Localities. Occurs in alkali pegmatites at the Cardiff uranium mine, Wilberforce, Ontario; Langesundfjord, Norway; and elsewhere.
Treatment. Melanocerite should be cleaned with distilled water.

Tritomite $(Y, Ca, La, Fe)_5$-$(Si, B, Al)_3(O, OH, F)_{13}$
Tritomite crystallizes in the hexagonal system as compact masses or grains coloured dark green, reddish-brown or black. H 3.5–6.5; SG 3.0–3.4
Localities. Occurs with magnetite and zircon in a granite pegmatite in Sussex Co., New Jersey and associated with red apatite, diopside and purple fluorite at Cardiff, Ontario.
Treatment. Clean with distilled water.

Kolbeckite $ScPO_4.2H_2O$
Kolbeckite crystallizes in the orthorhombic system as prismatic crystals coloured yellow or blue or colourless. They have a vitreous lustre. H 5; SG 2.4
Localities. Occurs with wardite in altered variscite nodules at Fairfield, Utah, and with silver ore from Felsöbánya, Romania.
Treatment. Clean with distilled water.

Viséite
$NaCa_5Al_{10}Si_3P_5O_{30}(OH, F)_{18}.16H_2O$
Viséite crystallizes in the cubic system in masses, white or bluish-white in colour. Vitreous lustre. H 3–4; SG 2.2
Localities. Occurs with delvauxite at Visé, Belgium.
Treatment. Clean with distilled water.

Lomonosovite
$Na_8(Mn, Fe, Ca, Mg)Ti_3Si_4P_2O_{24}$
Lomonosovite crystallizes in the triclinic system as laminated tabular crystals with a perfect cleavage. The colour is dark brown to black or reddish-violet, the lustre is vitreous. H 3–4; SG 3.1
Localities. Occurs with hackmanite and other minerals in pegmatites in syenite, Kola Peninsula, USSR.
Treatment. Clean with distilled water.

Steenstrupine $CeNaMnSi_3O_9$
Steenstrupine crystallizes in the hexagonal system as rhombohedral crystals or masses coloured reddish-brown or black. H 5; SG 3.1–3.6
Localities. Occurs in nepheline-sodalite syenite pegmatites at Julianehaab, Greenland; Kola Peninsula, USSR.
Treatment. Clean with distilled water.

Britholite
$(Ca, Y)_5(SiO_4, PO_4)_3(OH, F)$
Britholite crystallizes in the hexagonal system as hexagonal crystals black in colour, with an adamantine lustre. H 5; SG 4.2
Localities. Occurs in pegmatite in the Abukuma Range, Iisaka, Fukushima Prefecture, Japan, and in the USSR.
Treatment. Clean with distilled water.

Wilkeite
$Ca_5(SiO_4, PO_4, SO_4)_3(O, OH, F)$
Wilkeite crystallizes in the hexagonal system and forms masses or rounded crystals coloured pale pink or yellow. Vitreous lustre. H about 5; SG 3.1–3.2
Localities. Occurs in blue calcite in marble at Crestmore, Riverside Co., California, etc.
Treatment. Clean with distilled water.

Ardennite
$Mn_5Al_5(As, V)O_4Si_5O_{20}(OH)_2.2H_2O$
Ardennite crystallizes in the orthorhombic system as crystalline aggregates. It has a perfect cleavage. The colour is yellow to brown or black. H 6–7; SG 3.6
Localities. Occurs with calcite and cuprosklodowskite in a limestone at Grants, New Mexico; the Ardennes, Belgium; Madhya Pradesh, India; etc.
Treatment. Clean with distilled water.

Macgovernite
$Mn_9Mg_4Zn_2As_2Si_2O_{17}(OH)_{14}$
Macgovernite crystallizes in the hexagonal system, forming light- to reddish-brown granular masses with a pearly lustre. Cleavage perfect. H not known; SG 3.7
Localities. Occurs with franklinite and other minerals at Franklin, New Jersey.
Treatment. Macgovernite should be cleaned with distilled water.

Schallerite
$(Mn, Fe)_8Si_6As(O, OH, Cl)_{26}$
Schallerite crystallizes in the hexagonal system as reddish-brown granular masses, with a waxy or pearly lustre. H about 5; SG 3.3
Localities. Occurs with calcite and other minerals at Franklin, New Jersey.
Treatment. Clean with distilled water.

Yeatmanite
$(Mn, Zn)_{13}Sb_2Si_4Zn_2O_{28}$
Yeatmanite crystallizes in the triclinic system as brown pseudo-orthorhombic crystals with a perfect cleavage. H 4; SG 5.0
Localities. Occurs embedded in willemite at Franklin, New Jersey.
Treatment. Yeatmanite should be cleaned with distilled water.

Långbanite $Mn_4^{2+}Mn_9^{3+}Sb^{5+}Si_2O_{24}$
Långbanite crystallizes in the hexagonal system as prismatic crystals which are

black with a metallic lustre. H 6.5; SG 4.6–4.8
Localities. Occurs in ores at Långban, Sweden.
Treatment. Clean with distilled water.

Chapmanite $Fe_2Sb(SiO_4)_2(OH)$
Chapmanite crystallizes in the orthorhombic system as compact olive-green or deep yellow masses with a greenish-yellow streak. H up to 2.5; SG 3.6–3.7
Localities. Occurs with silver at the Keeley Mine, Cobalt, Ontario; from Durango, Mexico; with quartz at Braunsdorf, E. Germany.
Treatment. Clean with distilled water.

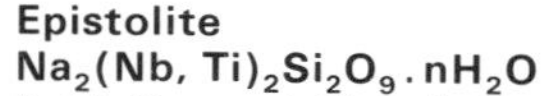

Epistolite
$Na_2(Nb, Ti)_2Si_2O_9 . nH_2O$
Epistolite crystallizes in the monoclinic system as rectangular plates with a perfect cleavage or as lamellar aggregates. The colour is white, grey, yellow or brown; the lustre is pearly. H 1–1.5; SG 2.6–2.8
Localities. Occurs in alkali pegmatites in the Julianehaab area, Greenland, and the Kola Peninsula, USSR.
Treatment. Clean with distilled water.

Murmanite
$Na_2(Ti, Nb)_2Si_2O_9 . H_2O$
Murmanite crystallizes in the monoclinic system as aggregates of platy crystals with a perfect cleavage and bright pink in colour. H 2–3; SG 2.7–2.8
Localities. Occurs in alkali pegmatites and nepheline-syenites, Kola Peninsula, USSR.
Treatment. Clean with distilled water.

Left: Murmanite from the Kola Peninsula, USSR, a classic mineral locality

Fersmanite
$(Na, Ca)_2(Ti, Nb)Si(O, F)_6$
Fersmanite crystallizes in the monoclinic system as pseudotetragonal crystals coloured light to dark brown with a white streak. They have a vitreous lustre. H 5–5.5; SG 3.4
Localities. Occurs in nepheline-syenite pegmatites, Kola Peninsula, USSR.
Treatment. Clean with distilled water.

Nenadkevichite
$(Na, Ca)(Nb, Ti)Si_2O_7 . 2H_2O$
Nenadkevichite probably crystallizes in the orthorhombic system and forms dark brown to rose-coloured masses. The streak is rose. H about 5; SG 2.8
Localities. Occurs in alkali pegmatites, Kola Peninsula, USSR.
Treatment. Clean with distilled water.

Wöhlerite
$NaCa_2(Zr, Nb)Si_2O_8(O, OH, F)$
Wöhlerite crystallizes in the monoclinic system as prismatic crystals or as grains. The colour is light to dark yellow or brown with a yellowish-white streak. H 5.5–6; SG 3–4
Localities. Occurs in alkali rocks and pegmatites in the Langesundfjord area, Norway; Kola Peninsula, USSR.
Treatment. Clean with distilled water.

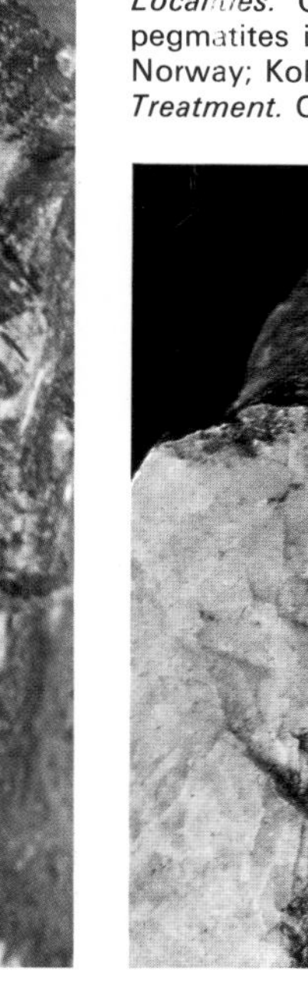

Far left: Ardennite from the Ardennes, Belgium

Left: Wöhlerite from the Langesundfjord, Norway

Right: Crystals of fergusonite in quartz from Itterby, Sweden

Låvenite
$(Na, Ca, Mn)_3(Zr, Ti, Fe)(SiO_4)_2F$
Låvenite crystallizes in the monoclinic system as prismatic crystals with a perfect cleavage; sometimes as radial aggregates. It is colourless or light yellow to brown with a vitreous lustre. H 6; SG 3.5
Localities. Occurs in alkali rocks and pegmatites in the Langesundfjord area, Norway; Kola Peninsula, USSR; and in other localities.
Treatment. Clean with distilled water.

Roeblingite
$Pb_2Ca_7Si_6O_{14}(OH)_{10}(SO_4)_2$
Roeblingite crystallizes in the monoclinic system; cleavage is perfect. It forms compact white or pink masses with a dull lustre. H 3; SG 3.4
Localities. Occurs with garnet, axinite and other minerals at Franklin, New Jersey, and in Sweden.
Treatment. Clean with distilled water.

Afghanite **$(Na, Ca, K)_{12}$-$(Si, Al)_{16}O_{34}(Cl, SO_4, CO_3)_4 . H_2O$**
Afghanite crystallizes in the hexagonal system; it has a perfect cleavage. It forms blue masses. H 5.5–6; SG 2.5
Localities. Occurs in the lapis-lazuli mine at Sar-e-Sang, Badakshan, Afghanistan.
Treatment. Clean with distilled water.

Thaumasite
$Ca_3Si(OH)_6(CO_3)(SO_4) . 12H_2O$
Thaumasite crystallizes in the hexagonal system as compact white or colourless masses with a vitreous lustre. Crystals are acicular. H 3.5; SG 1.9
Localities. Occurs with spurrite at Crestmore, Riverside Co., California; Paterson, New Jersey; Långban, Sweden; etc.
Treatment. Clean with distilled water.

Niobates and Tantalates

Lueshite $NaNbO_3$
Lueshite crystallizes in the orthorhombic system as cubes with striated faces. The colour is black. H 5.5; SG 4.4
Localities. Occurs on a yellow mica at Lueshe, Zaire.

Rankamaite **$(Na, K, Pb, Li)_3$-$(Ta, Nb, Al)_{11}(O, OH)_{30}$**
Rankamaite crystallizes in the orthorhombic system and is found as water-worn pebbles coloured white. H 3–4; SG 5.5
Localities. Occurs with simpsonite, cassiterite and other minerals at Mumba, Kivu, Zaire.
Treatment. Clean with distilled water.

Microlite
$(Ca, Na, Fe)_2(Ta, Nb)_2(O, OH, F)_7$
Microlite, a member of the pyrochlore group, crystallizes in the cubic system as octahedra or masses. The colour is pale yellow to reddish-brown, sometimes emerald-green; the lustre is vitreous. H 5–5.5; SG 4.3–5.7
Occurrence. Occurs in pegmatites with lepidolite and spodumene.
Localities. From Greenland; Norway; Malagasy Republic; fine crystals at the Rutherford mine, Amelia, Virginia; southern California; and elsewhere.

Treatment. Clean with dilute acid.
Fashioning. Uses: faceting or cabochons; *cutting angles*: crown 40°–50°, pavilion 37°–40°; *heat sensitivity*: low.

Pandaite
$(Ba, Sr)_2(Nb, Ti)_2(O, OH)_7$
Pandaite, a member of the pyrochlore group, crystallizes in the cubic system as small octahedra with cube faces also present. The colour is yellowish to light green or grey. H 4.5–5; SG 4.0
Localities. Occurs in a biotite-rich rock at Panda Hill, Tanzania.
Treatment. Clean with distilled water.

Simpsonite $Al_4Ta_3O_{13}OH$
Simpsonite crystallizes in the hexagonal system as tabular crystals or crystalline masses. The colour is yellow, brown or colourless; a yellow, blue or white fluorescence may be observed under ultra-violet light. There is a vitreous lustre. H 7–7.5; SG 5.9–6.8
Occurrence. Occurs in granite pegmatites.
Localities. From Rio Grande do Norte, Brazil; Kola Peninsula, USSR; the Mdara mine, Bikita, Rhodesia; and from some other localities.
Treatment. Clean with oxalic acid to remove iron stains.

Thoreaulite $SnTa_2O_7$
Thoreaulite crystallizes in the monoclinic system as prismatic crystals, sometimes with lamellar twinning. They have a perfect cleavage and an adamantine to resinous lustre. The colour is brown or yellow with a yellow or greenish streak. H 5.5–6; SG 7.6–7.9
Occurrence. Occurs in granite pegmatites.
Localities. From Maniema, Kivu, Zaire; Sebeia, Rwanda; USSR.
Treatment. Clean with distilled water.

Stibiotantalite $Sb(Ta, Nb)O_4$
Stibiotantalite crystallizes in the orthorhombic system as prismatic crystals sometimes with polysynthetic twinning. The colour is dark brown to yellowish-brown or red, and the lustre is vitreous. H 5–5.5; SG 7.3
Localities. Occurs with pink tourmaline, lepidolite, morganite and other minerals in a pegmatite at Mesa Grande, San Diego Co., California; fine crystals from the Muiane pegmatite, Ribaue-Alto Ligonha pegmatite, Mozambique; etc.
Treatment. Clean with dilute acid.
Fashioning. Uses: faceting or cabochons; *cleavage*: perfect, macropinacoidal; *cutting angles*: crown 30°–40°, pavilion 37°–40°; *heat sensitivity*: fairly high.

Bismutotantalite
$(Bi, Sb)(Ta, Nb)O_4$
Bismutotantalite crystallizes in the orthorhombic system as prismatic crystals; also found as water-worn pebbles. There is a perfect cleavage. The colour is brown to black with a yellow-brown streak; there is an adamantine lustre. H 5; SG 8.5
Localities. Occurs in pegmatites with tourmaline and cassiterite at Gamba Hill, south-west Uganda; as pebbles from Acari, Campina Grande, Brazil.
Treatment. Clean with dilute acid.

Tantalite $(Fe, Mn)(Ta, Nb)_2O_6$
Tantalite crystallizes in the orthorhombic system as tabular crystals or aggregates. The colour is black or brown with a reddish-brown streak and a metallic lustre. H 6–6.5; SG 8.2
Occurrence. Occurs in granite pegmatites with quartz, spodumene and other minerals.
Localities. From Canada; USSR; Rhodesia; Finland; South Dakota; etc.
Treatment. Clean with dilute acid.

Columbite $(Fe, Mn)(Nb, Ta)_2O_6$
Columbite crystallizes in the orthorhombic system as tabular crystals or as aggregates or compact masses. The colour is brown or black with a streak of the same colour, and the lustre is metallic. H 6; SG 5.1
Occurrence. Occurs in granite pegmatites.
Localities. From Canada; Brazil; Argentina; California; Virginia; and elsewhere.
Treatment. Clean with dilute acid.

Tapiolite
$(Fe, Mn)(Ta, Nb)_2O_6$
Tapiolite crystallizes in the tetragonal system as prismatic crystals with some polysynthetic twinning. The colour is black with a black or brown streak and there is a metallic lustre. H 6–6.5; SG 7.8
Occurrence. Occurs in granite pegmatites.
Localities. From Morocco; USSR; Western Australia; South Dakota; etc.
Treatment. Clean with dilute acid.

Fergusonite
$(Y, Er, Ce, Fe)(Nb, Ta, Ti)O_4$
Fergusonite crystallizes in the tetragonal system as prismatic crystals or as grains. They have a vitreous lustre. The colour is brown to black; the streak may be yellow, grey or brown. H 5.5–6.5; SG 5.4
Occurrence. Occurs in granite pegmatites.
Localities. From Greenland; USSR; San Diego Co., California.

Pyrochlore
$(Na, Ca, U)_2(Nb, Ta, Ti)_2O_6(OH, F)$
Pyrochlore crystallizes in the cubic system as octahedra or granular masses. The colour is yellow, red or brown, with a vitreous lustre. H 5–5.5; SG 4.5
Occurrence. Occurs in carbonatites, pegmatites and in nepheline-syenites.
Localities. From St. Peter's Dome, El Paso Co., Colorado; Oka, Quebec; the variety ellsworthite (containing uranium) from Hybla, Hastings Co., Ontario, and the variety hatchettolite, also with uranium, from the Woodcox mine in the same area. Good crystals from Mbeya, Tanzania, and elsewhere.
Treatment. Clean with dilute acid.

Fersmite **$(Ca, Ce, Na)(Nb, Ti, Fe, Al)_2(O, OH, F)_6$**
Fersmite crystallizes in the orthorhombic system as prismatic crystals with striations on the prism faces. The colour is dark brown or black with a greyish-brown streak; there is a resinous lustre. H 4–4.5; SG 4.7–4.8
Localities. Occurs in a marble in Ravalli Co., Montana; from pegmatites in the Central Urals, USSR.
Treatment. Clean with dilute acid.

Polymignite
$(Ca, Fe, Y, Th)(Nb, Ti, Ta)O_4$
Polymignite crystallizes in the orthorhombic system as prismatic crystals with vertical striations and a metallic lustre. The colour is black with a dark brown streak. H 6.5; SG 4.8
Localities. Occurs in a pegmatite at Fredriksvärn, Norway; Beverly, Massachusetts.
Treatment. Clean with dilute acid.

Aeschynite
$(Y, Ca, Fe, Th)(Ti, Nb)_2(O, OH)_6$
Aeschynite crystallizes in the orthorhombic system as prismatic crystals or compact masses. The colour is brown to black with a reddish-yellow streak and there is an adamantine lustre. H 5–6; SG 4.9
Localities. Occurs in pegmatites; also in placer deposits from the Malagasy Republic; Ural Mts, USSR; and some other localities.
Treatment. Clean with dilute acid.

Betafite **$(Ca, Fe, U)_{2-x}(Nb, Ti, Ta)_2O_6(OH, F)_{1-z}$**
Betafite, a member of the pyrochlore group, crystallizes in the cubic system as octahedra coloured black, brown, greenish or yellow, with a metallic lustre. H 3–5.5; SG 4.1
Occurrence. Occurs in pegmatites with zircon and other minerals.
Localities. From the Silver Crater mine, Bancroft, Ontario; the Malagasy Republic; Gunnison Co., Colorado; San Bernardino Co., California; and some other localities.
Treatment. Do not try to remove brown coating that occurs on specimens from the Malagasy Republic.

Yttrotantalite $(Y, U, Fe)(Ta, Nb)O_4$
Yttrotantalite crystallizes in the monoclinic system as prismatic crystals or compact masses coloured brown or black with a metallic or greasy lustre. H 5–5.5; SG 5.7
Occurrence. Occurs in pegmatites with gadolinite, mica and other minerals.
Localities. From Hattevick, Norway; Ytterby and Falun, Sweden.
Treatment. Clean with dilute acid.

Euxenite **$(Y, Er, Ce, La, U)(Nb, Ti, Ta)_2(O, OH)_6$**
Euxenite crystallizes in the orthorhombic system as prismatic crystals or compact masses. The colour is black tinged with brown or green, with a metallic lustre. The streak is yellow, brown or grey. H 5.5–6.5; SG 4.3–5.9
Occurrence. Occurs in granite pegmatites with rare-earth minerals.
Localities. From San Bernardino Co., California; Malagasy Republic; Brazil; etc.
Treatment. Do not try to remove brown coatings.
Fashioning. Uses: faceting or cabochons; *cleavage*: not significant; tough; *cutting angles*: crown 30°–40°, pavilion 37°–40°; *heat sensitivity*: low.

Left: Euxenite from Ambatofotskyely, Malagasy Republic

Right: Crust of cornetite crystals from the Empire Nevada mine, Yerington, Nevada

Polycrase (Y, Er, Ce, La, U)-(Ti, Nb, Ta)$_2$(O, OH)$_6$
Polycrase crystallizes in the orthorhombic system as prismatic crystals or compact masses. The colour is black with a grey, brown or yellow streak and a metallic lustre. H 5.5–6.5; SG 4.3–5.9
Occurrence. Occurs in granite pegmatites with other rare-earth minerals.
Localities. From Henderson Co., North Carolina; Canada; Brazil; USSR; etc.
Treatment. Do not try to remove brown coatings.

Samarskite
(Fe, Y, U)$_2$(Nb, Ti, Ta)$_2$O$_7$
Samarskite crystallizes in the orthorhombic system as prismatic crystals or compact masses. The colour is black to brown with a black or reddish-brown streak. The lustre is resinous. H 5–6; SG 5.2–5.7
Occurrence. Occurs in granite pegmatites associated with columbite.
Localities. From Brazil; Malagasy Republic; Riverside Co., California; North Carolina; and elsewhere.
Treatment. Do not try to remove coatings.
Fashioning. Uses: faceting or cabochons; *cleavage*: imperfect, brachypinacoidal; brittle; *cutting angles*: crown 30°–40°, pavilion 37°–40°; *heat sensitivity*: low.

Phosphates

Stercorite
$Na(NH_4)H(PO_4).4H_2O$
Stercorite crystallizes in the triclinic system as masses and nodules, which are white, colourless, yellow or brown. They have a vitreous lustre. H 2; SG 1.5
Localities. Occurs in guano deposits off the coast of Peru and South-West Africa.

Lithiophilite
Li(Mn^{2+}, Fe^{2+})PO$_4$
Lithiophilite crystallizes in the orthorhombic system as large crystals, but usually in cleavable masses, coloured reddish-brown. The streak is white or grey. There is a resinous lustre. H 4–5; SG 3.3
Occurrence. Occurs in granite pegmatites.
Localities. From Pala, San Diego Co., California; Namaqualand, South Africa; Argentina; etc.

Triphylite Li(Fe^{2+}, Mn^{2+})PO$_4$
Triphylite crystallizes in the orthorhombic system as masses coloured greenish or bluish-grey with a colourless or grey streak. They have a resinous lustre. H 4–5; SG 3.4
Occurrence. Occurs in granite pegmatites.
Localities. From the Black Hills, South Dakota; Pala, California; France; Germany; Finland; etc.

Sicklerite Li(Mn^{2+}, Fe^{3+})PO$_4$
Sicklerite crystallizes in the orthorhombic system as masses coloured yellow or brown, with a dull lustre. H about 4; SG 3.4
Occurrence. Occurs as a secondary mineral in granite pegmatites.
Localities. From Custer Co., South Dakota; Pala, San Diego Co., California; Sweden; Finland; etc.
Treatment. Clean with distilled water.

Tavorite LiFe^{3+}PO$_4$OH
Tavorite crystallizes in the triclinic system as prismatic crystals or granular masses. The colour is green or yellow, with a vitreous lustre. H not known; SG 3.3
Localities. Occurs in aggregates lining cavities in altered triphylite at the Tip-Top mine, Custer Co., South Dakota (some crystals are emerald-green); Malagasy Republic; Minas Gerais, Brazil; and elsewhere.
Treatment. Clean with distilled water.

Pseudomalachite
$Cu_5(PO_4)_2(OH)_4.H_2O$
Pseudomalachite crystallizes in the monoclinic system as prismatic crystals or as botryoidal masses; it has a perfect cleavage. The colour is dark green, with a vitreous lustre. H 4.5–5; SG 4.3
Localities. Occurs as a secondary mineral associated with quartz, malachite, etc, in the Rhineland, W. Germany; France; USSR; Canada; USA; and elsewhere.

Cornetite $Cu_3PO_4(OH)_3$
Cornetite crystallizes in the orthorhombic system as prismatic crystals or crusts. The colour is dark green or blue, and the lustre is vitreous. H 4.5; SG 4.1
Localities. Occurs with libethenite and malachite and other minerals at Yerington, Nevada; on an argillaceous sandstone at Bwana Mkubwa, Zambia; Etoile du Congo mine, Shaba, Zaire.
Treatment. Clean with distilled water.

Libethenite Cu_2PO_4OH
Libethenite crystallizes in the orthorhombic system as prismatic crystals or as crusts. The colour is light to dark green, with a vitreous lustre. H 4; SG 4.0
Occurrence. Occurs in the oxidation zone of copper-bearing ore deposits.
Localities. Found at Libethen, Romania; USSR; Zaire; San Benito Co., California; etc.

Nissonite
$Cu_2Mg_2(PO_4)_2(OH)_2.5H_2O$
Nissonite crystallizes in the monoclinic system as tabular and diamond-shaped crystals or as crusts. The colour is bluish-green. H 2.5; SG 2.7
Localities. Occurs with malachite, azurite and other minerals in metamorphic rocks in the Panoche Valley, California.
Treatment. Clean with distilled water.

Turquoise
$CuAl_6(PO_4)_4(OH)_8.4–5H_2O$
Turquoise crystallizes in the triclinic system as short prismatic crystals, rarely found; more commonly as masses, stalactites, veins or crusts. There is a perfect cleavage. The colour is bright blue to greenish-blue, green or grey. There is a waxy lustre. H 5–6; SG 2.6–2.8
Occurrence. Formed by the action of surface water on rocks containing aluminium.
Localities. Micro-crystals from Lynch Station, Campbell Co., Virginia; Arizona; Egypt; Iran, where the finest gem material comes from. Rashleighite, an iron-containing variety, from Cornwall.
Treatment. Clean with distilled water and a wetting agent.
Fashioning. Uses: cabochons, beads, carvings, tumbling, etc; *cleavage*: none in massive form; *heat sensitivity*: fairly high.

Chalcosiderite
$CuFe_6(PO_4)_4(OH)_8.4H_2O$
Chalcosiderite crystallizes in the triclinic system as prismatic crystals or crusts. There is a perfect cleavage. The colour is dark green, with a vitreous lustre. H 4.5; SG 3.2
Occurrence. Occurs as a secondary mineral in the oxidation zone of copper deposits.
Localities. From the Phoenix mine, Liskeard, Cornwall; and Seigen, Westphalia, W. Germany.
Treatment. Clean with distilled water.

Left: Libethenite crystals from Rheinbreitbach, Germany

Andrewsite
$(Cu, Fe^{2+})Fe_3^{3+}(PO_4)_3(OH)_2$
Andrewsite crystallizes in the orthorhombic system as botryoidal aggregates coloured blue or green, with a silky lustre. H 4; SG 3.5
Localities. Occurs with limonite and other minerals at the Phoenix mine, Liskeard, Cornwall.
Treatment. Clean with distilled water.

Beryllonite $NaBePO_4$
Beryllonite crystallizes in the monoclinic system as tabular prismatic crystals with polysynthetic twinning occurring. There is a perfect cleavage and a vitreous lustre. The colour is white, colourless or yellow. H 5.5–6; SG 2.8
Occurrence. Occurs in granite pegmatites.
Localities. From Newry and McKean Mountain, Maine.
Treatment. Clean with distilled water.
Fashioning. Uses: faceting or cabochons; *cleavage*: perfect // to basal pinacoid, distinct macropinacoidal; *cutting angles*: crown 40°–50°, pavilion 43°; *heat sensitivity*: fairly high.

Moraesite $Be_2PO_4OH.4H_2O$
Moraesite crystallizes in the monoclinic system as spherulitic masses. The colour and streak are white. H not known; SG 1.8
Localities. From vugs with beryl and with other phosphates at the Sapucaia pegmatite mine, Minas Gerais, Brazil.

Faheyite
$(Mn, Mg, Na)Be_2Fe_2^{3+}(PO_4)_4.6H_2O$
Faheyite crystallizes in the hexagonal system as tufts or botryoidal masses of fibres with a perfect cleavage. The colour is white to brown. H not known; SG 2.7
Localities. Occurs as a coating of quartz and other minerals at the Sapucaia pegmatite mine, Minas Gerais, Brazil.

Hurlbutite $CaBe_2(PO_4)_2$
Hurlbutite crystallizes in the orthorhombic system as prismatic crystals with striations, and a vitreous lustre. The colour is greenish white, yellow or colourless with a white streak. H 6; SG 2.9
Localities. Occurs with muscovite, smoky quartz and other minerals in pegmatite at the Smith mine, Newport, New Hampshire.
Treatment. Clean with distilled water.

Roscherite
$(Ca, Mn, Fe)_3Be_3(PO_4)_3(OH)_3.2H_2O$
Roscherite crystallizes in the monoclinic system as tabular crystals or platy aggregates. The colour is dark green or brown. H 4.5; SG 2.9
Occurrence. Occurs in pegmatites.
Localities. From Newry, Maine; Sapucaia, Minas Gerais, Brazil; Greifenstein, E. Germany.
Treatment. Clean with distilled water.

Newberyite $MgHPO_4.3H_2O$
Newberyite crystallizes in the orthorhombic system as prismatic crystals coloured grey, brown or colourless. They have a vitreous lustre. There is a perfect cleavage. H 3–3.5; SG 2.1
Occurrence. Occurs with hannayite in bat guano at the Skipton Caves, Ballarat, Victoria, Australia.
Treatment. Clean with distilled water.

Monetite $CaHPO_4$
Monetite crystallizes in the triclinic system as massive aggregates coloured white, yellow or colourless with a vitreous lustre. H 3.5; SG 2.9
Localities. Occurs with apatite on Ascension Island and elsewhere.
Treatment. Clean with distilled water.

Whitlockite $(Ca, Mg)_3(PO_4)_2$
Whitlockite crystallizes in the hexagonal system as rhombohedral crystals. The

Left: A large specimen of turquoise in matrix from Mongolia, together with some polished samples

Right: Hopeite crystals from Broken Hill, Zambia

Far right: Variscite showing characteristic distribution of colour

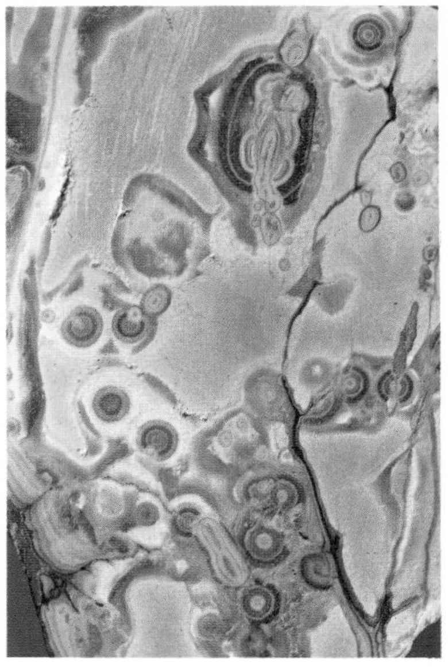

colour is pink, white, yellow or grey, and the lustre is vitreous. H 5; SG 3.1
Occurrence. In granite pegmatites.
Localities. From Oran, Algeria; Custer Co., South Dakota; etc.
Treatment. Clean with distilled water.

Hydroxyapatite $Ca_5(PO_4)_3OH$
Hydroxyapatite crystallizes in the hexagonal system as prismatic crystals or granular masses; they have a vitreous lustre. The colour is white, grey, yellow, green, violet, purple, red or brown. The streak is white. H 5; SG 3.0
Localities. Occurs in talc-schist, Cherokee Co., Georgia; in diallage-serpentine rock, Rossa, Ossola, Italy.
Treatment. Clean with distilled water.

Goyazite $SrAl_3(PO_4)_2(OH)_5 . H_2O$
Goyazite crystallizes in the hexagonal system as rhombohedral or pseudocubic crystals with a perfect cleavage. May also occur as pebbles. The colour is yellow, pink or colourless. The lustre is resinous or pearly. H 4.5–5; SG 3.3
Localities. Occurs near Diamantina, Minas Gerais, Brazil; Simplon Tunnel, Switzerland; pegmatites in Maine; the tungsten area of Boulder Co., Colorado.
Treatment. Clean with distilled water.

Palermoite
$(Li, Na)_2(Sr, Ca)Al_4(PO_4)_4(OH)_4$
Palermoite crystallizes in the orthorhombic system as vertically striated prisms with a perfect cleavage coloured white or colourless. The streak is white and the mineral exhibits a white fluorescence under x-rays. There is a vitreous lustre. H 5.5; SG 3.2
Localities. Occurs with siderite and other minerals at the Palermo pegmatite, North Gordon, New Hampshire.
Treatment. Clean with distilled water.

Gorceixite $BaAl_3(PO_4)_2(OH)_5 . H_2O$
Gorceixite crystallizes in the hexagonal system as botryoidal aggregates; also occurs as grains and pebbles. The colour is white or brown. H 6; SG 3.3
Localities. In novaculite at Hot Springs, Arkansas; in the diamond-bearing sands of Brazil, Guyana and Africa.

Hopeite $Zn_3(PO_4)_2 . 4H_2O$
Hopeite crystallizes in the orthorhombic system as prismatic crystals or compact masses; it has a perfect cleavage. Parahopeite is dimorphous with hopeite, but crystallizes in the triclinic system. Hopeite is yellow, white, colourless or grey, with a white streak. There is a vitreous lustre. H 3.5; SG 3.0
Occurrence. Occurs in zinc ore deposits.
Localities. From Kabwe, Zambia; Hudson Bay mine, Salmo, British Columbia; Altenberg, W. Germany. Parahopeite from the same localities.

Tarbuttite Zn_2PO_4OH
Tarbuttite crystallizes in the triclinic system as prismatic crystals, often with deep striations or as crusts. There is a perfect cleavage, and a vitreous lustre. The colour is white, colourless, yellow, red, green or brown with a white streak. H 3.5; SG 4.1
Localities. Occurs with limonite and other minerals in the oxidation zone of the zinc ores, Shaba, Zambia.
Treatment. Clean with distilled water.

Spencerite $Zn_4(PO_4)_2(OH)_2 . 3H_2O$
Spencerite crystallizes in the monoclinic system as tabular crystals or stalactitic masses. There is a perfect cleavage. The colour is white, and the lustre is vitreous. H 3; SG 3.1
Localities. In oxidized zinc ores at the Hudson Bay mine, British Columbia.

Scholzite $CaZn_2(PO_4)_2 . 2H_2O$
Scholzite crystallizes in the orthorhombic system as prismatic or platy crystals. The colour is white or colourless. There is a vitreous lustre. H 3–3.5; SG 3.1
Localities. Occurs with quartz and other minerals in pegmatite at Hagendorf, Bavaria, W. Germany.

Faustite
$(Zn, Cu)Al_6(PO_4)_4(OH)_8 . 5H_2O$
Faustite crystallizes in the triclinic system as compact apple-green masses with a waxy lustre. The streak is white or pale yellow. H 5.5; SG 2.9
Localities. Occurs at the Copper King mine, Eureka Co., Nevada.

Phosphophyllite
$Zn_2(Fe, Mn)(PO_4)_2 . 4H_2O$
Phosphophyllite crystallizes in the monoclinic system as prismatic crystals with a vitreous lustre; they are often twinned and have a perfect cleavage. The colour is bright blue-green to colourless. H 3–3.5; SG 3.1
Localities. Occurs in vugs in massive sulphides at San Luis Potosi, Bolivia; in granite pegmatites at Hagendorf, Bavaria, W. Germany.
Fashioning. Uses: faceting or cabochons; *cleavage*: perfect orthopinacoidal, distinct clinopinacoidal; *cutting angles*: crown 40°, pavilion 40°; *heat sensitivity*: fairly high.

Berlinite $AlPO_4$
Berlinite crystallizes in the hexagonal system as granular masses. The colour is grey or pale pink. H 6.5; SG 2.6
Localities. Occurs with other phosphates at the Westanå iron mine, Näsum, Sweden.
Treatment. Clean with distilled water.

Variscite $AlPO_4 . 2H_2O$
Variscite crystallizes in the orthorhombic system as octahedra but more commonly as masses, veins or nodules. The colour is pale to emerald-green and the lustre is waxy. The variety metavariscite crystallizes in the monoclinic system. H 3.5–4.5; SG 2.6
Occurrence. Occurs in aluminous rocks that have been altered by phosphatic waters.
Localities. From Fairfield, Utah; California, Nevada, Arizona; Germany; Austria; Spain; Brazil and elsewhere.
Treatment. Clean with distilled water only.
Fashioning. Uses: cabochons, carvings, beads, tumbling, etc; *cleavage*: none, material generally aggregated or massive; *heat sensitivity*: fairly high.

Augelite $Al_2PO_4(OH)_3$
Augelite crystallizes in the monoclinic system as tabular crystals or as triangular plates or as masses. There is a perfect cleavage. It is white, colourless, yellow, blue or pink with a white streak. There

is a vitreous lustre. H 4.5–5; SG 2.7
Localities. Occurs with andalusite at White Mountain, Mono Co., California; South Dakota; the Palermo mine, New Hampshire; Bolivia; and some other localities.
Treatment. Clean with distilled water.

Bolivarite
$Al_2(PO_4)(OH)_3.4\text{–}5H_2O$
Bolivarite is amorphous to 1,050°C and occurs as crusts and botryoidal masses, with a vitreous lustre. The colour is bright yellowish-green and there is a bright green fluorescence under ultraviolet light. H 2.5–3.5; SG 2.0
Localities. Occurs in granite near Pontevedra, Spain; from the pegmatite at Kobokobo, Kivu, Zaire.

Wavellite $Al_3(OH)_3(PO_4)_2.5H_2O$
Wavellite crystallizes in the orthorhombic system as radiating aggregates of acicular crystals, or as crusts. There is a perfect cleavage. The colour is green or white, yellow, brown, blue or colourless, with a vitreous lustre. H 3–4; SG 2.4
Occurrence. Occurs in hydrothermal veins or in phosphate rocks.
Localities. Fine specimens from Hot Spring Co., Arkansas; El Dorado Co., California; Barnstaple, Devon; and elsewhere.
Treatment. Wavellite should be cleaned ultrasonically.

Evansite $Al_3PO_4(OH)_6.6H_2O$
The formula quoted is not certain, and the crystal system is not known. Evansite occurs in massive stalactitic or botryoidal forms with a yellow green or blue colour; it may also be colourless. There is a vitreous lustre. The streak is white. H 3–4; SG 1.8–2.2
Occurrence. Occurs with allophane and limonite.
Localities. From the Coosa coal field, Alabama; Custer Co., South Dakota; Malagasy Republic.
Treatment. Clean with distilled water.

Left: Radiating acicular wavellite crystals

Below: Brazilianite, discovered in this century. It is sometimes cut as a gemstone. This specimen is from Brazil

Brazilianite $NaAl_3(PO_4)_2(OH)_4$
Brazilianite crystallizes in the monoclinic system as short prismatic crystals with striations or as globules. The colour is pale yellow to yellowish-green or colourless with a colourless streak. There is a vitreous lustre. H 5.5; SG 3.0; RI 1.60–1.62
Occurrence. Occurs in pegmatites.
Localities. From Conselheira Pena, Minas Gerais, Brazil; Smith mine, Newport, New Hampshire.
Treatment. Clean with dilute oxalic acid.
Fashioning. Uses: faceting or cabochons; *cleavage*: perfect pinacoidal; *cutting angles*: crown 40°, pavilion 40°; *heat sensitivity*: low.

Gordonite
$MgAl_2(PO_4)_2(OH)_2.8H_2O$
Gordonite crystallizes in the triclinic system as sheaf-like aggregates: it has a perfect cleavage. The colour is white, pink, green or colourless with a white streak. H 3.5; SG 2.2
Occurrence. Occurs with crandallite at Fairfield, Utah.

Overite $CaMg(H_2O)_4Al(OH)(PO_4)_2$
Overite crystallizes in the orthorhombic system as plates or platy aggregates;

Left: A fine crystal of brazilianite with an unusually long prismatic form

Right: Tabular crystals of torbernite from Bois Noir, France.

it has a perfect cleavage. The colour is light green to colourless with a vitreous lustre. H 3.5–4; SG 2.5
Localities. Occurs with crandallite at Fairfield, Utah.

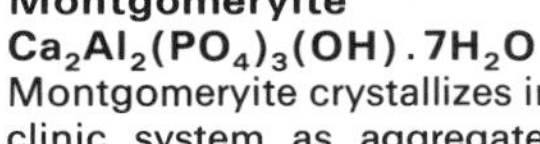

Montgomeryite $Ca_2Al_2(PO_4)_3(OH).7H_2O$
Montgomeryite crystallizes in the monoclinic system as aggregates of plates coloured deep to pale green and with a perfect cleavage. There is a vitreous lustre. H 4; SG 2.5
Localities. Occurs with crandallite at Fairfield, Utah.

Crandallite $CaAl_3(PO_4)_2(OH)_5.H_2O$
Crandallite crystallizes in the triclinic system; it forms nodular masses and has a perfect cleavage. The colour is yellow or white. The lustre may be vitreous or dull. H 5; SG 2.8–2.9
Localities. Occurs with other phosphate minerals at Fairfield, Utah; Brazil; Senegal; Guatemala; etc.

Wardite $NaAl_3(PO_4)_2(OH)_4.2H_2O$
Wardite crystallizes in the tetragonal system as pyramidal crystals or fibrous aggregates. There is a perfect cleavage. The colour is pale green, bluish, white or colourless with a vitreous lustre. H 5; SG 2.8
Localities. Occurs with crandallite at Fairfield, Utah; with amblygonite at Pala, San Diego Co., California; and elsewhere.
Treatment. Clean with distilled water.

Englishite $K_2Ca_4Al_8(PO_4)_8(OH)_{10}.9H_2O$
Englishite is thought to crystallize in the monoclinic system; it forms layers and aggregates of curved plates with a perfect cleavage. It is colourless, with a vitreous lustre. H about 3; SG about 2.6
Localities. Occurs with crandallite at Fairfield, Utah.

Below: Monazite, which crystallizes in the monoclinic system

Xenotime YPO_4
Xenotime crystallizes in the tetragonal system as prismatic crystals or as rosettes. There is a perfect cleavage. The colour is yellow to reddish-brown, greenish or grey, and the lustre is vitreous. H 4–5; SG 4.4–5.1
Occurrence. Occurs in alkaline and igneous rocks and pegmatites.
Localities. From California; Colorado; Brazil; Malagasy Republic; etc.
Treatment. Clean with distilled water.

Monazite (La, Ce, Y, Th)PO_4
Monazite crystallizes in the monoclinic system as tabular crystals or granular masses with a resinous lustre. The colour is reddish-brown, brown, pink, yellow, green, white with a white or grey streak. H 5–5.5; SG 4.6–5.4
Localities. Occurs in placer deposits around the world. Fine crystals from Encampment, Wyoming, where they are found in association with euxenite; Norway; Finland; Switzerland; etc.

Rhabdophane (Ce, Y, La, Di)$(PO_4).H_2O$
Rhabdophane crystallizes in the hexagonal system as stalactitic encrustations with a greasy lustre. The colour is white, yellow or pink. H 3.5; SG 3.9–4.0
Localities. Occurs in a limonite deposit at Salisbury, Connecticut; Cornwall.

Churchite (Y, Er)$PO_4.2H_2O$
Churchite, sometimes known as weinschenkite, crystallizes in the monoclinic system as crusts or rosettes. There is a perfect cleavage. The colour is white, grey or colourless. H 3; SG 3.3
Localities. Occurs in a manganese-rich limonite deposit at Rockbridge Co., Virginia; with copper in Cornwall.

Florencite $CeAl_3(PO_4)_2(OH)_6$
Florencite crystallizes in the hexagonal system as rhombohedral pink or pale yellow crystals with a resinous lustre. H 5–6; SG 3.5–3.7
Localities. Occurs in mica schists at Ouro Preto, Minas Gerais, Brazil; in pegmatite at Klein Spitzkopje, South-West Africa; Malawi; etc.
Treatment. Clean with distilled water.

Plumbogummite $PbAl_3(PO_4)_2(OH)_5.H_2O$
Plumbogummite crystallizes in the hexagonal system as compact masses or stalactitic crusts. The colour is white, grey, yellow or greenish-blue, reddish-brown with a white or colourless streak. There is a resinous or dull lustre. H 4.5–5; SG 4.0
Occurrence. Occurs in the oxidation zone of lead ores.
Localities. From Inyo Co., California; Cumbria; Minas Gerais, Brazil; etc.

Sincosite Ca $(VO)_2(PO_4)_2.5H_2O$
Sincosite crystallizes in the tetragonal system as tabular crystals or compact masses. There is a perfect cleavage. The colour is green or yellow with a green streak. There is a vitreous lustre. H not known; SG about 2.8
Localities. Occurs in siliceous gold ore at Lead, Lawrence Co., South Dakota; Junin, Peru.
Treatment. Clean with distilled water.

Meta-ankoleite $K_2(UO_2)_2(PO_4)_2.6H_2O$
Meta-ankoleite crystallizes in the tetragonal system as plates; it has a perfect cleavage and is coloured yellow. It has a vitreous lustre. There is a yellow-green fluorescence in ultra-violet light. H not known; SG 3.5
Localities. Occurs with muscovite and quartz in the Mungenyi pegmatite, Uganda; Sebungive, Rhodesia.

Uramphite $NH_4UO_2PO_4 . 3H_2O$
Uramphite's crystal system is not known. It forms rosettes and plates coloured dark or pale green with a yellow-green fluorescence under ultra-violet light. They have a vitreous lustre. H not known; SG 3.7
Localities. From a uranium-coal deposit, USSR.

Torbernite
$Cu(UO_2)_2(PO_4)_2 . 8–12H_2O$
Tobernite crystallizes in the tetragonal system as tabular crystals or lamellar aggregates. There is a perfect cleavage. The colour is emerald-green and lighter shades with a pale green streak. There is a vitreous lustre. H 2–2.5; SG 3.2
Localities. Occurs in pegmatites at Spruce Pine, North Carolina; Cornwall; Puy-de-Dôme, France; E. Germany; and elsewhere.
Treatment. Remove iron stains with weak solution of sodium dithionate; otherwise use distilled water.

Metatorbernite
$Cu(UO_2)_2(PO_4)_2 . 8H_2O$
Metatorbernite crystallizes in the tetragonal system as thin tablets or lamellar aggregates. There is a perfect cleavage. The colour is pale to dark green, with a vitreous lustre. H 2.5; SG 3.7–3.8
Occurrence. Occurs in the oxidized zone of veins with uraninite and copper.
Localities. From Redruth, Cornwall; Lake Athabasca, Saskatchewan, Canada; Puy-de-dôme, France; Erzgebirge, E. Germany; etc.
Treatment. Clean with distilled water.

Saléeite $Mg(UO_2)_2(PO_4)_2 . 8–10H_2O$
Saléeite crystallizes in the tetragonal system as rectangular plates or platy aggregates. There is a perfect cleavage. The colour is yellow to yellowish-green, with a vitreous or dull lustre. There is a bright yellow-green fluorescence in ultra-violet light. H 2.5; SG 3.3
Localities. In a sandstone at Gull mine, Fall River Co., South Dakota; Schneeberg, E. Germany; Shinkolobwe, Shaba, Zaire; Rum Jungle, Northern Territory, Australia; etc.

Autunite
$Ca(UO_2)_2(PO_4)_2 . 10–12H_2O$
Autunite crystallizes in the tetragonal system as tabular crystals or crusts and aggregates. There is a perfect cleavage. The colour is bright yellow, pale or dark green, with a vitreous or dull lustre. There is a bright yellow-green fluorescence under ultra-violet light. Meta-autunite is a dehydration product. H 2–2.5; SG 3.0–3.2
Occurrence. Formed by the alteration of uraninite in pegmatites or granites.
Localities. From Rum Jungle, Northern Territory, Australia; E. Germany; St. Austell and Redruth, Cornwall, England; and other localities.

Phosphuranylite
$Ca\ (UO_2)_4(PO_4)_2(OH)_4 . 7H_2O$
Phosphuranylite crystallizes in the orthorhombic system as scales or plates, aggregates and crusts. There is a perfect cleavage. The colour is light golden

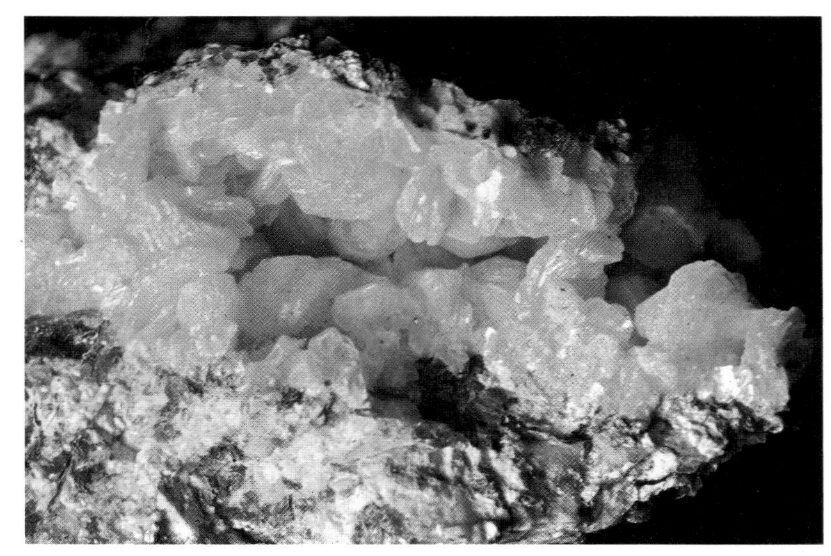

Left: Autunite. Bright yellows are characteristic of radioactive minerals

Left: Autunite plates from Peveragno, Piedmont, Italy

yellow to deep yellow. H about 2.5; SG 4.1
Occurrence. Occurs in the weathered zone of uraninite-bearing pegmatites.
Localities. From Rio Grande do Norte, Brazil; Margnac, Haute-Vienne, France; Palermo mine, New Hampshire; Mitchell Co., North Carolina; etc.

Uranocircite
$Ba(UO_2)_2(PO_4)_2.10–12H_2O$
The crystal system is probably tetragonal and the mineral is found as tabular crystals or as aggregates. There is a perfect cleavage. The colour is yellow, with a green fluorescence under ultra-violet light. There is a pearly lustre. H 2–2.5; SG 3.5
Localities. Occurs at Farnwitte, Black Forest, W. Germany.

Meta-uranocircite
$Ba(UO_2)_2(PO_4)_2.6–8H_2O$
Meta-uranocircite crystallizes in the tetragonal system as thin plates and aggregates with a perfect cleavage. They have a pearly lustre. The colour is yellow-green with a green fluorescence under ultra-violet light. H 2–2.5; SG 3.9
Localities. Occurs at Menzenschwand, W. Germany; the Banat, Romania; Harding Co., South Dakota; in an alluvial deposit near Antsirabe, Malagasy Republic; etc.

Sabugalite
$HAl(UO_2)_2(PO_4)_4.16H_2O$
Sabugalite crystallizes in the tetragonal system as thin plates and crusts; it has a perfect cleavage. The colour is bright yellow with a similarly-coloured fluorescence under ultra-violet light. There is a vitreous lustre. H 2.5; SG 3.2
Occurrence. Occurs in sandstone-type uranium deposits and in uraninite veins.
Localities. From Mina da Quarta Feira, Beira Alta, Portugal; Margnac, Haute-Vienne, France; Arizona; Wyoming; etc.

Parsonsite $Pb_2UO_2(PO_4)_2.2H_2O$
Parsonsite crystallizes in the triclinic system as prismatic crystals or tufts and fibrous masses. The colour is pale yellow to brown or pink, with an adamantine or greasy lustre. H 2.5–3; SG 5.7
Localities. Occurs with autunite at the Ruggles mine, Grafton Center, New Hampshire; with torbernite at Lachaux, Puy-de-Dôme, France; with metatorbernite at Shinkolobwe, Shaba, Zaire; etc.

Renardite
$Pb(UO_2)_4(PO_4)_2(OH)_4.7H_2O$
Renardite crystallizes in the orthorhombic system as rectangular plates and lamellar masses with a perfect cleavage. The colour is light yellow to brown, with a vitreous lustre. H 3.5; SG 4.3
Localities. Occurs with torbernite at Shinkolobwe, Shaba, Zaire; Lignol, Morbihan, and Lachaux, Puy-de-Dôme, France; etc.

Dumontite
$Pb_2(UO_2)_3(PO_4)_2(OH)_4.3H_2O$
Dumontite crystallizes in the monoclinic system as minute elongated crystals coloured pale to golden yellow. There is a weak green fluorescence under ultra-violet light. H not known; SG 5.6
Localities. Occurs with autunite at Nogales, Arizona; with parsonsite at Shinkolobwe, Shaba, Zaire.
Treatment. Clean with distilled water.

Hureaulite
$H_2(Mn,Fe)_5(PO_4)_4.4H_2O$
Hureaulite crystallizes in the monoclinic system as prismatic crystals or fibrous masses. The colour is rose, red, yellow, orange, brown or colourless with a white streak. There is a vitreous lustre. H 3.5; SG 3.1
Occurrence. Occurs in granite pegmatites with other phosphate minerals.
Localities. From the Black Hills, South Dakota; the Palermo Mine, New Hampshire; Hagendorf, W. Germany.

Stewartite
$MnFe_2(OH)_2(PO_4)_2.8H_2O$
Stewartite crystallizes in the triclinic system as tufts coloured yellow or brown. H not known; SG 2.9
Occurrence. Occurs in granite pegmatites with other phosphate minerals.
Localities. From the Stewart mine, Pala, San Diego Co., California; Alto Boqueiro, Rio Grande do Norte, Brazil; etc.

Natrophilite $NaMnPO_4$
Natrophilite crystallizes in the orthorhombic system as prismatic crystals and more commonly as granular masses. The colour is deep yellow, with a resinous lustre. H 4.5–5; SG 3.4
Localities. Occurs in a granite pegmatite at Branchville, Connecticut.

Griphite
$(Na,Al,Ca,Fe)_3Mn_2(PO_4)_{2.5}(OH)_2$
Griphite crystallizes in the cubic system as compact masses coloured brown to black, with a resinous lustre. H 5.5; SG 3.4
Occurrence. Occurs in granite pegmatites.
Localities. From Mt. Ida, Northern Territory, Australia; Pennington Co., South Dakota; etc.

Reddingite $(Mn,Fe)_3(PO_4)_2.3H_2O$
Reddingite crystallizes in the orthorhombic system as octahedra or fibrous masses. The colour is pink, yellow, reddish-brown or white, and the lustre is vitreous. H 3–3.5; SG 3.2
Occurrence. Occurs as an alteration of lithiophilite in granite pegmatites.
Localities. From Fairfield Co., Connecticut; Buckfield, Maine.
Treatment. Clean with distilled water.

Switzerite $(Mn,Fe)_3(PO_4)_2.4H_2O$
Switzerite crystallizes in the monoclinic system as flakes or bladed crystals with a perfect cleavage. The colour is pink or golden brown, with a pearly lustre. H about 2.5; SG 2.9
Localities. Occurs in a pegmatite at the spodumene mine, Kings Mountain, North Carolina.

Beusite $(Mn, Fe, Ca, Mg)_3(PO_4)_2$
Beusite crystallizes in the monoclinic system as prismatic crystals intermingled with lithiophilite. The colour is reddish-brown. H 5; SG 3.7
Occurrence. Occurs in granite pegmatites.
Localities. From Los Aleros, San Luis Province, Argentina.

Triploidite $(Mn^{2+}, Fe^{2+})_2PO_4OH$
Triploidite crystallizes in the monoclinic system as prismatic crystals with vertical striations or as fibrous aggregates. The colour is pink, yellow or brown to green. There is a vitreous lustre. The streak is white. H 4.5–5; SG 3.7
Localities. Occurs in granite pegmatites at the Ross mine, Custer Co., South Dakota; Branchville, Fairfield Co., Connecticut; etc.

Salmonsite
$Mn^{2+}_9, Fe^{3+}_2(PO_4)_8.14H_2O$
Salmonsite probably crystallizes in the orthorhombic system as fibrous masses coloured light brown. H 4; SG 2.8
Localities. Occurs in granite pegmatite at the Stewart mine, Pala, California.

Right: Uranocircite blades from Bergen, East Germany

Laueite $MnFe_2(PO_4)_2(OH)_2.8H_2O$
Laueite crystallizes in the triclinic system as wedge-shaped crystals with a perfect cleavage. The colour is yellow, orange or brown, with a vitreous lustre. H 3; SG 2.4–2.5
Localities. Occurs in a pegmatite in eastern Bavaria, W. Germany; Palermo quarry, New Hampshire; Black Hills, South Dakota.
Treatment. Clean with distilled water.

Strunzite
$MnFe_2(PO_4)_2(OH)_2.8H_2O$
Strunzite crystallizes in the monoclinic system as tufts and coatings coloured yellow to brown. H not known; SG 2.5–2.6
Localities. Occurs in pegmatites containing triphylite at Hagendorf, Bavaria, W. Germany; Palermo, New Hampshire; Custer Co., South Dakota; and some other localities.

Landesite
$Mn_{10}Fe_3^{3+}(PO_4)_8(OH)_5.11H_2O$
Landesite crystallizes in the orthorhombic system as octahedra pseudomorphous after reddingite. The colour is brown. H 3–3.5; SG 3.0
Occurrence. In granite pegmatites.
Localities. From Poland, Maine.
Treatment. Clean with distilled water.

Frondelite
$(Mn^{2+}, Fe^{2+}(Fe_4^{3+}(PO_4)_3(OH)_5$
Frondelite crystallizes in the orthorhombic system as crusts and fibrous masses. There is a perfect cleavage. The colour is olive green to black. H 4.5; SG 3.5
Occurrence. Occurs in granite pegmatites as an alteration product of the mineral triphylite.
Localities. From the Black Hills, South Dakota; Minas Gerais, Brazil.

Rockbridgeite
$(Fe^{2+}, Mn)Fe_4^{3+}(PO_4)_3(OH)_5$
Rockbridgeite crystallizes in the orthorhombic system as prismatic crystals, more commonly as botryoidal crusts. There is a perfect cleavage. The colour is light to dark green becoming brown on oxidation. H 4.5; SG 3.3–3.5
Occurrence. Found in limonite, and in granite pegmatites with iron-manganese phosphate minerals.
Localities. From the Black Hills, South Dakota; Brazil; Germany; and from some other localities.

Wolfeite $(Fe^{2+}, Mn^{2+})_2PO_4OH$
Wolfeite crystallizes in the monoclinic system as prismatic crystals with vertical striations; also found as fibrous aggregates. There is a vitreous lustre. The colour is reddish or dark brown and the streak is white. H 4.5–5; SG 3.8
Occurrence. Occurs in granite pegmatites.
Localities. From Custer Co., South Dakota; Hagendorf, Bavaria, W. Germany; etc.

Heterosite $(Fe^{3+}, Mn^{3+})PO_4$
Heterosite crystallizes in the orthorhombic system in masses coloured red to purple with a similarly-coloured streak. Dull lustre. H 4–4.5; SG 3.4
Occurrence. Occurs as an alteration of triphylite in granite pegmatites.
Localities. From Custer Co., South Dakota; Germany; Australia; Erongo, South-West Africa; etc.
Treatment. Remove dark alteration coating with dilute hydrochloric acid.

Purpurite $(Mn^{3+}, Fe^{3+})PO_4$
Purpurite crystallizes in the orthorhombic system as masses coloured dark red to purple with a similarly-coloured streak. There is a dull lustre. H 4–4.5; SG 3.7
Occurrence. Occurs in granite pegmatites as an alteration of lithiophilite.
Localities. From Custer Co., South Dakota; Chanteloube, France; Wodgina, Western Australia; etc.
Treatment. Remove dark alteration coating with dilute hydrochloric acid.

Dickinsonite $H_2Na_6(Mn, Fe, Ca, Mg)_{14}(PO_4)_{12}.H_2O$
Dickinsonite crystallizes in the monoclinic system as tabular crystals or foliated masses with a perfect cleavage. The colour is yellow to dark green with a white streak. There is a vitreous lustre. H 3.5–4; SG 3.4
Occurrence. Occurs in granite pegmatities in association with other phosphate minerals.
Localities. From Branchville, Connecticut; Poland, Maine.

Fillowite
$H_2Na_6(Mn, Fe, Ca)_{14}(PO_4)_{12}.H_2O$
Fillowite crystallizes in the hexagonal system as pseudorhombohedral crystals or granular masses; it has a perfect cleavage. The colour is yellow to brown with a white streak. There is a greasy lustre. H 4.5; SG 3.4
Localities. Occurs in granite pegmatites with reddingite, etc., at Branchville, Connecticut.

Bermanite
$Mn^{2+}, Mn_2^{3+}(PO_4)_2(OH)_2.4H_2O$
Bermanite crystallizes in the monoclinic system as tabular crystals, frequently twinned, or as lamellar masses; it has a perfect cleavage. The colour is pale red to brown, with a vitreous lustre. H 3.5; SG 2.8
Occurrence. Occurs in pegmatites with hureaulite and other phosphates.
Localities. From Custer Co., South Dakota; Sapucaia, Minas Gerais, Brazil· Malagasy Republic.
Treatment. Clean with distilled water.

Left: Eosphorite. The vitreous lustre is clearly visible in this fragment

Below left: Graftonite in a pegmatitic matrix

Graftonite
$(Fe^{2+}, Mn^{2+}, Ca)_3(PO_4)_2$
Graftonite crystallizes in the monoclinic system as masses with a pink to brown colour and a vitreous lustre. H 5; SG 3.7–3.8
Occurrence. Occurs in granite pegmatites.
Localities. From Custer Co., South Dakota; Brissago, Switzerland; Olgiasca, Lake Como, Italy; etc.
Treatment. Clean with distilled water.

Fairfieldite
$Ca_2(Mn^{2+}, Fe^{2+})(PO_4)_2.2H_2O$
Fairfieldite crystallizes in the triclinic system as prismatic crystals or lamellar masses; it has a perfect cleavage. The colour is white or yellow to green with a white streak. There is a vitreous lustre. H 3.5; SG 3.1
Occurrence. In granite pegmatites.
Localities. From Branchville, Fairfield Co., Connecticut; Hühnerkobel, Bavaria, W. Germany.
Treatment. Clean with distilled water.

Arrojadite
$(Na, Ca)_2(Fe^{2+}, Mn^{2+})_5(PO_4)_4$
Arrojadite crystallizes in the monoclinic system as dark green masses with a vitreous lustre. H 5; SG 3.5
Occurrence. Occurs in granite pegmatites.
Localities. From Pennington Co., South Dakota; Serra Blanca, Picuhy, Brazil.

Hagendorfite
$(Na, Ca)(Fe^{2+}, Mn^{2+})_2(PO_4)_2$
Hagendorfite crystallizes in the monoclinic system as greenish-black masses. H 3.5; SG 3.7
Localities. Occurs in pegmatite at Hagendorf, Bavaria, W. Germany; and at Nörro, Sweden.

Eosphorite
$(Mn, Fe)AlPO_4(OH)_2.H_2O$
Eosphorite crystallizes in the monoclinic system as prismatic crystals or radial

Right: Translucent vivianite from Uncía in Bolivia. Vivianite is pleochroic

aggregates. The colour is pink, yellow or colourless, red, brown or black; there is a vitreous lustre. H 5; SG 3.0
Occurrence. Occurs in granite pegmatites with other manganese phosphates.
Localities. From Minas Gerais, Brazil; Hagendorf, Bavaria, W. Germany; Branchville, Connecticut; Mt. Mica, Maine; Palermo mine, New Hampshire.
Treatment. Clean with distilled water.

Childrenite
$(Fe, Mn)AlPO_4(OH)_2.H_2O$
Childrenite crystallizes in the orthorhombic system as tabular crystals or as plates. The colour is brown or yellow, with a vitreous lustre. H 5; SG 3.2
Occurrence. Found in hydrothermal vein deposits and in granite pegmatites.
Localities. From Cornwall, England; Custer Co., South Dakota; Minas Gerais, Brazil.
Treatment. Clean with distilled water.

Vivianite $Fe_3(PO_4)_2.8H_2O$
Vivianite crystallizes in the monoclinic system as flexible laminae; they have a perfect cleavage and are sectile. The mineral also occurs as crusts and earthy coatings. The colour is colourless, darkening on exposure to green or blue, purple or black. There is a pearly lustre. The streak is colourless altering to brown or dark blue. H 1.5–2; SG 2.7
Occurrence. In metallic ore veins and as an alteration product of phosphate minerals in pegmatites; also in sedimentary deposits.
Localities. From Lemhi Co., Idaho; Black Hills, South Dakota; Poopó, Bolivia; Cameroons; St. Agnes, Cornwall; USSR; Australia; and elsewhere.
Treatment. Clean with distilled water.

Ludlamite $Fe_3(PO_4)_2.4H_2O$
Ludlamite crystallizes in the monoclinic system as tabular crystals or granular masses. There is a perfect cleavage. The colour is bright to apple green, or colourless, with a vitreous lustre. H 3.5; SG 3.2
Occurrence. Occurs in the oxidation zone of ore deposits and as an alteration product of iron bearing phosphate minerals in granite pegmatites.
Localities. From Lemhi Co., Idaho; Custer Co., South Dakota; Wheal Jane mine, Truro, Cornwall; Hagendorf, Bavaria, W. Germany.
Treatment. Clean with distilled water.

Lipscombite
$(Fe^{2+}, Mn^{2+})Fe^{3+}_2(PO_4)_2(OH)_2$
Lipscombite crystallizes in the tetragonal system as aggregates of tiny crystals coloured dark green to black. H not known; SG 3.7
Localities. Occurs with metastrengite in frondelite at the Sapucaia pegmatite, Minas Gerais, Brazil, etc.

Laubmannite
$Fe^{2+}_3Fe^{3+}_6(PO_4)_4(OH)_{12}$
Laubmannite crystallizes in the orthorhombic system as botryoidal aggregates coloured bright yellow to green, with a vitreous lustre. H 3.5–4; SG 3.3
Localities. Occurs in novaculite on Buckeye Mountain, Polk Co., Arkansas; Amberg-Auerbach, Bavaria, W. Germany; Leveäniemi, Sweden.

Barbosalite $Fe^{2+}Fe^{3+}_2(PO_4)_2(OH)_2$
Barbosalite crystallizes in the monoclinic system as prismatic crystals or masses, or as crusts. The colour is greenish-blue to black, with a dull or vitreous lustre. H 5.5–6; SG 3.6
Localities. Occurs in pegmatites with pyrite, Custer Co., South Dakota; and elsewhere.

Phosphosiderite $FePO_4.2H_2O$
Phosphosiderite crystallizes in the monoclinic system as tabular crystals or botryoidal crusts; interpenetration twins are common. The colour is red to violet or yellow; the lustre is vitreous. H 3.5–4; SG 2.8
Localities. Occurs with strengite in the Bull Moose mine, Custer Co., South Dakota; Pala, San Diego Co., California; Chanteloube, France; Hagendorf, W. Germany; etc.

Strengite $FePO_4.2H_2O$
Strengite crystallizes in the orthorhombic system as octahedra or spherical aggregates and crusts. The colour is pale to deep violet-red with a white streak. There is a vitreous lustre. H 3.5–4.5; SG 2.9
Occurrence. Formed by the alteration of phosphate minerals rich in iron.
Localities. From Custer Co., South Dakota; Pala, San Diego Co., California; Hagendorf, Bavaria, W. Germany; Mangualde, Portugal; etc.
Treatment. Clean with distilled water.

Beraunite
$Fe^{2+}Fe^{3+}_5(OH)_5(H_2O)_4(PO_4)_4.2H_2O$
Beraunite crystallizes in the monoclinic system as tabular crystals with vertical striations; more commonly as fibrous aggregates or crusts. The colour is red or brown with a yellow-brown streak. There is a vitreous lustre. H 3.5–4; SG 3.0
Occurrence. In iron ore deposits and in granite pegmatites.
Localities. From Pennington Co., South Dakota; Palermo mine, New Hampshire; Ireland; Germany; etc.

Dufrenite
$Fe^{2+}Fe^{3+}_4(PO_4)_3(OH)_5.2H_2O$
Dufrenite crystallizes in the monoclinic system as botryoidal masses or crusts; it has a perfect cleavage and is coloured dark green to black. H 3.5–4.5; SG 3.1–3.3
Localities. Occurs as a secondary mineral at Wheal Phoenix, Cornwall, England; Thüringer Wald, E. Germany; Cherokee Co., Alabama.

Cacoxenite
$Fe^{3+}_4(PO_4)_3(OH)_3.12H_2O$
Cacoxenite crystallizes in the hexagonal system as radial or tufted aggregates coloured golden yellow to brown. H 3–4; SG 2.3
Occurrence. Occurs as a secondary mineral associated with other phosphate minerals and with limonite.

Localities. From France; Germany; Sweden; Polk Co., Arkansas; Antwerp, New York, on hematite.

Cyrilovite
$NaFe^{3+}_3(PO_4)_2(OH)_4.2H_2O$
Cyrilovite crystallizes in the tetragonal system as very small squat crystals coloured orange to brown. H not known; SG 3.1
Localities. Occurs with metastrengite and leucophosphite at the Sapucaia pegmatite, Minas Gerais, Brazil; also from Cyrilov, Czechoslovakia; etc.

Anapaite $Ca_2Fe^{2+}(PO_4)_2.2H_2O$
Anapaite crystallizes in the triclinic system as tabular crystals or aggregates coloured green or white, with a vitreous lustre. H 3.5; SG 2.8
Localities. Occurs in phosphatic geodes at Bellaver de Cerdena, Spain; in a bituminous clay at Messel, W. Germany; from the Lewis Well, Kings Co., California.
Treatment. Clean with distilled water.

Mitridatite
$Ca_3Fe^{3+}_4(OH)_6(H_2O_3)(PO_4)_4$
Mitridatite crystallizes in the monoclinic system as tabular crystals or as crusts and veinlets. The colour is deep red with an olive-green streak, and the lustre is dull. H 3.5; SG 3.2
Localities. Occurs with jahnsite and collinsite in an altered triphylite in pegmatite, Custer Co., South Dakota; from Central Africa; etc.
Treatment. Clean with distilled water.

Jahnsite $CaMn^{2+}Mg_2(H_2O)_8Fe^{3+}_2(OH)_2(PO_4)_4$
Jahnsite crystallizes in the monoclinic system as well-developed crystals and as parallel aggregates, often twinned. The colour is brown, purple, yellow, orange or green. There is a vitreous lustre. H 4; SG 2.7
Occurrence. Occurs in granite pegmatites with other phosphate minerals.
Localities. From Custer Co., South Dakota; Palermo, New Hampshire.
Treatment. Clean with distilled water.

Messelite
$Ca_2(Fe^{2+}, Mn^{2+})(PO_4)_2.2H_2O$
Messelite crystallizes in the triclinic system as prismatic crystals or lamellar masses; it has a perfect cleavage. The colour is white, colourless, green or grey with a white streak. There is a vitreous lustre. H 3.5; SG 3.1
Occurrence. Occurs as a late hydrothermal mineral in granite pegmatites.
Localities. From Custer Co., South Dakota; Palermo, New Hampshire; King's Mountain, North Carolina.
Treatment. Clean with distilled water.

Collinsite
$Ca_2(Mg, Fe)(PO_4)_2.2H_2O$
Collinsite crystallizes in the triclinic system as prismatic crystals or as bundles; more commonly as crusts. There is a perfect cleavage. The colour is white, brown or colourless, with a vitreous lustre. H 3.5; SG 3.0
Localities. Occurs in altered phosphate nodules in a granite pegmatite at the Tip Top mine, Custer Co., South Dakota; and elsewhere.

Phosphoferrite
$(Fe, Mn)_3(PO_4)_2.3H_2O$
Phosphoferrite crystallizes in the orthorhombic system as octahedra or as granular masses. It is green or colourless, with a vitreous lustre. H 3–3.5; SG 3.3
Occurrence. Occurs in granite pegmatites as an alteration product of the mineral triphylite.
Localities. From Custer Co., South Dakota; Hagendorf, Bavaria, W. Germany.
Treatment. Clean with distilled water.

Metavauxite
$Fe^{2+}Al_2(PO_4)_2(OH)_2.8H_2O$
Metavauxite crystallizes in the monoclinic system as prismatic crystals or as radial aggregates. The colour is white, green or colourless, and the lustre is vitreous. H 3; SG 2.3
Localities. Occurs with paravauxite and wavellite with quartz in the tin mines of Llallagua, Bolivia.
Treatment. Clean with distilled water.

Above: Pyramidal crystals of lazulite, from Lincoln County, Georgia

Paravauxite
$Fe^{2+}Al(PO_4)_2(OH)_2.10H_2O$
Paravauxite crystallizes in the triclinic system as prismatic crystals or radial aggregates. There is a perfect cleavage. The colour is greenish-white or colourless with a white streak, and there is a vitreous lustre. H 3; SG 2.4
Localities. Occurs in tin deposits at Llallagua, Bolivia.
Treatment. Clean with distilled water.

Vauxite $Fe^{2+}Al_2(PO_4)_2(OH)_2.6H_2O$
Vauxite crystallizes in the triclinic system as tabular crystals and as radial aggregates coloured pale to dark blue. The streak is white. There is a vitreous lustre. H 3.5; SG 2.4
Localities. Occurs with paravauxite and quartz in the tin deposits at Llallagua, Bolivia.
Treatment. Clean with distilled water.

Leucophosphite
$KFe^{3+}_2(OH)(H_2O)(PO_4)_2.H_2O$
Leucophosphite crystallizes in the monoclinic system as prismatic crystals or as lamellar masses. The colour is brown in pegmatite and greenish in sedimentary occurrences; there is a vitreous lustre. H 3.5; SG 2.9
Localities. Occurs in pegmatites at the Sapucaia mine, Minas Gerais, Brazil; in sedimentary deposits, Bomi Hill, Liberia; and elsewhere.

Lazulite $(Mg,Fe^{2+})Al_2(PO_4)_2(OH)_2$
Lazulite crystallizes in the monoclinic system as pyramidal crystals or as granular masses with twinning commonly found. The colour is deep or light blue, with a vitreous lustre. H 5.5–6; SG 3.1
Occurrence. Occurs in granite pegmatites, quartz veins and quartzites associated with garnet, andalusite and other minerals.
Localities. From Bolivia; Brazil; Malagasy Republic; Lincoln Co., Georgia; Mono Co., California; Hörrsjoberg, Sweden; etc.
Treatment. Clean with dilute hydrochloric acid only.

Left: Strengite in matrix from Bavaria

Right: Tiny green crystals of euchroite from Lubietová, Czechoslovakia. The name means 'beautifully coloured'

Below right: Clinoclase from Gwennap, Cornwall

Fashioning. Uses: usually cabochons or beads; *cleavage*: indistinct prismatic; *pleochroism*: strong; *heat sensitivity*: fairly high.

Cassidyite
$Ca_2(Ni, Mg)(PO_4)_2 . 2H_2O$
Cassidyite crystallizes in the triclinic system as crusts and spherules which are coloured bright green. H not known; SG 3.1
Localities. Occurs in weathered meteorites from Wolf Creek Crater, Western Australia

Arsenates

Olivenite Cu_2AsO_4OH
Olivenite crystallizes in the orthorhombic system as prismatic crystals and as globular masses. The colour is olive green to yellow and brown. The lustre is vitreous or silky. H3; SG 4.4
Occurrence. Occurs in the oxidized zone of ore deposits with azurite and malachite.
Localities. From Tsumeb, South-West Africa; Majuba Hill, Nevada; and elsewhere.
Treatment. Olivenite should be cleaned with distilled water.

Euchroite $Cu_2AsO_4OH . 3H_2O$
Euchroite crystallizes in the orthorhombic system as prismatic crystals coloured bright emerald green. They have a vitreous lustre. H 3.5–4; SG 3.4
Localities. Occurs with olivenite in a sericite schist at Libethen, Hungary; from the Zapachitsa copper deposit, Bulgaria.

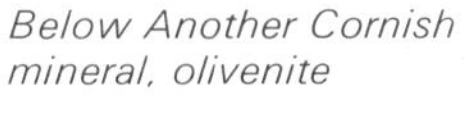

Below Another Cornish mineral, olivenite

Cornwallite $Cu_5(AsO_4)_2(OH)_4 . H_2O$
Cornwallite crystallizes in the monoclinic system as botryoidal crusts coloured light to dark green, with a dull lustre. H 4.5; SG 4.5
Localities. Occurs in Cornwall; Utah; Majuba Hill, Nevada; Germany; and other localities.

Cornubite $Cu_5(AsO_4)_2(OH)_4$
Cornubite crystallizes in the triclinic system as fibrous masses coloured light apple green. H not known; SG 4.6
Localities. Occurs in Cornwall, Devon and Cumbria, with malachite and olivenite.

Clinoclase $Cu_3AsO_4(OH)_3$
Clinoclase crystallizes in the monoclinic system as tabular crystals or rosettes; there is a perfect cleavage and the colour is dark green, blue or black. The streak is bluish-green. There is a vitreous lustre. H 2.5–3; SG 4.3
Localities. Occurs with olivenite in Cornwall; Tavistock, Devon; Majuba Hill, Nevada; Utah; etc.
Treatment. Clean with distilled water.

Chlorotile $Cu_3(AsO_4)_2 . 6H_2O$
Chlorotile crystallizes in the hexagonal system as prismatic crystals, more commonly as fibrous masses. The colour is pale to emerald green, with a dull lustre. H soft; SG 3.7
Occurrence. Occurs in the oxidation zone of ore deposits.
Localities. From Schneeberg, E. Germany; Dome Rock copper mine, South Australia; etc.

Conichalcite $CuCaAsO_4OH$
Conichalcite crystallizes in the orthorhombic system as prismatic crystals or as reniform masses and crusts. The colour is emerald-green or yellow. H 4.5; SG 4.3
Occurrence. Occurs in the oxidized zone of copper deposits with limonite and other copper minerals.
Localities. From South-West Africa; Mexico; Poland; Germany; Spain; Chile; South Dakota; Arizona; etc.

Liroconite
$Cu_2Al(As, P)O_4(OH)_4 . 4H_2O$
Liroconite crystallizes in the monoclinic system as wedge-shaped crystals or granular masses. The colour is blue to green, and the crystals have a

vitreous lustre. H 2–2.5; SG 2.9–3.0
Occurrence. Occurs in the oxidized zone of copper deposits with azurite and malachite.
Localities. From Cornwall and Devon; Germany; Inyo Co., California; and some other localities.

Ceruléite
$CuAl_4(AsO_4)_2(OH)_8 . 4H_2O$
Ceruléite's crystal system is not known. It is found as compact masses, blue in colour, resembling turquoise. SG 2.8
Localities. From Wheal Gorland, Cornwall; and from Chile.

Chenevixite
$Cu_2Fe_2(AsO_4)_2(OH)_4 . H_2O$
Chenevixite crystallizes in the orthorhombic system as compact masses with a greasy or dull lustre. They are coloured dark green with a greenish-yellow streak. H 3.5–4.5; SG 3.9
Localities. Occurs with azurite and chrysocolla at Klein Spitzkopje, South-West Africa; Tamanrasset, Algeria; Broken Hill, New South Wales; and some other localities.
Treatment. Clean with distilled water.

Arthurite $Cu_2Fe_4(AsO_4, PO_4, SO_4)_4(O, OH)_4 . 8H_2O$
Arthurite crystallizes in the monoclinic system as prismatic crystals or globular aggregates. The colour is apple green or emerald green, with a vitreous lustre. H not known; SG 3.0
Occurrence. Occurs as thin crusts with pharmacosiderite.
Localities. From Calstock, Cornwall; Majuba Hill, Nevada; Atacama Province, Chile.

Rösslerite $MgH(AsO_4) . 7H_2O$
Rösslerite crystallizes in the monoclinic system as fibrous crusts coloured white. H 2–3; SG 1.9
Occurrence. Occurs in arsenic-bearing ore deposits.
Localities. From Bieber, Hanau, W. Germany; Jachymov, Czechoslovakia; etc.

Pharmacolite $CaHAsO_4 . 2H_2O$
Pharmacolite crystallizes in the monoclinic system as botryoidal clusters or acicular crystals. There is a perfect cleavage. They are white or colourless. They have a vitreous or silky lustre. H 2–2.5; SG 2.7
Localities. Occurs with erythrite at San Gabriel Canyon, Los Angeles Co., California; Wittichen, Black Forest, W. Germany; Jachymov, Czechoslovakia; etc.
Treatment. Pharmacolite should be cleaned with distilled water.

Sainfeldite $H_2Ca_5(AsO_4)_4 . 4H_2O$
Sainfeldite crystallizes in the monoclinic system as rosettes coloured light pink or colourless. H not known; SG 3.0
Localities. Occurs in the Gabe Gottes vein, Ste. Marie-aux-Mines, Alsace, France; etc.

Picropharmacolite
$H_2Ca_4Mg(AsO_4)_4 . 12H_2O$
Picropharmacolite crystallizes in the monoclinic system as needles and encrustations. There is a perfect cleavage

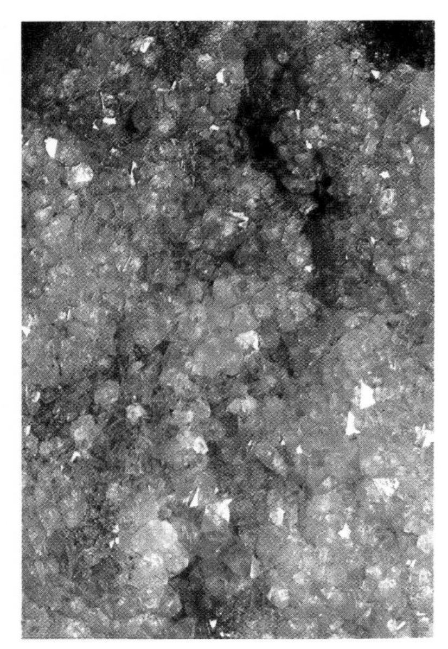

Left: Adamite crystals from the Ojuela mine, Mapimi, Mexico

Far left: Liroconite from Cornwall

Below left: Pharmacolite from Ste. Marie-aux-Mines, Alsace, France

and the mineral is white or colourless. H not known; SG 2.6
Localities. Occurs on dolomite at Joplin, Missouri; also from Ste. Marie-aux-Mines, Alsace, France; etc.

Köttigite $Zn_3(AsO_4)_2 . 8H_2O$
Köttigite crystallizes in the monoclinic system as prismatic crystals or as crusts. There is a perfect cleavage. The colour is dark red to brown with a reddish-white streak. H 2.5–3; SG 3.3
Localities. Occurs at the Daniel mine, Schneeberg, E. Germany; Mapimi, Durango, Mexico.
Treatment. Clean with distilled water.

Legrandite $Zn_2(OH)AsO_4 . H_2O$
Legrandite crystallizes in the monoclinic system as prismatic crystals or as aggregates, coloured yellow to colourless. They have a vitreous lustre. H 4.5; SG 4.0
Localities. Occurs in a limonite matrix with adamite at Mapimi, Durango, Mexico.
Treatment. Clean with distilled water.

Adamite Zn_2AsO_4OH
Adamite crystallizes in the orthorhombic system as tabular crystals or as druses of interlocked crystals. The colour is bright yellowish-green, blue, violet or light red. There may be a fluorescence under ultra-violet light. The lustre is vitreous. H 3.5; SG 4.3–4.5
Occurrence. Occurs in the oxidation zone of ore deposits with limonite, calcite and other minerals.
Localities. From San Bernardino Co., California; Mapimi, Durango, Mexico; Laurion, Greece; and some other localities.
Treatment. Clean with distilled water.

Paradamite Zn_2AsO_4OH
Paradamite crystallizes in the triclinic system as aggregates or as rounded crystals. There is a perfect cleavage and the colour is pale yellow. There is a vitreous lustre. H 3.5; SG 4.5
Localities. Occurs with mimetite and adamite on limonite at Mapimi, Durango, Mexico.
Treatment. Clean with distilled water.

Veszelyite
$(Cu, Zn)_3(PO_4)(OH)_3 2H_2O$
Veszelyite crystallizes in the monoclinic system as prismatic crystals or granular aggregates. The colour is green to blue, the lustre vitreous. H 3.5–4; SG 3.4

Right: Bayldonite from Tsumeb, South-West Africa

Localities. Kabwe, Zambia; Kipushi, Shaba, Zaire; Banat, Romania; and other localities.
Treatment. Clean with distilled water.

Austinite $CaZnAsO_4OH$
Austinite crystallizes in the orthorhombic system as prismatic crystals or drusy crusts coloured white, pale yellow or bright green. They have an adamantine or silky lustre. H 4–4.5; SG 4.1
Localities. Occurs with adamite on limonite at Tooele Co., Utah; Mapimi, Durango, Mexico; Tsumeb, South-West Africa; etc.

Chudobaite $(Na, K, Ca)(Mg, Zn, Mn)_2H(AsO_4)_2 . 4H_2O$
Chudobaite crystallizes in the triclinic system as small crystals coloured pink. H 2.5–3; SG 2.9
Localities. Occurs with conichalcite at Tsumeb, South-West Africa.

Holdenite $(Mn, Zn)_6(AsO_4)(OH)_5O_2$
Holdenite crystallizes in the orthorhombic system as tabular crystals coloured pink or yellowish-red to deep red. They have a vitreous lustre. H 4; SG 4.1
Localities. Occurs at Franklin, New Jersey. The only known specimen is found in a vein in franklinite.

Chlorophoenicite $(Mn, Zn)_5AsO_4(OH)_7$
Chlorophoenicite crystallizes in the monoclinic system as needles and fibres coloured light green in daylight and purple in artificial light. H 3.5; SG 3.5
Localities. Occurs with pyrochroite and other minerals in zinc-manganese-iron oxides in marbles at Franklin, New Jersey.
Treatment. Clean with distilled water.

Retzian $Mn_2Y(AsO_4)(OH)_4$
Retzian crystallizes in the orthorhombic system as prismatic crystals coloured dark brown with a light brown streak. They have a vitreous lustre. H 4; SG 4.1
Localities. Occurs in dolomite at Nordmark, Sweden.
Treatment. Clean with distilled water.

Agardite $(Y, Ca)_2Cu_{12}(AsO_4)_6(OH)_{12} . 6H_2O$
Agardite crystallizes in the hexagonal system as acicular crystals, coloured blue to green. H 3–4; SG 3.7
Localities. From the oxidation zone of the copper deposit at Bou-Skour, Morocco; also from Utah; etc.
Treatment. Clean with distilled water.

Cafarsite $Ca_3(Fe, Ti)_3Mn(AsO_4)_6 . 2H_2O$
Cafarsite crystallizes in the cubic system as well-formed crystals coloured dark or yellowish-brown. The streak is yellowish-brown. H 5.5–6; SG 3.9
Localities. Occurs on gneisses in Valais, Switzerland; Italy.

Schultenite $PbHAsO_4$
Schultenite crystallizes in the monoclinic system as rhombohedra, often with striated faces. They are colourless, with a vitreous lustre. H 2.5; SG 5.9
Localities. Occurs in association with anglesite and bayldonite in the Tsumeb area of South-West Africa.

Bayldonite $(Pb, Cu)_3(AsO_4)_2(OH)_2$
Bayldonite crystallizes in the monoclinic system as granular masses or as crusts. The colour is yellow to green; the lustre is resinous. H 4.5; SG 5.5
Occurrence. Occurs in the oxidation zone of copper ore deposits with azurite and other minerals.
Localities. From St. Day, Cornwall; Tsumeb, South-West Africa; etc.
Treatment. Clean with distilled water.
Fashioning. Uses: usually cabochons; *cleavage*: not significant; *heat sensitivity*: low.

Duftite $CuPbAsO_4OH$
Duftite crystallizes in the orthorhombic system as tiny crystals or as crusts coloured apple to dark green. H 3; SG 6.4
Localities. Occurs with wulfenite and malachite at Tsumeb, South-West Africa.

Caryinite $(Ca, Na, Pb, Mn)_3(Mn, Mg)_2(AsO_4)_3(OH)$
Caryinite crystallizes in the monoclinic system as granular masses with a yellow to brown colour. H 4; SG 4.3
Localities. Occurs in skarn at Långban, Sweden.

Carminite $PbFe_2(AsO_4)_2(OH)_2$
Carminite crystallizes in the orthorhombic system as minute crystals or as tufted aggregates. The colour is red to brown. H 3.5; SG 5.2
Localities. Occurs with anglesite and other minerals at Mapimi, Durango, Mexico; Calstock, Cornwall; Horhausen, W. Germany; Utah; Colorado; etc.

Ludlockite $(Fe, Pb)As_2O_6$
Ludlockite crystallizes in the triclinic system its crystals sometimes show lamellar twinning. The mineral is sectile and flexible. The colour is red with a light brown streak, and the lustre is adamantine. H 1.5–2; SG 4.4
Localities. Occurs with zinc in siderite at Tsumeb, South-West Africa.

Atelestite $Bi_8(AsO_4)_3O_5(OH)_5$
Atelestite crystallizes in the monoclinic system as tabular crystals or spherical crystalline aggregates. The colour is bright yellow to green with an adamantine lustre. H 4.5–5; SG 6.8
Localities. Occurs with bismutite at Schneeberg, E. Germany.
Treatment. Clean with distilled water.

Mixite $Cu_{12}Bi_2(AsO_4)_6(OH)_{12} . 6H_2O$
Mixite crystallizes in the hexagonal system as needle-like crystals or as tufted aggregates coloured emerald green to white. The streak is lighter than the colour. H 3–4; SG 3.8
Localities. Occurs in Germany; Inyo Co., California; El Carmen mine, Durango, Mexico; etc.

Walpurgite $(BiO)_4UO_2(AsO_4)_2 . 3H_2O$
Walpurgite crystallizes in the triclinic system as thin crystals, commonly twinned, or as aggregates. There is a perfect cleavage. The colour is yellow, and the lustre is adamantine. H 3.5; SG 5.9
Localities. Occurs with torbernite and other minerals in the oxidized zone of a uraninite vein, in the Walpurgis vein at the Weisser Hirsch mine, Schneeberg, E. Germany; etc.

Trögerite $H_2(UO_2)_2(AsO_4)_2 . 8H_2O$
Trögerite crystallizes in the tetragonal system as tabular crystals or as aggregates. There is a perfect cleavage. The colour is lemon yellow and there is a similarly-coloured fluorescence under ultra-violet light. The lustre is vitreous. H 2–3; SG 3.5
Localities. Occurs with walpurgite in the Walpurgis vein, Weisser Hirsch mine, Schneeberg, E. Germany.
Treatment. Clean with distilled water.

Abernathyite $KUO_2AsO_4 . 3H_2O$
Abernathyite crystallizes in the tetragonal system as tabular crystals or as coatings. There is a perfect cleavage; the colour is yellow and the streak pale yellow. There is a yellow-green fluorescence under ultra-violet light. The lustre is vitreous. H 2.5; SG 3.3
Localities. Occurs with scorodite in

sandstone at Temple Mountain, Utah; Harding Co., South Dakota.

Heinrichite
$Ba(UO_2)_2(AsO_4)_2.10–12H_2O$
Heinrichite crystallizes in the tetragonal system as tabular crystals coloured yellow-green and with a perfect cleavage. They have a vitreous lustre. There is a bright green fluorescence under ultra-violet light. H 2.5; SG 3.6
Localities. Occurs in an altered rhyolite tuff in Lake Co., Oregon; as an alteration of pitchblende at Wittichen in the Black Forest, W. Germany.

Uranospinite
$Ca(UO_2)_2(AsO_4)_4.10H_2O$
Uranospinite crystallizes in the tetragonal system as rectangular plates with a perfect cleavage. The colour is yellow to green with a pearly lustre. There is a bright yellow fluorescence under ultra-violet light. H 2–3; SG 3.4
Occurrence. Formed by the alteration of uraninite.
Localities. From the Weisser Hirsch mine, Schneeberg, E. Germany; Kane Co., Utah; etc.

Zeunerite $Cu(UO_2)_2(AsO_4)_2.12H_2O$
Zeunerite crystallizes in the tetragonal system as tabular crystals with a perfect cleavage. The colour is yellow to emerald-green, with a vitreous lustre. H 2.5; SG 3.4
Occurrence. Occurs in the oxidized zone of uraninite and arsenic-bearing deposits.
Localities. From the Weisser Hirsch mine, Schneeberg, E. Germany; etc.
Treatment. Clean with distilled water.

Uranospathite
$Cu(UO_2)_2[(As, P)O_4]_2.12H_2O$
It is not certain that uranospathite is a separate species. It probably crystallizes in the tetragonal system as groups of rectangular plates with a perfect cleavage. The colour is yellow to green or bluish-green. H not known; SG 2.5
Localities. Occurs with bassetite at Redruth, Cornwall.

Sarkinite Mn_2AsO_4OH
Sarkinite crystallizes in the monoclinic system as tabular crystals or as spherical granular masses. The colour is dark red or yellow with a red or yellow streak. There is a greasy lustre. H 4–5; SG 4.1–4.2
Localities. Occurs with hausmannite in Sweden at Långban and with jacobsite and other minerals at Pajsberg, Sweden.
Treatment. Clean with distilled water.

Eveite $Mn_2(OH)(AsO_4)$
Eveite crystallizes in the orthorhombic system as tabular crystals coloured apple-green with a white streak. H 4; SG 3.8
Localities. Occurs in fractures in hausmannite from Långban, Sweden.

Brandtite $Ca_2Mn(AsO_4)_2.2H_2O$
Brandtite crystallizes in the monoclinic system as prismatic crystals or as reniform masses and radial groups. There is a perfect cleavage and the colour is white or colourless. There is a vitreous lustre. H 3.5; SG 3.7
Localities. Occurs with rhodochrosite at Sterling Hill, New Jersey; etc.
Treatment. Clean with distilled water.

Berzeliite
$(Ca, Na)_3(Mg, Mn)_2(AsO_4)_3$
Berzeliite crystallizes in the cubic system as trapezohedra; more commonly as grains or in masses. The colour is orange-yellow. H 4.5–5; SG 4.1
Localities. Occurs in limestone with rhodonite and other minerals at Långban, Sweden.

Hematolite
$Mn_4Al(OH)_2(AsO_4)(AsO_3)_2$
Hematolite crystallizes in the hexagonal system as tabular or rhombohedral crystals with horizontal striations on the rhombohedral faces. There is a perfect cleavage. The colour is brownish-red to black, and the lustre is vitreous. H 3.5; SG 3.5
Localities. Occurs in crystalline limestone with jacobsite and other minerals at Nordmark, Sweden.
Treatment. Clean with distilled water.

Symplesite $Fe_3(AsO_4)_2.8H_2O$
Symplesite crystallizes in the triclinic system as small crystals or as spherical aggregates; it has a perfect cleavage. The colour is light blue or green to deep blue, with a vitreous lustre. H 2.5; SG 3.0
Localities. Occurs as a powder filling cavities in löllingite in a pegmatite at Custer Mountain Lode, Custer Co., South Dakota; from Lobenstein, E. Germany; Italy; Tasmania; etc.

Scorodite $FeAsO_4.2H_2O$
Scorodite crystallizes in the orthorhombic system as pyramidal crystals or as crusts or masses. The colour is pale green, yellow to brown or violet. The lustre is vitreous. H 3.5–4; SG 3.3
Occurrence. Formed by the oxidation of arsenic-bearing minerals.
Localities. From Ouro Preto, Minas Gerais, Brazil; Ontario; California; Mexico; and elsewhere.
Treatment. Clean ultrasonically.

Pharmacosiderite
$Fe_3(AsO_4)_2(OH)_3.5H_2O$
Pharmacosiderite crystallizes in the cubic system as cubes with diagonal striations or tetrahedra. The mineral is sectile and is coloured olive to emerald-green or reddish-brown. It has an adamantine or a greasy lustre. H 2.5; SG 2.8
Occurrence. As an oxidation product of arsenic minerals.
Localities. From Cornwall; Utah; South Dakota; Germany; and elsewhere.
Treatment. Clean with distilled water.

Arseniosiderite
$Ca_3Fe_4^{3+}(OH)_6(H_2O)_3(AsO_4)_4$
Arseniosiderite crystallizes in the monoclinic system as masses or fibrous aggregates with a yellow or brown colour. The streak is yellow. H 4.4; SG 3.6
Localities. Occurs at Mapimi, Durango, Mexico; Tooele Co., Utah; France; Germany; etc.
Treatment. Clean with distilled water.

Left: Bladed crystals of erythrite from Bou Azzer, Morocco

Liskeardite
$(Al, Fe)_3AsO_4(OH)_6.5H_2O$
Liskeardite crystallizes in the monoclinic or orthorhombic systems as aggregates and crusts. The colour is white or greenish to black. H not known; SG 3.0
Localities. Occurs coating quartz at Liskeard, Cornwall; and other places.

Erythrite $Co_3(AsO_4)_2.8H_2O$
Erythrite crystallizes in the monoclinic system as prismatic crystals or as bladed aggregates; there is a perfect cleavage and the mineral is sectile. The colour is deep purple to pale pink. The streak is lighter than this. The lustre is adamantine or dull. H 1.5–2.5; SG 3.2
Occurrence. Occurs in the oxidation zone of cobalt ore deposits.
Localities. From Schneeberg, E. Germany; England; Sweden; Bou Azzer, Morocco; Cobalt, Ontario; California; Nevada; New Mexico; etc.
Treatment. Clean with distilled water.

Left: Cubic crystals of pharmocosiderite from Cornwall

Right: Lavendulan from Meskani, Iran

Lavendulan $(Ca, Na)_2Cu_5(AsO_4)_4Cl.4–5H_2O$
Lavendulan crystallizes in the orthorhombic system as flaky aggregates coloured lavender-blue, with a vitreous lustre. H 2.5; SG 3.5
Localities. Occurs with olivenite at Tooele Co., Utah; with erythrite at Jachymov, Czechoslovakia; etc.

Roselite $Ca_2(Co, Mg)(AsO_4)_2.2H_2O$
Roselite crystallizes in the monoclinic system as prismatic crystals or as spherical aggregates. There is a perfect cleavage. The colour is dark red to pink, with a vitreous lustre. H 3.5; SG 3.5–3.7
Localities. Occurs with erythrite at Bou Azzer, Morocco; Schneeberg, E. Germany; Shapbach, W. Germany.
Treatment. Clean with distilled water.

Annabergite $Ni_3(AsO_4)_2.8H_2O$
Annabergite crystallizes in the monoclinic system as prismatic crystals or as crusts or powders. There is a perfect cleavage. The colour is white to intense yellow-green with the streak a lighter shade than the colour. There is an adamantine lustre. H 1.5–2.5; SG 3.1
Occurrence. Occurs in the oxidation zone of nickeliferous ore deposits.
Localities. From Schneeberg, E. Germany; Cobalt, Ontario; Sierra Cabrera, Spain; Laurion, Greece; Humboldt Co., Nevada; Inyo Co., California; etc.
Treatment. Clean with distilled water.

Vanadates

Volborthite $Cu_3(VO_4)_2.3H_2O$
Volborthite crystallizes in the monoclinic system as compact masses; it has a perfect cleavage and frequently displays lamellar twinning. The colour is dark green, yellow or brown, with a vitreous lustre. H 3.5; SG 3.5–3.8
Localities. Occurs in sandstone at Richardson, Utah; in a sedimentary rock at Menzies Bay, Vancouver Island, Canada; Syssersk, Urals, USSR.
Treatment. Clean with distilled water.

Pascoite $Ca_2V_6O_{17}.11H_2O$
Pascoite crystallizes in the monoclinic system as granular crusts coloured orange-yellow with a yellow streak and soluble in water. H 2.5; SG 1.9
Localities. Occurs in the vanadium deposit at Minas Ragra, Peru; Mesa Co., Colorado; etc.

Hewettite $CaV_6O_{16}.9H_2O$
Hewettite probably crystallizes in the orthorhombic system as aggregates and coatings coloured deep red. H not known; SG 2.5
Localities. Occurs in the oxidized zone of the vanadium deposit at Minas Ragra, Peru; etc.

Hummerite $KMgV_5O_{14}.8H_2O$
Hummerite crystallizes in the triclinic system as tabular crystals or as crusts with a bright orange colour. H not known; SG 2.5
Localities. Occurs in grey clay at the Hummer mine, Montrose Co., Colorado; also from South Dakota.

Sherwoodite $Ca_3(V^{4+}O)_2V^{5+}_6O_{20}.15H_2O$
Sherwoodite crystallizes in the tetragonal system as polycrystalline aggregates coloured dark blue to black, altering to yellow-green. H about 2; SG 2.8
Localities. Vanadium mines in Colorado.
Treatment. Clean with distilled water.

Hendersonite $Ca_2V^{4+}_{1+x}V^{5+}_{8-x}(O, OH)_{24}.8H_2O$
Hendersonite crystallizes in the orthorhombic system as platy crystals or as fibrous aggregates coloured dark green to black with a brownish-green streak. H 2.5; SG 2.8
Localities. Occurs in oxidized ore with sherwoodite at vanadium mines in Colorado and in New Mexico.
Treatment. Clean with distilled water.

Steigerite $AlVO_4.3H_2O$
Steigerite's system is not known. It forms plates of a yellow colour, with a waxy lustre. H and SG not known.
Localities. Occurs with gypsum in fractures in corvusite, San Miguel Co., Colorado.
Treatment. Clean with distilled water.

Vanalite $NaAl_8V_{10}O_{38}.30H_2O$
Vanalite's crystal system is not known; it forms bright yellow to orange crystals with a waxy or dull lustre. SG 2.3–2.4
Localities. Occurs as encrustations with clay minerals at Kara-Tau, Kazakhstan, USSR.
Treatment. Clean with distilled water.

Wakefieldite YVO_4
Wakefieldite crystallizes in the tetragonal system as yellow masses. H 5; SG 4.2
Localities. Occurs with quartz and other minerals at Wakefield, Quebec.
Treatment. Clean with distilled water.

Chervetite $Pb_2V_2O_7$
Chervetite crystallizes in the monoclinic system as twinned crystals or as pseudomorphs after francevilleite. The colour is grey to brown or colourless with a white streak. There is an adamantine

lustre. H about 2.5; SG 6.3–6.5
Occurrence. Occurs with francevillite in the oxidation zone of uranium deposits.
Localities. From Mounana, Gabon.
Treatment. Clean with distilled water.

Mottramite $Pb(Cu, Zn)VO_4OH$
Mottramite crystallizes in the orthorhombic system as prismatic crystals or as aggregates or crusts. The colour is light to dark green to black; the lustre is vitreous. H 3.0–3.5; SG 5.9
Occurrence. Occurs in the oxidation zone of ore deposits.
Localities. From Mottram St. Andrew, Cheshire; Crestmore, Riverside Co., California; Sardinia; Bolivia; Chile; and elsewhere.
Treatment. Clean with distilled water.

Descloizite $Pb(Zn, Cu) VO_4OH$
Descloizite crystallizes in the orthorhombic system as prismatic crystals or as aggregates or crusts. The colour is orange-red to brown or green, with a vitreous lustre. H 3.5; SG 6.2
Occurrence. Occurs in the oxidation zone of ore deposits with pyromorphite, calcite and other minerals.
Localities. From Tsumeb and Otavi, South-West Africa; Rhodesia; Mexico; Galena, South Dakota; and some other localities.
Treatment. Descloizite should be cleaned with distilled water.

Left: A fragment of amblygonite

Below left: A group of descloizite crystals from Tsumeb, South-West Africa

Pucherite $BiVO_4$
Pucherite crystallizes in the orthorhombic system as small crystals with curved faces, or as masses. There is a perfect cleavage. The colour is yellow to brown with a yellow streak. There is a vitreous lustre. H 4; SG 6.2
Localities. Occurs in the oxidized portion of a bismuth-bearing vein in the Pucher shaft of the Wolfgang mine, Schneeberg, E. Germany; Brejauba, Minas Gerais, Brazil; etc.
Treatment. Clean with distilled water.

Carnotite
$K_2(UO_2)_2(VO_4)_2 . 1\text{–}3H_2O$
Carnotite crystallizes in the monoclinic system as tiny crystals or as compact masses or aggregates. There is a perfect cleavage. The colour is bright golden yellow, with a pearly or dull lustre. H not known; SG 4.7
Occurrence. In sedimentary rocks.
Localities. From Kambove, Shaba, Zaire; El Borouj, Morocco; Radium Hill, South Australia; Wyoming; and some other localities.

Sengierite $Cu(UO_2)_2(VO_4)_2 . 6H_2O$
Sengierite crystallizes in the monoclinic system as thin plates with a perfect cleavage coloured green, with a vitreous lustre. H 2.5; SG 4.0
Localities. Occurs with malachite and oxides of copper and other minerals at Shaba, Zaire; Bisbee, Arizona.

Tyuyamunite
$Ca(UO_2)_2(VO_4)_2 . 5\text{–}8H_2O$
Tyuyamunite crystallizes in the orthorhombic system as scales or compact masses with a perfect cleavage. They have an adamantine or dull lustre. The colour is green or yellow, sometimes with a weak fluorescence of a similar colour under ultra-violet light. H 2; SG 3.3–3.6
Occurrence. Occurs as a secondary mineral with other uranium and vanadium minerals.
Localities. From Ferghana, USSR; Wyoming; Texas; Colorado; Nevada.

Vanuralite
$Al(UO_2)_2(VO_4)_2(OH) . 11H_2O$
Vanuralite crystallizes in the monoclinic system as plates with a perfect cleavage and coloured yellow. H 2; SG 3.6.
Localities. Occurs with francevillite at Mounana, Gabon.

Curienite $Pb(UO_2)_2(VO_4)_2 . 5H_2O$
Curienite crystallizes in the orthorhombic system as powders coloured yellow. SG 4.9
Localities. Occurs in sandstones at Mounana, Gabon.

Schubnelite $Fe_2(VO_4)_2 . 2H_2O$
Schubnelite crystallizes in the triclinic system as tiny crystals coloured black; yellow in thin fragments. H not known; SG 3.3
Localities. From Mounana, Gabon.

Phosphates, Arsenates and Vanadates with other anions

Amblygonite $(Li, Na)AlPO_4(F, OH)$
Amblygonite crystallizes in the triclinic system as prismatic crystals or as masses; it has a perfect cleavage. The colour is white to yellow or pink, with a vitreous lustre. H 5.5–6; SG 3.1; RI 1.61, 1.64
Occurrence. In granite pegmatites.
Localities. From Newry, Oxford Co., Maine; Brazil; France; Sakangyi, Burma; etc.
Treatment. Clean with distilled water.
Fashioning. Uses: faceting or cabochons; *cleavage*: perfect // to basal pinacoid, distinct macropinacoidal; *cutting angles*: crown 40°, pavilion 40°; *heat sensitivity*: fairly high.

Herderite $CaBePO_4(F, OH)$
Herderite crystallizes in the monoclinic system as prismatic or tabular crystals or as fibrous aggregates. The colour is pale yellow or green with a vitreous lustre. H 5–5.5; SG 3.0
Occurrence. In granite pegmatites.
Localities. From Newry and other places in Oxford Co., Maine; Minas Gerais, Brazil; Bavaria, W. Germany and Saxony, E. Germany; Mursinsk, USSR.
Treatment. Remove iron stains with dilute oxalic acid.

Wagnerite $(Mg, Fe, Mn, Ca)_2PO_4F$
Wagnerite crystallizes in the monoclinic system as prismatic crystals or

masses. The colour is yellow, grey to brown, with a vitreous lustre. H5–5.5; SG 3.1
Localities. Occurs in quartz veins with magnesite near Salzburg, Austria; in pegmatite at Mangualde, Portugal; Mt. Vesuvius, Italy; etc.
Treatment. Clean with distilled water.

Apatite $Ca_5(PO_4)_3F$
Apatite crystallizes in the hexagonal system as prismatic crystals or as compact masses. The colour is white, colourless, green, yellow, dark blue, purple, brown or reddish-brown. The streak is white. There is a vitreous lustre. H 5; SG 3.1; RI 1.63–1.64
Occurrence. Occurs in igneous rocks and pegmatites; also in hydrothermal veins and as detrital deposits.
Localities. From Durango, Mexico; South Dakota; Ontario; Quebec; Panasqueira, Portugal; Spain; Germany; Norway; Malagasy Republic; Burma, etc.
Treatment. Clean with distilled water.
Fashioning. *Uses*: faceting or cabochons; *cleavage*: imperfect basal; very brittle; *cutting angles*: crown 40°, pavilion 40°; *dichroism*: strong in Burmese stones; weak in other stones, except in deeper colours; *heat sensitivity*: high, use low-temperature dop wax.

Right: Violet apatite crystals showing tabular form. From Ehrenfriedersdorf, East Germany

Right: A hexagonal prism of apatite from Mexico

Triplite
$(Mn^{2+}, Fe^{2+}, Mg, Ca)_2(PO_4)(F, OH)$
Triplite crystallizes in the monoclinic system as masses coloured reddish-brown. The streak is white or brown. There is a vitreous lustre. H 5–5.5; SG 3.5–3.9
Occurrence. Occurs in granite pegmatites and in high-temperature vein deposits.
Localities. From Rhodesia; Germany; France; South-West Africa; Arizona; Colorado; and elsewhere.
Treatment. Clean with distilled water.

Svabite $Ca_5(AsO_4)_3(F, Cl, OH)$
Svabite crystallizes in the hexagonal system as prismatic crystals or as masses. It is yellow or colourless, with a vitreous lustre. H 4–5; SG 3.5–3.8
Localities. Occurs in zinc ores at Franklin, New Jersey; Långban, Sweden.
Treatment. Clean with distilled water.

Durangite $NaAlAsO_4F$
Durangite crystallizes in the monoclinic system as pyramidal crystals coloured light orange to red. There is a vitreous lustre. The streak is yellow. H 5; SG 3.9–4.1
Localities. Occurs in tin mines with cassiterite and topaz, etc, at Durango, Mexico.
Treatment. Clean with distilled water.

Pyromorphite $Pb_5(PO_4)_3Cl$
Pyromorphite crystallizes in the hexagonal system as hexagonal prisms or more commonly as reniform or globular masses. The colour is green, yellow, brown and the streak is white. Its lustre is adamantine. H 3.5–4; SG 7.0
Occurrence. Occurs in the oxidation zone of lead ores.
Localities. From England; Scotland; France; Broken Hill, New South Wales;

Bad Ems, W. Germany; Burma; Zaire; and elsewhere.
Treatment. Clean with distilled water.

Mimetite $Pb_5(AsO_4)_3Cl$
Mimetite crystallizes in the monoclinic system as acicular crystals or as granular masses with a vitreous lustre. The colour is bright yellow, orange or colourless. H 3.5–4; SG 7.3
Occurrence. Occurs in the oxidation zone of lead ore deposits.
Localities. From Tsumeb, South-West Africa; Mapimi, Durango, Mexico; Leadhills, Scotland; California; Utah; Nevada; Arizona; Germany; and elsewhere.
Treatment. Clean with distilled water.
Fashioning. Uses: usually cabochons; *cleavage*: imperfect pyramidal; brittle; *heat sensitivity*: high, dop with care or use low-temperature wax.

Vanadinite $Pb_5(VO_4)_3Cl$
Vanadinite crystallizes in the hexagonal system as prismatic crystals or as globular masses. The colour is bright red or orange to yellow with a white or yellow streak. The lustre is resinous. H 3; SG 6.9
Occurrence. Occurs in the oxidation zone of lead ore deposits.
Localities. From Scotland; Sardinia; Mibladen, Morocco; Arizona; New Mexico; Colorado; etc.
Treatment. Clean with distilled water.

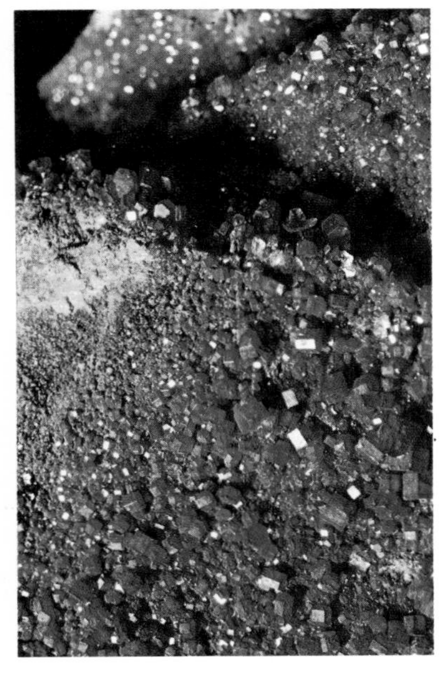

Left: Vanadinite crystals from the Old Yuma mine, Arizona

Far left: Mimetite from Johanngeorgenstadt, East Germany

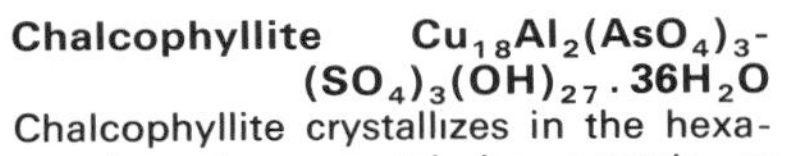

Chalcophyllite $Cu_{18}Al_2(AsO_4)_3(SO_4)_3(OH)_{27}.36H_2O$
Chalcophyllite crystallizes in the hexagonal system as tabular crystals or foliated masses and rosettes with a perfect cleavage. The colour is emerald green to bluish-green. The streak is lighter than the colour. There is a vitreous lustre. H 2; SG 2.7
Occurrence. Chalcophyllite is found in

Far left: Hexagonal prism of pyromorphite from Czechoslovakia

Below: Fine hexagonal tabular crystals of chalcophyllite from Cornwall

Right: Fornacite with dioptase from Renéville, Zaire

the oxidation zone of copper ore deposits.
Localities. From Cornwall; France; Germany; Bisbee, Arizona; Utah; Nevada; etc.
Treatment. Clean with distilled water.

Woodhouseite $CaAl_3PO_4SO_4(OH)_6$
Woodhouseite crystallizes in the hexagonal system as pseudocubic crystals with curved or striated faces. There is a perfect cleavage and the colour is white to pink, with a vitreous lustre. H 4.5; SG 3.0
Localities. Occurs in veins with quartz and topaz in an andalusite deposit near Bishop, California.
Treatment. Clean with distilled water.

Hinsdalite $(Pb, Sr)Al_3PO_4SO_4(OH)_6$
Hinsdalite crystallizes in the hexagonal system as pseudocubic crystals or as granular masses. There is a perfect cleavage. The colour is yellow, green or colourless; the lustre is vitreous. H 4.5; SG 3.6
Localities. Occurs on covelline at Butte, Montana; Hinsdale Co., Colorado.
Treatment. Clean with distilled water.

Hidalgoite $PbAl_3(AsO_4)(SO_4)(OH)_6$
Hidalgoite crystallizes in the hexagonal system as masses coloured white, grey, or light green, with a dull lustre. H 4.5; SG 4.0
Localities. Occurs with iron oxides in ores at Tooele Co., Utah; Cap Garonne, France; etc.

Beudantite $PbFe_3AsO_4SO_4(OH)_6$
Beudantite crystallizes in the hexagonal system as rhombohedral crystals coloured dark green to black with a green streak; there is a vitreous lustre. Pseudocubic forms are also found. H 3.5–4.5; SG 4–4.3
Occurrence. Occurs in the oxidation zone of ore deposits.
Localities. From Tiger, Arizona; Laurion, Greece; Western Australia; etc.
Treatment. Clean with distilled water.

Destinezite
$Fe_2PO_4(SO_4)(OH).5H_2O$
Destinezite crystallizes in the triclinic system as crusts or masses, sometimes stalactitic. The colour is deep yellow to brown. H 3–4; SG 2.0–2.4
Localities. Occurs in mine workings in San Benito Co., California; Black Hills, South Dakota; and elsewhere.

Below: Tyrolite. The foliated structure of the mineral can be clearly seen in these crystals from Penamellera, Spain

Pitticite near
$Fe^{3+}_2AsO_4SO_4OH.nH_2O$
Pitticite's crystal system is not yet established; it occurs as crusts or as stalactitic masses light or dark brown in colour. They have a vitreous lustre. The streak is yellow to white. H 2–3; SG about 2.3
Localities. Occurs as a deposit from springs; found as an alteration product of arsenopyrite at Devil's Gulch, California; Nevada; England; Romania; Germany; etc.

Tsumebite $Pb_2Cu(PO_4)(SO_4)(OH)$
Tsumebite crystallizes in the monoclinic system as tabular crystals or crusts. The crystals are twinned as trillings or as more complex groups. The colour is emerald green, with a vitreous lustre. H 3.5; SG 6.1
Localities. Occurs with smithsonite and other minerals at Tsumeb, South-West Africa; and other places.
Treatment. Clean with distilled water.

Tyrolite
$Ca_2Cu_9(AsO_4)_4(OH)_{10}.10H_2O$
Tyrolite crystallizes in the orthorhombic system as scales or as reniform masses with a foliated structure. There is a perfect cleavage. Pseudohexagonal aggregates are found. The colour is pale green to blue, with a vitreous lustre. H 2; SG 3.2
Occurrence. In the oxidation zone of copper deposits.
Localities. Found with chalcophyllite at Majuba Hill, Nevada; England; France; Italy; Austria and elsewhere.

Vauquelinite
$Pb_2Cu(CrO_4)(PO_4)(OH)$
Vauquelinite crystallizes in the monoclinic system as wedge-shaped crystals and as fibrous aggregates. The colour is brown to black, and the lustre adamantine. H 2.5–3; SG 6.0–6.1
Localities. Occurs with mimetite at Beresov, USSR, and other places.
Treatment. Clean with distilled water.

Embreyite
$Pb_5(CrO_4)_2(PO_4)_2.H_2O$
Embreyite crystallizes in the monoclinic system as crystalline crusts made up of tabular crystals showing multiple twinning. The colour is orange with a yellow streak. H 3.5; SG 6.4
Localities. Occurs with crocoite and vauquelinite at Beresov, USSR.

Fornacite
$(Pb, Cu)_3[(Cr, As)O_4]_2(OH)$
Fornacite crystallizes in the monoclinic system as prismatic crystals coloured dark green with a yellow streak. H not known; SG 6.3
Localities. Occurs with dioptase at Renéville, Congo and other places.

Arsenites

Trippkeïte $CuAs_2O_4$
Trippkeïte crystallizes in the tetragonal system as prismatic crystals coloured brilliant greenish-blue. There is a perfect cleavage. H not known; SG 4.8
Localities. Occurs in copper deposits at Copiapó, Chile.
Treatment. Clean with distilled water.

Heliophyllite $Pb_3AsO_{4-n}Cl_{2n+1}$
Heliophyllite crystallizes in the orthorhombic system as pyramidal crystals with horizontal striations; more commonly found as granular masses. The colour is yellow to green. The lustre is vitreous. H about 2; SG 6.9
Localities. Occurs with ekdemite at Jacobsberg, Sweden.
Treatment. Clean with distilled water.

Trigonite
$Pb_3Mn[AsO_3]_2[AsO_2(OH)]$
Trigonite crystallizes in the monoclinic system as tabular crystals with a perfect cleavage and a vitreous lustre. The colour ranges from yellow to brown or black. H 3; SG 6.1–7.1
Localities. Occurs with native lead in a

dolomite-haussmannite ore found at Långban, Sweden.
Treatment. Clean with distilled water.

Magnussonite
$(Mn, Mg, Cu)_5(AsO_3)_3(OH, Cl)$
Magnussonite crystallizes in the cubic system as encrustations coloured emerald or grass-green with a white streak. H 3.5–4; SG 4.3
Localities. Occurs in fine-grained hematite at Långban, Sweden.

Reinerite $Zn_3(AsO_3)_2$
Reinerite crystallizes in the orthorhombic system as blue to yellow-green crystals with a vitreous lustre. H 5–5.5; SG 4.3
Localities. Found with chalcocite and bornite at Tsumeb, South-West Africa.

Antimonates and Antimonites

Stetefeldtite
$Ag_{2-y}Sb_{2-x}(O, OH, H_2O)_{6-7}$
Stetefeldtite crystallizes in the cubic system as brown masses with a shining streak. H 3.5–4.5; SG 4.6
Localities. Found with chalcocite and pyrite at Belmont, Nevada, etc.

Roméite
$(Ca, Fe, Na)_2(Sb, Ti)_2(O, OH)_7$
Roméite crystallizes in the cubic system as octahedra or as masses. The colour is yellow to brown with a pale yellow streak; there is a vitreous lustre. Lewisite is a titanium-rich variety. H 5.5–6.5; SG 4.7–5.4
Occurrence. Found in manganese ores and with cinnabar.
Localities. From Minas Gerais, Brazil; Långban, Sweden; etc.
Treatment. Clean with distilled water.

Nadorite $PbSbO_2Cl$
Nadorite crystallizes in the orthorhombic system as tabular crystals or divergent groups with a perfect cleavage. The colour is brown to yellow, and the lustre is resinous. H 3.5–4; SG 7
Localities. Occurs with smithsonite and bindheimite at Djebel Nador, Constantine, Algeria; Långban, Sweden; as an alteration of jamesonite at the Bodannon mine, St. Endellion, Cornwall.
Treatment. Clean with distilled water.

Bindheimite $Pb_2Sb_2O_6(O, OH)$
Bindheimite crystallizes in the cubic system as cryptocrystalline masses or as crusts or pseudomorphs. The colour is yellow to reddish-brown or may be greenish to white. H 4.5; SG 4.6–7.3
Occurrence. Occurs in the oxidation zone of lead-antimony deposits.
Localities. From San Bernardino Co., California; Black Hills, South Dakota; England; Bolivia; USSR; Australia; and elsewhere.

Sulphates

Thenardite Na_2SO_4
Thenardite crystallizes in the orthorhombic system as tabular dipyramidal crystals with a perfect cleavage. Also found as crusts. The colour is white, brown or reddish. There is a vitreous lustre and a salty taste. H 2.5–3; SG 2.7
Occurrence. Occurs in salt lake deposits and around fumaroles.
Localities. From Searles Lake, San Bernardino Co., California; Nevada; Spain; USSR; Sicily; etc.

Mirabilite $Na_2SO_4 . 10H_2O$
Mirabilite crystallizes in the monoclinic system as prismatic or acicular crystals or as fibrous masses or crusts. There is a perfect cleavage. It is colourless or white, has a bitter taste with a vitreous lustre. H 1.5–2; SG 1.4
Occurrence. Occurs as a saline lake deposit or as a deposit from hot springs.
Localities. From the Great Salt Lake, Utah; Albany Co., Wyoming; Spain; Sicily; USSR; etc.
Treatment. Store in a sealed container.

Aphthitalite $K_3Na(SO_4)_2$
Aphthitalite crystallizes in the hexagonal system as tabular crystals or as pseudohexagonal twins or as masses or crusts. The colour is colourless or white with a vitreous lustre. There is a bitter taste. H 3; SG 2.7
Localities. Occurs with borax at Searles Lake, San Bernardino Co., California; Carlsbad, New Mexico; Stassfurt, E. Germany; in fumaroles at Mt. Vesuvius and Mt. Etna, Sicily; etc.

Chalcanthite $CuSO_4 . 5H_2O$
Chalcanthite crystallizes in the triclinic system as prismatic crystals or as stalactites or veins, also as granular masses. The colour is pale to dark blue and the lustre is vitreous. There is a colourless streak. The taste is metallic. H 2.5; SG 2.3
Occurrence. Occurs in the oxidation zone of copper ore deposits.
Localities. From England; Ireland; Germany; Chile; California; New Mexico; Arizona; and elsewhere.
Treatment. Store in a sealed container.

Boothite $CuSO_4 . 7H_2O$
Boothite crystallizes in the monoclinic system as masses coloured light blue, with a vitreous or silky lustre. Metallic taste. H 2–2.5; SG about 2.1
Localities. Occurs with copper sulphates at Alameda Co., California; Sain-Bel, Rhône, France.

Above: Brochantite from Chuquicamata, Chile

Antlerite $Cu_3SO_4(OH)_4$
Antlerite crystallizes in the orthorhombic system as tabular crystals or as fibrous aggregates. There is a perfect cleavage. The colour is emerald green to dark green with a pale green streak and a vitreous lustre. H 3.5; SG 3.9
Occurrence. Occurs in the oxidation zone of copper deposits especially in hot dry regions.
Localities. From Chuquicamata, Chile; Coahuila, Mexico; Bisbee, and the Antler mine, Mohave Co., Arizona; etc.

Brochantite $Cu_4SO_4(OH)_6$
Brochantite crystallizes in the monoclinic system as prismatic or acicular crystals or as crystalline aggregates, commonly with twinning. There is a perfect cleavage. The colour is emerald green to black with a vitreous lustre. There is a pale green streak. H 3.5–4; SG 4.0
Occurrence. Occurs in the oxidation zone of copper deposits.
Localities. From Bisbee, Arizona; Chile; Mexico; England; Zaire; USSR; and elsewhere.
Treatment. Clean with distilled water.

Left: Thenardite from Borax Lake, California

Above: Epsomite from Hérault, France

Above right: Anhydrite from Aussee, Styria

Langite $Cu_4SO_4(OH)_6.2H_2O$
Langite crystallizes in the orthorhombic system as small twinned crystals or as crusts. There is a perfect cleavage. The colour is blue to green, with a vitreous or silky lustre. H 2.5–3; SG 3.3
Localities. Occurs with gypsum and copper sulphate at St. Just and St. Blazey, Cornwall; with chalcopyrite at Mollau, Haut Rhin, France; etc.
Treatment. Clean with distilled water.

Devilline
$CaCu_4(SO_4)_2(OH)_6.3H_2O$
Devilline crystallizes in the monoclinic system as six-sided plates or as crusts. There is a perfect cleavage. The colour is emerald-green with a pale green streak. H 2.5; SG 3.1
Occurrence. Occurs with copper ores.
Localities. From Cornwall; Montgomery Co., Pennsylvania; Czechoslovakia; USSR.
Treatment. Clean with distilled water.

Cyanotrichite
$Cu_4Al_2SO_4(OH)_{12}.2H_2O$
Cyanotrichite crystallizes in the orthorhombic system as coatings or aggregates with a pale to dark blue colour. The streak is pale blue. H not known; SG 2.7–2.9
Occurrence. Occurs in the oxidation zone of copper ore deposits.
Localities. From Coconino Co., Arizona; Nevada; Utah; Greece; Scotland; USSR; South Africa; etc.

Woodwardite
$Cu_4Al_2SO_4(OH)_{12}.4–6H_2O$
Woodwardite's crystal system is not determined. It is found as rounded concretions coloured blue to green. H not known; SG 2.4
Localities. Woodwardite occurs in Cornwall; Nantlle, Caernarvonshire, Wales; Trentino, Italy.

Kieserite $MgSO_4.H_2O$
Kieserite crystallizes in the monoclinic system as granular masses with a perfect cleavage. The colour ranges from white to grey, yellow or colourless. H 3.5; SG 2.6
Occurrence. Occurs in sedimentary deposits in association with halite.
Localities. From Carlsbad, New Mexico; Texas; USSR; Sicily; Poland; Germany; Austria; etc.
Treatment. Store in a sealed container.

Hexahydrite $MgSO_4.6H_2O$
Hexahydrite crystallizes in the monoclinic system as tabular crystals, more commonly as fibrous or columnar masses; it has a perfect cleavage. The colour is white to pale green with a vitreous lustre. There is a salty taste. H not known; SG 1.8
Localities. Occurs as a dehydration product of epsomite from Saki salt lakes, Crimea, USSR; Bonaparte River, British Columbia; Oroville, Washington.

Epsomite $MgSO_4.7H_2O$
Epsomite crystallizes in the orthorhombic system as fibrous crusts; it has a perfect cleavage and a vitreous lustre. It is colourless, white, greenish to pinkish with a salty taste. H 2–2.5; SG 1.7
Occurrence. Occurs as an efflorescence in mine workings and in salt lakes.
Localities. From Epsom, Surrey; Oroville, Washington; Mt. Vesuvius, Italy; and elsewhere.
Treatment. Clean with alcohol and store in a sealed container.

Vanthoffite $Na_6Mg(SO_4)_4$
Vanthoffite crystallizes in the monoclinic system as aggregates or grains. The mineral is colourless. H 3.5; SG 2.7
Localities. Occurs in potash deposits at Stassfurt, E. Germany; Carlsbad, New Mexico.

Blödite $Na_2Mg(SO_4)_2.4H_2O$
Blödite crystallizes in the monoclinic system as prismatic crystals or as compact masses, blue-green, colourless or grey. There is a salty taste. H 2.5–3; SG 2.2
Occurrence. Occurs in salt deposits and in the Chilean nitrate deposits.
Localities. From Atacama, Chile; Soda Lake, California; Germany; USSR; Austria; Poland; India.
Treatment. Clean with alcohol. Store in sealed container.

Pickeringite $MgAl_2(SO_4)_4.22H_2O$
Pickeringite crystallizes in the monoclinic system as acicular crystals or fibrous masses or as encrustations. It is colourless, white or yellowish to reddish. H 1.5; SG 1.8
Occurrence. Occurs as a product of the weathering of pyrite rocks; from coal veins.
Localities. From Inyo Co., California; Colorado; New Mexico; Germany; Austria; and elsewhere.
Treatment. Store in sealed container.

Anhydrite $CaSO_4$
Anhydrite crystallizes in the orthorhombic system as tabular crystals or as granular masses; it has a perfect cleavage. It is white, grey, bluish, or colourless. There is a vitreous lustre. The streak is white. H 3.5; SG 3.0
Occurrence. Occurs as a rock-forming mineral in salt beds, in limestone or dolomite or in hydrothermal vein deposits.
Localities. From Carlsbad, New Mexico; Bancroft, Ontario; Germany; France; India; England; etc.
Fashioning. Uses: faceting or cabochons; *cleavage*: distinct to perfect // to the macro-, brachy- and basal pinacoids; very brittle; *cutting angles*: crown 40°–50°, pavilion 43°; *pleochroism*: weak; *heat sensitivity*: high; *mechanical sensitivity*: diamond saws and normal grinding techniques should not be used.

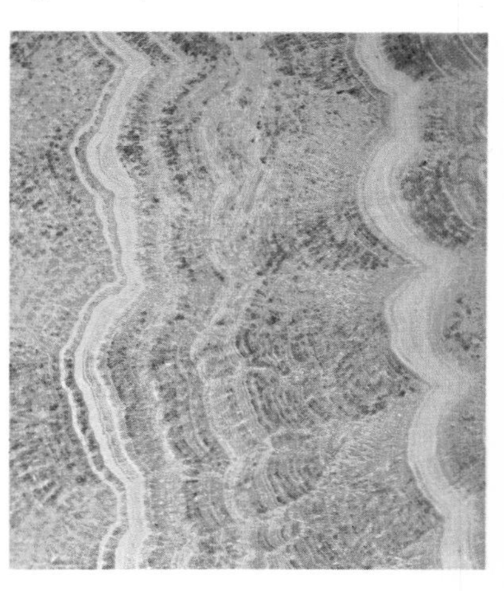

Right: A section of gypsum in its alabaster form

Gypsum $CaSO_4 . 2H_2O$
Gypsum crystallizes in the monoclinic system as tabular crystals or granular masses with twinning common and a perfect cleavage. There is a vitreous lustre. The colour is white, colourless, greenish or yellowish to brownish. There may be a greenish fluorescence and phosphorescence under ultra-violet light. H 2; SG 2.3
Occurrence. Occurs in sedimentary deposits, in saline lakes and in volcanic deposits.
Localities. From the London area, Sussex and Kent; California; Colorado; New Mexico; Chihuahua, Mexico; Chile; France; USSR; and elsewhere.
Treatment. Remove iron stains with dilute oxalic acid.
Fashioning. Uses: cabochons, carvings, bowls, etc; *cleavage*: perfect, clinopinacoidal; *heat sensitivity*: low, but avoid overheating during dopping, etc.

Above: Tabular gypsum crystal from Wiesloch, West Germany

Above left: Twinned gypsum crystals that have taken sand into their structure during growth

Left: Acicular gypsum crystals from Sicily

Right: Celestine with sulphur (bright yellow) from Sicily

Glauberite $Na_2Ca(SO_4)_2$
Glauberite crystallizes in the monoclinic system as tabular crystals or groups. There is a perfect cleavage. The colour is yellow, grey or colourless with a white streak and a salty taste. There is a vitreous lustre. H 2.5–3; SG 2.8
Occurrence. Occurs in salt deposits.
Localities. From Spain; France; Chile; Canada; Searles Lake, San Bernardino Co., California; and elsewhere.

Polyhalite $K_2MgCa_2(SO_4)_4 \cdot 2H_2O$
Polyhalite crystallizes in the triclinic system as fibrous or foliated masses; it has a perfect cleavage. It is white or grey and may be coloured pink or red from inclusions of iron oxide. Its lustre is vitreous. H 3.5; SG 2.8
Occurrence. Occurs in sedimentary deposits with anhydrite, halite and other salts.
Localities. From Carlsbad, New Mexico; England; France; USSR; Germany; and elsewhere.

Ettringite $Ca_6Al_2(SO_4)_3(OH)_{12} \cdot 24H_2O$
Ettringite crystallizes in the hexagonal system as hexagonal bipyramids or as fibres. There is a perfect cleavage and a vitreous lustre. It is colourless. H 2–2.5; SG 1.8
Localities. Occurs in metamorphosed limestones near Ettringen, Germany; Franklin, New Jersey; Arizona; Scawt Hill, Moyle, Ireland.

Celestine $SrSO_4$
Celestine crystallizes in the orthorhombic system as tabular crystals or as nodules or granular masses. There is a perfect cleavage. The colour is white, light blue, yellow, orange, red to brown. There is a vitreous lustre. Some specimens may fluoresce under ultra-violet light. H 3–3.5; SG 4.0
Occurrence. Occurs in limestones, and in hydrothermal vein deposits.
Localities. From Matehuala, San Luis Potosi, Mexico; in geodes, Malagasy Republic; Yate, Gloucestershire; Germany; Switzerland; Italy; USSR; Egypt; and elsewhere.
Treatment. Clean with dilute acid.
Fashioning. Uses: faceting or cabochons; *cleavage*: perfect prismatic and // to basal pinacoid; very brittle; *cutting angles*: crown 40°, pavilion 40°; *pleochroism*: weak; *heat sensitivity*: very high; *mechanical sensitivity*: very fragile.

Baryte $BaSO_4$
Baryte crystallizes in the orthorhombic system as tabular crystals or as aggregates; lamellar or fibrous masses, sometimes stalactitic. There is a perfect cleavage. It is white, grey, colourless, yellow, bluish, brown, with a vitreous lustre. It may fluoresce and phosphoresce under ultra-violet light. H 3–3.5; SG 4.5
Occurrence. Occurs in hydrothermal vein deposits; in sedimentary rocks, cavities in igneous rocks and as a deposit from hot springs.
Localities. From Meade and Pennington Counties, South Dakota; Cumbria and Derbyshire; Germany; France; Romania; and elsewhere.
Treatment. Remove iron stains with dilute oxalic acid.
Fashioning. Uses: faceting or cabochons; *cleavage*: perfect, prismatic and // to basal pinacoid; very brittle; *cutting angles*: crown 40°, pavilion 40°; *pleochroism*: weak; *heat sensitivity*: very high; *mechanical sensitivity*: very fragile.

Goslarite $ZnSO_4 \cdot 7H_2O$
Goslarite crystallizes in the orthorhombic system as granular masses; it has a perfect cleavage. It is white, colourless, greenish to brown, with a vitreous lustre. Metallic taste. H 2–2.5; SG 2.0
Occurrence. Occurs as an efflorescence in mines containing zinc minerals.
Localities. From Mexico; Argentina; Peru; New Mexico; California; Sweden; Germany; France; etc.
Treatment. Store in sealed container.

Ktenasite $(Cu, Zn)_3(SO_4)(OH)_4 \cdot 2H_2O$
Ktenasite crystallizes in the monoclinic system as tabular groups coloured blue to green, with a vitreous lustre. H 2–2.5; SG 3.0
Localities. Occurs with serpierite and smithsonite at Laurion, Greece.
Treatment. Clean with distilled water.

Right: Orthorhombic baryte crystal from Piacenza, Italy

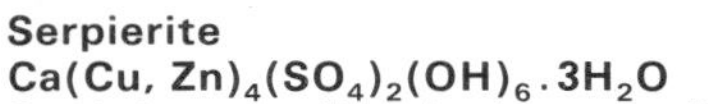

Serpierite
$Ca(Cu, Zn)_4(SO_4)_2(OH)_6 . 3H_2O$
Serpierite crystallizes in the monoclinic system as crusts and rounded masses. There is a perfect cleavage. The colour is blue. H not known; SG 3.1
Localities. Occurs with smithsonite at Ross Island, Killarney, Ireland; Laurion, Greece; etc.
Treatment. Clean with distilled water.

Alunogen $Al_2(SO_4)_3 . 16H_2O$
Alunogen crystallizes in the triclinic system as prismatic crystals, more commonly as crusts, efflorescences or masses. There is a perfect cleavage. The colour is white, yellow or colourless, with a vitreous lustre. There is an acid taste. H 1.5–2; SG 1.8
Occurrence. Occurs as an efflorescence in formations with iron sulphides, and in the oxidation zone of iron ore deposits.
Localities. From North Carolina; Utah; Colorado; New Mexico; Canada; Chile; Peru; France and elsewhere.
Treatment. Clean with alcohol.

Aluminite $Al_2SO_4(OH)_4 . 7H_2O$
Aluminite crystallizes in the monoclinic or orthorhombic systems as nodular masses coloured white to grey, with a dull lustre. H 1–2; SG 1.7–1.8
Localities. From Newhaven, Sussex; Green River, Utah; Halle, E. Germany; Salt Range, Punjab, India; etc.
Treatment. Clean with distilled water.

Felsöbanyite $Al_4SO_4(OH)_{10} . 5H_2O$
Felsöbanyite probably crystallizes in the orthorhombic system, with a perfect cleavage. It forms spherulitic aggregates with a yellow colour and a vitreous lustre. H 1.5; SG 2.3
Localities. Occurs with marcasite and quartz at Felsöbanya, Romania.
Treatment. Clean with distilled water.

Basaluminite $Al_4SO_4(OH)_{10} . 5H_2O$
Basaluminite probably crystallizes in the hexagonal system as microcrystalline masses coloured white with yellow or orange staining. H not known; SG 2.1
Localities. Occurs at the Lodge Pit siderite deposit, Irchester, Northamptonshire; Clifton Hill, Brighton, Sussex; Epernay, France; Utah; Kansas; etc.

Ammonia alum
$NH_4Al(SO_4)_2 . 12H_2O$
Ammonia alum crystallizes in the cubic system as octahedra or as fibrous masses. The colour is white or colourless and there is a sweetish taste. It has a vitreous lustre. H 1.5; SG 1.6
Localities. From brown coal at Tschermig, Czechoslovakia; Lake Co., California; Italy; Zaire; etc.
Treatment. Clean with alcohol.

Potash alum $KAl(SO_4)_2 . 12H_2O$
Potash alum crystallizes in the cubic system as masses of cubic or octahedral crystals. There is a vitreous lustre. It is white or colourless. Taste sweet. H 2–2.5; SG 1.7
Occurrence. Occurs in argillaceous rocks or in brown coals.
Localities. From England; Scotland; France; Germany; South Dakota; etc.
Treatment. Clean with alcohol.

Left: An aggregate of alunogen from Czechoslovakia

Soda alum $NaAl(SO_4)_2 . 12H_2O$
Soda alum crystallizes in the cubic system as octahedra. They are colourless. Vitreous lustre. H about 3; SG 1.7
Localities. From California; Colorado; Argentina.
Treatment. Clean with alcohol.

Anglesite $PbSO_4$
Anglesite crystallizes in the orthorhombic system as tabular crystals or as masses. The colour is white, colourless, yellowish, pale green or blue with a colourless streak. There is a vitreous lustre. May fluoresce yellow under UV light. H 2.5–3; SG 6.4– RI 1.87, 1.89
Occurrence. Occurs as a secondary mineral in lead deposits, and formed by the oxidation of galena.
Localities. Matlock, Derbyshire; Leadhills, Lanarkshire, Scotland; Anglesey, Wales; Chihuahua, Mexico; Tsumeb, South-West Africa; Australia; Wheatley Mine, Chester Co., Pennsylvania; Idaho; New Mexico; and elsewhere.
Treatment. Clean with distilled water.
Fashioning. Uses: faceting or cabochons; *cleavage*: distinct, prismatic and // to basal pinacoid; *cutting angles*: crown 40°, pavilion 40°; *heat sensitivity*: very high, with a cold setting cement; *mechanical sensitivity*: very fragile.

Left: Anglesite crystals from Sardinia showing a large number of cleavage planes

Right: Linarite crystals from Leadhills, Scotland

Lanarkite Pb_2SO_5
Lanarkite crystallizes in the monoclinic system as prismatic crystals or as masses. There is a perfect cleavage. The colour is grey to yellow with an adamantine lustre. There is a white streak. May fluoresce yellow under ultra-violet light. H 2–2.5; SG 6.9
Occurrence. In the oxidation zone of lead ore deposits.
Localities. From Leadhills, Lanarkshire, Scotland; Harz Mountains and the Black Forest, Germany; France; Chile; etc.
Treatment. Clean with distilled water.

Linarite $PbCuSO_4(OH)_2$
Linarite crystallizes in the monoclinic system as tabular crystals or as aggregates, frequently showing twinning. There is a perfect cleavage. The colour is dark blue, with a vitreous lustre. H 2.5; SG 5.3
Occurrence. In the oxidation zone of lead and copper deposits.
Localities. From Leadhills, Lanarkshire, Scotland; Mammoth mine, Tiger, Arizona; Spain; USSR; Canada; Argentina; and elsewhere.
Treatment. Clean with distilled water.

Beaverite
$Pb(Cu, Fe, Al)_3(SO_4)_2(OH)_6$
Beaverite crystallizes in the hexagonal system as friable masses of hexagonal plates coloured yellow. H not known; SG 4.4
Occurrence. Occurs in the oxidation zone of lead-copper deposits.
Localities. From Beaver Co. and Salt Lake Co., Utah; also from Nevada.

Plumbojarosite $PbFe_6(SO_4)_4(OH)_6$
Plumbojarosite crystallizes in the hexagonal system as compact masses coloured yellow to brown, with a dull lustre. H not known, but soft; SG 3.6
Occurrence. Occurs in the oxidation zone of lead ore deposits.
Localities. From western USA; Laurion, Greece; Bolivia; Anatolia, Turkey; and other localities.

Zippeite $(UO_2)_2(SO_4)(OH)_2 . 4H_2O$
Zippeite crystallizes in the orthorhombic system as compact masses, often with twinning. The colour is orange-yellow to brown, with a dull lustre. There is a variable fluorescence in ultraviolet light. H not known; SG 3.7
Occurrence. As an efflorescence in mines.
Localities. From Cornwall; Colorado; Utah; Příbram, Czechoslovakia; etc.
Treatment. Clean with distilled water.

Meta-uranopilite
$(UO_2)_6SO_4(OH)_{10} . 5H_2O$
Meta-uranopilite crystallizes in the orthorhombic system as needles coloured green or yellow. There is a yellow-green fluorescence under ultra-violet light. H and SG not known.
Localities. Occurs at Jachymov, Czechoslovakia; Cornwall.

Uranopilite
$(UO_2)_6SO_4(OH)_{10} . 12H_2O$
Uranopilite probably crystallizes in the monoclinic system as crusts coloured bright yellow, fluorescing a strong green under ultra-violet light. They have a silky lustre. H not known; SG 4.0
Localities. Occurs with johannite and zippeite, San Juan Co., Utah; Wheal Owles, Cornwall; Příbram, Czechoslovakia; Grury, France; Northwest Territory, Canada; etc.

Johannite
$Cu(UO_2)_2(SO_4)_2(OH)_2 . 6H_2O$
Johannite crystallizes in the triclinic system as tabular crystals or as aggregates. The colour is emerald green to yellow with a pale green streak. There is a vitreous lustre. H 2–2.5; SG 3.3
Occurrence. Occurs in the oxidized zone of uranium-bearing veins.
Localities. From Great Bear Lake, Canada; Jachymov, Czechoslovakia; Colorado; USSR; etc.

Mooreite
$(Mg, Zn, Mn)_8SO_4(OH)_{14} . 3\text{–}4H_2O$
Mooreite crystallizes in the monoclinic system as tabular or platy crystals with a perfect cleavage. It is colourless, with a vitreous lustre. H 3; SG 2.5
Localities. Occurs with calcite at Franklin, New Jersey.
Treatment. Clean with distilled water.

Torreyite
$(Mg, Zn, Mn)_7SO_4(OH)_{12} . 4H_2O$
Torreyite crystallizes in the monoclinic system as granular masses coloured blue. H 3; SG 2.7
Localities. Occurs with calcite at Franklin, New Jersey.
Treatment. Clean with distilled water.

Melanterite $FeSO_4 . 7H_2O$
Melanterite crystallizes in the monoclinic system as prismatic crystals or stalactitic crusts. The colour is green or blue with a colourless streak. There is a vitreous or silky lustre. H2; SG 1.9
Occurrence. Occurs as a weathering product of pyrite.
Localities. From Rio Tinto, Spain; South Dakota; Colorado; etc.
Treatment. Store in sealed container.

Butlerite $Fe^{3+}SO_4OH . 2H_2O$
Butlerite crystallizes in the monoclinic system as tabular crystals or as octahedra. There is a perfect cleavage. The colour is orange, the lustre vitreous. The streak is pale yellow. H 2.5; SG 2.5
Localities. Occurs with copiapite at Calf Mesa, Utah; Jerome, Arizona; La Alcaparrosa, Argentina; and elsewhere.

Below right: An encrustation of zippeite, a typically brightly coloured uranium mineral

Below: A coating of johannite crystals

Amarantite $Fe^{3+}SO_4OH.3H_2O$
Amarantite crystallizes in the triclinic system as prismatic crystals or as aggregates of needle-like crystals. There is a perfect cleavage. The colour is red or orange-yellow with a vitreous lustre. There is a yellow streak. H 2.5; SG 2.2
Localities. Occurs with copiapite at Blythe, California; South Dakota; Chuquicamata, Chile.

Copiapite
$(Fe, Mg)Fe_4^{3+}(SO_4)_6(OH)_2.20H_2O$
Copiapite crystallizes in the triclinic system as tabular crystals or as aggregates. There is a perfect cleavage; the colour is golden yellow or orange to green with a pearly lustre. H 2.5–3; SG 2.1–2.2
Occurrence. Formed by the oxidation of pyrite or other sulphides.
Localities. From Chuquicamata, Chile; Utah; California; Nevada; France; Spain; Germany; etc.
Treatment. Store in a sealed container.

Natrojarosite $NaFe_3^{3+}(SO_4)_2(OH)_6$
Natrojarosite crystallizes in the hexagonal system as minute tabular crystals or as crusts. There is a perfect cleavage. The colour is yellow or brown with a pale yellow streak. H 3; SG 3.2
Occurrence. Occurs as a secondary mineral in iron-bearing rocks, particularly ore deposits.
Localities. From the Black Hills, South Dakota; Nevada; Chile; Mexico; USSR; and elsewhere.
Treatment. Clean with distilled water.

Voltaite $K_2Fe_5^{2+}Fe_4^{3+}(SO_4)_{12}.18H_2O$
Voltaite crystallizes in the cubic system as octahedra or dodecahedra and as granular masses. The colour is dark green to black with a resinous lustre. There is a grey-green streak. H 3; SG 2.7
Localities. Occurs with sulphates at Bisbee, Arizona; Mt. Vesuvius, Italy; Bolivia; Chile; Japan; etc.

Botryogen $MgFe^{3+}(SO_4)_2OH.7H_2O$
Botryogen crystallizes in the monoclinic system as prismatic crystals or as botryoidal masses. There is a perfect cleavage. The colour is orange, the lustre vitreous. H 2.5; SG 2.1
Localities. Occurs with copiapite at the Redington mine, Napa Co., California; Chuquicamata, Chile; Santa Elena mine, San Juan, Argentina; France; Germany; Sweden; etc.
Treatment. Clean with distilled water.

Halotrichite $Fe^{2+}Al_2(SO_4)_4.22H_2O$
Halotrichite crystallizes in the monoclinic system as acicular crystals or as aggregates. They are white, colourless, yellow to greenish; the lustre is vitreous. There is a sharp taste. H 1.5; SG 1.9
Occurrence. Occurs as a weathering product of pyrite and in coal veins.
Localities. From Chile; France; Germany; Utah; and elsewhere.
Treatment. Store in sealed container.

Bieberite $CoSO_4.7H_2O$
Bieberite crystallizes in the monoclinic system as stalactites and crusts; it has a perfect cleavage. The colour is red; becomes opaque on dehydration. There is a vitreous lustre. H 2; SG 1.9
Localities. Occurs on pyrrhotine, Trinity Co., California; Siegen, Germany; Chalanches, France; etc.

Morenosite $NiSO_4.7H_2O$
Morenosite crystallizes in the orthorhombic system as stalactitic crusts. The colour is white or green with a greenish-white streak; there is a vitreous lustre. H 2.5; SG 2.0
Occurrence. Occurs as an alteration product of nickel-bearing sulphides.
Localities. From the Copper King mine, Boulder Co., Colorado; with erythrite at Julian, San Diego Co., California; Peru; France; Italy; Ontario, Canada; etc.
Treatment. Store in a sealed container.

Retgersite $NiSO_4.6H_2O$
Retgersite crystallizes in the tetragonal system as tabular crystals or as crusts. There is a perfect cleavage; the colour is emerald-green and the streak greenish-white. There is a vitreous lustre. The taste is bitter. H 2.5; SG 2.1
Occurrence. Formed as a secondary mineral from other nickel-bearing minerals.
Localities. From Lancaster Co., Pennsylvania; Minas Ragra, Peru; Lobenstein, Thuringia, Germany.
Treatment. Store in a sealed container.

Connellite
$Cu_{19}Cl_4SO_4(OH)_{32}.3H_2O$
Connellite crystallizes in the hexagonal system as groups of blue or blue-green acicular crystals with a vitreous lustre. The streak is pale blue. H 3; SG 3.4
Occurrence. Occurs in the oxidation zone of copper ore deposits.
Localities. From Cornwall; Bisbee, Arizona; South Africa; etc.
Treatment. Clean with distilled water.

Spangolite
$Cu_6AlSO_4(OH)_{12}Cl.3H_2O$
Spangolite crystallizes in the hexagonal system as prismatic or tabular crystals

Left: *Amarantite crystals from Sierra Gorda, Chile*

Below left: An aggregate of botryogen crystals from Knoxville, California

Far left: Copiapite from Copiapó, Chile

with a perfect cleavage. The colour is dark or emerald-green with a pale green streak. There is a vitreous lustre. H 3; SG 3.1
Occurrence. Occurs in the oxidized zone of copper ore deposits.
Localities. From the Blanchard mine, Socorro Co., New Mexico; St. Day, Cornwall; Arenas, Sardinia.
Treatment. Spangolite should be cleaned with distilled water.

Creedite
$Ca_3Al_2SO_4(F, OH)_{10}.2H_2O$
Creedite crystallizes in the monoclinic system as prismatic or acicular crystals with a perfect cleavage. It is white or colourless and has a vitreous lustre. H 4; SG 2.7
Localities. From the tin veins of Colquiri, Bolivia; Colorado; Chihuahua, Mexico.
Treatment. Clean with distilled water.

Chromates and Molybdates

Tarapacaite K_2CrO_4
Tarapacaite crystallizes in the orthorhombic system as tabular pseudohexagonal crystals coloured bright yellow. H not known; SG 2.7
Localities. Occurs with lopezite in nitrate deposits at Antofagasta Province, Chile.

Lopezite $K_2Cr_2O_7$
Lopezite crystallizes in the triclinic system as spherical aggregates; it has a perfect cleavage, and is orange-red in colour and has a vitreous lustre. H 2.5; SG 2.7
Localities. In a nitrate rock at the Maria Elena mine Antofagasta, Chile.

Right: The characteristic colour of crocoite. This specimen comes from its classic locality of Dundas, Tasmania

Below left: Wulfenite showing tabular form

Below right: Stolzite from Broken Hill, Australia

Crocoite $PbCrO_4$
Crocoite crystallizes in the monoclinic system as prismatic crystals or as masses or aggregates. The colour is bright red or orange with an orange-yellow streak. Adamantine lustre. H 2.5–3; SG 6.0
Occurrence. Occurs in the oxidation zone of lead and chromium ores.
Localities. From Inyo Co., California; Mammoth mine, Tiger County, Arizona; Dundas, Tasmania; Beresov, USSR; Minas Gerais, Brazil; etc.
Treatment. Clean ultrasonically.
Fashioning. Uses: faceting or cabochons; *cleavage*: distinct, prismatic; sectile; *cutting angles*: crown 30°–40°, pavilion 37°–40°, *heat sensitivity*: very high, stones should be dopped with a cold-setting cement; *mechanical sensitivity*: very fragile.

Powellite $Ca(Mo, W)O_4$
Powellite crystallizes in the tetragonal system as pyramidal crystals or as crusts or masses. The colour is yellow, brown, grey, blue or black, the lustre is adamantine or greasy. There is a golden yellow fluorescence. H 3.5–4; SG 4.2
Occurrence. Occurs in the oxidation zone of ore deposits.
Localities. From Isle Royale, Michigan; Utah; Arizona; Turkey; USSR; Morocco.
Treatment. Clean with distilled water.

Wulfenite $PbMoO_4$
Wulfenite crystallizes in the tetragonal system as tabular crystals or granular masses. The colour is yellow or orange to brown with a white streak. There is a resinous lustre. H 2–3; SG 6–7
Occurrence. Occurs in the oxidation

zone of ore deposits.
Localities. From Morocco; Tsumeb, South-West Africa; South Dakota; Utah; Austria; Mexico; and elsewhere.
Treatment. Clean ultrasonically.
Fashioning. Uses: faceting or cabochons; *cleavage*: distinct, pyramidal; brittle; *cutting angles*: crown 30°–40°, pavilion 37°–40°; *heat sensitivity*: high.

Tungstates

Cuprotungstite $Cu_2WO_4(OH)_2$
Cuprotungstite's crystal system is not known; it forms masses and crusts coloured emerald-green or brown, with a vitreous lustre. H not known; SG 5.4
Localities. Occurs in a copper deposit at South Peacock mine, Adams Co., Idaho; Chile; Transvaal, South Africa; etc.
Treatment. Clean with distilled water.

Scheelite $CaWO_4$
Scheelite crystallizes in the tetragonal system as octahedral or tabular crystals or as granular masses. They are white, colourless, orange-yellow, green or purple. The lustre is adamantine or vitreous. There is a bright bluish-white fluorescence under short-wave ultraviolet light. The streak is white. H 4.5–5; SG 6.1; RI 1.92, 1.93
Occurrence. Occurs in contact metamorphic deposits and in hydrothermal veins and pegmatites.
Localities. From Cochise Co., Arizona; Inyo Co., California; Peru; England; Japan; Korea; and elsewhere.
Treatment. Clean with dilute acid.
Fashioning. Uses: faceting or cabochons; *cleavage*: distinct, pyramidal; brittle; *cutting angles*: crown 40°, pavilion 40°; *heat sensitivity*: low.

Stolzite $PbWO_4$
Stolzite crystallizes in the tetragonal system as tabular or dipyramidal crystals coloured yellow, red, brown or green with a colourless streak. There is a resinous lustre. H 2.5–3; SG 7.9–8.3
Occurrence. Occurs in the oxidation zone of tungsten ore deposits.
Localities. From England; Brazil; Broken Hill, New South Wales; Utah; Arizona; Massachusetts; etc.
Treatment. Clean with distilled water.

Raspite $PbWO_4$
Raspite crystallizes in the monoclinic system as tabular crystals coloured light yellow or brown with a perfect cleavage. They have an adamantine lustre. H 2.5–3; SG 8.5
Localities. Occurs with stolzite at Broken Hill, New South Wales; Minas Gerais, Brazil; in tin veins at Cerro Estano, Guanajuato, Mexico.
Treatment. Clean with distilled water.

Russellite Bi_2WO_6
Russellite crystallizes in the tetragonal system as compact masses coloured yellow-green. H 3.5; SG 7.3
Localities. Occurs with native bismuth at Castle-an-Dinas, St. Columb Major, Cornwall.
Treatment. Clean with distilled water.

Hübnerite $MnWO_4$
Hübnerite crystallizes in the monoclinic system as prismatic crystals and as groups of parallel crystals. There is a

Left: An aggregate of wolframite crystals, each prismatic but tending towards a tabular habit

perfect cleavage. The colour is yellow or reddish-brown with a reddish-brown or greenish-grey streak. The lustre is resinous. H 4–4.5; SG 7.2
Occurrence. In hydrothermal ore veins.
Localities. From Silverton, San Juan Co., Colorado; Idaho; Nevada; New Mexico; France; Peru; etc.
Treatment. Clean with distilled water.

Wolframite (Fe, Mn)WO_4
Wolframite crystallizes in the monoclinic system as prismatic crystals or as groups of crystals or lamellar masses. There is a perfect cleavage. The colour is grey or black with a reddish-brown streak. The lustre is metallic. H 4–4.5; SG 7.4
Occurrence. Occurs in high-temperature ore veins and quartz veins in granite.
Localities. From Panesqueira, Portugal; South Dakota; New Mexico; Arizona; England; France; and elsewhere.
Treatment. Clean with dilute acid.

Chalcomenite $CuSeO_3 . 2H_2O$
Chalcomenite crystallizes in the orthorhombic system as prismatic or acicular crystals of minute size. The colour is bright blue, with a vitreous lustre. H 2–2.5; SG 3.3
Localities. Formed by the oxidation of copper and lead selenides at Sierra Famatina, Argentina; also from Bolivia.
Treatment. Clean with distilled water.

Tellurites, Iodates and Oxalates

Cliffordite UTe_3O_8
Cliffordite crystallizes in the cubic system as octahedra coloured bright yellow with an adamantine lustre. H 4; SG 6.6

Left: Pyramidal scheelite crystal from the Traversella mine, near Ivres, Italy

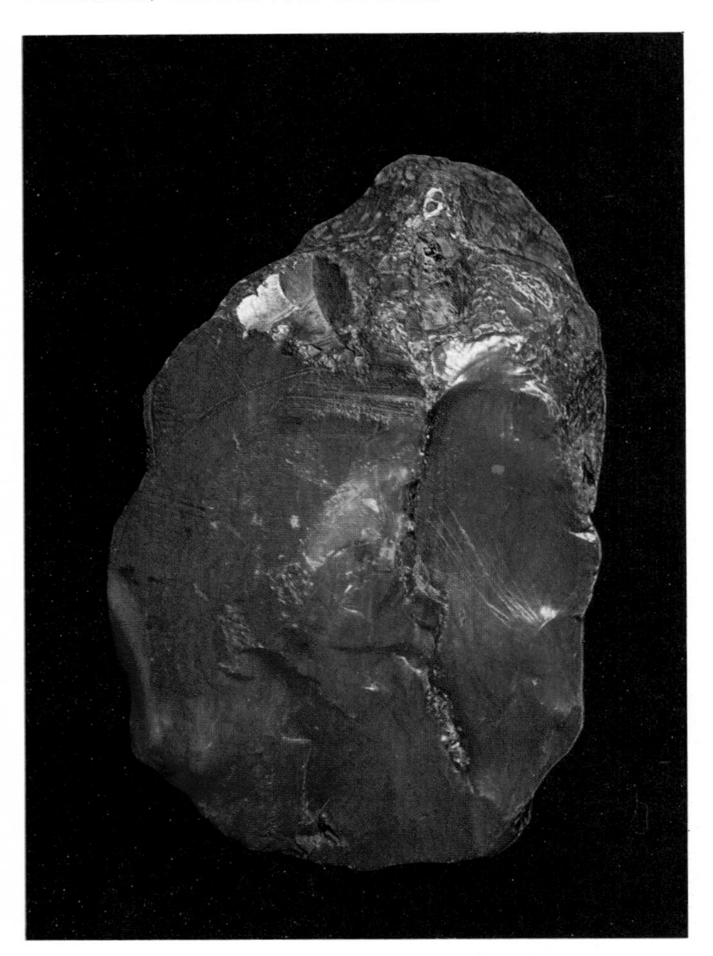

Above: A large specimen of amber from the Baltic coast

Localities. Occurs with tellurium at Moctezuma, Sonora, Mexico.
Treatment. Clean with dilute acid.

Emmonsite $Fe_2(TeO_3)_3 . 2H_2O$
Emmonsite crystallizes in the monoclinic system as compact masses; it has a perfect cleavage. The colour is yellowish-green, the lustre is vitreous or dull. H 5; SG 4.5
Occurrence. Occurs in the oxidation zone of deposits with native tellurium.
Localities. From Cripple Creek, Colorado; Nevada; New Mexico; Honduras.
Treatment. Clean with distilled water.

Salesite $CuIO_3OH$
Salesite crystallizes in the orthorhombic system as prismatic crystals with a perfect cleavage and bluish-green in colour. They have a vitreous lustre. H 3; SG 4.8
Localities. Occurs in oxidized ore at Chuquicamata, Chile.
Treatment. Clean with distilled water.

Whewellite $CaC_2O_4 . H_2O$
Whewellite crystallizes in the monoclinic system as crystalline masses coloured white, colourless, yellow or brown. There is a vitreous lustre. H 2.5–3; SG 2.2
Occurrence. Occurs in coal seams and also in sedimentary nodules and concretions.
Localities. From Zwickau and Burgk, E. Germany; San Juan Co., Utah; Alsace, France; USSR.
Treatment. Clean with distilled water.

Appendix: some organic materials

Amber
A fossil resin with a chemical composition close to $C_{10}H_{16}O$ and containing some H_2S. It is most commonly brown or yellow but may also be green, reddish or close to white. It commonly displays a bluish-white fluorescence under long-wave ultraviolet light; under short-wave radiations the fluorescent colour inclines to green. H about 2; SG about 1.08; RI 1.54
Occurrence. From pine trees of the Oligocene period, notably *Pinus succinifera*.
Localities. The major locality is the Kaliningrad area of the USSR; also from the Baltic coast; the east coast of England; Burma; Sicily; Romania; etc.
Treatment. Clean with soap and water.
Fashioning. Uses: faceting, cabochons, carvings, beads, etc; *cleavage*: none; splintery; *cutting angles*: crown 40°–50°, pavilion 43°; *heat sensitivity*: very high, softens at about 150°C.

Jet
A fossil wood close in composition to brown coal. It is black and has a shining conchoidal fracture. It burns with a smell of coal. H about 2.5; SG about 1.33; RI about 1.66
Localities. From the Upper Lias of east Yorkshire, particularly from the Whitby area; Asturias, Spain; Aude, France; Utah.
Treatment. Clean with soap and water.
Fashioning. Uses: cabochons, carvings, beads, tumbling, etc; *cleavage*: none; *heat sensitivity*: low, but avoid overheating during dopping.

Ivory
Ivory is mainly a phosphate of calcium and includes both organic and mineral matter. Most of the worked ivory seen comes from the elephant; in transverse section it shows curved lines, the lines of Retzius, resembling the arcs formed by engine-turning. Other ivories come from the walrus, hippopotamus and whale. Ivory fluoresces under ultraviolet light with a bluish-white fluorescence. H just above 2; SG 1.7–1.9; RI 1.54

Right: A branch of red coral, Corallium rubrum

Localities. The best elephant ivory comes from Africa, particularly from the Cameroons.
Treatment. Clean gently with methylated spirits.

Vegetable ivory
From the nut of the Ivory Palm (*Phytelephas macrocarpa*). The composition is close to $C_6H_{10}O_5$. It fluoresces bluish-white under ultraviolet light, but much more faintly than elephant ivory. It shows no lines of Retzius. H about 2.5; SG 1.42; RI 1.54
Localities. Ivory Palm is found in Peru and Colombia.

Bone
The commonest simulant of ivory. Like ivory, it fluoresces bluish-white under ultraviolet light. In transverse section it displays cavities, surrounded by dark dots, called the Haversian systems and quite different from ivory's lines of Retzius. H about 2.5; SG about 2; RI about 1.54
Treatment. Clean gently with methylated spirits.

Tortoise-shell
The composition of tortoise-shell is close to the protein keratin, containing oxygen, carbon, nitrogen, sulphur and hydrogen. The brown mottling seen on the shell is shown by the microscope to consist of dots of colour. It is sectile. H about 2.5; SG 1.29; RI 1.55
Localities. From the shell of the hawk's-bill turtle, *Chelone imbricata,* an inhabitant of the Malay archipelago, the West Indies and other tropical centres.
Treatment. Clean in tepid water.

Coral
Coral is made up of the skeletons of coral polyps, which belong to the class Anthozoa. Chemically, coral is almost completely calcium carbonate with some magnesium carbonate, and also contains some iron oxide and organic matter. The colour may be a fine red, pink or white, but there are also black and blue varieties. H about 3.5; SG 2.6–2.7; RI about 1.5
Localities. Red varieties from the Mediterranean; black from the Indian Ocean and off Hawaii; blue from the Cameroons.

Pearl
Pearls are a product of pearl oysters, bivalve molluscs that are not true oysters. They are found in both sea water and fresh water. The chemical composition of pearl is largely calcium carbonate in the form of aragonite, with some conchiolin ($C_{32}H_{48}N_2O_{11}$) and water. Natural and artificial pearls are distinguished with x-rays. SG 2.6–2.7
Localities. Pearls are found in the Persian Gulf, the Gulf of Manaar between Ceylon and India, the Red Sea, the Gulfs of Mexico and Panama, and the waters off Australia and Venezuela. Fresh-water pearls are found in many rivers, including those of Scotland and North America – in particular the Mississippi river.
Treatment. Pearls should not be stored in too dry an atmosphere.

Living coral.
When dried this species is one of those suitable for carving

Glossary

acicular: needle-like

amorphous: lacking crystalline atomic structure

angstrom: a unit of measurement of wavelength equal to a hundred-millionth of a centimetre

asterism: an optical effect in a gem in which a star with four or six rays is seen

birefringence: see *double refraction*

cabochon: a rounded, polished but not faceted gemstone

cat's-eye effect: see *chatoyancy*

chatoyancy: an optical effect in gemstones in which a single band of light traverses the stone

cleavage: the propensity of a crystal to split along a direction parallel to a possible crystal face

concretion: a spherical or disc-shaped mass found in sedimentary rocks, formed by the local cementation of the rock's constituent minerals

covalent bond: atomic bond involving electron-sharing

crown: the upper part of a faceted gemstone

cryptocrystalline: consisting of individual crystals that are not discernible except under magnification

density: the mass per unit volume of a substance

dichroism: see *pleochroism*

dop stick: a stick to which a gemstone is attached with wax (dopped) to aid grinding or polishing

double refraction: the formation of two refracted rays from a single incident ray; occurring in all crystal systems except the cubic

efflorescence: a fluffy deposit of crystalline material on rock surfaces

enantiomorphous: having left- and right-handed forms, which are mirror-images of each other

euhedral: having a well-developed crystal shape

fluorescence: the emission of visible light under illumination by radiation (visible or otherwise) of a shorter wavelength, such as ultraviolet light and x-rays; if the emission persists when the stimulating radiation ceases, it is called phosphorescence

fracture: breakage of a material that is not along a crystallographic plane (that is, not parallel to an actual or possible crystal face); fracture leaves a surface that may be smooth, hackly, uneven, etc, and is characteristic of the mineral

geode: a cavity in a rock lined with crystals growing towards the centre

habit: the shape of a crystal arising from the relative development of its faces; it may be described as tabular, pyramidal, etc, and is often characteristic of the mineral

hydrothermal processes: geological processes involving heated water, associated with igneous activity

ionic bond: a bond between two atoms involving the transfer of electrons from one to the other

isomorphous: having identical crystal structure at the atomic level, but differing chemical compositions

lamellae: thin mineral layers resembling the pages of a book; due to multiple *twinning*

lustre: quality of the brilliance or shine of a mineral's surface; it may be metallic, adamantine, etc

matrix: the rock in which a specimen is embedded

metamorphism: the alteration of rocks by heat, pressure and chemical processes after their original formation

mole: that weight of a substance that, when expressed in grams, is numerically equal to the molecular weight; a mole of carbon (molecular weight 12) weighs 12 grams

monochromatic light: light of one wavelength only (that is, of a single pure colour)

Nicol prism: a filter used to produce *polarized* light, made from Iceland spar, a variety of calcite

pavilion: the lower half of a faceted stone

phosphorescence: see *fluorescence*

pinacoid: a pair of parallel crystal faces arising from the presence of a centre or plane of symmetry; specific types include the basal pinacoid, having faces at right angles to the crystal's vertical axis; brachypinacoid, having faces parallel to the shortest axis; macropinacoid, having faces parallel to the y-axis where this is longer than the x-axis

pleochroic: showing several different colours when viewed in different directions; crystals displaying only two such colours are called dichroic

polarized light: light whose vibrations are in one plane only (at right angles to the direction of travel)

polymorphous: (of a given substance) taking several different crystal forms

porphyry: any igneous rock in which relatively large crystals, called phenocrysts, occur in a finer-grained background

prism: in crystallography, a crystal form whose faces meet in edges that are all parallel

pseudomorph: crystalline material that has taken on the shape of another crystal or body which it has replaced; for example, opalized shell

refraction: the bending of a light ray's path on leaving one medium and entering another

specific gravity: the ratio of a material's *density* to the density of water

streak: the colour of a mineral in its powdered form; revealed in the trace left by a specimen when drawn across a piece of unglazed porcelain

thermoluminescence: emission of light by a substance when heated to less than red-heat

twinned crystal: one having the appearance of two or more individual crystals, in contact or interpenetrating, that are identical or are mirror-images of each other

unit cell: the basic unit of pattern in a crystal; it consists of the smallest grouping of atoms that gives rise to the total crystal when repeated in all directions

Identification Tables

Mineral specimens are often difficult to identify because their crystal system is not immediately apparent, their colour is unusual because of surface coatings, and so forth. The customary form of mineral identification table, in which one property such as hardness is taken as a standard and the others related to it, has therefore not been adopted in this book. Instead, each major property has been allocated a separate table. This method has the advantage of making available alternative paths to identification. For example, a specimen whose hardness is not easily determined may be referred to the table of colour; a specimen whose crystal form is distorted beyond immediate recognition may be referred to the table of specific gravities; and so on.

Not all minerals included in the Mineral Kingdom section are included in these identification tables. This is partly to keep the size of the tables within reasonable bounds and partly because many of the minerals described in The Mineral Kingdom are rare or too difficult to identify without more sophisticated methods of testing being employed. The minerals listed in the tables therefore comprise only a part of the whole, but do include the commoner materials that are most likely to be encountered by the collector in the field.

Sometimes a mineral shows more than one set of characteristics, and then appears in several parts of a given table; for example, blende occurs in a variety of colours. Where numerical constants are given they represent general figures; values of some properties can vary widely within a particular mineral species.

Explanations of terms used in the tables and of how to determine quantities are given in the Glossary and in Identifying Minerals.

Table 1
Form and habit
A form is a basic shape that a crystal can take, which has all the symmetries of the crystal's system. Several forms may be combined in a crystal. The overall shape of a crystal or group of crystals is largely due to the relative development of the forms present and is called the crystal's habit

System	Form/habit	Minerals
Cubic system	cubes	acanthite boracite diamond pyrite galena cuprite fluorite uraninite halite cerargyrite pharmacosiderite
	dodecahedra	diamond garnet magnetite copper amalgam cuprite sodalite
	icositetrahedra	garnet leucite
	interpenetrant twins	fluorite
	masses or wires	gold silver copper platinum bornite acanthite smaltite skutterudite sodalite lazurite uraninite chloanthite
	octahedra	diamond gold magnetite cuprite spinel franklinite fluorite chromite periclase pyrochlore uraninite
	pyritohedra	pyrite cobaltite gersdorffite hauerite
	tetrahedra	tetrahedrite blende boracite diamond (rarely)
Tetragonal system	dipyramids	cassiterite scheelite wulfenite zircon apophyllite
	masses	cassiterite pyrolusite
	prismatic	zircon idocrase scapolite apophyllite phosgenite
	tabular	torbernite autunite wulfenite
Orthorhombic system	bladed	diaspore strontianite goethite
	masses	iolite variscite prehnite olivine
	prismatic	arsenopyrite goethite chrysoberyl aragonite stibnite löllingite bournonite cerussite baryte anglesite danburite natrolite enargite celestine adamite olivenite wavellite
	prismatic dipyramidal	topaz andalusite staurolite olivine
	pseudohexagonal twins	chrysoberyl chalcocite cerussite aragonite
	wedge-shaped	azurite gypsum orpiment wolframite sphene monazite

Form and habit
continued

System	Form	Minerals
	tabular	sulphur chalcocite marcasite hemimorphite columbite tantalite
Monoclinic system	bladed	erythrite vivianite sphene tremolite
	masses	jadeite nephrite actinolite malachite serpentine
	prismatic	gypsum actinolite tremolite azurite realgar manganite spodumene epidote clinozoisite pyroxenes amphiboles orthoclase
	pseudohexagonal tabular	muscovite lepidolite phlogopite biotite
Triclinic system	bladed	kyanite albite
	masses	ulexite turquoise rhodonite microcline
	wedge-shaped	axinite
Hexagonal system	hexagonal and trigonal prismatic	corundum tourmaline willemite dioptase apatite mimetite quartz beryl vanadinite calcite pyrargyrite millerite proustite pyromorphite
	rhombohedra	hematite cinnabar siderite calcite chabazite dolomite
	scalenohedra	calcite rhodochrosite smithsonite
	tabular	phenakite benitoite graphite molybdenite pyrrhotine covelline corundum—variety ruby ilmenite beryl apatite

Table 2
Lustre and colour
The colour of a mineral specimen is the easiest property to observe. Unfortunately, a mineral's colour is often highly variable from specimen to specimen. It is a more valuable guide to identity when taken in conjunction with lustre, the quality of the reflected light

Lustre	Colour	Minerals
adamantine	white and colourless	diamond zircon cassiterite scheelite cerussite anglesite
	brown	pyromorphite zircon blende sphene diamond cerussite wulfenite monazite scheelite
	black	diamond rutile zircon cassiterite cerussite blende
	red	cinnabar realgar proustite rutile cuprite vanadinite
	reddish-brown	rutile monazite blende cuprite zircon descloizite
	orange	zircon crocoite wulfenite scheelite mimetite
	yellow	sphene wulfenite blende rutile orpiment mimetite
	green	sphene vanadinite blende andradite zircon
metallic	white (silver- or tin-white)	silver arsenic antimony mercury arsenopyrite bismuth cobaltite skutterudite chloanthite smaltite
	greyish-white	galena stibnite
	steel-grey	platinum arsenopyrite
	blue-grey	galena stibnite tetrahedrite molybdenite
	purplish-grey	covelline bornite
	dark-grey	chalcocite platinum graphite acanthite galena enargite tennantite
	golden-yellow	gold chalcopyrite pyrite
	reddish-brown to black	goethite rutile tantalite

	black	rutile hematite ilmenite uraninite pyrolusite manganite wolframite goethite columbite tantalite acanthite enargite graphite magnetite chalcocite
	red	copper niccolite
	brownish-red	bornite
	yellow	pyrrhotine millerite pyrite marcasite
pearly (usually seen on cleavage surfaces)	white or colourless	gypsum stilbite muscovite albite apophyllite
	brown or reddish-brown	stilbite muscovite phlogopite biotite erythrite
	yellow, yellow-green or green	torbernite autunite kyanite chlorite microcline vivianite
	blue or purple	kyanite vivianite
resinous	white or colourless	nepheline halite gypsum apatite
	brown to black	axinite willemite wurtzite pyromorphite serpentine
	orange-yellow	orpiment sulphur willemite serpentine blende
	green	serpentine apatite mottramite
	blue	sodalite
vitreous	white or colourless	fluorite corundum grossular calcite baryte quartz orthoclase danburite topaz spodumene stilbite beryl idocrase leucite celestine gypsum apophyllite natrolite scapolite dolomite hemimorphite phenakite datolite petalite
	brown	corundum tourmaline calcite chrysoberyl baryte quartz topaz staurolite andalusite stilbite epidote clinozoisite idocrase
	black	quartz hornblende olivine spinel andradite
	red and pink	corundum spinel fluorite tourmaline rhodochrosite rhodonite almandine pyrope spodumene erythrite strengite topaz
	orange	spessartine topaz corundum calcite grossular
	yellow	corundum topaz chrysoberyl calcite fluorite orthoclase smithsonite beryl andradite quartz spodumene apatite aragonite colemanite brazilianite amblygonite
	green	corundum tourmaline chrysoberyl fluorite beryl olivine spodumene jadeite orthoclase apatite epidote idocrase diopside dioptase andalusite andradite quartz tremolite actinolite adamite
	blue	corundum spinel fluorite smithsonite topaz baryte celestine benitoite sodalite beryl lazulite apatite tourmaline kyanite vivianite cordierite phosphophyllite
	purple	corundum fluorite cordierite spodumene spinel almandine kyanite
waxy	blue or blue-green	turquoise variscite

Table 3
Cleavage and parting
Cleavage is the tendency shown by many minerals to split readily in directions parallel to possible or actual crystal faces. The cleavage is described in terms of the form defined by the cleavage faces. Only minerals with a perfect cleavage are listed here. Where twinning is present, even if not apparent, specimens may also part readily in certain preferred directions

System	Cleavage	Minerals
cubic system	cubic	galena halite cobaltite periclase sylvine
	dodecahedral	blende sodalite
	octahedral	diamond fluorite
tetragonal system	basal	anatase apophyllite boléite
orthorhombic system	basal	topaz celestine baryte
	prismatic	anthophyllite enargite erythrosiderite hemimorphite natrolite wavellite
monoclinic system	basal	muscovite phlogopite biotite chlorite lepidolite epidote clinozoisite
	prismatic	amphibole spodumene glaucophane
triclinic system	pinacoidal	kyanite rhodonite ulexite amblygonite
hexagonal system	rhombohedral	calcite dolomite rhodochrosite

Table 4
Fluorescence in ultraviolet light
Some minerals emit visible light when they are illuminated with radiation of shorter wavelength, such as ultraviolet light. Different colours may be emitted according to the wavelength band of the ultraviolet light

Colour	Ultraviolet	Minerals
whitish	long-wave	amblygonite colemanite quartz (some chalcedonies) aragonite ulexite
	short-wave	anglesite scheelite
red	long-wave	beryl (a few chrome rich emeralds) corundum (variety ruby) spinel (red variety)
	short-wave	calcite spinel (red variety)
orange	long-wave	corundum (yellow varieties from Ceylon) diamond lapis-lazuli (in patches) sodalite (in patches) spodumene (variety kunzite) zircon
	short-wave	corundum (colourless varieties) diamond
yellow	long-wave	anglesite diamond apatite phosgenite scapolite zircon
	short-wave	opal powellite zircon
greenish-yellow	long-wave	autunite uranocircite adamite
	short-wave	autunite uranocircite adamite
green	long-wave	adamite willemite autunite
	short-wave	opal willemite chalcedony (some)
blue	long-wave	diamond fluorite danburite witherite
	short-wave	benitoite fluorite scheelite diamond
violet	long-wave	fluorite scapolite
	short-wave	fluorite

Table 5
Phosphorescence following illumination with ultraviolet light
The emission of light under the stimulation of shorter-wave radiation is called phosphorescence when it persists after the stimulus has ceased

Colour	Minerals
white	gypsum opal ulexite
brown	albite—after short-wave irradiation (rare)
pink	calcite diopside
red	wollastonite
orange	blende spodumene (variety kunzite)
orange-yellow	scapolite
green	adamite willemite
bluish-white	celestine zircon
pale blue	amblygonite cerussite diamond
blue or violet	fluorite

Table 6
Thermoluminescence
These minerals may emit visible light when heated to a temperature below red heat

Colour	Minerals
white	calcite oligoclase
greenish-white	willemite
pink	corundum (variety ruby)
red	smithsonite
orange	calcite topaz tourmaline (some red varieties)
yellow	calcite

Table 7
Hardness according to Mohs' scale

Mohs' original scale was defined by ten selected minerals. Intermediate values are now employed. Each mineral will scratch those below it on the scale

Hardness	Minerals
1	talc
1–2	graphite covelline realgar orpiment vivianite
1.5–2.5	sulphur erythrite stibnite gypsum
2–2.5	cinnabar proustite argentite bismuth autunite torbernite chlorite galena pyrargyrite halite ulexite
2.5–3	copper gold silver chalcocite anglesite crocoite vanadinite bournonite phlogopite biotite lepidolite muscovite serpentine wolfenite
3	enargite bornite cerussite olivenite
3–3.5	calcite antimony millerite celestine descloizite mottramite
3–4	tetrahedrite arsenic strontianite adamite chalcopyrite blende siderite rhodochrosite dolomite malachite wurtzite cuprite mimetite wavellite pyromorphite strengite variscite pyrrhotine
4	aragonite fluorite stilbite chabazite manganite
4–4.5	platinum smithsonite wolframite
4.5–5	pectolite turquoise colemanite hemimorphite scheelite apophyllite kyanite
5	apatite dioptase analcime
5–5.5	goethite lazurite niccolite monazite
5.5–6	ilmenite uraninite actinolite tremolite cobaltite löllingite datolite willemite arsenopyrite lazulite skutterudite natrolite opal leucite
6	orthoclase sodalite scapolite magnetite hornblende diopside rhodonite olivine sphene
6–6.5	columbite tantalite pyrite marcasite rutile benitoite pyrolusite
6.5–7	cassiterite zircon hematite jadeite nephrite idocrase andradite prehnite diaspore spodumene axinite
7	quartz danburite epidote clinozoisite grossular staurolite
7–7.5	pyrope spessartine tourmaline almandine
7.5–8	phenakite beryl
8–8.5	spinel topaz chrysoberyl
9	corundum
10	diamond

Table 8
Streak

The colour of a mineral in its powdered form is called its streak. It can be observed by drawing a specimen across a piece of unglazed porcelain to leave a trace

Specimen	Streak	Minerals
Specimen with metallic lustre	silver	arsenic bismuth silver
	grey	platinum
	golden-yellow	gold
	red	copper
Specimen with non-metallic lustre	grey	bornite chalcocite galena covelline stibnite pyrrhotine cobaltite löllingite marcasite arsenopyrite enargite bournonite graphite antimony molybdenite
	brown	blende wurtzite tetrahedrite rutile
	black	pyrolusite tetrahedrite magnetite columbite
	red to reddish-brown	cinnabar cuprite hematite manganite pyrargyrite proustite tennantite tantalite
	orange	realgar descloizite
	orange-yellow	crocoite
	yellow	orpiment vanadinite autunite torbernite
	green	malachite vivianite mottramite torbernite
	blue	azurite lazurite
	purple	vivianite

Table 9
Refractive index and birefringence

In most cases the upper and lower limits of a range are quoted. Where significant, the birefringence (the difference between the two indices in a doubly refractive material) is given in brackets

Mineral	Refractive index
fluorite	1.43
opal	1.45
borax	1.46
sodalite	1.48
chabazite	1.48
natrolite	1.49
stilbite	1.49–1.50
calcite	1.49–1.66 (0.172)
lazurite	1.50
dolomite	1.50–1.76 (0.179)
leucite	1.51
orthoclase	1.52–1.53 (0.006)
strontianite	1.52–1.66
gypsum	1.53
plagioclase feldspars	1.53–1.59 (0.051)
quartz (chalcedony)	1.53–1.54
cordierite	1.53–1.55
lepidolite	1.55
aragonite	1.53–1.69 (0.155)
nepheline	1.53–1.55
apophyllite	1.54
scapolite	1.54–1.58
phlogopite	1.54–1.58
biotite	1.54–1.58
halite	1.54
muscovite	1.56
quartz	1.54–1.55 (0.009)
serpentine	1.56–1.60
beryl	1.56–1.59 (0.004–0.007)
chlorite	1.57–1.59
vivianite	1.58–1.65
variscite	1.58
colemanite	1.59–1.61
rhodochrosite	1.60–1.82 (0.219)
tremolite	1.60–1.68
actinolite	1.60–1.68
nephrite	1.62
lazulite	1.60–1.64
topaz	1.62–1.64

Refractive index and birefringence
continued

Mineral	Refractive index (birefringence)
hemimorphite	1.62–1.64
tourmaline	1.62–1.65 (0.014–0.020)
smithsonite	1.63–1.84 (0.228)
turquoise	1.62–1.63
celestine	1.62–1.63
hornblende	1.63–1.65
erythrite	1.63–1.70
datolite	1.64–1.67
prehnite	1.63
danburite	1.63–1.64
apatite	1.63–1.64
siderite	1.63–1.87 (0.240)
andalusite	1.63–1.65
baryte	1.64–1.65
phenakite	1.65–1.67
jadeite	1.66
spodumene	1.65–1.67
peridot	1.65–1.69 (0.036)
dioptase	1.65–1.71
axinite	1.68–1.69
willemite	1.69–1.73
idocrase	1.70–1.73
clinozoisite	1.71–1.72
spinel	1.71–1.80
augite	1.71–1.75
kyanite	1.72–1.74
adamite	1.73–1.75
rhodonite	1.73–1.75
pyrope	1.74
azurite	1.73–1.84 (0.107)
grossular	1.74–1.75
epidote	1.74–1.77
staurolite	1.74–1.76
chrysoberyl	1.75–1.76
corundum	1.76–1.78 (0.008)
malachite	1.77
almandine	1.77–1.83
monazite	1.79–1.85
spessartine	1.79–1.81
cerussite	1.80–2.08 (0.275)
zircon	1.84–1.99 (0.059)
sphene	1.84–2.09 (0.142)
anglesite	1.88–1.89
andradite	1.89
scheelite	1.92–1.94
sulphur	1.96–2.25 (0.29)
cassiterite	2.0–2.1
pyromorphite	2.05–2.06
mimetite	2.13–2.14
descloizite	2.19–2.35
mottramite	2.21–2.33
wulfenite	2.28–2.40
crocoite	2.29–2.66
blende	2.37
orpiment	2.40–3.02 (0.62)
vanadinite	2.35–2.42
diamond	2.42
realgar	2.54–2.70
rutile	2.61–2.69 (0.287)
proustite	2.79–3.09 (0.295)
cuprite	2.85
pyrargyrite	2.88–3.08

Table 10
Specific gravity
A material's specific gravity is the ratio of its density (weight per unit volume) to the density of water. It is numerically equal to the material's density in grams per cubic centimetre. Only the specific gravities of the commoner minerals are given here, and mean values only are quoted

Mineral	Specific gravity
borax	1.7
ulexite	1.9
opal	2.1
chabazite	2.0–2.1
sulphur	2.0
stilbite	2.1
graphite	2.1
halite	2.1
natrolite	2.2
sodalite	2.2–2.3
gypsum	2.3
apophyllite	2.3
wavellite	2.4
lazurite	2.4–2.5
leucite	2.4–2.5
colemanite	2.4
serpentine	2.5–2.6
variscite	2.5
nepheline	2.5–2.6
cordierite	2.5–2.7
microcline	2.5–2.6
orthoclase	2.5–2.6
scapolite	2.7
turquoise	2.6–2.9
albite	2.6
plagioclase feldspars	2.6–2.7
quartz	2.6
oligoclase	2.6
beryl	2.6–2.9
vivianite	2.6–2.7
chlorite	2.7–2.9
calcite	2.7
labradorite	2.7
muscovite	2.8–2.9
lepidolite	2.8–2.9
prehnite	2.8–2.9
datolite	2.8–3.0
dolomite	2.8–3.0
biotite	2.8–3.4
phlogopite	2.8–3.4
aragonite	2.9–3.0
phenakite	2.9–3.0
tremolite	2.9–3.1
danburite	3.0
erythrite	3.0
nephrite	3.0
tourmaline	3.0–3.3
hornblende	3.0–3.4
lazulite	3.1–3.2
apatite	3.1–3.2
autunite	3.1–3.2
andalusite	3.1–3.2
actinolite	3.1–3.3
spodumene	3.1–3.2
fluorite	3.1
axinite	3.2–3.3
olivine	3.2–4.3
torbernite	3.2
jadeite	3.3
diopside	3.2–3.3
dioptase	3.2–3.3
diaspore	3.3–3.5
clinozoisite	3.3–3.4
idocrase	3.3–3.6
goethite	3.3–4.3
hemimorphite	3.4–3.5
rhodochrosite	3.4–3.6
epidote	3.4–3.6
sphene	3.4–3.5
augite	3.4–3.5
orpiment	3.4
diamond	3.5
topaz	3.5
rhodonite	3.5–3.7
realgar	3.5
spinel	3.5–3.9
grossular	3.6
malachite	3.6–4.0
benitoite	3.6
kyanite	3.6–3.7
pyrope	3.6–3.8
staurolite	3.7–3.8
chrysoberyl	3.7
strontianite	3.7
azurite	3.7
siderite	3.8–3.9
andradite	3.8–3.9
willemite	3.8–4.1
blende	3.9–4.1
zircon	3.9–4.7
almandine	3.9–4.3
celestine	3.9
corundum	3.9–4.0
smithsonite	4.0–4.5
chalcopyrite	4.1–4.3
spessartine	4.1–4.2
rutile	4.2
manganite	4.3
adamite	4.3–4.5
baryte	4.5
enargite	4.4–4.5
pyrolusite	4.4–5.0
pyrrhotine	4.6
molybdenite	4.6–4.7
covelline	4.6–4.7
tennantite	4.6–4.9
monazite	4.6–5.4
stibnite	4.6
ilmenite	4.7
marcasite	4.8
tetrahedrite	4.9–5.1
pyrite	5.0
bornite	5.0
columbite	5.1–6.5
magnetite	5.1
hematite	5.3
millerite	5.5
chalcocite	5.5–5.8
proustite	5.5
arsenic	5.7
bournonite	5.8
pyrargyrite	5.8
mottramite	5.9
scheelite	5.9–6.1
crocoite	6.0
arsenopyrite	6.0
skutterudite	6.1–6.9
cuprite	6.1
descloizite	6.2
cobaltite	6.3
anglesite	6.3
wulfenite	6.5–7.0
vanadinite	6.5–7.1
uraninite	6.5–9.7
cerussite	6.5
tantalite	6.5–8.0
antimony	6.7
cassiterite	6.8–7.1
pyromorphite	7.0
argentite	7.2–7.4
mimetite	7.2
wolframite	7.3
löllingite	7.4
galena	7.5
niccolite	7.7
cinnabar	8.1
bismuth	9.7
silver	10–11
platinum	14–19
gold	15–19

Conversion Tables

Linear measure

0.001 inch	= 0.0254	millimetre
1 inch	= 2.54	centimetres
12 inches = 1 foot	= 0.3048	metre
3 feet = 1 yard	= 0.9144	metre
5,280 feet = 1 (statute) mile	= 1.6093	kilometres

	1 millimetre	=	0.03937	inch
10 millimetres	= 1 centimetre	=	0.3937	inch
10 centimetres	= 1 decimetre	=	3.937	inches
10 decimetres	= 1 metre	=	39.37	inches or 3.2808 feet
10 metres	= 1 decametre	=	393.7	inches
10 decametres	= 1 hectometre	=	328.08	feet
10 hectometres	= 1 kilometre	=	0.621	mile or 3,280.8 feet
10 kilometres	= 1 myriametre	=	6.21	miles

Square measure

	1 square inch	= 6.452	square centimetres
144 square inches	= 1 square foot	= 929.03	square centimetres
9 square feet	= 1 square yard	= 0.8361	square metre
	1 square mile	= 259.00	hectares or 2.590 square kilometres

	1 square millimetre	= 0.00155	square inch
100 square millimetres	= 1 square centimetre	= 0.15499	square inch
100 square centimetres	= 1 square decimetre	= 15.499	square inches
100 square decimetres	= 1 square metre	= 1,549.9	square inches or 1.196 square yards
100 square metres	= 1 square decametre	= 119.6	square yards
100 square decametres	= 1 square hectometre	= 2.471	acres
100 square hectometres	= 1 square kilometre	= 0.386	square mile or 247.1 acres

Volume measure

	1 cubic inch	= 16.387	cubic centimetres
1,728 cubic inches	= 1 cubic foot	= 0.0283	cubic metre
27 cubic feet	= 1 cubic yard	= 0.7646	cubic metre

1,000 cubic millimetres	= 1 cubic centimetre	= 0.06102	cubic inch
1,000 cubic centimetres	= 1 cubic decimetre	= 61.023	cubic inches or 0.0353 cubic foot
1,000 cubic decimetres	= 1 cubic metre	= 35.314	cubic feet or 1.308 cubic yards

Weight

437.5 grains	= 1 ounce	= 28.3495	grams
16 ounces or 7,000 grains	= 1 pound	= 453.59	grams
100 pounds	= 1 US hundredweight	= 45.36	kilograms
112 pounds	= 1 Imp. hundredweight	= 50.80	kilograms
2,000 pounds	= 1 US ton	= 907.18	kilograms
2,240 pounds	= 1 Imp. ton	= 1,016	kilograms

10 milligrams	= 1 centigram		
10 centigrams	= 1 decigram		
10 decigrams	= 1 gram	= 0.035274	ounce
10 grams	= 1 decagram	= 0.3527	ounce
10 decagrams	= 1 hectogram	= 3.5274	ounces
10 hectograms	= 1 kilogram	= 2.2046	pounds
10 kilograms		= 22.046	pounds
1,000 kilograms	= 1 metric ton	= 2,204.6	pounds
		= 1.1023	US tons
		= 0.9842	Imp. ton

Important Addresses

Australia

Gemmological Association of Australia, GPO Box 149, Sydney, New South Wales 2001
Geological and Mining Museum, 36 George Street North, Sydney, New South Wales 2000

Canada

Royal Ontario Museum, 100 Queen's Park, Toronto, Ontario
Geological Survey of Canada, 601 Booth Street, Ottawa, Ontario

Czechoslovakia

Lapidarium Historické Múzeum, Park Kultury a oddechu Julia Fučika, Praha

France

Musée de l'Ecole Supérieure des Mines, 60 Boulevard Saint-Michel, Paris 75006

Iran

Bank Melli, Tehran (in the custody of the Bank Markazi, Supervisory Board for the Crown Jewels Collection)

Switzerland

Geologische und Mineralogisch-Petrographische Sammlung der Technischen Hochschule, 8000 Zürich

UK

Institute of Geological Sciences and Geological Museum, Exhibition Road, London SW7 2DE
British Museum (Natural History), Cromwell Road, London SW7 5BD
Mineralogical Society of Great Britain and Ireland, 41 Queen's Gate, London SW7 5HR
Gemmological Association of Great Britain, St. Dunstan's House, Carey Lane, London EC2V 8AB
Geological Society of London, Burlington House, Piccadilly, London W1V 0JU
Geologists' Association, 278 Fir Tree Road, Epsom, Surrey

USA

Smithsonian Institution, 1000 Jefferson Drive, Washington, DC 20560
American Museum of Natural History, Central Park West, New York, NY 10024
Geological Survey of the United States, 12201 Sunrise Valley Drive, Reston, Virginia 22092
Gemological Institute of America, 11940 San Vincente Boulevard, Los Angeles, California 90049
Mineralogical Society of America, 1707 L Street, NW, Washington, DC 20036
American Federation of Mineralogical Societies, 704 SW 31st Street, Pendleton, Oregon 97801
International Mineralogical Association, 2018 Luzerne Avenue, Silver Spring, Maryland 20910

USSR

Mining Museum of the G V Plekhanov Mining Institute, Vasilievsky ostrov, Leningrad V 26

West Germany

Bayrisches Geologisches Landesamt, Prinzregentenstrasse 28,8 München 22
Geologisches Landesamt in Baden Württemberg, Albertstrasse 5,Freiburg Ei.Br.
Geologisches Landesamt Nordrhein-Westfalen, Postschliessfach 1080, 415 Krefeld
Geologisches Landesamt Schleswig-Holstein, Mecklenburgerstrasse 22/24, 23 Kiel-Wik
Geologisches Landesamt Rheinland-Pfalz, Flachsmarktstrasse 9, 65 Mainz
Hessisches Landesamt für Bodenforschung, Leberberg 9, 62 Wiesbaden
Deutsches Edelsteinmuseum, D658 Idar-Oberstein
Deutsche Gemmologische Gesellschaft, Gewerbehalle, Postfach 2260, D658 Idar-Oberstein

Bibliography

Chemistry, Crystallography, Geochemistry, Physics

Bishop, A C, *An Outline of Crystal Morphology*, Hutchinson, London, 1967.
Intended for students of geology.

Burns, R G, *Mineralogical Applications of Crystal Field Theory*, University Press, Cambridge, 1970.
A lucid book on the nature of chemical bonding applied to mineralogy and geochemistry.

Cotton, F A and Wilkinson, G, *Advanced Inorganic Chemistry*, 3rd edition, Interscience Publishers, New York and London, 1972.
The standard text.

Ditchburn, R W, *Light*, 3rd edition, Blackie, London, 1963.
First-year undergraduate-level textbook.

Donnay, J D H, *Crystal Data: Determinative Tables*, 3rd edition; vol 2, *Inorganic Compounds*, National Bureau of Standards, Washington, 1973.

Evans, R C, *An Introduction to Crystal Chemistry*, 2nd edition, Cambridge University Press, 1964.

Gleason, S, *Ultraviolet Guide to Minerals*, Ultra-Violet Products, San Gabriel, California, 1972.
A useful guide with coloured illustrations.

Goldschmidt, V M, *Geochemistry* (edited by Alex Muir), Clarendon Press, Oxford, 1958.
The standard work on geochemistry; supplements, closely linked to this original work, are also published.

Hartshorne, N H and Stuart, A, *Crystals and the Polarizing Microscope*, 4th edition, Edward Arnold, London, 1970.
The standard work in the field.

Lonsdale, K, *Crystals and X-Rays*, G Bell & Sons, London, 1948.
A most lucid account of X-ray diffraction by crystals. Out of print but worth a search.

Sidgwick, N V, *The Chemical Elements and their Compounds*, Clarendon Press, Oxford, 1950. Two volumes.

Geology and Mineralogy —general

Blyth, F G H, *Geology for Engineers*, 6th edition, Edward Arnold, London, 1975.
An excellent applied geology.

Chalmers, R O, *Australian Rocks, Minerals and Gemstones*, Angus and Robertson, Sydney, 1967.
The best guide to Australian gems and minerals.

Dana, E S, *A Textbook of Mineralogy*, 4th edition, revised by W E Ford, John Wiley, New York and London, 1932.
The instructional part of the great Dana series of mineralogical texts. Somewhat overtaken by new minerals, occurrences and research but still useful.

Dana, J D and E S, *The System of Mineralogy*, 7th edition by Palache, Berman and Frondel, 1944–. John Wiley and Sons, New York.
Three volumes so far published. Most works on minerals are based on the classification adopted for this book on its original publication in 1837.

Dana, J D, *Manual of Mineralogy*, John Wiley & Sons, New York. Various editions.
A simplified guide based on the textbook by E S Dana and others.

Deer, W A, Howie, R A and Zussman, J, *Rock-forming Minerals*, Longman, London, 1962. Five volumes.
Full details are given of the common minerals of igneous, sedimentary and metamorphic rocks. *An Introduction to the Rock-forming Minerals* is a digest in one volume.

Desautels, P E, *The Mineral Kingdom*, Paul Hamlyn, London, 1969.
Based on the collections of the Smithsonian Institution this book is a lucid introduction to the beauty of the mineral world.

Hamilton, W R, Woolley, A R and Bishop, A C, *The Hamlyn Guide to Minerals, Rocks and Fossils*, Paul Hamlyn, London, 1974.
The best of the pocket-sized field guides to minerals.

Hey, M H and Embrey, P, *Index of Mineral Species arranged Chemically*, British Museum (Natural History), London. 1955–1974.
Includes two supplements. The spine reads *Chemical Index of Minerals*.

Hurlbut, C S, *Minerals and Man*, Thames and Hudson, London, 1969.
An account of the economic use of minerals.

Institute of Geological Sciences, *British Regional Geology*, London, 1935–.
Continuously revised.

Michele, V de, *The World of Minerals*, Orbis Publishing Ltd, London, 1976.
Introduction to minerals, illustrated with high-quality colour photographs.

Moorhouse, W W, *The Study of Rocks in Thin Section*, Harper and Row, New York, 1964.
An introductory text.

Nature Conservancy (Great Britain), *List of Protected Geological Sites*.

Orbis Publishing Ltd, *Crystals*, London, 1976.
High-quality colour photographs, with an introduction.

Roberts, W L, Rapp, G R and Weber, J, *Encyclopedia of Minerals*, Van Nostrand Reinhold Co., New York and London, 1974.
Very useful up-to-date references for each mineral.

Rodgers, Peter R, *Agate Collecting in Britain*, Batsford, London, 1975.

Sinkankas, J, *Gemstone and Mineral Data Book*, Winchester Press, New York, 1972.
Contains all possible data needed for the identification and preservation of gems and minerals.
—*Prospecting for Gemstones and Minerals*, Van Nostrand Reinhold, New York and London, 1970.
Contains information on the recognition of gem-bearing geological formations, handling specimens, etc.

Traill, R J, *A Catalogue of Canadian Minerals*, Geological Survey of Canada, 1970 (Paper 69–45).
Updates R A A Johnston's Memoir 74 of 1915.

Vlasov, K, *Mineralogy of Rare Elements*, (Vol. 2 of the series *Geochemistry and Mineralogy of Rare Elements and Genetic Types of their Deposits*, translated from the Russian), Israel Program for Scientific Translations, 1966.

Wells, A F, *Structural Inorganic Chemistry*, 3rd edition, Clarendon Press, Oxford, 1962.
A standard reference-work of great clarity.

Winchell, A N and Winchell, H, *Elements of Optical Mineralogy*, 4th edition, John Wiley, New York, 1951.
The best work in this field.

Periodicals and Serials

Acta Crystallographica, Munksgaard, Copenhagen, for the International Union of Crystallography, 1948–.
In 2 sections: A—theoretical; B—structural.

American Mineralogist, Journal of the Mineralogical Society of America.
The official American mineralogical journal.

Australian Gemmologist, Gemmological Association of Australia, Sydney.
Reports from university and government research bodies on Australian gem materials, particularly on opal.

Bulletin Association Française de Gemmologie, Paris.
The official French gemmological journal. Contains reports of new occurrences of gem materials, especially from the Malagasy Republic and Brazil.

Bulletin of the Geological Survey of Great Britain, HMSO, London.
Irregular.

Canadian Mineralogist, Mineralogical Association of Canada, 555 Booth St., Ottawa. Semi-annual.

Gemmological Newsletter, 7, Hillingdon Avenue, Sevenoaks, Kent.
A weekly bulletin of new gem finds, new testing methods, book reviews and other material.

Gems and Gemology. Gemological Institute of America, Los Angeles.
Especially noteworthy for reports of unusual gems tested at the GIA laboratories in Los Angeles and New York.

Geochimica et Cosmochimica Acta. Pergamon, New York, 1950–.

Geological Magazine, Cambridge University Press, six times a year.

Geologists' Association Guides, Benham & Co., Colchester, Essex, 1958–.

Geoserials, Geosystems, London, 1969–.
A world list of current geoscience serial publications. Supplements published in Geoscience Documentation.

Geoscience Documentation, Geosystems, London, 1969–.
Articles, news and bibliographies.

Journal of Gemmology, Gemmological Association of Great Britain.
The official British gemmological journal with material of a good standard.

Journal of the Geological Society of London, Geological Society of London, Burlington House, Piccadilly, London W1V 0JV. Six issues a year.

Journal of Petrology, Clarendon Press, Oxford, 1960–.

Lapidary Journal, PO Box 80937, San Diego, California 92138.
The best journal devoted to fashioning and also containing up-to-date reports on the occurrence of new gem materials.

Memoirs, Geological Society of Great Britain, London, 1958–.

Memoirs, Geological Survey of Great Britain, London, 1864–. In sheets.

Mineral Digest, Mineral Digest Ltd, PO Box 341, Murray Hill Station, New York, NY 10016.
A serial issued irregularly though with periodical-type numeration. Contains magnificent coloured illustrations. Seven issues extant at the time of writing.

Mineralogical Abstracts, Mineralogical Society of Great Britain, London SW7, and the Mineralogical Society of America.
Reviews all relevant literature in the fields of descriptive mineralogy, gemmology, crystallography, etc.

Mineralogical Magazine, Mineralogical Society of Great Britain, London, SW7.
Essential for the serious student of mineralogy.

Mineralogical Record, John S White, Jr, PO Box 783, Bowie, Maryland 20715. Six issues a year.

Monthly List of New Publications [of the US Geological Survey], US Dept. of the Interior, Geological Survey, National Center, Stp 329, Reston, Virginia 22092.

Synthetic Crystals Newsletter, 7, Hillingdon Avenue, Sevenoaks, Kent.
An occasional bulletin of information on new synthetic materials, particularly those of gemmological interest.

Gemmology

Anderson, B W, *Gem Testing*, 8th edition, Butterworths, London, 1971.
The classic manual on the scientific identification of gemstones. Admirably lucid.

Gübelin, E, *Internal World of Gemstones*, ABC Edition, Zürich, 1974.
The best book extant on the inclusions found in gemstones; fine-quality coloured illustrations and a good bibliography.

Kalokerinos, A, *Australian Precious Opal*, Nelson, Melbourne, 1971.
Illustrated with fine-quality coloured plates. The author's nomenclature is sometimes non standard.

Liddicoat, R T, *Handbook of Gem Identification*, 10th edition, Gemological Institute of America, Los Angeles, 1975.
An excellent American textbook.

Rutland, E H, *An Introduction to the World's Gemstones*, Country Life, London, 1974.
The best simple introduction available in English.

Sinkankas, J, *Gemstones of North America*, D Van Nostrand Co., Princeton and London, 1959.
The best guide to the gem materials that occur in North America.

—*Van Nostrand's Standard Catalog of Gems*, D Van Nostrand Co., Princeton and London, 1968.
Gives tables of values, now dated, of rough and cut gem material; still very useful for the criteria upon which the values are based.

Smith, G F Herbert, *Gemstones*, 14th edition, revised by F Coles Phillips, Chapman and Hall, London, 1972.
One of the first and best scientific studies of the gem.

Webster, R, *Gems*, Newnes-Butterworths, London, 1975.
A comprehensive survey of the gem materials.

—*Practical Gemmology*, NAG Press, London. Various editions.
Based on the examination syllabus of the Gemmological Association of Great Britain and including questions.

—*The Gemmologists' Compendium*, NAG Press, London. Various editions.
For use as an adjunct to the laboratory examination of gemstones. Contains useful tables.

Gem-cutting

Sinkankas, J, *Gem Cutting*, 2nd edition, Van Nostrand Reinhold Co., New York and London, 1962.
An excellent work on the fashioning of gemstones.

Vargas, G and M, *Faceting for Amateurs*, published by the authors, Thermal, California, 1969.
Another well written guide to gem fashioning.

Index

References in *italic* type are to illustration captions. In the Mineral Kingdom section principal references to minerals are indexed in **bold** type; passing references are not indexed. The Identification Tables are likewise referred to by bold type; mineral names appearing in the tables are not indexed. A page number followed by ff denotes a topic discussed on several consecutive pages

The authors and publishers wish to thank J A D Cotton for the loan of equipment and J Winter for assistance with the text

The photographs in this book were taken by Carlo Bevilacqua and supplied by courtesy of Istituto Geografico de Agostini, Novara, except those on the following pages:

8 C Dani. **9** (left) C Dani, (right) Ronan Picture Library. **14** Paul Brierley. **15** Barnaby's Picture Library. **23** IGDA. **26** from Materials Research Bulletin Vol II, p 1 by Cooke and Nye. **27** Cambridge Instruments. **35** C Dani. **42** N Cirani. **45** (left) Crown copyright, reproduced by permission of the Controller HMSO, (right) American Association of Petroleum Geologists. **47** (right) Peter R Rodgers. **49** (top) G Tomsich, (bottom) Picturepoint. **51** A Rittman–V Gottini. **61** (bottom) IGDA. **63** Titus. **68** (bottom) D Fernandez. **72** N Cirani. **75** Australian Information Service, London. **76** Kernowcraft. **77** Amey Roadstone Corporation. **78** Photair/A Perceval. **79** SEF. **80** S E Hedin-bild. **84** Archivio Foto B. **92–95** Angelo Hornak/Orbis. **98–100** Angelo Hornak/Orbis. **104** NERC copyright, reproduced by permission of the Director, Institute of Geological Sciences. **108–9** Angelo Hornak/Orbis. **120–1** De Beers. **132** Paul Brierley. **143** Smithsonian Institution. **144–7** Angelo Hornak/Orbis. **154** (centre right) De Beers. **156** (top left) Angelo Hornak/Orbis. **186** (bottom) Archivio Foto B. **195** (top) Angelo Hornak/Orbis. **210** (both) Angelo Hornak/Orbis. **222** (bottom right) Archivio Foto B. **229** (top) NERC copyright, reproduced by permission of the Director, Institute of Geological Sciences. **265** (top) Angelo Hornak/Orbis. **273** (top) Angelo Hornak/Orbis. **286** (bottom) G Mazza. **287** C Annunziata